Wireless Networks

Series Editor
Xuemin Sherman Shen, University of Waterloo, Waterloo, ON, Canada

The purpose of Springer's Wireless Networks book series is to establish the state of the art and set the course for future research and development in wireless communication networks. The scope of this series includes not only all aspects of wireless networks (including cellular networks, WiFi, sensor networks, and vehicular networks), but related areas such as cloud computing and big data. The series serves as a central source of references for wireless networks research and development. It aims to publish thorough and cohesive overviews on specific topics in wireless networks, as well as works that are larger in scope than survey articles and that contain more detailed background information. The series also provides coverage of advanced and timely topics worthy of monographs, contributed volumes, textbooks and handbooks.

Xiaohua Tian • Xinyu Tong • Xinbing Wang

Wireless Localization Techniques

Springer

Xiaohua Tian
Shanghai Jiao Tong University
Shanghai, China

Xinyu Tong
Tianjin University
Tianjin, China

Xinbing Wang
Shanghai Jiao Tong University
Shanghai, China

ISSN 2366-1186 ISSN 2366-1445 (electronic)
Wireless Networks
ISBN 978-3-031-21177-5 ISBN 978-3-031-21178-2 (eBook)
https://doi.org/10.1007/978-3-031-21178-2

This Springer imprint is published by the registered company Springer Nature Switzerland AG
The registered company address is: Gewerbestrasse 11, 6330 Cham, Switzerland

Preface

Location service is a prerequisite for numerous IoT applications. While most modern mobile devices are equipped with global navigation satellite system (GNSS) receivers, the satellite system–based location service could be unavailable in certain circumstances, for example, in indoor spaces and "urban canyon" areas, where the satellite signals could be blocked by walls and buildings. It is, therefore, necessary to perform localization with other technical means, which are supposed to meet the following needs: (1) universality—can use commercial off-the-shelf (COTS) devices for localization without hardware modification; (2) effectiveness—can provide services with certain accuracy and reliability; (3) deployability—can be easily deployed with existing infrastructure and limited investments.

Wireless localization can be a very appropriate candidate solution, which estimates a wireless terminal's location by analyzing features of wireless communication signals, such as Wi-Fi, LTE, and BLE. This technique has been attracting interest from both academia and industry since the past decades; however, most of the research work has been conducted empirically, that is, observing experimental results and phenomena, then coming up with a new idea, and then verifying the new idea with experiments again. A systematic theoretical study has been long pending. Answers to some fundamental issues of the wireless localization technique are still unknown.

In this book, we provide systematic theoretical studies and design approaches for wireless localization systems, which tentatively answer the following fundamental questions:

(1) How good the wireless localization system could possibly achieve?
(2) How the system could scale up with the number of users to be served?
(3) How to make key design decisions in implementing the system?
(4) How to mitigate human efforts in deploying the wireless localization system?

Such theoretical insights could help researchers and engineers understand the essence of wireless localization methodology and develop practical systems.

We hope this book can help in advancing research and the practice of wireless localization. This book is primarily intended for anyone who wants to learn wireless localization technology. It is an ideal guidebook for students or beginners who are the first to walk into this area, as well as professionals, engineers, and researchers working in related fields.

Shanghai, China Xiaohua Tian
Tianjin, China Xinyu Tong
Shanghai, China Xinbing Wang

Introduction

Wireless localization is a technique used to estimate a wireless terminal's location by utilizing characteristics of wireless communication signals, such as Wi-Fi, LTE, and BLE. This technique has been attracting interest from both academia and industry since the past decades; however, most of the research work has been conducted using the methodology of "empirical study." A systematic theoretical study on the topic has been long pending. In this book, we present a systematic theoretical study of the wireless localization technique. Guided by theoretical results, we will provide design approaches for improving performance of the localization system and making its deployment more convenient.

Contents

Chapter 1
Why and What?

1.1 Wireless Localization

1.1.1 Why Not GPS?

The localization of mobile terminal has become a prerequisite for numerous IoT applications. Recently, most mobile devices are equipped with Global Navigation Satellite System (GNSS) such as GPS and BDS for location services; however, satellite systems are not always able to provide qualified localization services in diverse environment. For example, as satellite signals would be blocked by obstacles, we can hardly realize satisfying localization accuracy indoors or in "urban canyon" environments with tall buildings. Besides, in the scene of emergency calls for cellular networks, such as E911 in North America [260] and E112 in Europe [262], the government mandates that network providers have accurate and reliable localization capabilities for all mobile terminals in order to localize the caller in emergency situations [231]. However, the callers are likely to use unintelligent terminals with only basic communication functions. According to the statistics of Pew Research Center in the United States in 2018, the smartphone ownership rate of Americans is 77%, and those of German and French smartphones are 72% and 62%, respectively.

As these special environments and terminal constraints make the satellite system unable to effectively perform localization, it is necessary to obtain location information through other technical methods. As for large-scale practical scenarios, the alternative solution should meet the following three basic requirements: (1) **Universality:** We can use commercial off-the-shelf (COTS) mobile devices to locate targets without hardware modification; (2) **Effectiveness:** The localization method should be accurate and reliable for given applications; (3) **Deployability:** The system should be easily deployable on the existing infrastructure without any hardware or firmware modification at the access points (APs).

X. Tian et al., *Wireless Localization Techniques*, Wireless Networks,
https://doi.org/10.1007/978-3-031-21178-2_1

1.1.2 What Is Wireless Localization?

The wireless fingerprinting localization system can meet these requirements simultaneously. In particular, fingerprinting localization is independent of satellite systems and can locate targets based on wireless signal characteristics, which consist of two phases. In offline phase, we associate the signal characteristics with their observed locations and record them in the database, where the signal characteristics are also called "wireless fingerprints." In online phase, the user reports observed wireless fingerprints to the server. In order to localize users, the server compares these fingerprints with fingerprint database called "fingerprint map" to find the likely location where the fingerprints are collected. In practice, we usually select the received signal strength (RSS) as fingerprints [221]. The fingerprints can be detected by mobile devices from wireless infrastructure including Wi-Fi APs, Bluetooth low-energy (BLE) nodes [253, 254, 259, 275], and LTE base stations [261, 263–269] that have been widely deployed. Benefiting from the spatial difference of RSS fingerprints, we can distinguish the physical location of measurement devices. Since almost all wireless devices can provide RSS information acquisition interfaces, there is no need to redesign the hardware for localization.

1.1.3 Issues to Be Studied in the Book

Among the basic requirements of the localization system, universality and deployability are contradictory to effectiveness. Consequently, there are three major challenges when building a large-scale fingerprinting localization system.

Accuracy and Reliability The contradiction between effectiveness and universality is reflected in the characteristics of RSS fingerprints. Specifically, while RSS is easily available, it can just reflect coarse-grained features of wireless signals. With such information, we can hardly realize accurate and reliable localization.

To this end, we explored current wireless localization systems that yield decimeter-level accuracy. They all rely on fine-grained signal characteristics, e.g., CSI, that are just available with open communication chip interface. These fine-grained information inevitably contain trade secrets and are not opened by most manufacturers. In the academic field, Intel 5300 and Atheros [237] network interface controllers (NICs) are most used, where we have to embed the NICs into microcomputers or routers to set up special APs. Moreover, as these dedicated devices only support the Wi-Fi protocol, it is difficult to apply them to diverse scenarios. In a cellular system, massive users and complex channel environment increase the cost of network infrastructure for accurate CSI or ToF localization. In contrast, RSS is coarse-grained but available for almost all wireless devices. Consequently, the system must make full use of coarse-grained information to realize high performance as much as possible.

Scalability It is difficult to achieve large-scale collection even for coarse-grained RSS fingerprints [232, 233]. The scalability of the fingerprinting localization system depends on the "quality" and "quantity" of the fingerprints collected in the offline phase. Nevertheless, we can hardly ensure "quality" and "quantity" at the same time due to the contradiction between effectiveness and deployability.

Recently, we have observed the trend of utilizing crowdsensing methods to collect wireless fingerprints. In particular, early fingerprinting localization systems require professionals to survey the area to construct the fingerprint map. This method can guarantee the "quality" of fingerprints, but it cannot ensure the "quantity" of fingerprints especially applied in large scenarios. The crowdsensing method [239–246] can construct the fingerprint map in a non-participatory manner and increase the amount of fingerprints at a low cost. However, non-participating users will not perform accurate measurements like professionals, so it is difficult to guarantee "quality." From the perspective of practicability, the "quality" of fingerprints should be guaranteed first. Therefore, we must propose the corresponding mechanism in the crowdsensing process and filter out high-quality fingerprints.

Coverage Crowdsensing can expand the fingerprint map at a low cost but is limited by the number, behavior patterns, and a range of activities of participants. Therefore, it cannot guarantee full coverage of all locations in the service area, which is a problem of system implementation caused by the contradiction between the above pair of requirements.

The data collected by Crowdsensing not only are uneven in quality but also have a distinct geographical imbalance. Some areas have a lot of sampled data, while some areas have scarce data or even are uncovered. The basic reason is the uneven distribution and various behavior modes of participants. In light of rigid requirements of localization such as E911 emergency call, the target may appear anywhere in the service area, so full coverage of the fingerprint map is needed. In a broad area of the city level, full coverage cannot be guaranteed even if the wireless fingerprint collection of Crowdsensing is adopted.

1.2 Background and Literature Review

1.2.1 Early Localization System

RSSI indoor localization has long been an active research field, which enables a vast range of mobile computing applications. As a typical fingerprinting based indoor localization system in 2000, RADAR [32, 234] infers the device's location using the technique of nearest neighbor(s) in signal space (NNSS). With the development of indoor localization technique, the researchers devote their efforts to studying the property of wireless signals [31, 140], so that we can benefit from higher accuracy. The related work is shown in Table 1.1; in contrast to previous work [39] that is

Table 1.1 Early localization system

Research	Description
[33]	Bayesian networks
[34]	Bayesian networks and Gaussian model
[35]	Probability distributions and locations clustering
[36]	Sample correlations
[37]	Statistical learning
[38]	Correlation, clustering, and continuous space estimator

based on propagation model, Wi-Fi fingerprinting localization [33–35] utilizes the probability model such as Bayesian networks [21] and Gaussian model [215, 235] to improve the quality of RSSI fingerprints.

With the development of IoT, more and more smart devices support Wi-Fi communication, which promotes the further development of Wi-Fi localization technology. More comprehensive experiments [40, 113, 138] are conducted to verify the performance of Wi-Fi localization. These experiments show the limitations of fingerprinting based localization. In particular, since RSS fingerprints are based on site survey in real environment, we can realize higher accuracy in contrast to model-based localization system. However, fingerprint collection requires a lot of labor, which limits the deployment of fingerprinting localization. Next, we will introduce the current solutions to the following questions: (1) how to relieve the deployment cost of fingerprinting localization? (2) how to improve the accuracy of fingerprinting localization?

1.2.2 Mobile Crowdsensing and Wi-Fi Fingerprints

The sensing paradigm of utilizing the crowd power to fulfill large-scale sensing tasks at a low cost is termed as mobile crowdsensing (MCS). A MCS application of particular interest is that the sensors embedded in mobile phones are leveraged for various sensing tasks [42]. Compared with the tradition paradigm relying on wireless sensor networks, MCS has a broader coverage, lower deployment cost, and higher scalability [43]. Consequently, some researches study how to use MCS to relieve the time-consuming cost of fingerprinting localization. The related work can be divided into two categories: (1) how to design an incentive mechanism to encourage users to collect high-quality fingerprints? (2) how to reduce the overall workload for fingerprints collection?

High-Quality Fingerprints Collection The purpose of MCS is to attract more users to collect high-quality fingerprints. Consequently, it is necessary to design a mechanism to evaluate the quality of fingerprints and realize quality-aware pricing. For this purpose, Table 1.2 lists some recent work that optimizes the fingerprinting based localization system. In particular, [45–48] explore the method to select

Table 1.2 High-quality fingerprints collection

Research	Description
[45]	Extract accurate fingerprint values from short RSS measurement times
[47]	An on-demand approach to optimally select representative fingerprints
[46]	Evaluates the performance of various radio map construction methods
[48]	Hybrid fingerprint quality evaluation
[49]	Quality-aware pricing and total expected payment minimizing
[50]	Online quality-aware pricing and expected regret minimizing.

Table 1.3 Automatic label for fingerprints

Research	Landmarks?	Floor Plan?	PDR or Odometry?
CARLOC [52]	Roadway landmarks	Y	Y
Zee [53]	N	Y	Y
UnLoc [128]	Magnetic, elevator and Wi-Fi	N	Y
LiFS [55, 219]	N	Y	Y
WILL [56]	N	Y	Y
[58]	Static node	N	Y

optimized fingerprints or evaluate the quality of fingerprints. These methods can prevent malicious users from submitting low-quality fingerprints. Next, [49, 50] design quality-aware pricing for crowdsensed fingerprints to ensure cost and quality at the same time. Moreover, the recent work [51] also takes the users' preferences into consideration and designs a personalized task matching mechanism.

Fingerprints Without Pain We can also facilitate the deployment of fingerprinting localization through relieving the overall workload for fingerprints collection. To realize such goal, the state-of-the-art researches consist of two categories:

First, since traditional MCS requires the user to report both fingerprints and locations, we can save cost through automatic labeling. In particular, the fingerprints are usually collected in mobile devices equipped with inertial measurement unit (IMU) [210–212, 216]. We can utilize pedestrian dead reckoning (PDR) method to transform these IMU readings into user's relative trajectory. Next, it is possible to leverage the indoor floor plan to localize the such relative trajectory in the map. Naturally, we can establish the relationship between fingerprints and real locations (Table 1.3).

Second, since there is a correlation between fingerprints at different locations, we can infer those missing fingerprints with limited fingerprints collection [213, 214, 220]. In this way, we can figure out the overall radio map with a few human labor [225, 226]. For example, [59, 60] set up the mathematical model to infer the fingerprints at missing locations, while [58, 64] leverage the PDR to assist the fingerprints recovery. Besides the indoor localization, the above proposed method is also suitable for outdoor localization such as [63] and [61]. Table 1.4 summarizes some recent work about radio map generation.

Table 1.4 Radio map generation

Research	Indoor or Outdoor?	Mathematical model	PDR?
[65]	Indoor	Unsupervised manifold alignment	N
[111]	Indoor	Reliability bounds analysis	N
[66]	Indoor	Approximate volume sampling	N
[60]	Indoor	Random forests	N
[67]	Indoor	Maximum likelihood and stochastic gradient descent	N
[68]	Indoor	Signal propagation model and autonomous fingerprints	N
[69]	Indoor	Tensor algebraic methods	N
[58]	Indoor	Peer-to-peer distance measurement	Y
[63]	Outdoor	Support vector machines	Y
[64]	Indoor	Hidden Markov model	Y

Table 1.5 Fine-grained self-calibrating indoor localization system comparison

System name	FineLoc [73]	Walkie-Markie [70]	PiLoc [71]	SenseWit [74]
Technology	BLE	Wi-Fi	Wi-Fi	only INS
Pathway floor plan	Yes	Yes	Yes	Yes
Sub-room classification	Yes	No	No	Yes
Obstacle recognition	Yes	No	No	No
Non-calibration	Yes	Yes	Yes	Yes
Unique calibration	Yes	Yes	Yes	No

Indoor Plan Construction Recently, some researches found there is a relationship between indoor plan and radio map. Consequently, they begin to explore whether it is possible to generate the indoor plan with such radio map, so that we can localize targets even without the prior knowledge about indoor map. Self-calibrating indoor localization systems are capable of constructing the radio map and even the floor plan automatically [70–72], where the opportunistically sensed data from users are smartly combined. The core idea of such systems is to exploit the notable feature of ambient Wi-Fi signals serving either as landmarks in the pathway environment [70] or as labels of users' trajectories [71, 72]. However, since previous works just provide coarse-grained indoor map, FineLoc designs a novel method to construct fine-grained map based on Bluetooth low-energy (BLE) nodes. We list these typical systems in Table 1.5, where we can see that some systems [70–72] can only generate pathway floor plan without any detailed information.

1.2.3 *The Accuracy of RSSI Localization*

Though Wi-Fi localization can yield a higher accuracy than GPS localization systems, the meter-level accuracy is not sufficient for some indoor applications. The root reason is that RSSI fingerprints are influenced by many factors including time, location, and obstacles in the environment. In order to improve the performance of RSSI localization, recent researchers have designed a series of methods, and these methods can be divided into the following categories:

(1) Model Optimization. We can optimize the model and strategy of indoor localization to remove noise and address temporal fluctuation problem in [5, 75, 76].
(2) Peer Devices. In a typical localization system, there might be more than one user, and these users' devices can measure RSSI with each other. Hence, some work [77, 78, 228–230] proposes to improve the accuracy with these measurements among peer devices.
(3) New Features. Besides RSSI measurements, we can also squeeze more features related to indoor localization such as the frequency of RSSI detection and multiple channels. So, some researches [79, 80] begin to dig out more features and design the corresponding algorithm for localization.
(4) Multi-modality Features. With more sensors embedded in smart devices, we can obtain multi-modality data to depict the state of users. Accordingly, it is possible to fuse RSSI with more information, so that we can witness a higher accuracy as presented in [15, 81].
(5) Anchor Placement. In practice, the placement of anchors is also an important factor that influences the accuracy. That is, unreasonable placement will make it difficult for us to reach the upper bound of localization accuracy. To address this issue, there are a lot of systems to study the best strategy for anchor placement such as [9] and [108]. To better understand these methods, we summarize and classify the related researches in Table 1.6.

1.2.4 Theoretical Analyses

Though there are many researches improving the accuracy of RSSI localization, there still lacks theoretical guidance on what accuracy indoor localization can achieve. Consequently, some fundamental theoretical questions have not been substantively answered. For example, what is the upper bound of wireless localization performance under ideal conditions? Is there a general method, independent of the deployment environment and equipment heterogeneity, that can optimize the wireless localization system fundamentally? Our book answers these questions to a certain extent.

1.2.5 From RSSI to CSI

Since the previous active localization systems require the user to carry intelligent devices anywhere, the recently proposed works also explore to localize the users

Table 1.6 The accuracy of Wi-Fi localization

Research	Category	Description
[75]	Model Optimization	Order-k max and min Voronoi Diagrams
[76]	Model Optimization	Principal component localization
[82]	Model Optimization	Fuzzy grid-prediction scheme
[30]	Model Optimization	Geometry-assisted localization
Wi-Dist [29]	Model Optimization	Indoor localization framework fusing noisy fingerprints
[28]	Model Optimization	Historical beacons assistant
[27]	Model Optimization	Information-theoretic measure
[26]	Model Optimization	Hypothesis testing
[25]	Model Optimization	Grey model predicting Kalman filter
[24]	Model Optimization	Supervised machine learning and Cramer–Rao lower bound
[23]	Model Optimization	Cramer–Rao lower bound
[6]	Model Optimization	Temporal diversity and spatial dependency of fingerprints
[124]	Model Optimization	Time-varying path-shadowing model
ViVi [4]	Model Optimization	Fingerprint spatial gradient
EZ [77]	Peer Devices	Wireless propagation and proposed EZ model
[78, 105]	Peer Devices	Peer-assisted localization approach
[79]	New Features	Time Synchronization Function (TSF) time stamps
[48]	New Features	Signal strength difference and hyperbolic location fingerprint
MuD [80]	New Features	Multiple radio paths
TOC [22]	New Features	Unique time of charge sequences among wireless sensors
[163]	New Features	Multi-channel signal strength
[18]	New Features	Fine time measurement (FTM)
[20]	New Features	Fine time measurement (FTM)
Wi-Go [19]	New Features	Fine time measurement (FTM)
[13]	New Features	RSS, Path-loss and Data Rate fingerprints
[17]	Multi-modality Features	Anonymous and id-linked sensor
[83]	Multi-modality Features	Pedometer and Wi-Fi measurements
[84]	Multi-modality Features	Pedometer and Wi-Fi measurements
Centaur [16]	Multi-modality Features	Radio frequency and acoustic signals
ZiFind [15]	Multi-modality Features	ZigBee and Wi-Fi signals
WAIPO [81]	Multi-modality Features	Wi-Fi and magnetic fingerprints
SoiCP [12]	Multi-modality Features	Pedometer, GPS, and Wi-Fi fingerprinting
iVR [11]	Multi-modality Features	Vision and radio
[10]	Multi-modality Features	Pedometer, magnetic, and Wi-Fi fingerprints
[9]	Anchor Placement	MaxL–minE algorithm
[8]	Anchor Placement	Elegant convex optimization problem and CRLB
[108]	Anchor Placement	Source position estimation for anchor position uncertainty
[7]	Anchor Placement	Mixed integer linear programming

Table 1.7 The summary of CSI indoor localization systems

Research	Category	Physical features
ArrayTrack [86]	Active Localization	AoA (WARP)
Phaser [94]	Active Localization	AoA
ToneTrack [95]	Active Localization	ToF (WARP)
SpotFi [87]	Active Localization	AoA and ToF
Chronos [227]	Active Localization	AoA and ToF
UAT [90]	Active Localization	AoA
RIM [120]	Active Localization	Time-Reversal Resonating Strength (TRRS)
LocAP [14]	Active Localization	AoA
[93]	Active Localization	ADoA
LiFS [89]	Passive Localization	Raw CSI readings
WiDAR [127]	Passive Localization	DFS
MaTrack [92]	Passive Localization	AoA
IndoTrack [91]	Passive Localization	DFS and AoA
WiDAR 2.0 [88]	Passive Localization	AoA, ToF, and DFS
mD-Track [119]	Passive Localization	AoA, ToF, and DFS
WiPolar [96]	Passive Localization	Polarization, AoA, ToF, and DFS

who do not carry devices [2, 3, 125]. However, the RSSI readings are so unsteady that we can hardly obtain accurate results within one meter, and this limits the deployment of indoor localization in those scenarios that require a high accuracy. This motivates the researchers to think about whether we can find another reliable signal feature to replace such course-grained RSSI. Consequently, the CSI based localization system is gradually becoming popular [85].

The ability of CSI acquirement [97] stimulates the development of indoor localization systems, where we can realize decimeter-level accuracy even with commercial Wi-Fi devices. In particular, CSI can reflect how the communication channel influences the phase and amplitude of signals. According to the properties of electromagnetic waves, the phase changes 2π when the signal propagates the distance of wavelength. Since the wavelength of Wi-Fi signals is usually $5 \sim 12$ cm, we own the ability to locate the target accurately. In fact, due to the periodicity of phase, it is difficult for us to tell how many cycles the signal has propagated. This makes the solution to CSI localization not so straightforward. To address this issue, the recently proposed researches develop a series of methods to extract physical features from the raw CSI readings. These physical features include the angle of arrival (AoA), time of flight (ToF), and doppler frequency offset (DFS). Specifically, the AoA depicts the arrived direction of signals, the ToF can reflect the distance of propagation, while DFS is related to the movement velocity of the target. Now, we can summarize off-the-shelf CSI localization systems into two categories, i.e., passive localization and active localization in Table 1.7.

Passive localization [1, 91, 92, 126, 127, 223] requires no sensor attached to targets and even achieves a better performance than device-based localization; however, the application of passive localization is limited for the following two reasons: First, the existing passive localization systems can not accurately identify users identities [99]. Second, passive localization systems suffer more from environment noise and can hardly localize multiple targets.

Active localization requires targets equipped with Wi-Fi devices. Some recent systems [227] can localize targets with one single Wi-Fi; however, these systems require special targets that can retrieve CSI and therefore cannot be utilized to localize smart phones and most regular devices around us.

Actually, despite higher accuracy, CSI localization systems suffer even more from the labor-intensive site survey. In particular, the crux of these CSI localization systems [86, 87, 94, 227] is AoA measurement; however, it is necessary to calibrate phase offsets for feasible AoA. To address this problem, ArrayTrack [86] manually calibrates this phase offset utilizing RF cables. Since this phase offset might change after device restarted, manual calibration is not feasible. Thus, Phaser [94] proposes a method to automatically calibrate phase offsets with prior knowledge of the AP's position and antenna array direction; CUPID [98] utilizes smartphone sensors and CSI to localize the target. However, an automatic calibration method of the AP's position and antenna array direction is still unavailable. In contrast to RSSI localization systems where only AP's position needs to be calibrated, direction calibration is special for CSI localization systems and is more difficult to be achieved. Therefore, it is necessary to automatically calibrate these information. This book will design a method to realize auto-calibration and provide a theoretical analysis of the AoA based localization method.

Chapter 2
Theoretical Model for RSS Localization

2.1 Radio Propagation Model

2.1.1 Theoretical Model

RSS localization consists of model based and fingerprinting based methods. The basic idea of RSS model based localization is to map the RSS value into distance between the transmitter and the receiver. To this end, researchers have proposed several propagation models to depict such relationship, such as log-distance path loss (LDPL) model [85] and wall attenuation factor (WAF) [32]. Among the existing propagation models, the LDPL model is well known and has been employed in most RSS localization systems. Specifically, if we represent the average path loss at reference point d_0 as $\overline{PL_0}$, the measured path loss $PL(d)$ at distance d should satisfy the following equation:

$$PL(d) = \overline{PL_0} + 10\gamma lg\left(\frac{d}{d_0}\right), \tag{2.1}$$

where γ is the path loss exponent. However, the LDPL model does not consider the fact that environmental clutter might be vastly different at two different locations with the same transmitter–receiver distance. The measurement has shown that at any distance d, the path loss at a particular position is random and distributed log-normally about the mean value. As a result, the log-normal path loss (LNPL) model has been proposed to cover this issue. If we utilize P_T and $P_R(d)$ to, respectively, represent the transmitted and received power, we have

$$P_R(d) = P_T - \overline{PL_0} - 10\gamma lg\left(\frac{d}{d_0}\right) - \sigma\mathcal{Y}, \tag{2.2}$$

X. Tian et al., *Wireless Localization Techniques*, Wireless Networks,
https://doi.org/10.1007/978-3-031-21178-2_2

where $\mathcal{Y}$ is the normalized Gaussian random variable with $\mathcal{Y} \sim \mathcal{N}(0, 1)$ and σ is a constant representing the standard variance of the received signal.

2.1.2 Challenges and Motivations

The RSS model based localization systems rely on the assumption that RSS monotonically decreases with distance. Although this trend can be met on a relatively large scale, the shadowing σ will destroy local RSS trends especially in multipath rich indoor environment. Consequently, we can hardly distinguish some near locations because the large σ blurs the monotonic trend. In order to address such challenge in RSS model based localization systems, the RSS fingerprinting based methodology has been a seminal idea that induces many indoor localization systems with different flavors [85].

Most of the RSS fingerprinting based localization systems are implemented in IEEE 802.11 wireless local area network (WLAN) environment, where the RSS measured for frames sent from different access points (APs) is utilized to infer the user's location. A typical RSS fingerprint based method can be divided into two phases: *(1) Offline Phase:* the location fingerprints are collected in each measurement point to build the radio map by performing a site survey of RSS from multiple APs and *(2) Online Phase:* we estimate the user location by utilizing localization algorithm based on the user's RSS vector and the radio map. During the above process, there are the following challenges to be addressed.

Challenge 1: Accuracy and Reliability While efforts have been dedicated to the RSS fingerprinting based indoor localization in order to improve the accuracy and efficiency, performance of the RSS fingerprinting based methodology itself is still unknown in a theoretical perspective. Results of empirical studies are highly dependent on experimental environment and implementation [40, 41, 78, 100, 105, 280]. Theoretical analysis borrowed from ranging based cooperative localization in wireless sensor networks is based on ideal radio propagation model and unsuitable for fingerprinting based localization [40, 106–108]. The current lack of a theoretical insight into the RSS fingerprinting based methodology could incur the blindness for system designers: can we further improve the performance of the localization system with better implementations or this has been the best we can achieve with the methodology?

Challenge 2: Scalability of Fingerprinting Localization Systems There are two factors that limit the scalability of fingerprinting localization systems. The first one is the number of measurement points. As we know, the number of measurement points is large and RSS is time-variant. When numerous RSS fingerprints need to be recalibrated frequently, the site survey is both laborious and time-consuming.

Consequently, is it possible to propose a method to relieve the deployment cost of such a system? The second one is the number of users. In particular, when people are moving around in the indoor space, it continuously changes the radio propagation environment. Intuitively, the more human bodies are within the space, the more serious the consequent shadowing and multipath effects are, which results in more complicated radio propagation environment and more alienation of radio map constructed in the offline phase to the current reality. This motivates us to think about the following question: how the fingerprinting localization system scales with the number of users?

Challenge 3: Theoretical Guidance on Fingerprinting Localization Systems Wireless localization is a kind of technique to estimate the wireless terminal's location by utilizing characteristics of wireless communication signals. The technique has been attracting interests from both academia and industrial in the past decides; however, most of the research work is conducted using the methodology of "empirical study," that is, observing experimental results or phenomena, then coming up with a new idea, and then verifying the new idea with experiments again. A systematic theoretical study on the topic has been long needing but lacking. Answers to some fundamental issues of wireless localization technique are still unknown.

2.2 Accuracy and Reliability Analysis

This section presents a general probabilistic model to shed light on a fundamental issue: how good the RSS fingerprinting based indoor localization can achieve? Concretely, we present the probability that a user can be localized in a region with certain size. We reveal the interaction among accuracy, reliability, and the number of measurements in the localization process. Moreover, we present the optimal fingerprints reporting strategy that can achieve the best localization accuracy with given reliability and the number of measurements, which provides a design guideline for the RSS fingerprinting based indoor localization system. Furthermore, we analyze the influence of imperfect database information on the reliability of localization and find that the impact of imperfect information is still under control with reasonable number of samplings when building the database.

2.2.1 System Model

Consider an indoor space denoted by S, where the long and narrow space such as a corridor can be modelled as a one-dimensional Cartesian space with $S \subset \mathbb{R}$, and the ordinary space such as a room can be modelled as a two-dimensional Cartesian space with $S \subset \mathbb{R}^2$. We use $\mathbf{r}$ to denote a location in S, where $\mathbf{r} = x_1$ and $\mathbf{r} =$

(x_1, x_2) in the one-dimensional and two-dimensional Cartesian coordinate system, respectively. For both fingerprints collection and location determination, the mobile device reports the RSS readings obtained by measuring signals sent from each AP. The measured result is a random variable with respect to a specific location denoted by $\mathcal{P}(\mathbf{r})$

$$\mathcal{P}(\mathbf{r}) = \mu(\mathbf{r}) + \sigma\mathcal{Y}, \tag{2.3}$$

where $\mu(\mathbf{r})$ represents how the mean of RSS readings varies with respect to locations. $\mathcal{Y}$ is the normalized Gaussian random variable with $\mathcal{Y} \sim \mathcal{N}(0, 1)$, and σ is a constant representing the standard variance of the received signal.

Equation (2.3) is a generalized model derived from the LNPL model [77, 85]. If we let $\mu(\mathbf{r}) = P_T - \overline{PL_0} - 10\gamma lg(\frac{d}{d_0})$, where d is the distance between the location of the transmitter and $\mathbf{r}$, Eq. (2.3) degenerates to the LNPL model, where $\sigma\mathcal{Y}$ factually represents the shadowing effect. Extensive practical measurements and studies in the literature have revealed that the value of σ can be regarded as a constant in a certain region if the location of the transmitter is given, which has been widely adopted in the industrial standardization documents for radio propagation modeling [109, 110].

The LNPL model above only considers the path loss and shadowing but ignores the small-scale fading incurred by multi-path effect. The multi-path effect will result in changes in the received power, but it is extremely difficult to accurately predict how the multi-path effect will incur the change. To this end, we use a function $\mu(\mathbf{r})$ to denote the average aggregated effect of free space path loss and multi-path effect at location $\mathbf{r}$, and thus we generalize the original LNPL model and obtain the radio propagation model as shown in Eq. (2.3). Such modeling approach is also adopted by a number of work focusing on characterizing wireless communication channels such as in [102, 137, 138], and the Gaussian assumption is validated by a number of work on indoor localization [38, 39, 77, 101, 280].

As the real wireless environment is very unpredictable, we are unable to know exactly what the mean value of the measured RSSes $\mu(\mathbf{r})$ is like. However, it is observed from many previous experiments [38, 77, 100, 280] that the mean of measured RSSes changes in a non-dramatic manner with the small change of locations, which makes it reasonable to assume that $\mu(\mathbf{r})$ is continuous. Thus we can have the following approximation:

$$\mu(\mathbf{r}') \approx \mu(\mathbf{r}) + \nabla\mu(\mathbf{r})(\mathbf{r}' - \mathbf{r}), \tag{2.4}$$

where $\mathbf{r}'$ is an estimated location of the user given the actual location of the user $\mathbf{r}$. Note that all the analysis based on Eq. (2.3) can be applied to the scenario using the LNPL model, which itself has been widely adopted in many indoor localization systems [38, 39, 77, 280]. The experimental results to be shown in Sect. 2.2.7 also support the modeling above.

In the training phase, the mobile device is randomly assigned a point in the indoor space, and the device randomly chooses an AP and measures the RSS fingerprint once. As the result of each measurement is a random variable as shown in Eq. (2.3), all possible outcomes for the measurement form a sample space Ω. We define a σ-algebra $\mathcal{F}$ that is the collection of all events, where each event is a set containing zero or more outcomes. Equation (2.3) gives an assignment of probabilities to events $\mathcal{P} : \Omega \rightarrow [0, 1]$, and thus we can construct a probability space $(\Omega, \mathcal{F}, \mathcal{P})$. Suppose that the mobile device performs the measurement n times at a given assigned point, and then the Cartesian product of probability spaces induced by RSS measurements forms an n-dimensional probability space $(\Omega^n, \mathcal{F}^n, \mathcal{P})$, where we abuse $\mathcal{P} : \Omega^n \rightarrow [0, 1]$ for the convenience of demonstration.

The location determination phase can be considered as a mapping $\mathcal{M} : \Omega^n \rightarrow S$, $\mathbf{r}' = \mathcal{M}(\mathbf{o})$, where $\mathbf{o}$ is an outcome of Ω^n. It means that the localization system outputs an estimated location $\mathbf{r}'$ for a series of measurements of RSSes from randomly chosen APs. Since RSS measurement results are independently and identically distributed, the induced sample spaces of RSS measurements are orthogonal to each other, and Ω^n is homeomorphic to n-dimensional Cartesian space. In the coordinate system of n-dimensional Cartesian space, we can obtain a presentation of $\mathbf{o}$ denoted as $\mathbf{P} = [P_1, P_2, \ldots, P_n]^T$, in which P_i is the reading of each measurement result. Consequently, the original measurement $\mathcal{P} : \Omega^n \rightarrow [0, 1]$ becomes $f : \mathbb{R}^n \rightarrow [0, 1]$.

We use Q to denote the area that is centered at the user's actual location with radius δ in the physical space. $E(\delta)$ is used to denote the event in the sample space, which makes the localization system estimates the user's location to be in Q. We will use E to represent $E(\delta)$ in the following discussion to avoid tedious mathematical presentation. The size of δ determines the *accuracy* of the localization system. We use R to denote the probability that the user's estimated location is within the area Q, which is defined as the *reliability* of the localization system.

With the definition of accuracy and reliability, we can see that the probability of the event that the system correctly estimates the user's location is

$$R(E) = \int_E f(\mathbf{P})d\mathbf{P} = \int_Q g(\mathbf{r}', \mathbf{r})d\mathbf{r}', \tag{2.5}$$

where $f(\mathbf{P})$ is the possible measurement of the sample space in the n-dimensional Cartesian coordinate system, and $g(\mathbf{r}', \mathbf{r})$ is the probability distribution function that the user is localized at $\mathbf{r}'$ given that the real location of the user is $\mathbf{r}$ in the physical space. Equation (2.5) indicates that the reliability can be interpreted as either the probability that the measurement falls into the event region E or the user is localized in an area that is centered at $\mathbf{r}$ and with radius δ. Based on such a model, we are to calculate the one-dimensional localization reliability for the case of one-time measurement for a single AP and then extend the result to multiple-time measurements for multiple APs in the following part.

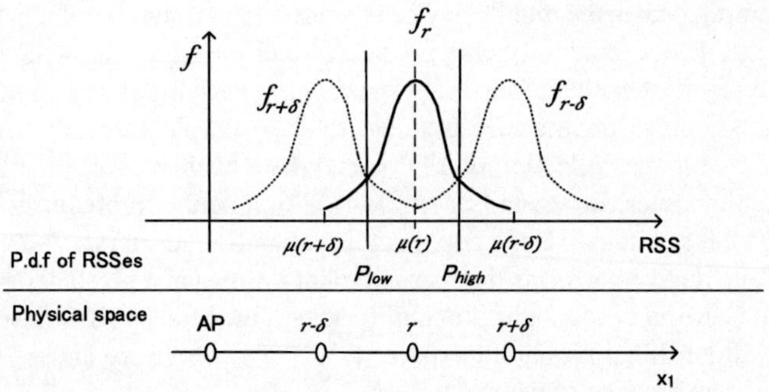

Fig. 2.1 Integration domain for one-dimensional localization with single measurement for single AP

2.2.2 *Localization in One-Dimensional Space*

2.2.2.1 One-Time Measurement for Single AP

We set the origin of the spatial coordinate system at the location of the sole AP in the one-dimensional physical space; the corresponding probability density function (PDF) for each location in the sample space could be represented as shown in Fig. 2.1.

The location in one-dimensional physical space is a scalar, and the δ neighborhood of the user's actual location r is the line segment from $r-\delta$ to $r+\delta$. Note that the farther the location is from the AP, the smaller mean value could be observed at the location; therefore, $\mu(r+\delta)$ is less than $\mu(r-\delta)$. The probability the user is localized in the δ neighborhood of r (denoted by Q) is equivalent to that the reported RSSes fall within the range between P_{high} and P_{low} according to the principle of maximum likelihood estimation (MLE), which is the event E in this case. Due to the symmetry of the PDF according to Eq. (2.3), it is straightforward that $f_{r-\delta}(P_{high}) = f_r(P_{high})$ and $f_{r+\delta}(P_{low}) = f_r(P_{low})$. We have

$$\begin{cases} P_{high} = \frac{\mu(r-\delta)+\mu(r)}{2}, \\ P_{low} = \frac{\mu(r+\delta)+\mu(r)}{2}, \end{cases} \tag{2.6}$$

thus the reliability

$$R(\delta, r, \sigma) = \int_{P_{low}}^{P_{high}} f_r(P)dP = erf\left(|\frac{-\mu'(\mathbf{r})\delta}{2\sqrt{2}\sigma}|\right), \tag{2.7}$$

where $erf(\cdot)$ is the error function defined as $\text{erf}(x) = \frac{2}{\sqrt{\pi}} \int_0^x e^{-t^2}\,\text{d}t$.

2.2.2.2 Multiple Measurements for Multiple APs

According to the probability theory, the average of n i.i.d. Gaussian variables is equivalent to a Gaussian variable with a standard deviation $\frac{\sigma}{\sqrt{n}}$. Two measurements to a single AP can be regarded as measurements for two identical APs located at the same place, which is to be confirmed by our result shown in Eq. (2.16). If several measurements are performed on a single AP, the RSS fingerprint is set to be a new random variable with the standard deviation $\frac{\sigma}{\sqrt{n}}$.

The probability the user is localized in Q is equivalent to the probability the user's measurement of the RSS P_i falls into the event E:

$$E = \left\{ o | \prod_{i=1}^{n} f_r(P_i) \geq \prod_{i=1}^{n} f_{r+\delta}(P_i), \prod_{i=1}^{n} f_r(P_i) \geq \prod_{i=1}^{n} f_{r-\delta}(P_i) \right\}. \tag{2.8}$$

According to the radio propagation model, Eq. (2.3), the outcomes in the event E satisfy the following inequality:

$$\prod_{i=1}^{n} \frac{1}{\sigma_i \sqrt{2\pi}} e^{-\frac{(P_i - \mu_i(r))^2}{2\sigma^2}} \geq \prod_{i=1}^{n} \frac{1}{\sigma_i \sqrt{2\pi}} e^{-\frac{(P_i - \mu_i(r \pm \delta))^2}{2\sigma^2}}. \tag{2.9}$$

After simplification, it is equivalent to

$$\sum_{i=1}^{n} \frac{\mu_i(r \pm \delta) - \mu_i(r)}{\sigma_i} \left(Y_i - \frac{\mu(r \pm \delta) - \mu(r)}{2\sigma_i} \right) \leq 0, \tag{2.10}$$

where $\mu_i(r)$ is the average RSS of AP i at r.

We normalize RSS readings P_i with respect to $Y_i = \frac{P_i - \mu_i}{\sigma_i}$, where σ_i might differ among different APs. For a given location of the receiver, the received signal propagated from different APs can be through different paths. A simple example is that one AP's signal is propagated through the line-of-sight (LOS) channel and another AP's signal could be propagated through the non-line-of-sight (NLOS) channel to the receiver; therefore, the observed values of σ for different APs can be different [109]. The distance between two APs is usually hundreds of meters in practice to save the infrastructure investment, since one AP could cover an area of radius about 100 m; moreover, densely deploying APs could incur interference. Consequently, it is reasonable to assume that the distance between two APs for localization is larger than the dimension of δ neighborhood of a receiver's real location. This is why we assume that the value of σ is constant for a given AP as shown in Eq. (2.3), but the values of σ with respect to different APs are different. The experiment results to be shown in Sect. 2.2.7 also support such modeling.

We use $\boldsymbol{\mu}(r)$ to denote the mean RSS outcome at r. Now that it refers to the outcome itself, this notation does not depend on any coordinate system. Given a specific average RSS $\boldsymbol{\mu}(r)$, the set of $\{Y_i\}$ forms a coordinate basis for the sample space, where the origin is $\boldsymbol{\mu}(r)$, and each dimension is suppressed by a factor of σ_i. We use this coordinate system to characterize the event E and its probability in the following several sections due to its simplicity for the small-scale analysis.

Apparently, the two constraints shown in Eq. (2.10) are two non-parallel hyperplanes in the sample space. Vectors $\mathbf{h_1} = [\frac{1}{2\sigma_1}(-\mu_1(r)+\mu_1(r-\delta)), \frac{1}{2\sigma_2}(-\mu_2(r)+\mu_2(r-\delta)), \ldots, \frac{1}{2\sigma_n}(-\mu_n(r)+\mu_n(r-\delta))]^T$ and $\mathbf{h_2} = [\frac{1}{2\sigma_1}(-\mu_1(r)+\mu_1(r+\delta)), \frac{1}{2\sigma_2}(-\mu_2(r)+\mu_2(r+\delta)), \ldots, \frac{1}{2\sigma_n}(-\mu_n(r)+\mu_n(r+\delta))]^T$ together span a plane $\mathcal{W}$. As restrictions to the event E, Eq. (2.10) can then be rewritten in the vector form:

$$\begin{cases} 2\mathbf{h_1}(\mathbf{o}-\mathbf{h_1}) \leq 0, \\ 2\mathbf{h_2}(\mathbf{o}-\mathbf{h_2}) \leq 0. \end{cases} \tag{2.11}$$

It is important to note that $\mathbf{h_1}$ and $\mathbf{h_2}$ can denote both the normal vectors to each hyperplane and the two points on each hyperplane that are closest to the origin. This fact will be helpful to deal with the two-dimensional issue. Now that the PDF and the constraint conditions are all normalized, we can rotate the coordinate system $\{Y_i\}$ to another orthonormal basis $\{\mathbf{e_i}\}$, $i = 1, 2, \ldots, n$, where $\mathbf{e_1}$ is parallel to $\mathbf{h_1}$ and $\mathbf{e_2} \in \mathcal{W}$. Consequently, $\mathcal{W}$ is spanned by only two coordinate axes. The rest coordinate axes are therefore all orthogonal to the plane. There exists an orthonormal basis for subspace $\bar{\mathcal{W}}$, i.e., $\{\mathbf{e_i}\}$, $i = 3, \ldots, n$. Any outcome $\mathbf{o}$ in the sample space can be decomposed into $\mathbf{o} = \sum_i c_i \mathbf{e_i}$, where coefficients c_i are determined and unique for any given vector $\mathbf{o}$ and orthonormal basis $\{\mathbf{e_i}\}$. Equation (2.10) can then again be rewritten in the component form of the $\{\mathbf{e_i}\}$ basis:

$$\begin{cases} 2|\mathbf{h_1}\mathbf{e_1}|(c_1 - |\mathbf{h_1}\mathbf{e_1}|) \leq 0, \\ 2|\mathbf{h_2}\mathbf{e_1}|(c_1 - |\mathbf{h_2}\mathbf{e_1}|) + 2|\mathbf{h_2}\mathbf{e_2}|(c_2 - |\mathbf{h_2}\mathbf{e_2}|) \leq 0. \end{cases} \tag{2.12}$$

Thus the probability that the system correctly estimates the user's location is

$$\begin{aligned} R(E) &= \int_E f_r(\mathbf{P}) de^n \\ &= \int_{-\infty}^{c_1 \leq |\mathbf{h_1}|} de_1 \int_{-\infty}^{\frac{|\mathbf{h_2}\mathbf{e_1}|^2 + |\mathbf{h_2}\mathbf{e_2}|^2 - |\mathbf{h_2}\mathbf{e_1}|c_1}{|\mathbf{h_2}\mathbf{e_2}|}} \frac{1}{2\pi} e^{-\frac{e_1^2+e_2^2}{2}} de_2. \end{aligned} \tag{2.13}$$

Note that $f_r(\mathbf{P})$ is an n-variable Gaussian PDF. As of now, we successfully reduce the multiple integral to a much simpler two-dimensional one. Multivariate Gaussian integral equation (2.15) is integrated on the area indicated in Fig. 2.2.

By Eq. (2.4), the dimension of E can be further reduced, for that $\mathbf{h_1}$ and $\mathbf{h_2}$ will now be parallel to each other, though in different directions.

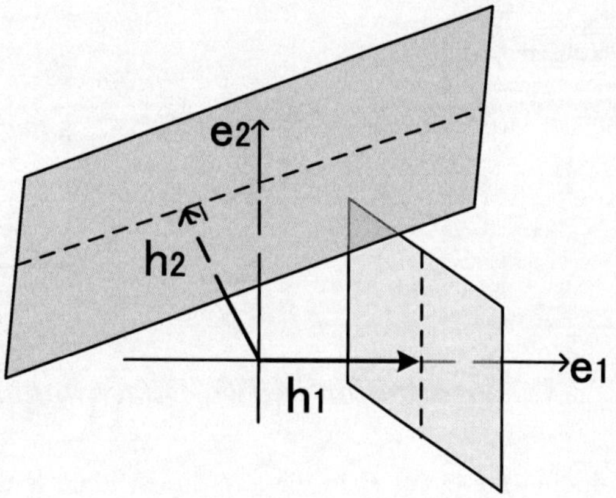

Fig. 2.2 Integration domain for one-dimensional localization with multiple measurements for multiple APs

$$\begin{cases} \mathbf{h}_1 = \left[-\frac{1}{2\sigma_1}\mu_1'(r)\delta, \ldots, -\frac{1}{2\sigma_n}\mu_n'(r)\delta\right]^T, \\ \mathbf{h}_2 = \left[\frac{1}{2\sigma_1}\mu_1'(r)\delta, \ldots, \frac{1}{2\sigma_n}\mu_n'(r)\delta\right]^T. \end{cases} \tag{2.14}$$

Thus the two constraint conditions shown in Eq. (2.10) are parallel to each other. $R(E)$ can then be simplified as

$$R(E) = \int_{|\mathbf{h}_2|}^{|\mathbf{h}_1|} de_1 \int_{-\infty}^{-\infty} \frac{1}{(\sqrt{2\pi})^2} e^{-\frac{e_1^2+e_2^2}{2}} de_2 \tag{2.15}$$

$$\approx erf\left(\frac{\sqrt{\sum_{i=1}^{n}(\frac{\mu_i'(r)\delta}{2\sigma_i})^2}}{\sqrt{2}}\right). \tag{2.16}$$

2.2.2.3 Discussions

We can see that Eq. (2.16) is equivalent to Eq. (2.7) when $n = 1$ meaning that there is a single AP in the room, which corroborates our analysis. Moreover, the analysis above reveals some insight into the design of the indoor localization. First, the more data are reported to the system, the more reliable the location estimation is, since $\sqrt{\sum_i(\frac{\mu_i'(r)\delta}{2\sigma_i})^2} > \sqrt{(\frac{\mu_i'(r)\delta}{2\sigma_i})^2}$. Second, if $\frac{\mu_i'(r)}{\sigma_i} \geq \frac{\mu_j'(r)}{\sigma_j}, \forall i, j \leq n$, the reliability of the result can be improved if the user reports AP_i's RSS rather than the other. This means that reporting a measurement of a cleaner channel (smaller σ) is more effective, and the sharper the signal varies around the user's location (greater $\mu'(r)$), the easier it is for the system to pinpoint the user's location.

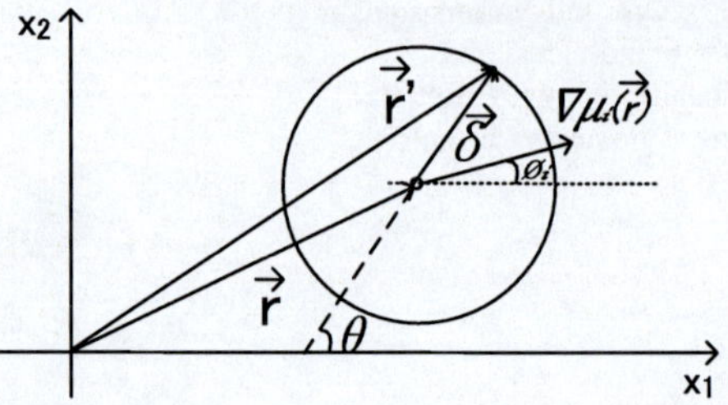

Fig. 2.3 Two-dimensional localization with multiple measurements for multiple APs

2.2.3 *Localization in Two-Dimensional Space*

Finding the event E in the two-dimensional localization is more challenging. To simplify the modeling, we first present a mathematical expression of Q in the physical space, based on which we try to find the shape of the event E in the sample space. We are to prove that E is a hyper-cylinder and then prove that the intersection between the hyper-cylinder and the cross-section plane is in the shape of an ellipse, which makes it possible to obtain the reliability through integration.

2.2.3.1 Multiple Measurements for Multiple APs

Figure 2.3 illustrates how to represent a location in the two-dimensional physical space, where the location of the user is $\mathbf{r}$ and the location of any point on the boundary of the area Q is $\mathbf{r}'$. We use $\boldsymbol{\delta} = \mathbf{r}' - \mathbf{r}$ to denote a two-dimensional vector with the direction from the user's actual location to any point on the boundary of Q. We use θ to denote the angle between $\boldsymbol{\delta}$ and the horizontal axis and use ϕ_i to denote the angle between $\nabla\mu_i(\mathbf{r})$ and the horizontal axis. By Eq. (2.4), we have

$$\mu_i(\mathbf{r}') - \mu_i(\mathbf{r}) = \nabla\mu_i(\mathbf{r})\boldsymbol{\delta} = \delta|\nabla\mu_i(\mathbf{r})|cos(\theta - \phi_i), \tag{2.17}$$

where $\delta = |\boldsymbol{\delta}|$.

We want to find the event that the user is localized within the area Q. According to MLE, this is equivalent to find E, where the probability density of the user's appearing on the boundary of Q is no greater than that of the user's appearing at $\mathbf{r}$:

$$E = \left\{ \mathbf{o} | \prod_{i=1}^{n} f_{\mathbf{r}}(P_i) \geq \prod_{i=1}^{n} f_{\mathbf{r}+\boldsymbol{\delta}}(P_i) \right\}. \tag{2.18}$$

All outcomes in E follow the inequality

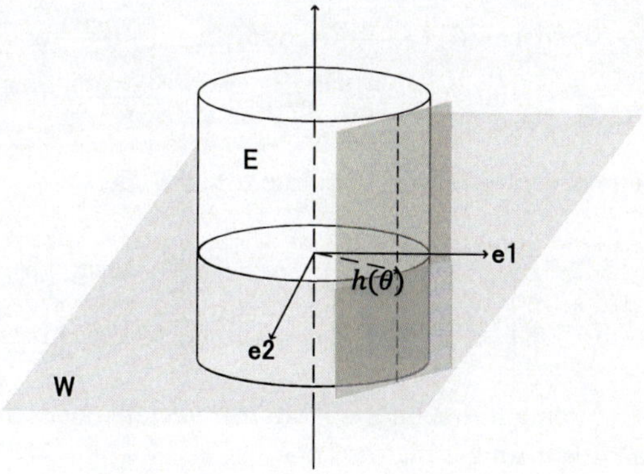

Fig. 2.4 Integral area of two-dimensional space

$$\sum_{i=1}^{n} \frac{\mu_i(\mathbf{r}+\boldsymbol{\delta}) - \mu_i(\mathbf{r})}{\sigma_i} \left(Y_i - \frac{\mu_i(\mathbf{r}+\boldsymbol{\delta}) - \mu_i(\mathbf{r})}{2\sigma_i} \right) \leq 0. \tag{2.19}$$

Substituting Eq. (2.17) into Eq. (2.19), we have the specific description of E:

$$\sum_{i=1}^{n} \frac{\delta|\nabla\mu_i(\mathbf{r})|}{\sigma_i} \cos(\theta - \phi_i) \left(Y_i - \frac{\delta|\nabla\mu_i(\mathbf{r})|}{2\sigma_i} \cos(\theta - \phi_i) \right) \leq 0. \tag{2.20}$$

Constraint condition equation (2.20) should hold true for any θ, and thus there will be an infinite set of hyperplanes surrounding event E in the sample space, as shown in Fig. 2.4. We define the n-dimensional normal vector of the hyperplane to be a function of θ:

$$\mathbf{h}(\theta) = \left[\frac{\delta|\nabla\mu_1(\mathbf{r})|}{2\sigma_1} \cos(\theta - \phi_1), \ldots, \frac{\delta|\nabla\mu_n(\mathbf{r})|}{2\sigma_n} \cos(\theta - \phi_n) \right]^T . \tag{2.21}$$

Theorem 1 *The orbit of* $\{\mathbf{h}(\theta)\}$ *and the origin are coplanar, i.e., on the same two-dimensional plane in the sample space.*

Proof 1 This is equivalent to prove that there exists a rank $n - 2$ complementary subspace $\bar{\mathcal{W}}$ of $\mathcal{W}$, where $\mathcal{W}$ is spanned by $\{\mathbf{h}(\theta)\}$. Formally, $\forall \mathbf{g} \in \bar{\mathcal{W}}, \forall \theta, \mathbf{h}(\theta) \cdot \mathbf{g} \equiv 0$, that is,

$$\left[\frac{\delta|\nabla\mu_1(\mathbf{r})|}{2\sigma_1} \{\cos\theta\cos\phi_1 + \sin\theta\sin\phi_1\}, \ldots, \right.$$
$$\left. \frac{\delta|\nabla\mu_n(\mathbf{r})|}{2\sigma_n} \{\cos\theta\cos\phi_n + \sin\theta\sin\phi_n\} \right] [g_1, \ldots, g_2]^T \equiv 0.$$

Consequently, we need to prove

$$\begin{cases} \cos(\theta)\left[\frac{\delta|\nabla\mu_1(\mathbf{r})|}{2\sigma_1}\cos\phi_1,\ldots,\frac{\delta|\nabla\mu_n(\mathbf{r})|}{2\sigma_n}\cos\phi_n\right][g_1,\ldots,g_n]^T \equiv 0, \\ \sin(\theta)\left[\frac{\delta|\nabla\mu_1(\mathbf{r})|}{2\sigma_1}\sin\phi_1,\ldots,\frac{\delta|\nabla\mu_n(\mathbf{r})|}{2\sigma_n}\sin\phi_n\right][g_1,\ldots,g_n]^T \equiv 0. \end{cases}$$

The equations above hold true for all θ, thus

$$\begin{cases} \left[\frac{\delta|\nabla\mu_1(\mathbf{r})|}{2\sigma_1}\cos\phi_1,\ldots,\frac{\delta|\nabla\mu_n(\mathbf{r})|}{2\sigma_n}\cos\phi_n\right][g_1,\ldots,g_n]^T \equiv 0, \\ \left[\frac{\delta|\nabla\mu_1(\mathbf{r})|}{2\sigma_1}\sin\phi_1,\ldots,\frac{\delta|\nabla\mu_n(\mathbf{r})|}{2\sigma_n}\sin\phi_n\right][g_1,\ldots,g_n]^T \equiv 0. \end{cases} \tag{2.22}$$

Adding $n-2$ lines of zero row vectors under the row vector in Eq. (2.22) makes an $n \times n$ square matrix $\mathcal{H}$:

$$\mathcal{H} = \begin{pmatrix} \frac{\delta|\nabla\mu_1(\mathbf{r})|}{2\sigma_1}\cos(\phi_1) & \ldots & \frac{\delta|\nabla\mu_n(\mathbf{r})|}{2\sigma_n}\cos(\phi_n) \\ \frac{\delta|\nabla\mu_1(\mathbf{r})|}{2\sigma_1}\sin(\phi_1) & \ldots & \frac{\delta|\nabla\mu_n(\mathbf{r})|}{2\sigma_n}\cos(\phi_n) \\ 0 & .. & 0 \end{pmatrix}.$$

Then $\bar{\mathcal{W}}$ is the solution space to the linear formula: $\mathcal{H}\mathbf{g} = 0$, where $rank(\mathcal{H}) \leq 2$, so $rank(\bar{\mathcal{W}}) \geq n-2$; therefore, $rank(W) = n - rank(\bar{\mathcal{W}}) \leq 2$. The straight line connecting $\mathbf{h}(\theta)$ and $\mathbf{h}(-\theta)$ will always come across the origin, and thus the origin is also in plane $\mathcal{W}$.

Equation (2.17) means that $\mathcal{W}$ is a tangent plane of surface M at $\boldsymbol{\mu}(\mathbf{r})$, where $M = \{\boldsymbol{\mu}(\mathbf{r}')|\mathbf{r}' \in S\}$ is the mean surface of RSS readings. Repeat the technique we used in Sect. 2.2.3, and we will be again able to reduce the multivariate integral in the whole event to a two variable integral on a subset of plane $\mathcal{W}$. The next is to determine the domain of probability integration. By definition, it is the area inside the envelop of $\{\mathbf{h}(\theta)\}$.

Theorem 2 *The orbit of* $\{\mathbf{h}(\theta)\}$ *is an ellipse.*

Proof 2 Theorem 1 states that the orbit of $\{\mathbf{h}(\theta)\}$ is in a plane. To prove Theorem 2 is equivalent to prove that $\forall\theta, \exists\psi$,

$$\mathbf{h}(\theta) = \mathbf{U}\cos(\psi) + \mathbf{V}\sin(\psi), \tag{2.23}$$

where $\mathbf{U}$ and $\mathbf{V}$ are constant vectors and $\mathbf{UV} = 0$. If there exists such a constant $\alpha = \theta - \psi$ satisfying Eq. (2.23), then Theorem 2 is proven.

$$\begin{aligned} \mathbf{h}(\theta) &= \sum_{i=1}^{n} \frac{\delta|\nabla\mu_i(\mathbf{r})|}{2\sigma_i}\{\cos(\psi + \alpha - \phi_i)\}Y_i \\ &= \sum_{i=1}^{n} \frac{\delta|\nabla\mu_i(\mathbf{r})|}{2\sigma_i}\{\cos(\alpha - \phi_i)\cos\psi - \sin(\alpha - \phi_i)\sin\psi\}Y_i. \end{aligned} \tag{2.24}$$

If **U** and **V** are assigned as follows, then Eq. (2.24) is satisfied.

$$\begin{cases} \mathbf{U} = \sum_i \left\{ \frac{\delta |\nabla \mu_i(\mathbf{r})|}{2\sigma_i} \cos(\alpha - \phi_i) Y_i \right\}, \\ \mathbf{V} = \sum_i \left\{ -\frac{\delta |\nabla \mu_i(\mathbf{r})|}{2\sigma_i} \sin(\alpha - \phi_i) Y_i \right\}. \end{cases} \tag{2.25}$$

Thus $\mathbf{UV} = 0$ is equivalent to

$$f(\alpha) = \sum_{i=1}^{n} \left(\frac{\delta |\nabla \mu_i(\mathbf{r})|}{2\sigma_i} \right)^2 \sin(2\alpha - 2\phi_i) = 0. \tag{2.26}$$

Apparently, formula (2.26) has four different solutions of α in the interval between 0 and 2π because $f(\alpha) = -f(\alpha + \pi/2)$ and $f(\alpha)$ is a continuous function. There will be four zero points within each 2π period. The four solutions actually correspond to four different assignments of vector **U** to the semi-major axes and semi-minor axes. However, as we are only interested in the length of the semi-major axis and semi-minor axis, all four kinds of assignments are the same. We will use **U** as the semi-major axis in the following sections.

$$\sum_i \left(\frac{\delta |\nabla \mu_i(\mathbf{r})|}{2\sigma_i} \right)^2 (\sin(2\alpha)\cos(2\phi_i) - \cos(2\alpha)\sin(2\phi_i)) = 0, \tag{2.27}$$

where

$$\tan 2\alpha = \frac{\sum_{i=1}^{n} \left(\frac{\delta |\nabla \mu_i(\mathbf{r})|}{2\sigma_i} \right)^2 \sin(2\phi_i)}{\sum_{i=1}^{n} \left(\frac{\delta |\nabla \mu_i(\mathbf{r})|}{2\sigma_i} \right)^2 \cos(2\phi_i)}.$$

Thus we have

$$\begin{cases} |\mathbf{U}| = \sqrt{\sum_{i=1}^{n} \left\{ \frac{\delta |\nabla \mu_i(\mathbf{r})|}{2\sigma_i} \cos(\alpha - \phi_i) \right\}^2}, \\ |\mathbf{V}| = \sqrt{\sum_{i=1}^{n} \left\{ \frac{\delta |\nabla \mu_i(\mathbf{r})|}{2\sigma_i} \sin(\alpha - \phi_i) \right\}^2}. \end{cases} \tag{2.28}$$

Consequently, the reliability of the location estimation in the two-dimensional space is

$$
\begin{aligned}
R(E) &= \int_{\frac{e_1^2}{|\mathbf{U}|^2}+\frac{e_2^2}{|\mathbf{V}|^2}=1} \frac{1}{2\pi} e^{-\frac{e_1^2+e_2^2}{2}} de_1 de_2 \\
&= |\mathbf{U}||\mathbf{V}| \frac{1}{2\pi} \int_0^{2\pi} \frac{1-e^{-\frac{\cos^2\psi|\mathbf{U}|^2+\sin^2\psi|\mathbf{V}|^2}{2}}}{\cos^2\psi|\mathbf{U}|^2+\sin^2\psi|\mathbf{V}|^2} d\psi.
\end{aligned} \tag{2.29}
$$

2.2.3.2 Discussions

Most of the conclusions in the one-dimensional situation still hold true in the two-dimensional case. More data will yield higher reliability; however, there are some distinguishable properties in the two-dimensional case worth of mentioning. First, if the $\nabla\mu_i(\mathbf{r})$ for all APs are the same, which means that $\phi_i = \phi_j, \forall i, j$, then it is impossible to determine the user's location. This is because $|\mathbf{V}| = 0$ in this case, thus $R(E) = 0$. It means that there should be at least two APs, and the corresponding directions of $\nabla\mu_i(\mathbf{r})$ are different from each other. Second, if the user observes that $\nabla\mu_i(\mathbf{r})$ and $\nabla\mu_j(\mathbf{r})$ for two APs i and j are either in the same direction or in the opposite direction, then it is just like in the one-dimensional case, and thus if $\frac{|\nabla\mu_i(r)|}{\sigma_i} \geq \frac{|\nabla\mu_j(r)|}{\sigma_j}, \forall i, j \leq n$, reporting the RSS reading from AP i is more effective than reporting that of AP j for location determination.

2.2.4 *Best Strategy for Location Determination*

The analysis above shows that the utilities for reporting RSS fingerprints from different APs are different in the location determination process. A natural question to ask is: which fingerprints should the user report to the system so that the most accurate location estimation can be obtained? Before revealing the answer to such a question, we first present the fundamentals of the location determination.

2.2.4.1 Fundamentals of Location Determination

The fundamental issue of location determination is that can every outcome in the sample space be mapped into a location in the physical space. The mean of RSSes $\mu(\cdot)$ is a continuous mapping from the physical space to the mean surface of RSSes M. According to Eq. (2.4), each small area around $\mu(\mathbf{r})$ can be approximated as a plane. Recall that the event E in the two-dimensional case is a hyper-cylinder. According to Theorem 2, the intersection of the hyper-cylinder and M forms an orbit, which is the same two-dimensional plane as $\mu(\mathbf{r})$; therefore, if we shrink δ to zero, then the hyper-cylinder will shrink to an $n-2$-dimensional body $\bar{\mathcal{W}}$ and intersect with M at $\mu(\mathbf{r})$. We use Fig. 2.5a to illustrate a simple example of 3D

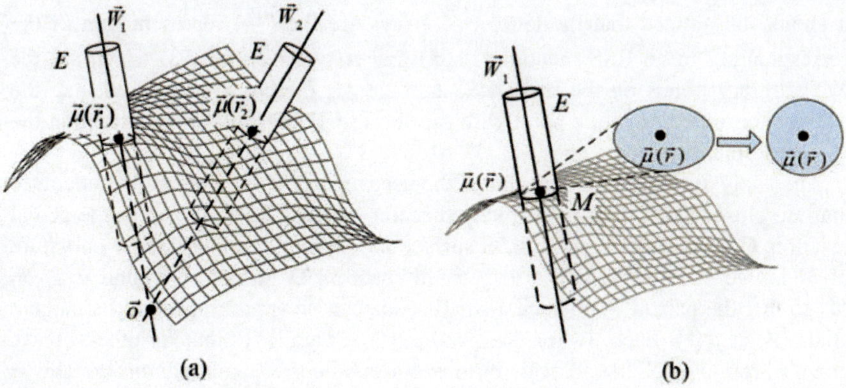

Fig. 2.5 Sample space of RSSes. (**a**) Fundamentals of location determination. (**b**) Best RSS reporting strategy

sample space. $\bar{\mathcal{W}}$ is the event that the user is estimated to be most likely appearing at $\mathbf{r}$ because E is the event that the system estimates the user's location in the area with a radius no more than δ.

Consequently,

$$\mathbf{r} = \mathcal{M}(\bar{\mathcal{W}}), \tag{2.30}$$

where $\mathcal{M} : \Omega^{n-2} \rightarrow S$ is a mapping from the set of outcomes $\bar{\mathcal{W}}$ to the user's most likely location $\mathbf{r}$.

However, it is worth noting that we in fact abuse the notation $\mathbf{r}$ here, since the location obtained from the mapping $\mathcal{M}$ is not necessarily the actual location of the user. To see this, recall that $\bar{\mathcal{W}}$ is an $(n-2)$-dimensional body perpendicular to the tangent plane of M at $\boldsymbol{\mu}(\mathbf{r})$ in the $\{\frac{P_i}{\sigma_i}\}$ coordinate system, and M is a surface with curvature, and thus it may happen that $\bar{\mathcal{W}}_1$ and $\bar{\mathcal{W}}_2$ that induced by two tangent planes of M intersect at an outcome, and this outcome can be mapped into two different points on M. This scenario is illustrated as in Fig. 2.5a, where the outcome $\mathbf{o}$ could be mapped into both $\mathbf{r_1}$ and $\mathbf{r_2}$. That is, the same set of RSSes can result in different localizations.

If this happens, we also want to use MLE to derive which location the user is more likely to appear. By the definition of MLE, the intersection outcome should be mapped to the point $\mathbf{r}'$ on M, if the inequality $f_{\mathbf{r}'}(\mathbf{P}) > f_{\mathbf{r}''}(\mathbf{P})$ is satisfied, where $f_{\mathbf{r}'}(\mathbf{P}) = \prod_{i=1}^{n} \frac{1}{\sigma_i\sqrt{2\pi}} e^{-(P_i-\mu_i(\mathbf{r}'))^2/2\sigma_i^2}$, $f_{\mathbf{r}''}(\mathbf{P}) = \prod_{i=1}^{n} \frac{1}{\sigma_i\sqrt{2\pi}} e^{-(P_i-\mu_i(\mathbf{r}''))^2/2\sigma_i^2}$, and $\mathbf{r}'$ and $\mathbf{r}''$ can be any points in the physical space. If $f_{\mathbf{r}'}(\mathbf{P}) > f_{\mathbf{r}''}(\mathbf{P})$, then $(\frac{P_i-\mu_i(\mathbf{r}')}{\delta_i})^2 < (\frac{P_i-\mu_i(\mathbf{r}'')}{\delta_i})^2$. This means that in the $\{\frac{P_i}{\sigma_i}\}$ coordinate system, the Euclidean distance from the outcome $\mathbf{o} = [\frac{P_1}{\sigma_1}, \frac{P_2}{\sigma_2}, \ldots, \frac{P_n}{\sigma_n}]^T$ to the point $[\frac{\mu_1(\mathbf{r}')}{\sigma_1}, \frac{\mu_2(\mathbf{r}')}{\sigma_2}, \ldots, \frac{\mu_n(\mathbf{r}')}{\sigma_n}]^T$ should be less than that to $[\frac{\mu_1(\mathbf{r}'')}{\sigma_1}, \frac{\mu_2(\mathbf{r}'')}{\sigma_2}, \ldots, \frac{\mu_n(\mathbf{r}'')}{\sigma_n}]^T$.

It should be noticed that the latter two arrays are the $\{\frac{P_i}{\sigma_i}\}$ representation of the corresponding mean RSS outcomes $\boldsymbol{\mu}(\mathbf{r}')$ and $\boldsymbol{\mu}(\mathbf{r}'')$, respectively, which are the two arbitrary points on the mean RSS surface M; therefore, we should map the intersection outcome to the point with the shortest Euclidean distance to M in the $\{\frac{P_i}{\sigma_i}\}$ coordinate system.

The analysis above indicates that if the user reports to the system the outcomes that are closer to M, it is more likely that the user can be localized to the actual location. Considering the mean RSS surface M, no matter how great the curvature of M is, we can always find a very small space in Ω^n which is around $\boldsymbol{\mu}(\mathbf{r})$ on M, so that the part of M in such a small space can be approximated to its tangent plane $\mathcal{W}$ at $\boldsymbol{\mu}(\mathbf{r})$. Each $\bar{\mathcal{W}}$ for the given $\boldsymbol{\mu}(\mathbf{r}')$ is parallel to that for others. If we move $\mathbf{r}$ around on S, the point $\boldsymbol{\mu}_{\mathbf{i}}(\mathbf{r})$ moves around correspondingly on $\mathcal{W}$. This is equivalent to say that $\bar{\mathcal{W}}$ scans the entire small space in Ω^n.

Every point within the small space is on a unique $\bar{\mathcal{W}}$, and there must be a mapping from Ω^n to S for every point in the sample space around M. It is interesting to find that if the space is very small, the tangent plane approximation is more accurate, and thus mapping the outcome into the surface is almost the same to find the Euclidean distance. If the outcome is far from M, it may happen that there is no such a $\bar{\mathcal{W}}$ so that the outcome can be mapped to a location.

In conclusion, if M is a plane, there must exist a function from Ω^n to S; if M is with curvature, the nearer the reported outcomes to M, the more likely the system will return a reliable location estimation with accuracy δ.

2.2.4.2 Best Strategy

With revealing the fundamentals of location determination, we now derive which AP users should measure so that they can be localized with the highest accuracy with the given reliability. Let $\mathbb{U} = \{AP_i\}, i = 1, \ldots, m$, be the set of all APs that can be sensed by the user's mobile device. A measurement strategy is defined as a sequence of measurements on APs and is denoted by $\mathcal{V}_n = (s^1, \ldots, s^n), s^j \in \mathbb{U}$. Note that the superscript of s^j is the index of the measurement in the sequence, and it does not necessarily mean that the measurement is performed on AP_j. One AP can be measured more than once in the sequence. The whole set of strategies is denoted as $\mathbb{U}^n$, where the size of the set is m^n. The optimal strategy is denoted by $\mathcal{V}_n^*, \mathcal{V}_n^* \in \mathbb{U}^n$.

Recall that the event E is a hyper-cylinder in the sample space and the intersection between the hyper-cylinder and M is an ellipse centered at $\boldsymbol{\mu}(\mathbf{r})$, as shown in Fig. 2.5b. We now consider another event $\mathcal{E}(c)$, which is also a hyper-cylinder in the sample space; however, we let the intersection between such a hyper-cylinder and M be a circle centered at $\boldsymbol{\mu}(\mathbf{r})$ and with radius c, where $\mathbf{r}$ is the actual location of the user as shown in Fig. 2.5b. In another perspective, $\mathcal{E}(c)$ denotes the event that the outcomes for localizing a user at $\mathbf{r}$ fall in the newly defined hyper-cylinder. Thus the reliability of the location estimation is in fact the probability of the event $\mathcal{E}(c)$, which is similar to the previous analysis:

$$\begin{aligned} R(\mathcal{E}(c)) &= \int_0^{2\pi} \int_0^c \frac{1}{2\pi} e^{-\frac{\rho^2 \cos^2 \psi + \rho^2 \sin^2 \psi}{2}} \rho d\rho d\psi \\ &= 1 - e^{-c^2/2}. \end{aligned} \tag{2.31}$$

Let us switch our attention to the physical space. We consider the vicinity of $\mathbf{r}$, where each point on the boundary of the vicinity represents an outcome in Ω^n. The vicinity is denoted as $\mathcal{U}$ and it must satisfy that the outcomes for localizing those location points on the boundary of $\mathcal{U}$ just fall on the circle on M. The point on the circle is denoted as $\mu(\mathbf{r}')$. Thus

$$\sum_{i=1}^{n} \frac{(\mu_i(\mathbf{r}') - \mu_i(\mathbf{r}))^2}{(2\sigma_i)^2} = c^2. \tag{2.32}$$

Put $\mathbf{r}$ and $\mathbf{r}'$ in the polar coordinate system with the origin at $\mathbf{r}$, then Eq. (2.32) can be transformed into $\sum_{i=n}^{n} (\rho(\theta)|\nabla\mu_i(\mathbf{r})|cos(\theta - \phi_i))^2/(2\sigma_i)^2 = c^2$, and thus we have

$$\rho^2(\theta) = \frac{4c^2}{\sum p_i \cos^2(\theta - \phi_i)}, \tag{2.33}$$

where $p_i = (|\nabla\mu_i(\mathbf{r})|/\sigma_i)^2$. For Eq. (2.33), if we let $Q_1 = \sum p_i \cos^2\phi_i$, $Q_2 = \sum p_i \sin^2\phi_i$, and $Q_3 = \sum 2 p_i \cos\phi_i \sin\phi_i$, we have

$$Q_1 \rho^2 \cos^2\theta + Q_2 \rho^2 \sin^2\theta + Q_3 \rho^2 \cos\theta \sin\theta = 4c^2, \tag{2.34}$$

which means that $\mathcal{U}$ is in fact an ellipse.

Define a complex parameter Z_i characterizing AP_i, where $Z_i = p_i e^{2i\phi_i}$, $\sum Z_i = \sum p_i e^{2i\phi_i}$, and $\sum Z_i^* = \sum p_i e^{-2i\phi_i}$. The area of $\mathcal{U}$ is denoted by u, where $u = 8\pi c^2/\sqrt{4Q_1Q_2 - Q_3^2}$. The area of the ellipse u profiles the accuracy of the localization. Recall that the event $\mathcal{E}(c)$ determines the area of $\mathcal{U}$ in the physical and $\mathcal{E}(c)$ is determined by outcomes the user has submitted. This means that which RSS fingerprints the user submitted determines the localization accuracy. To maximize the accuracy is equivalent to minimize u, and thus the best strategy for the user is to adopt the measurement sequence $\mathcal{V}_n^*$, where

$$\mathcal{V}_n^* = \arg\max\nolimits_{\mathcal{V}_n \in \mathbb{U}^n} \left\{ \left(\sum_{i \in \mathcal{V}_n} |Z_i| \right)^2 - |\sum_{i \in \mathcal{V}_n} Z_i|^2 \right\}. \tag{2.35}$$

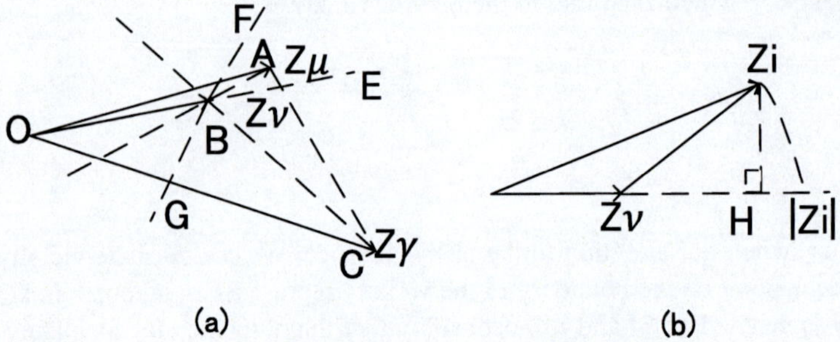

Fig. 2.6 Proof of Theorem 3. **(a)** Z_ν in the area of the triangle. **(b)** $|Z_i| - |Z_\nu| > |Z_i - Z_\nu| cos(\theta_i)$

It is indicated by Eq. (2.35) that the location determination system needs to search over the entire strategy profile $\mathbb{U}^n$ to find the optimal strategy. In the following discussion, we are to prove that we can narrow down the searching space by eliminating APs with small $|Z_i|$ from the set of all visible APs.

Theorem 3 *Suppose that a user can choose to measure AP_ν, AP_μ, and AP_γ, where the measurement for each AP is denoted by Z_ν, Z_μ, and Z_γ, respectively. If Z_ν falls inside the $\triangle OZ_\mu Z_\gamma$ in the complex plane, then $Z_\nu \notin \mathcal{V}_n{}^*$.*

Proof 3 Figure 2.6a shows an arbitrary Z_ν falls inside $\triangle OZ_\mu Z_\gamma$ in the complex plane. O is the origin. We use A and C to represent the corresponding point of Z_μ and Z_γ. We are to prove that any measurement such as Z_ν that falls in the area of the triangle must not be an element of $\mathcal{V}_n^*$ using contradiction.

If $Z_\nu \in \mathcal{V}_n^*$, let $T = \sum_{\mathcal{V}_n^*/\{Z_\nu\}} |Z_i|$ and $G = \sum_{\mathcal{V}_n^*/\{Z_\nu\}} Z_i$, where $\mathcal{V}_n^*/\{Z_\nu\}$ stands for the difference sequence of $\mathcal{V}_n^*$ eliminating an arbitrary measurement to AZ_ν. If there are multiple measurements to AP_ν, it makes no difference to eliminate any one of them, since the measurement order does not matter. $\mathcal{V}_n^*/\{Z_\nu\} \cup \{Z_i\}$ stands for the strategy $\mathcal{V}_n^*/\{Z_\nu\}$ plus a measurement to AP_i. T is a real number, while G could be a complex number. According to Eq. (2.35), we should have $u(\mathcal{V}_n^*) \leq u(\mathcal{V}_n), \forall \mathcal{V}_n \in \mathbb{U}$, where $u(\mathcal{V}_n)$ is the area of the ellipse in the physical space given the chosen $\mathcal{V}_n$. Then the following equations should hold true for both μ and γ:

$$\begin{cases} \left(\frac{8\pi c^2}{u(\mathcal{V}_n^*)}\right)^2 \geq \left(\frac{8\pi c^2}{u(\mathcal{V}_n^*/\{Z_\nu\}\cup\{Z_\mu\})}\right)^2, \\ \left(\frac{8\pi c^2}{u(\mathcal{V}_n^*)}\right)^2 \geq \left(\frac{8\pi c^2}{u(\mathcal{V}_n^*/\{Z_\nu\}\cup\{Z_\gamma\})}\right)^2. \end{cases} \tag{2.36}$$

This is equivalent to prove $(T + |Z_\nu|)^2 - |G + Z_\nu|^2 \geq (T + |Z_i|)^2 - |G + Z_i|^2$, for $i = \mu$ and γ. According to Eq. (2.35), we should prove that

$$\frac{G}{T}(Z_i - Z_\nu) \geq |Z_i| - |Z_\nu|. \tag{2.37}$$

Let θ_i be the angle between Z_ν and $Z_i - Z_\nu$, and then $|Z_i| - |Z_\nu| > |Z_i - Z_\nu| cos(\theta_i)$ as shown in Fig. 2.6b. This inequality still holds for the case $|Z_i| < |Z_\nu|$ or $\theta_i > \frac{\pi}{2}$, where the proof is straight and thus skipped due to the limitation of space. It is straightforward that $|G| < T$, and therefore $|\frac{G}{T}| < 1$. Thus if Z_ν were to be an element of $\mathcal{V}_n^*$, the following two equations would both be true:

$$e(Z_\mu - Z_\nu) > |Z_\gamma - Z_\nu| \cos(\theta_\mu), \tag{2.38}$$

$$e(Z_\gamma - Z_\nu) > |Z_\gamma - Z_\nu| \cos(\theta_\gamma), \tag{2.39}$$

where e is a unit vector.

In Fig. 2.6a, OB and OE are collinear. We draw two lines BF and BG so that $\angle EBA = \angle ABF$ and $\angle EBC = \angle CBG$. Equation (2.38) indicates the range of direction for e is from BE to BF (counterclockwise), and Eq. (2.39) indicates the range of direction for e is from BE to BG (clockwise); $\angle ABC < \pi$, which means that it is impossible for the two scopes to overlap, which means there is no such e that makes Eqs. (2.38) and (2.39) true at the same time, and the inequalities (2.36) cannot hold simultaneously. Consequently, Theorem 3 is proved by the contradiction.

Theorem 3 can be understood as follows: if we use a point on the convex plane to represent the measurement Z_i, then there will be many points on the plane representing all possible measurements. Only those points on the convex hull of all points are possible candidates of the best strategy. It is worth mentioning that parameters used for determining the best strategy can be derived by analyzing the fingerprints collected for each AP in the database. There is no need for information about the location of APs and no need for explicit efforts from users either. Finding the best strategy in a general case turns out to be non-trivial, and we present the details in another work [111].

2.2.5 *Impact on Localization in One-Dimensional Space*

2.2.5.1 Imperfect Information

The cornerstone underpinning our analysis on localization reliability above is the assumption: the distribution of the RSS at each location $\mathbf{r}$ is perfectly known. With such perfect information, we can construct a one-to-one mapping from the sample space to the physical space. In particular, we can always find a point on the mean surface of the RSS based on reported fingerprints, and the point on the mean surface $\boldsymbol{\mu}(\mathbf{r})$ corresponds to a location in the physical space $\mathbf{r}$.

Ideally, the mean of RSS readings at a given location can be perfectly known from the database, if the number of measurements at the location is large enough in the training phase; however, due to the cost of the training phase, the infor-

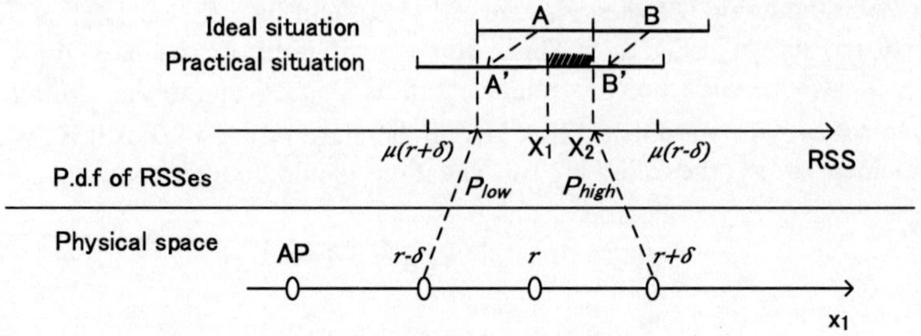

Fig. 2.7 One-dimensional localization with imperfect information

mation recorded at the fingerprints database is usually imperfect, and the current crowdsourcing based fingerprints collection is unable to guarantee the quality of submitted fingerprints. In particular, crowdsourcing workers submit the current location and corresponding RSS fingerprints observed to the localization server in an opportunistic sensing manner [70, 71, 77, 219], where the location of the reporting worker is estimated with dead reckoning by utilizing the inertial measurement unit (IMU) of the worker's mobile device such as accelerometer, magnetometer, and gyroscope [70, 71]. Due to the IMU error, the worker may incorrectly report the position where fingerprints are collected. As a result, the perfect information is usually unavailable in the database.

The consequence of the imperfect information is that the value of $\boldsymbol{\mu}(\mathbf{r})$ for each location in the physical space is inaccurate. A natural question to ask is: how the imperfect information will impact the reliability of location determination? In particular, what is the deviation from the true probability that a user can be correctly localized, which is incurred by the imperfect information? With a limited number of measurements in the training phase, what is the best localization reliability can be obtained? These important issues are to be addressed in the following.

2.2.5.2 One-Time Measurement for Single AP

Recall our investigation of the simple case where the fingerprint is measured only once with respect to a single AP. The domain E in the sample space corresponding to the δ neighborhood of $\mathbf{r}$ in the physical space is a line segment, where the endpoints of the segment are P_{high} and P_{low}, respectively. If the database has perfect information of fingerprints, sequential line segments in the physical space should be corresponding to sequential line segments in the sample space, as the ideal situation shown in Fig. 2.7, where A and B are midpoints of the two line segments, respectively.

However, the practical situation is that the information can be derived from the imperfect database, which means that the region E can migrate to somewhere else, as shown in the figure. We can use the midpoint to denote the line segment itself. If the values of submitted fingerprints fall in the shadow area as shown in Fig. 2.7, the server will determine that the user's physical location should be corresponding to the line segment B in the sample space with imperfect information; however, the user's actual location is in fact corresponding to the line segment A. The localization server can mistakenly determine the user's location due to the imperfect information.

Points on line segments in the practical situation could be regarded as points on line segments in the ideal situation after a random migration as shown in Fig. 2.7. For any point on the δ neighborhood of r in the physical space, there is a corresponding point in the sample space. We assume that the users appear on each point of the neighborhood with the same probability. We use x_0 to denote the corresponding point in the sample space for a given point in the physical space, X_1 to denote the right boundary of line segment A after the random migration, and X_2 to denote the right boundary of line segment A before the random migration.

In the user's perspective, the probability deviation of correct localization is the absolute value of the difference between the probability that the user should be localized in certain location with perfect information and that with imperfect information. If we define such a probability deviation as the probability error, then the probability error caused by the imperfect database in this particular case is

$$p_{e_1} = \int_{X_1}^{X_2} \frac{1}{L_0} \frac{1}{\sqrt{2\pi}\sigma} e^{-\frac{(x-x_0)^2}{2\sigma^2}} dx, \tag{2.40}$$

where $L_0 = 2\delta$.

We now consider a general case. Suppose that the length of the line segment A is L, and we set the origin of the horizontal axis to be at the left endpoint of line segment A, then $X_2 = L$, where $L = P_{high} - P_{low}$ as shown in Eq. (2.6). It is not straightforward to determine the coordinate of X_1 because points on the line segment A can migrate to anywhere in the sample space. A key observation is that X_1 is actually a P_{high} on the line segment A', thus $X_1 = L + \frac{r_{1x}+r_{2x}}{2}$, where r_{1x} and r_{2x} are deviations of the two midpoints A and B in the practical situation, respectively. Note that r_{1x} and r_{2x} are deviations along the horizontal axis, where deviating to the right is positive and to the left is negative. As a result, the probability error considering the general case as shown in the figure is

$$p_{e_2} = \int_{-\infty}^{\infty} \int_{-\infty}^{\infty} p_{e_1} \cdot \frac{N}{2\pi\sigma^2} e^{-\frac{N(r_{1x}^2+r_{2x}^2)}{2\sigma^2}} dr_{1x} dr_{2x}. \tag{2.41}$$

Since the error can also happen to the line segment A and the line segment left to A, the overall error probability is

$$P_e = 2\int_0^{L_0} p_{e_2} dr, \tag{2.42}$$

where r denotes the user's physical location in the coordinate system. Consider a very small δ, the mean of the RSS is not changing dramatically according to

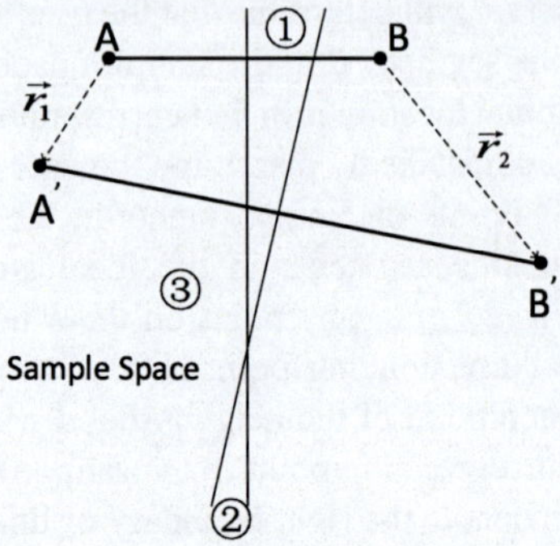

Fig. 2.8 Imperfect information with multiple measurements for multiple APs

Eq. (2.4), and thus we can apply local linearization to points in both sample space and physical space, which means that the length of a line segment in the sample space is proportional to that in the physical space:

$$\begin{aligned}\mu(\mathbf{r}') - \mu(\mathbf{r}) &\approx (\mathbf{r}' - \mathbf{r}) \cdot \nabla\mu(\mathbf{r}) \\ &\approx \left|(\mathbf{r}' - \mathbf{r})\right| \cdot |\nabla\mu(\mathbf{r})| \cdot \cos\varphi,\end{aligned} \tag{2.43}$$

where φ is the angle between two vectors: $\mathbf{r}' - \mathbf{r}$ and $\nabla\mu(\mathbf{r})$. In one-dimensional case, the angle φ would be either 0 or π, leading to $|\cos\varphi| = 1$ and

$$\left|\mu(\mathbf{r}') - \mu(\mathbf{r})\right| \approx \left|(\mathbf{r}' - \mathbf{r})\right| \cdot |\nabla\mu(\mathbf{r})| . \tag{2.44}$$

Then we have

$$P_e = 2\int_0^L p_{e2}\frac{L}{L_0}dx_0, \tag{2.45}$$

where every parameter can be obtained from the database in practice.

2.2.5.3 Multiple Measurements for Multiple APs

We now extend our analysis to the case where the database contains fingerprints measured multiple times with respect to multiple APs. Assume that the number of measurements is n, and then the sample space is n-dimensional. Suppose that nodes A and B are two points on the mean surface of the sample space. The challenge comes with the n-dimension is that the two nodes can migrate to any positions in the space, which results in that the drifted points may not be on the mean surface, thus making probability error analysis extremely complicated. This is illustrated in Fig. 2.8, where A' and B' denote the drifted means in the practical situation, respectively.

In Fig. 2.8, AB is a one-dimensional line segment and we could use a hyperplane to cut it in the middle. Since A and B denote means of two locations respectively, all reported fingerprints fall in the left side of the hyperplane should be determined to be at the location corresponding to A. All reported fingerprints fall in the right side of the hyperplane should be determined to be at the location corresponding to B. We use $\mathbf{r_1}$ and $\mathbf{r_2}$ to denote the deviation of A and B, respectively. Similarly, we can have hyperplane cut line segment $A'B'$ in the middle, and each side of the hyperplane represents those fingerprints that can entail two different localization results in the practical situation.

It is straightforward that if reported fingerprints fall in area 1, the location determination result with imperfect database is different from that with perfect database, which incurs error. The location of the user should be determined to be corresponding to the area of B, but it is determined to be at the location corresponding to A. If reported fingerprints fall in area 2, the location of the user should be determined to be corresponding to the area of A, but it is determined to be at that corresponding to B. Localization errors happen when any of the events happening because the imperfect information makes the serve believe that the boundary is the hyperplane intersecting with $A'B'$, while the real boundary is the one intersecting with AB.

The probability deviation that the user is correctly localized can be derived if area 1 and area 2 can be mathematically characterized; however, the challenge is that it is difficult to imagine the shape of areas in the n-dimensional sample space. Line segment AB is one-dimensional, so its bisecting hyperplane is $n-1$-dimensional. Similarly, the bisecting hyperplane of line segment $A'B'$ is also $n-1$-dimensional. Although AB and $A'B'$ are not in the same hyperplane, their bisecting hyperplanes intersect with each other, thus sharing $n-2$ dimensions. Consequently, we could always rotate the coordinate system, so that the projections of the two line segments are in the same plane while the rest of the shared $n-2$ dimensions orthogonal to the plane. That is, no matter how many dimensions the sample space has, we can always illustrate the situation in the sample space as shown in Fig. 2.8.

For any point in the δ neighborhood of r in the physical space, there is a corresponding point in the sample space. The point in the sample space can also be mapped into a two-dimensional surface after the coordinate system rotation as described above. We use (x_0, y_0) to denote the coordinate of the point in the system after rotation. Then the probability error for the particular case as shown in Fig. 2.8 is

$$
\begin{aligned}
p &= \int_{area1} \frac{1}{L_0} \frac{1}{2\pi\sigma^2} e^{-\frac{(x-x_0)^2+(y-y_0)^2}{2\sigma^2}} dxdy \\
&\quad - \int_{area2} \frac{1}{L_0} \frac{1}{2\pi\sigma^2} e^{-\frac{(x-x_0)^2+(y-y_0)^2}{2\sigma^2}} dxdy \\
&= \frac{1}{L_0}(p_1 - p_2),
\end{aligned}
$$

where $L_0 = 2\delta$. Note that fingerprints fall into area 3 will definitely make the system to localize the user at the location corresponding to A in both practical and ideal situation, and thus the deviation is only incurred by the difference between area 1 and area 2.

Consider the two hyperplanes bisecting AB and $A'B'$, and we now study the angle between the two hyperplanes so that the specific expression of probability error can be derived. It is straightforward that

$$\cos\theta = \frac{\mathbf{AB}\cdot\mathbf{A'B'}}{|AB|\,|A'B'|} = \frac{(L,0)\cdot(r_{2x}-r_{1x}+L, r_{2y}-r_{1y})}{L\times\sqrt{(r_{2x}-r_{1x}+L)^2+(r_{2y}-r_{1y})^2}},$$

$$\sin\theta = \frac{|r_{2y}-r_{1y}|}{L},$$

where r_{1x}, r_{1y}, r_{2x}, and r_{2y} denote deviations of A and B in two dimensions, respectively. Assume that N times of measurements are performed independently in the training stage to build up the fingerprints database, and then the drift distance of A and B follows the Gaussian distribution with mean 0 and variance value $\frac{\sigma}{\sqrt{N}}$, according to the law of large numbers. That is, $r_{1x} \sim N(0, \frac{\sigma}{\sqrt{N}})$, $r_{1y} \sim N(0, \frac{\sigma}{\sqrt{N}})$, $r_{2x} \sim N(0, \frac{\sigma}{\sqrt{N}})$, and $r_{2y} \sim N(0, \frac{\sigma}{\sqrt{N}})$. Consequently, the values of r_{1x}, r_{1y}, r_{2x}, and r_{2y} could appear in the range $\frac{-3\sigma}{\sqrt{N}} \sim \frac{3\sigma}{\sqrt{N}}$ with high probability. We are able to perform measurements many times so that r_{1x}, r_{1y}, r_{2x}, and r_{2y} are all small polynomial terms compared with L, and thus the following approximation can be obtained:

$$\cos\theta \approx \frac{|r_{2x}-r_{1x}+L|}{L}. \tag{2.46}$$

Since we have $\theta < \frac{|\mathbf{r_1}|+|\mathbf{r_2}|}{L}$, we can also make N big enough to have the following approximation

$$\sin\theta \approx \tan\theta \approx \theta. \tag{2.47}$$

Note that $\int_{area1} - \int_{area2} = \int_{area1+area3} - \int_{area2+area3}$. For the integration over area 1 and area 3, we have

$$p_1 = \int_{-\infty}^{\frac{L/2-x_0}{\cos\theta}} \frac{1}{2\pi\sigma^2} e^{-\frac{x^2}{2\pi\sigma^2}}\,dx, \tag{2.48}$$

$$p_2 = \int_{-\infty}^{\frac{L}{2}-x_0} \frac{1}{2\pi\sigma^2} e^{-\frac{x^2}{2\pi\sigma^2}}\,dx. \tag{2.49}$$

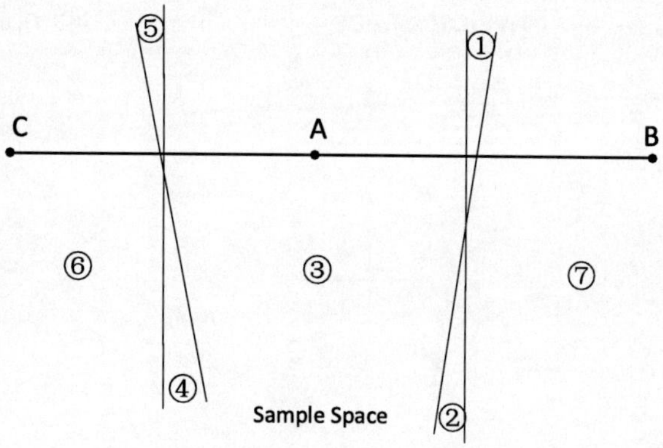

Fig. 2.9 One-dimensional localization with multiple measurements for multiple APs

Taking all possible situations for the data drift incurred by imperfect information into account, the probability error for location determination is

$$p = \left| \frac{1}{L_0} \left[erf\left(\frac{L/2 - x_0}{\cos\theta\sqrt{2}\sigma} \right) - erf\left(\frac{L/2 - x_0}{\sqrt{2}\sigma} \right) \right] \right|. \tag{2.50}$$

The error can also happen in the line segment AB and the line segment left to AB such as the line segment AC shown in Fig. 2.9. Consequently, the location determination error is determined by $area1 + area5 - area2 - area4$, which we could be put as $[(area1 + area3 + area4 + area5 + area6) - (area2 + area3 + area4 + area5 + area6)] + [(area5 + area1 + area2 + area3 + area7) - (area4 + area1 + area2 + area3 + area7)]$. Thus we get the final probability error for the multiple measurements over multiple APs

$$\begin{aligned} P_e &= 2\int_0^L \int_{-\infty}^{\infty} \int_{-\infty}^{\infty} \int_{-\infty}^{\infty} \int_{-\infty}^{\infty} \frac{N^2}{4\pi^2\sigma^4} e^{-\frac{N\left(r_{1x}^2 + r_{2x}^2 + r_{1y}^2 + r_{2y}^2\right)}{2\sigma^2}} p \\ &\quad \times dr_{1x} dr_{1y} dr_{2x} dr_{2y} \frac{L}{L_0} dx_0 \\ &= 2\int_0^L \int_{-\infty}^{\infty} \int_{-\infty}^{\infty} \int_{-\infty}^{\infty} \int_{-\infty}^{\infty} \frac{N^2}{4\pi^2\sigma^4} e^{-\frac{N\left(r_{1x}^2 + r_{2x}^2 + r_{1y}^2 + r_{2y}^2\right)}{2\sigma^2}} \\ &\quad \times \left[erf\left(\frac{L/2 - x_0}{\cos\theta\sqrt{2}\sigma} \right) - erf\left(\frac{L/2 - x_0}{\sqrt{2}\sigma} \right) \right] \\ &\quad \times dr_{1x} dr_{1y} dr_{2x} dr_{2y} \frac{L}{L_0^2} dx_0. \end{aligned} \tag{2.51}$$

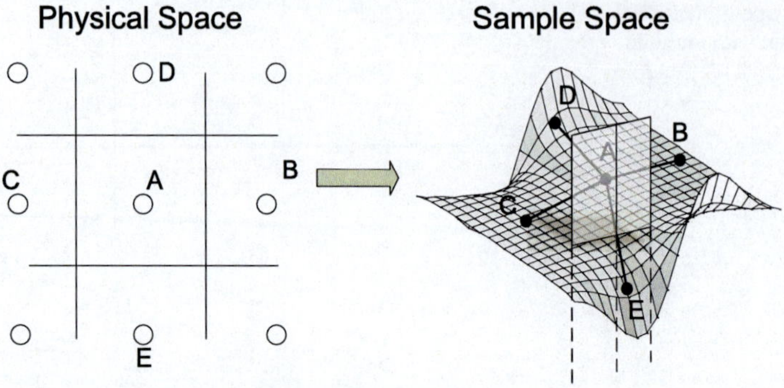

Fig. 2.10 Mapping from physical space to sample space

2.2.6 *Impact on Localization in Two-Dimensional Space*

The practical indoor localization system partitions the two-dimensional physical space into blocks [55, 219], such as the one shown in the left part of Fig. 2.10, where the center of each block represents the block itself. Let us first consider the ideal case with perfect information. Recall that the corresponding image in the sample space with respect to each point in the physical space is a point on the mean surface M as shown in the right part of Fig. 2.10. We use 4 hyperplanes to surround the point A on the mean surface M, where each hyperplane is orthogonally cutting the line segment between A and the neighboring node in the middle. According to the principle of MLE, if reported fingerprints fall in the surrounded area, then the system will localize the user's location to be at block A. It is worth mentioning that we here do not adopt the hyper-cylinder discussed in Sect. 2.2.3 as the boundary in the sample space because such a boundary can leave certain areas in the physical space uncovered.

We now provide the mathematical expression of such a surrounded area as shown in Fig. 2.10. Suppose that $A = [x_{1A}, x_{2A}, \cdots, x_{nA}]^T$ and $B = [x_{1B}, x_{2B}, \cdots, x_{nB}]^T$, and then the bisector plane of line segment $\mathbf{AB}$ is

$$\mathbf{h_1} = \frac{\mathbf{AB}}{2} = \frac{(x_{1A} - x_{1B}, x_{2A} - x_{2B}, \cdots, x_{nA} - x_{nB})}{2}. \tag{2.52}$$

The other three bisector planes could also be presented in a similar manner. If we use area 1 to denote the surrounded area, then area 1 can be determined by $\mathbf{h_1} \cdot \mathbf{r} \le |\mathbf{h_1}|^2$, $\mathbf{h_2} \cdot \mathbf{r} \le |\mathbf{h_2}|^2$, $\mathbf{h_3} \cdot \mathbf{r} \le |\mathbf{h_3}|^2$, and $\mathbf{h_4} \cdot \mathbf{r} \le |\mathbf{h_4}|^2$.

If we use area 0 to denote block A in the physical space and assume that the user will appear in any point of block A with identical probability, then the reliability for the ideal situation is

$$P_{e_1} = \iint_{area0} \mathrm{d}x_0 \mathrm{d}y_0 \int \cdots \int_{area2} \left(\frac{\sqrt{N}}{2\pi\sigma} \right)^n$$

$$\times e^{-\frac{N[(x-x_1)^2+...(x-x_n)^2]}{2\sigma^2}} \frac{1}{S_0} \mathrm{d}x_1 \mathrm{d}x_2 \cdots \mathrm{d}x_n, \quad (2.53)$$

where S_0 is the area of block A.

In the practical situation, the nodes A, B, C, D, and E migrate to A', B', C', D', and E' in the sample space. We can also use 4 hyperplanes to surround a corresponding area 2, so that if reported fingerprints fall in area 2, the user is localized in block A in the physical space. Similarly, the reliability for the practical situation is

$$P_{e_2} = \iint_{area0} \mathrm{d}x_0 \mathrm{d}y_0 \int \cdots \int_{area2} \left(\frac{\sqrt{N}}{2\pi\sigma} \right)^n$$

$$\times e^{-\frac{N\left[(x-x_1)^2+...(x-x_n)^2\right]}{2\sigma^2}} \frac{1}{S_0} \mathrm{d}x_1 \mathrm{d}x_2 \cdots \mathrm{d}x_n. \quad (2.54)$$

Consequently, the probability error is $P_e = \left| P_{e_1} - P_{e_2} \right|$.

It is extremely difficult to give a closed-form expression of P_e; however, the probability error analysis inspires us to consider a very special case when determining the surrounded area in the sample space mentioned above, which could potentially incur large localization error. That is, what if the nodes A, C, and D are on the same straight line, which means that the surrounded area is actually an open area. Although the general mathematical expressions also hold in the special case, the consequence in location determination is that large-scale localization error could happen. The physical meaning of the open area is that the user could be localized in physical locations corresponding to faraway areas in the sample space. In particular, if the reported fingerprint is $\mu(\mathbf{r}')$, based on which the user is most likely at location $\mathbf{r}'$, the system however could still localize the user to be at some location faraway from $\mathbf{r}'$.

Such a phenomenon can be avoided by utilizing the best fingerprints reporting strategy when constructing the database. We first illustrate why A, C, and D could be on the same straight line, as shown in Fig. 2.11. The left part of the figure shows the setting of the physical space, and the right part shows an example of two-dimensional sample space, which means the number of measurements is two. In the fingerprints collection phase, a site surveyor standing at A could measure AP_1 and AP_2 once, respectively, surveyors at C and D measure AP_1 and AP_2 twice, respectively, and then the corresponding nodes of these physical locations in the sample space is on the same straight line as shown in the right part of Fig. 2.11.

The way surveyors construct the database described above could be very possible if the best strategy is not considered. This is because surveyors usually prefer to measure APs with strongest signal strengths, such as AP_1 and AP_2 with respect

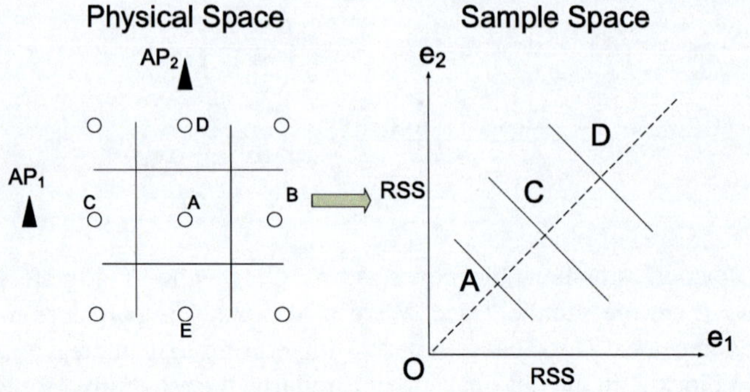

Fig. 2.11 Special case

to C and D, respectively. However, referring to the best strategy theory could reveal that such a survey can be of little avail for localization. Take location C for example, if the surveyor measures AP_1 twice, the corresponding complex parameter Z_1 are on the same straight line in the complex plane, which makes $(\sum_{i\in\mathcal{V}_n}|Z_i|)^2 - |\sum_{i\in\mathcal{V}_n} Z_i|^2 = 0$, where $\mathcal{V}_n = \{1, 1\}$ according to the surveying process. This could be better understood by reviewing Eq. (2.35). Although the best strategy presented earlier is for location estimation, it also provides guidance for the training phase.

However, if the number of measurements is small, it may happen that the migrated points in the sample space are collinear, which will incur large deviation from the reliability. Fortunately, our numerical analysis shows that it is very difficult for such subtle migration of points in the sample space to happen, and deviations of reliability incurred by imperfect information are usually very small.

2.2.7 *Experimental Results*

We conduct experiments with the trace data collected by the EVARILOS testbed [193], in order to verify our theory. The data are collected in an unmanned utility room with many metal objects termed as "Zwijnaarde," where there is almost no outside interference and no persons are present in the environment. The trace data contains 144557 combinations of access point (AP) and reference point (RP), and each (AP, RP) tuple contains a number of RSS raw measurements. A detailed description of the testbed and data could be found in [112–114].

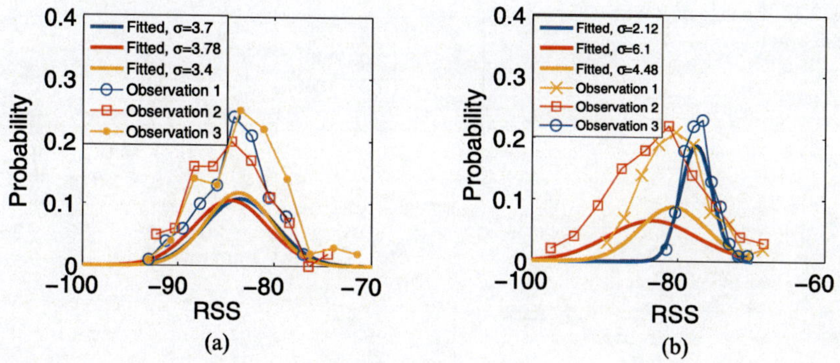

Fig. 2.12 Characteristics of radio propagation. (**a**) Adjacent locations W.R.T. the same AP. (**b**) The same location W.R.T. different APs

2.2.7.1 Verification of Main Assumptions

Although the rationale of our main assumption about the radio propagation has been explained in Sect. 2.3, where a number of work adopting the similar assumption is briefly surveyed, we still validate our assumption by performing analysis of the trace data from the EVARILOS testbed. We first filter out those unreliable measurements, where there is only 1 or 2 RSS readings recorded or all the RSS readings are exactly the same.

Figure 2.12a shows three adjacent RPs' RSS observations and fitted curves. The RSSes observed at the three locations (1290, 1980), (1290, 1270), and (1890, 1270) are with respect to the same AP at (1000, 1712), and the RPs are around 3.5 m from one another. There are totally 272, 146, and 98 observations at the three RPs, respectively. Observation curves represent the proportions of the observed RSS value. We fit the observation curves and find they are approximately to be Gaussian distribution, with skewness and kurtosis less than $|0.5|$ and $|3.3|$, respectively. It can be seen that the observed RSS readings at adjacent RPs have similar value of σ, and the mean does not change dramatically.

Figure 2.12b shows the RSS values observed at the same RP (2490, 1270) with respect to 3 different APs at (2644, 1500), (3998, 728), and (1000, 600), respectively. The total numbers of RSS records with respect to each AP are 173, 208, and 196; the fitted curves are with the skewness and kurtosis less than $|0.31|$ and $|3.2|$, respectively. The observations corroborate our assumption that the RSS values observed at the same location with respect to different APs are with quite different values of σ.

Figure 2.13 shows the change of σ's value with the distance from the AP. We show the trend of the change with respect to the 3 APs. We examine RSSes observed at all RPs that are less than 35 m from the AP, and we calculate the corresponding σ value at each RP. As shown in the figure, 45% of the locations' values of σ vary less than 2, and 81% of the values of σ vary less than 5, if the RP is less than 17 m

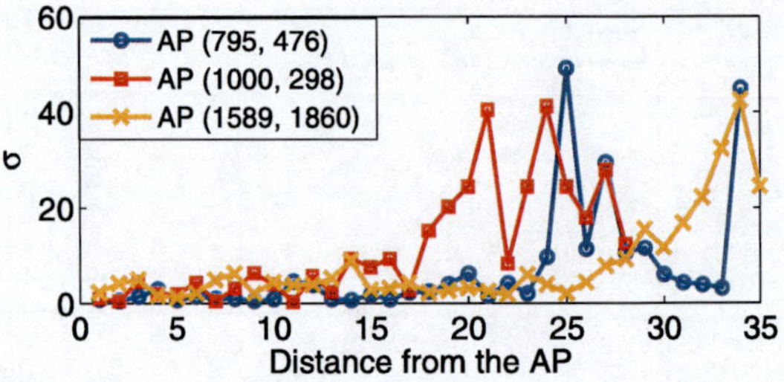

Fig. 2.13 Change of σ with the distance

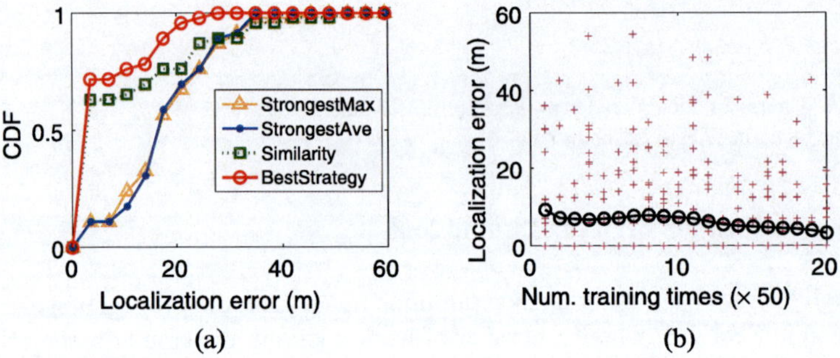

Fig. 2.14 Localization results. (**a**) Influence of AP selection. (**b**) Influence of training times

from the AP. If the distance exceeds 23 m, the change of the value of σ becomes dramatic. This observation corroborates the model in [109], which supports our modeling assumption.

2.2.7.2 Localization Performance

We use the data observed at a part of the RPs as the training set and that observed at the rest of the RPs as the test set to perform localization, with results illustrated in Fig. 2.14a. We compare the performance of the proposed best strategy based on Eqs. (2.35) and (2.29) with other three reporting strategies widely used in the previous work [77]. The best strategy is the logical result of our modeling and analysis, and thus if it outperforms other frequently used strategies, our modeling and analysis could be validated.

With *StrongestAvg*, the user measures APs with the strongest average RSSes can be observed at the to-be-determined location in the online phase. With *Strongest-Max*, the user measures APs with the strongest RSSes can be observed. With the *Similarity* strategy, APs are first clustered according to the similarity of their generated RSSes and the representative AP of each cluster is then selected. How to compute the similarity metric and how to select the representative AP in a cluster are described in [77]. As can be seen from Fig. 2.14a, it is obvious that the best strategy outperforms other three strategies. More detailed statistical results are tabulated in

Table 2.1 Localization errors

Metric	StrgstMax	StrgstAvg	Similarity	BestStrategy
Avg. err [m]	19.3	19.3	18.4	16.4
Min. err [m]	6.0	3.8	3.6	3.5
Max. err [m]	42.6	37.6	48.0	55.8
Med. err [m]	17.1	18.4	14.0	11.4

Table 2.1. Although the best strategy has the largest maximum error due to the randomness of fingerprinting based localization approach, the overall performance is better than the other three strategies.

Figure 2.14b illustrates the localization results influenced by the number of training times in the offline phase. We use reliable RSS records in all RPs to do the experiment in order to ensure that there are enough data for training. We use a part of RSS data for the offline phase and the rest of the data for testing. With the number of data used in the offline phase increasing, it is clear that the average localization error is decreasing as shown with the curve in the figure, which corroborates our theoretical analysis. Since for each RP, there are a part of RSS records used for training, and if the data for testing are with the same RP, it is possible that the minimum localization error equals 0 as shown in the figure.

2.3 Scalability with the Collocation of Measurement Points

This section focuses on measurement point collocation in different cases and their effects on localization accuracy [224]. We first study two simple preliminary cases under assumption that users are uniformly distributed: when measurement points are collocated regularly, we propose a collocation pattern which is most beneficial to localization accuracy; when measurement points are collocated randomly, we prove that localization accuracy is limited by a tight bound. Under the general case that users are distributed asymmetrically, we show the best allocation scheme of measurement points: measurement point density ρ is proportional to $(c\mu)^{2/3}$ in every part of the region, where μ is the user density and c is a constant determined by the collocation pattern. We also give some guidelines on collocation choice and perform extensive simulations to validate our assumptions and results.

2.3.1 *Case 1: Regular Collocation*

In crowdsourcing based RSS fingerprints collection method, crowdworkers devote various efforts in different incentive mechanisms. With high rewards, crowdworkers can be encouraged to update RSS information at certain measurement points which have been collocated regularly in a proper way. In this section, assuming that users are uniformly distributed, we prove that EQLE and MQLE can be minimized and DNN can be maximized when measurement points are at the intersecting locations of a mesh of equilateral triangles as shown in Fig. 2.15a. We also show specific values of EQLE, MQLE, and DNN in a mesh of equilateral triangles and a mesh of grids and compare their performance.

2.3.1.1 Background and Modeling

Chen et al. in [9] developed a simple iterative algorithm that finds an optimized indoor landmark deployment and showed deployment pattern with up to eight landmarks in the indoor environment. However, multiple indoor landmarks increase the computation complexity of this algorithm. Actually, as is shown in Fig. 2.15b, the measurement points are usually collocated at the intersecting locations of a mesh of grids like [126]. Kaemarungsi et al. in [280] showed specific analysis on how grid spacing influences the location error. A mesh of grids can be regarded as a mesh network that two groups of parallel lines with the identical spacing intersect vertically.

In this section, we study a more general case, the regular collocation, where measurement points are at the intersecting locations of a mesh network that two groups of parallel lines with the various spacing intersect at a certain angle as shown in Fig. 2.15b. The users are uniformly distributed in a certain region with area[1] S and N measurement points. Before theoretical analysis, we give the following definitions first.

Definition 1 (Neighboring Region) The neighboring region for a measurement point refers to the region where it is the nearest measurement point to any user located in.

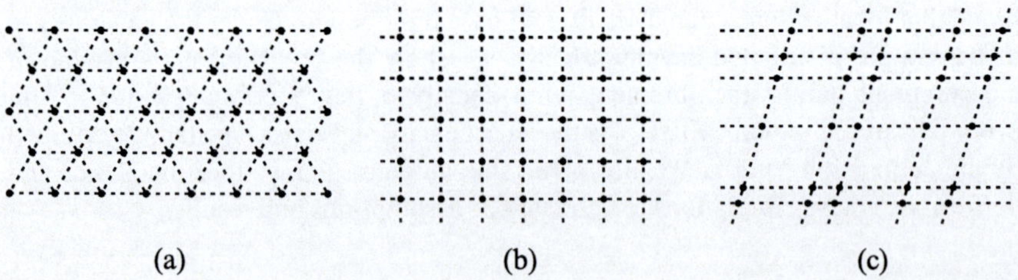

Fig. 2.15 (a) A mesh of equilateral triangles, (b) a mesh of grids, and (c) the regular collocation

[1] Throughout this work, with abuse of notation, we shall use S to represent both a certain region and its area.

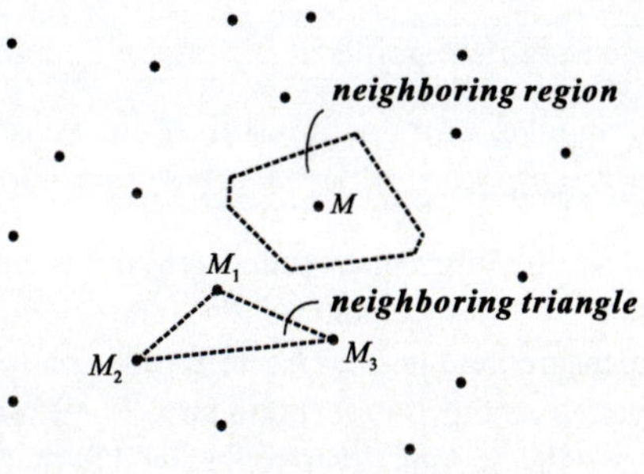

Fig. 2.16 Neighboring region and neighboring triangle

As shown in Fig. 2.16, the neighboring region for measurement point M is surrounded by dashed line. The nearest neighbor algorithm will return the estimated location M for any user located in its neighboring region. The whole region can be partitioned into neighboring regions in only one way.

Definition 2 (Neighboring Triangle) The neighboring triangle refers to the triangle combined with three measurement points with no other measurement points in.

As shown in Fig. 2.16, $\Delta M_1 M_2 M_3$ is a neighboring triangle combined with M_1, M_2, and M_3. Different from the neighboring region, the whole region can be partitioned into neighboring triangles in multiple ways.

Definition 3 (Measurement Point Density) If there are N measurement points in a certain region with the area S, the measurement point density $\mu = \frac{N}{S}$.

Intuitively, the reciprocal of measurement point density $\frac{S}{N}$ can be regarded as the average area of neighboring region and $\sqrt{\frac{S}{N}}$ can be regarded as the size scale of the neighboring region.

For mathematical tractability, we introduce the following assumption and approximation. Any indoor region has indoor infrastructures more or less, which indicates that parts of region are inaccessible to people and measurement points. Since the large open indoor environment usually has a large open area with sparse indoor infrastructures, in our theoretical analysis, we assume that there is no infrastructure and the whole region is accessible to people and measurement points. In Sect. 2.3.4, the simulation value in a large open area with sparse infrastructures has a small deviation from the theoretical value, which indicates that the influence of sparse indoor infrastructures is limited. In large open indoor environment, we also have the following proposition:

Proposition 1 *If measurement points are collocated regularly, the ratio of the number of measurement points at the region boundary can be scaled as* $\Theta\left(\frac{1}{\sqrt{N}}\right)$.

Proof 4 In the regular collocation, the perimeter can be scaled as $\Theta\left(\sqrt{S}\right)$. Since the spacing of measurement points can be scaled as $\Theta\left(\sqrt{\frac{S}{N}}\right)$, the number of measurement points at the region boundary can be scaled as $\Theta\left(\sqrt{N}\right)$. So the ratio of number of these points to N can be scaled as $\Theta\left(\frac{1}{\sqrt{N}}\right)$. This finishes the proof of Proposition 1.

In the large open indoor environment, both S and N are large, which indicates that the measurement points at the region boundary account for a small fraction of all measurement points. As an approximation, we will ignore the effect of measurement points at the region boundary. For example, the sum of the area of neighboring triangles is considered to be equal to that of neighboring regions. We will also verify this approximation by extensive simulations in Sect. 2.3.4.

2.3.1.2 Theoretical Analysis

Lemma 1 *The number of neighboring triangles is double of that of measurement points.*

Proof 5 Since the sum of three interior angles for a single neighboring triangle is π and a certain region can be partitioned into neighboring triangles, the number of neighboring triangles is

$$\frac{N \times 2\pi}{\pi} = 2N.$$

This finishes the proof of Lemma 1.

In the above proof, based on Proposition 1, we ignore the effect of measurement points at the region boundary.

Lemma 2 *If measurement points are regularly collocated, the whole region can be partitioned into acute or right neighboring triangles.*

Proof 6 The proof is straightforward. Since the regular collocation is the mesh of parallelograms and any parallelogram can be partitioned into two acute or right triangles, we can draw the above conclusion. This finishes the proof of Lemma 2.

Lemma 3 *The boundaries of neighboring regions in an acute or right triangle are three midperpendiculars of triangle edges.*

Proof 7 In Fig. 2.17, OA_1, OB_1, and OC_1 are three midperpendiculars on the BC, AC, and AB. Using the property of the midperpendicular, for any node in region $\square BC_1OA_1$, B is the nearest measurement point to it in A, B, and C. Hence, $\square BC_1OA_1$ is the neighboring region for B in $\triangle ABC$. Similarly, we can know that $\square AC_1OB_1$ and $\square CB_1OA_1$ are the neighboring regions for A and C, respectively. So three midperpendiculars OA_1, OB_1, and OC_1 are the boundaries of neighboring regions in $\triangle ABC$. This finishes the proof of Lemma 3.

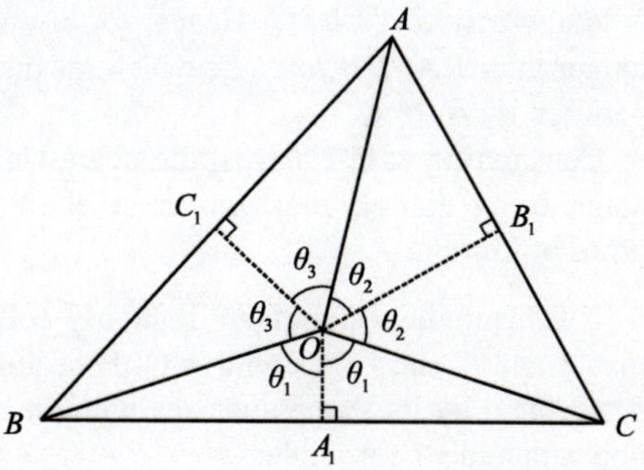

Fig. 2.17 Neighboring regions in the triangle

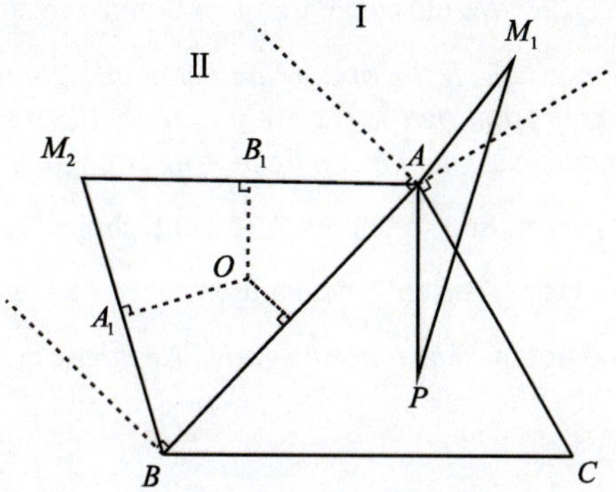

Fig. 2.18 Two possible positions of outside measurement points

From Lemma 3, it is apparent that the crossing point of three midperpendiculars O is the circumcenter of $\triangle ABC$. Moreover, the MQLE of an acute or right triangle is the radius of its circumcircle, where the corresponding position is the circumcenter O.

Lemma 4 *If a region can be partitioned into acute or right neighboring triangles, the nearest measurement point to any node inside a neighboring triangle must be one of the three triangle vertexes.*

Proof 8 We consider two cases according to positions of the measurement point outside the neighboring triangle as follows.

Case I: As is shown in Fig. 2.18, if the measurement point M_1 is in region I, for any node P in $\triangle ABC$, $\triangle APM_1$ is an obtuse triangle ($\triangle APM_1$ may reduce to a line segment when node P is in special positions). Hence, $PA < PM_1$, which indicates that compared to any other measurement point in region I, A is the nearest one to any node P inside $\triangle ABC$.

Case II: As is shown in Fig. 2.18, if the measurement point M_2 is in region II, $\triangle ABM_2$ is an acute or right triangle. Using the property of the midperpendicular, $\square B_1OA_1M_2$ is the neighboring region for M_2 in region $\square ACBM_2$. Since $\triangle ABM_2$ is an acute or right triangle, the crossing point of three midperpendiculars (circumcenter) is inside $\triangle ABM_2$ or on edge AB, which indicates that $\square B_1OA_1M_2$

is totally inside $\triangle ABM_2$. Hence, for any node P inside $\triangle ABC$, PM_2 is always longer than PA, PB, or PC, which means the nearest measurement point to P is between A, B, or C.

Considering case I and II, the nearest measurement point to any node inside a neighboring triangle must be one of the three triangle vertexes. This finishes the proof of Lemma 4.

If measurement points are regularly collocated in a certain region, using Lemmas 2 and 4, the EQLE and MQLE of any neighboring triangle in this region is determined by its shape and area without being influenced by other measurement points outside the triangle.

EQLE We introduce Lemmas 5 and 6 to simplify the proof.

Lemma 5 *If the area of an acute or right neighboring triangle is fixed, under the assumption that users are uniformly distributed in this triangle, the EQLE can be minimized when this neighboring triangle is an equilateral triangle.*

Proof 9 See Appendix A in our technical report [281].

Using Jensen's inequality, we can derive Lemma 6.

Lemma 6 *For n non-negative variables* $x_1, x_2, \ldots, x_n$,

$$\frac{\sum_{i=1}^{n} x_i^{3/2}}{n} \geq \left(\frac{\sum_{i=1}^{n} x_i}{n} \right)^{3/2}, \tag{2.55}$$

where the equality holds if and only if $x_1 = x_2 = \ldots = x_n$.

Proof 10 See Appendix B in our technical report [281].

Theorem 1 *EQLE can be minimized when measurement points are at the intersecting locations of a mesh of equilateral triangles.*

Proof 11 Using Lemma 2, the region in regular collocation can be partitioned into acute or right neighboring triangles. Using Lemma 1, the number of neighboring triangles is $2N$. So we can use $S_1, S_2, \ldots, S_{2N}$ to denote the area of these triangles. Equilateral triangles whose area is equal to $S_1, S_2, \ldots, S_{2N}$ are denoted by $S_{\triangle 1}, S_{\triangle 2}, \ldots, S_{\triangle 2N}$, respectively. Their edge length is denoted by $a_1, a_2, \ldots, a_{2N}$, respectively. Applying Lemma 5, we know that[2]

$$\mathbf{E_e}(S_i) \geq \mathbf{E_e}\left(S_{\triangle i}\right) = c_e a_i \quad (i = 1, 2, \ldots, 2N), \tag{2.56}$$

[2] $\mathbf{E_e}(\cdot)$ is a notation for calculating the EQLE in a region.

where c_e is a constant denoting the ratio of the EQLE to the edge length in equilateral triangles. Together with Lemmas 2 and 4 and noting that

$$S_i = S_{\Delta i} = \frac{\sqrt{3}}{4}a_i^2 \ (i = 1, 2, \ldots, 2N), \tag{2.57}$$

we have

$$\begin{aligned} \mathbf{E_e}(S) &= \left(\sum_{i=1}^{2N} S_i \mathbf{E_e}(S_i)\right)/S \\ &\geq \left(\sum_{i=1}^{2N} S_{\Delta i} \mathbf{E_e}(S_{\Delta i})\right)/S \\ &= \left(\sum_{i=1}^{2N} \frac{\sqrt{3}}{4} c_e a_i^3\right)/S. \end{aligned} \tag{2.58}$$

Since the whole area S equals the sum of area of neighboring triangles, noting Eq. (2.57),

$$S = \sum_{i=1}^{2N} S_i = \sum_{i=1}^{2N} S_{\Delta i} = \frac{\sqrt{3}}{4}\sum_{i=1}^{2N} a_i^2. \tag{2.59}$$

Using Lemma 6 and noting Eq. (2.59),

$$\begin{aligned} \left(\sum_{i=1}^{2N} \frac{\sqrt{3}}{4} c_e a_i^3\right)/S &= \frac{\sqrt{3}c_e N}{2S}\left(\frac{\sum_{i=1}^{2N}(a_i^2)^{3/2}}{2N}\right) \\ &\geq \frac{\sqrt{3}c_e N}{2S}\left(\frac{\sum_{i=1}^{2N} a_i^2}{2N}\right)^{3/2} \\ &= \frac{\sqrt{2}c_e}{\sqrt[4]{3}}\sqrt{\frac{S}{N}}. \end{aligned} \tag{2.60}$$

Together with Eq. (2.58),

$$\mathbf{E_e}(S) \geq \frac{\sqrt{2}c_e}{\sqrt[4]{3}}\sqrt{\frac{S}{N}}. \tag{2.61}$$

The equality in Eq. (2.61) holds when all localization triangles are equilateral triangles with the same edge length, which indicates that the measurement points are at the intersecting locations of a mesh of equilateral triangles. This finishes the proof of Theorem 1.

In the proof of Theorem 1, we use Lemmas 5 and 6 to reduce the optimal collocation of all measurement points to the optimal choice of neighboring triangles.

MQLE In reality, some application scenarios are sensitive to the MQLE. We will show that a mesh of equilateral triangles can also minimize MQLE. Lemma 7 is introduced to avoid formula redundancy in the proof.

Lemma 7 *If* $2\theta_1 + 2\theta_2 + 2\theta_3 = 2\pi$, *then*

$$\sin 2\theta_1 + \sin 2\theta_2 + \sin 2\theta_3 \leq \frac{3\sqrt{3}}{2}, \tag{2.62}$$

where the equality holds when $\theta_1 = \theta_2 = \theta_3 = \pi/3$.

Proof 12 See Appendix C in our technical report [281].

Theorem 2 *MQLE can be minimized when measurement points are at the intersecting locations of a mesh of equilateral triangles.*

Proof 13 From Lemma 3, MQLE of an acute or right neighboring triangle is the radius of its circumcircle, where the corresponding position is the circumcenter. Together with Lemmas 2 and 4, the MQLE of this region is the largest circumcircle radius of a certain neighboring triangle.

Using S_m and R_m to denote the largest triangle area and its corresponding radius of the circumcircle, it is obvious that[3]

$$\mathbf{E_m}(S) \geq R_m. \tag{2.63}$$

As is shown in Fig. 2.17,

$$S_m = \frac{1}{2} R_m^2 (\sin 2\theta_1 + \sin 2\theta_2 + \sin 2\theta_3). \tag{2.64}$$

Since S_m is the largest neighboring triangle area, using Lemma 1,

$$S_m \geq \frac{S}{2N}. \tag{2.65}$$

Together with Eqs. (2.63) and (2.64) and using Lemma 7,

[3] $\mathbf{E_m}(\cdot)$ is a notation for calculating the MQLE in a region.

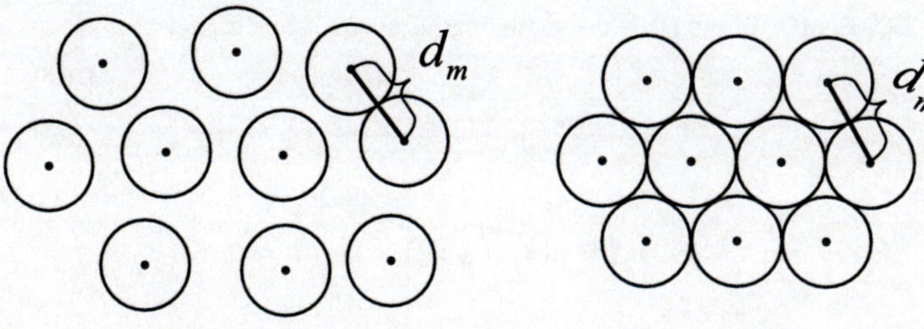

Fig. 2.19 Sparse packing and dense packing. (**a**) Sparse. (**b**) Dense

$$
\begin{aligned}
\mathbf{E_m}(S) &\geq R_m \\
&= \sqrt{\frac{2S_m}{\sin 2\theta_1 + \sin 2\theta_2 + \sin 2\theta_3}} \\
&\geq \sqrt{\frac{S}{N}} \cdot \sqrt{\frac{1}{\sin 2\theta_1 + \sin 2\theta_2 + \sin 2\theta_3}} \\
&\geq \frac{\sqrt{2\sqrt{3}}}{3}\sqrt{\frac{S}{N}},
\end{aligned} \tag{2.66}
$$

where the equality holds when all neighboring triangles are equilateral triangles with identical area. This finishes the proof of Theorem 2.

DNN The low pass loss of signal in large open indoor environment results in the similarity of RSS fingerprints between neighboring measurement points, which degrades the performance of RSS fingerprint based localization algorithm. Since large DNN can make RSS more spatially varying, it is beneficial to increase the performance of localization algorithm. Using Thue's theorem (the proof is shown in [282]), we will prove that a mesh of equilateral triangles can also maximize DNN.

Theorem 3 *DNN can be maximized when measurement points are at the intersecting locations of a mesh of equilateral triangles.*

Proof 14 This theorem is equivalent to the one that given the DNN d_m, the region area can be minimized when the measurement points are at the intersecting locations of a mesh of equilateral triangles.

As is shown in Fig. 2.19a, if we draw circles centering on each measurement point with radius $d_m/2$, there is no overlap in these circles. Recall Thue's theorem that regular hexagonal packing is the densest circle packing in the plane and the density of this circle configuration is $\pi/\sqrt{12}$ as shown in Fig. 2.19b. This indicates that

Table 2.2 EQLE, MQLE, and DNN in a mesh of grids or equilateral triangles

Collocation pattern	EQLE	MQLE	DNN
Equilateral triangles	$\left(\frac{\sqrt{2\sqrt{3}}}{9}+\frac{\sqrt{2\sqrt{3}}\ln 3}{12}\right)\sqrt{\frac{S}{N}}$ $\approx 0.377\sqrt{\frac{S}{N}}$	$\frac{\sqrt{2\sqrt{3}}}{3}\sqrt{\frac{S}{N}}$ $\approx 0.620\sqrt{\frac{S}{N}}$	$\frac{\sqrt{6\sqrt{3}}}{3}\sqrt{\frac{S}{N}}$ $\approx 1.075\sqrt{\frac{S}{N}}$
Grids	$\left(\frac{\sqrt{2}}{6}+\frac{1}{12}\ln\left(\frac{2+\sqrt{2}}{2-\sqrt{2}}\right)\right)\sqrt{\frac{S}{N}}\approx$ $0.383\sqrt{\frac{S}{N}}$	$\frac{\sqrt{2}}{2}\sqrt{\frac{S}{N}}\approx 0.707\sqrt{\frac{S}{N}}$	$\sqrt{\frac{S}{N}}$

$$\frac{S_c}{S}=\frac{\pi N(d_m/2)^2}{S}\leq\frac{\pi}{\sqrt{12}}, \tag{2.67}$$

where S_c is the sum of the area of all circles. Hence,

$$S\geq\frac{\sqrt{3}}{2}Nd_m^2. \tag{2.68}$$

The equality in Eq. (2.68) holds when these circles are in a regular hexagonal packing, which is equivalent to that the measurement points are at the intersecting locations of a mesh of equilateral triangles. This finishes the proof of Theorem 3.

Since the proof of Theorem 3 is not based on the regular collocation condition, Theorem 3 is a general theorem, which indicates that DNN in a mesh of equilateral triangles is the maximum in all collocation patterns.

2.3.1.3 Comparisons of Collocation Patterns

The specific values of EQLE, MQLE, and DNN in a mesh of grids or equilateral triangles are shown in Table 2.2. From this table, EQLE and MQLE in a mesh of grids are about 1.4% and 14.1% larger than those in a mesh of equilateral triangles. DNN in a mesh of grids is about 7.5% lower than that in a mesh of equilateral triangles. This indicates that a mesh of equilateral triangles is more beneficial to localization accuracy and algorithm performance than a mesh of grids.

However, the difficulty of measurement point deployment varies in different collocation patterns. Since the grid floor tiles are widely used in indoor environment, it is easy to locate measurement points in a mesh of grids.

Compared to this, locating measurement points in a mesh of equilateral triangles needs extra assistance such as markers on the floor.

2.3.2 Random Collocation

With low rewards, crowdworkers will not bother to update RSS information at certain measurement points. Actually, some indoor localization systems like Zee [53] can run in the background on a smartphone without requiring any explicit crowdworker participation. In this section, we still assume that the users are uniformly distributed in a region. Since users are also crowdworkers, the measurement points where they update RSS information will be randomly and dynamically collocated in the region. We will show that EQLE is lower bounded by $\frac{(2N)!!}{(2N+1)!!}\sqrt{\frac{S}{\pi}}$ and this lower bound becomes tight when the number of measurement points is large.

2.3.2.1 Background and Modeling

In a region with infinite area and infinite measurement points, we use r to denote the distance of a user located at a certain node to its closest measurement point. For a homogeneous two-dimensional Poisson point process where measurement points are uniformly randomly collocated, the probability density function (p.d.f.) of the distance to the nearest measurement point is (see [283], Chapter 8)

$$f(r) = 2\pi\mu r \cdot e^{-\mu\pi r^2}, \tag{2.69}$$

where μ is the measurement point density. Using the conclusion in [284], the expected distance of the user to its nearest measurement point is

$$\mathbf{E}(r) = \int_0^\infty rf(r)\,dr = \frac{1}{2\sqrt{\mu}}. \tag{2.70}$$

However, in the indoor environment, the region has a certain shape with finite area S and finite measurement points N, in which case the above conclusion cannot apply to.

For convenience of analysis, we give some definitions. In a region with an arbitrary shape, the area of a certain part of the region where the distance of the user to any node located in is less than R can be expressed as

$$S(R) = \int_0^R r\theta(r)dr, \tag{2.71}$$

where the function $\theta(r)$ denotes the sum of angle covered by the region at the distance r to the user node and $\theta(r) \le 2\pi$ as shown in Fig. 2.20. In this region, the largest distance to the user node is denoted by R_m and the total area S satisfies

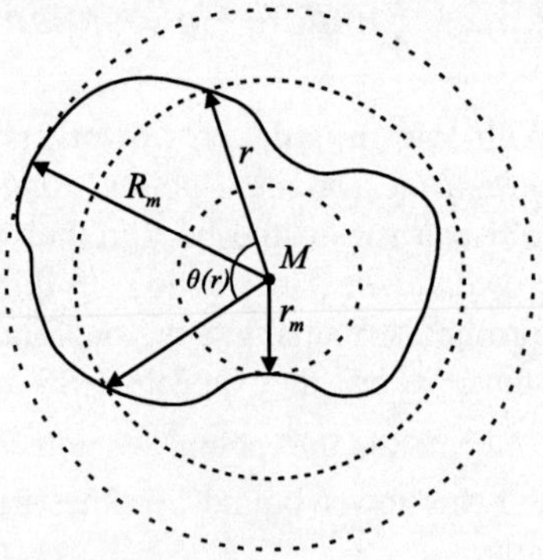

Fig. 2.20 R_m, $\theta(r)$, and r_m in a region with an arbitrary shape

$$S = S(R_m) = \int_0^{R_m} r\theta(r)dr. \tag{2.72}$$

2.3.2.2 Theoretical Analysis

Lemma 8 *In a region with an arbitrary shape, when the user is located at a certain node of this region, the EQLE of the region can be expressed as*

$$\mathbf{E_e}(S) = \int_0^{R_m} \left(1 - \frac{S(r)}{S}\right)^N dr. \tag{2.73}$$

Proof 15 See Appendix D in our technical report [281].

Lemma 9 *When the user is located at a center of a circular region, the EQLE is* $\frac{(2N)!!}{(2N+1)!!}\sqrt{\frac{S}{\pi}}$.

Proof 16 See Appendix E in our technical report [281].

Theorem 4 *If N measurement points are uniformly randomly collocated in the region, the EQLE is lower bounded by* $\frac{(2N)!!}{(2N+1)!!}\sqrt{\frac{S}{\pi}}$ *and this bound becomes tight when N is large.*

Proof 17 We will first show that when the user is located at a certain node of a region with an arbitrary shape, the EQLE is no smaller than that when the user is located at the center of a circular region which has identical area and an identical number of measurement points. We use $S_o(R)$ and R_{mo} to denote the area within the distance R to the user node and the radius in the circular region.

Since the area of two regions is identical,

$$S = S(R_m) = \int_0^{R_m} r\theta(r)dr \tag{2.74}$$

and

$$S = S_o(R_{mo}) = \int_0^{R_{mo}} r \cdot 2\pi \, dr. \tag{2.75}$$

Recalling that $\theta(r) \leq 2\pi$ and noting Eqs. (2.74) and (2.75), we know that $R_m \geq R_{mo}$ and $S(r) \leq S_o(r)$ when $r \leq R_{mo}$.

Using Lemma 8 and noting that $S(r) \leq S_o(r)$ and $R_m \geq R_{mo}$, the EQLE in a circular region satisfies

$$\begin{aligned}
\mathbf{E_e}(S_o(R_{mo})) &= \int_0^{R_{mo}} \left(1 - \frac{S_o(r)}{S}\right)^N dr \\
&\leq \int_0^{R_{mo}} \left(1 - \frac{S(r)}{S}\right)^N dr \\
&\leq \int_0^{R_m} \left(1 - \frac{S(r)}{S}\right)^N dr \\
&= \mathbf{E_e}(S(R_m)).
\end{aligned} \tag{2.76}$$

Using Lemma 9, we know

$$\mathbf{E_e}(S_o(R_{mo})) = \frac{(2N)!!}{(2N+1)!!}\sqrt{\frac{S}{\pi}}. \tag{2.77}$$

Substituting Eq. (2.77) into Eq. (2.76),

$$\mathbf{E_e}(S(R_m)) \geq \frac{(2N)!!}{(2N+1)!!}\sqrt{\frac{S}{\pi}}. \tag{2.78}$$

In the following, we will show that this lower bound becomes tighter with N increasing. We use r_m to denote the shortest distance of a user node to the region boundary as shown in Fig. 2.20. Noting that

$$\int_0^{r_m} \left(1 - \frac{S(r)}{S}\right)^N dr = \int_0^{r_m} \left(1 - \frac{S_o(r)}{S}\right)^N dr, \tag{2.79}$$

the difference of EQLE in two regions satisfies

$$
\begin{aligned}
&\mathbf{E_e}\left(S\left(R_m\right)\right)-\mathbf{E_e}\left(S\left(R_{mo}\right)\right)\\
&=\int_0^{R_m}\left(1-\frac{S\left(r\right)}{S}\right)^N dr-\int_0^{R_{mo}}\left(1-\frac{S_o\left(r\right)}{S}\right)^N dr\\
&=\int_{r_m}^{R_m}\left(1-\frac{S\left(r\right)}{S}\right)^N dr-\int_{r_m}^{R_{mo}}\left(1-\frac{S_o\left(r\right)}{S}\right)^N dr\\
&\le\int_{r_m}^{R_m}\left(1-\frac{S\left(r\right)}{S}\right)^N dr\\
&\le\int_{r_m}^{R_m}\left(1-\frac{S\left(r_m\right)}{S}\right)^N dr\\
&=\int_{r_m}^{R_m}\left(1-\frac{\pi r_m^2}{S}\right)^N dr.
\end{aligned}
\tag{2.80}
$$

From Eq. (2.80), when N increases, the difference of EQLE in two regions will decrease exponentially. So far we have proved that if the user is located at any node of a region with an arbitrary shape, the EQLE is all lower bounded by $\frac{(2N)!!}{(2N+1)!!}\sqrt{\frac{S}{\pi}}$ and this bound becomes tighter as N increases.

Hence, if users are uniformly distributed in this region, the EQLE is also lower bounded by $\frac{(2N)!!}{(2N+1)!!}\sqrt{\frac{S}{\pi}}$. This bound becomes tighter as N increases, which indicates that it is a tight bound when N is large. This finishes the proof of Theorem 4.

Proposition 2

$$
\lim_{N\to\infty}\frac{(2N)!!}{(2N+1)!!}\cdot 2\sqrt{\frac{N}{\pi}}=1. \tag{2.81}
$$

Proof 18 See Appendix F in our technical report [281].

From Proposition 2, we can conclude that when N is large, the lower bound approximates

$$
\frac{(2N)!!}{(2N+1)!!}\cdot\sqrt{\frac{S}{\pi}}\approx\frac{1}{2}\sqrt{\frac{\pi}{N}}\cdot\sqrt{\frac{S}{\pi}}=\frac{1}{2}\sqrt{\frac{S}{N}}.
$$

Together with Theorem 4, $\frac{1}{2}\sqrt{\frac{S}{N}}$ can be regarded as the approximate value for the EQLE in a region with arbitrary shape when the number of measurement points is large. We find that the approximate value $\frac{1}{2}\sqrt{\frac{S}{N}}$ is exactly equal to that in Eq. (2.70). Intuitively, when N is large, the nearest measurement point will extremely approach the user node, where the effect of the region shape and boundary can be ignored and the problem is reduced to the two-dimensional homogeneous Poisson point process. Compared with EQLE in regular collocation (see Table 2.2), EQLE in random collocation is much larger than that in regular collocation. However, random collocation does not require explicit deployment of measurement points and the budget is low.

2.3.3 Asymmetrical Distribution

In Sects. (2.3.1) and (2.3.2), we hold the strong assumption that users are uniformly distributed in the region. However, in reality, user density varies in different parts of the region, which results in an asymmetrical distribution. In this section, we show the best allocation scheme of measurement points when the number of measurement points is fixed.

2.3.3.1 Modeling

If users are asymmetrically distributed, the problem can be modeled as follows. The whole region S can be divided into different parts denoted by $S_1, S_2, \ldots, S_l$, where l is the number of parts. In each part, users are uniformly distributed, where the probability density is denoted by $\rho_1, \rho_2, \ldots, \rho_l$. Noting that the term $\rho_i S_i$ indicates the probability of users locating in the i_{th} region, the area and probability density of each part satisfy

$$\sum_{i=1}^{l} \rho_i S_i = 1. \tag{2.82}$$

Without generality, the collocation pattern of measurement points in each part can be different. Since the EQLE is proportional to the size scale of the neighboring region, the EQLE of the i_{th} region is $c_i\sqrt{\frac{S_i}{N_i}}$, where N_i denotes the number of measurement points allocated to the i_{th} region and c_i is a collocation pattern constant. As is shown in Table 2.2, the collocation pattern constant in a mesh of equilateral triangles and a mesh of grids is approximately 0.377 and 0.383, respectively. The collocation pattern constant in a random collocation is approximately 0.5 if the number of measurement points is large. Given the above definitions, we investigate on how to minimize EQLE of the whole region when the total number of measurement points N is fixed.

2.3.3.2 Theoretical Analysis

Using Hölder's inequality, we can derive Lemma 10.

Lemma 10 *For n non-negative variables $x_1, x_2, \ldots, x_n$ and n non-negative variables $y_1, y_2, \ldots, y_n$,*

$$\sum_{i=1}^{n} \frac{y_i}{\sqrt{x_i}} \geq \frac{\left(\sum_{i=1}^{n} y_i^{2/3}\right)^{3/2}}{\left(\sum_{i=1}^{n} x_i\right)^{1/2}}, \tag{2.83}$$

where the equality holds if and only if $\frac{y_1^{2/3}}{x_1} = \frac{y_2^{2/3}}{x_2} = \ldots = \frac{y_n^{2/3}}{x_n}$.

Proof 19 See Appendix G in our technical report [281].

Theorem 5 *EQLE of the whole region is minimized when measurement point density $\mu_i \propto (c_i \rho_i)^{2/3}$, in which case the minimum EQLE is $\left(\sum_{i=1}^{l} S_i (c_i \rho_i)^{2/3}\right)^{3/2} / N^{1/2}$.*

Proof 20 EQLE of the whole region can be expressed as

$$\mathbf{E_e}(S) = \sum_{i=1}^{l} \rho_i S_i \mathbf{E_e}(S_i) = \sum_{i=1}^{l} \rho_i S_i c_i \sqrt{\frac{S_i}{N_i}}. \tag{2.84}$$

Using Lemma 10, Eq. (2.84) satisfies

$$\begin{aligned} \mathbf{E_e}(S) &= \sum_{i=1}^{l} \frac{\rho_i c_i S_i^{3/2}}{\sqrt{N_i}} \\ &\geq \frac{\left(\sum_{i=1}^{l} \left(\rho_i c_i S_i^{3/2}\right)^{2/3}\right)^{3/2}}{\left(\sum_{i=1}^{l} N_i\right)^{1/2}} \\ &= \frac{\left(\sum_{i=1}^{l} S_i (c_i \rho_i)^{2/3}\right)^{3/2}}{N^{1/2}}. \end{aligned} \tag{2.85}$$

Equation (2.85) holds if and only if

$$\frac{S_i (c_i \rho_i)^{2/3}}{N_i} = k \ (i = 1, 2, \ldots, l),$$

where k is a constant. This equality condition is equivalent to

$$\frac{(c_i\rho_i)^{2/3}}{\mu_i} = k \ (i = 1, 2, \ldots, l)$$

or $\mu_i \propto (c_i\rho_i)^{2/3}$. This finishes the proof of Theorem 5.

Using the ratio constant k, the minimum EQLE can be simplified as

$$\frac{\left(\sum_{i=1}^{l} S_i (c_i\rho_i)^{2/3}\right)^{3/2}}{N^{1/2}} = \frac{\left(\sum_{i=1}^{l} S_i k u_i\right)^{3/2}}{N^{1/2}} = Nk^{3/2}.$$

As a special case in which the collocation pattern in each part is identical, we can draw a simpler conclusion that EQLE is minimized when $\mu_i \propto \rho_i^{2/3}$.

2.3.4 *Simulation*

In this section, we will perform extensive simulations to validate our assumptions and results in the above analysis.

2.3.4.1 Regular Collocation

We create a 1×1 square region with 1024 (1020) measurement points in grid (equilateral triangle) collocation pattern. Figure 2.21a, b shows collocation patterns without indoor infrastructures, while Fig. 2.21c, d shows collocation patterns with indoor infrastructures accounting for 16% of the total area. In each case, we compare the EQLE of a mesh of grids and that of a mesh of equilateral triangles. In each experiment, 100 users are distributed uniformly in the accessible region. Repeating each experiment for 1000 times, we obtain EQLE by calculating the mean value of distance of the nearest measurement points to users. Experiment results are illustrated in Table 2.3. As is shown in the table, when there is no indoor infrastructure, EQLE obtained by simulation is almost equal to theoretical EQLE. This validates the accuracy of approximation that the effect of measurement points at region boundary can be ignored. When there are indoor infrastructures which accounts for 16% of the total area, EQLE obtained by simulation is about 1.8% larger than theoretical EQLE, which indicates that the influence of sparse indoor infrastructures is limited Fig. 2.22.

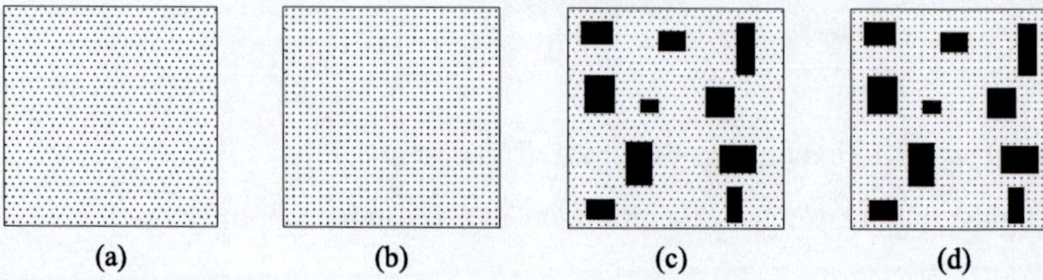

Fig. 2.21 Different collocation patterns in various environments. (**a**) A mesh of equilateral triangles without infrastructures. (**b**) A mesh of grids without infrastructures. (**c**) A mesh of equilateral triangles with infrastructures. (**d**) A mesh of grids with infrastructures

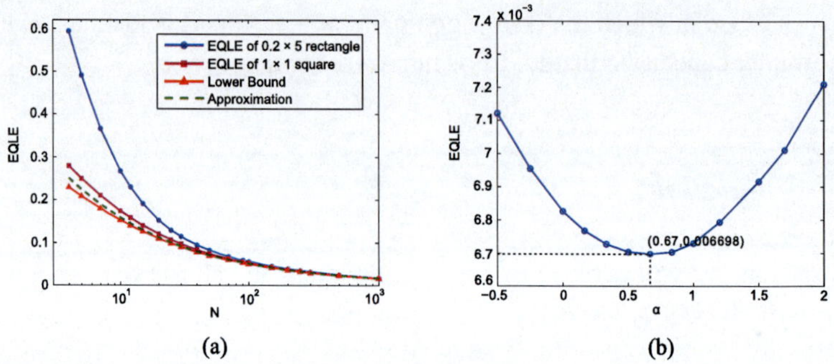

Fig. 2.22 (**a**) EQLE versus the number of measurement points. (**b**) EQLE of the region when measurement points are allocated following $\mu_i \propto \rho_i^{\alpha}$.

Table 2.3 EQLE of collocation patterns in various environments

Collocation pattern	Theoretical	No infrastructures	Infrastructures
Equilateral triangles	0.011810	0.011810	0.011997
Grids	0.011956	0.011955	0.012185

2.3.4.2 Random Collocation

In the simulation of random collocation, we evaluate the relationship between the EQLE and the number of measurement points in a 0.2×5 rectangular region and 1×1 square region, respectively, where the simulation results are shown in Fig. 2.21a. Noting that curves in this figure are logarithmic scale for the x-axis and linear scale for the y-axis, the EQLE in two regions approaches the lower bound fast with N increasing. EQLE of 1×1 square is always lower than that of 0.2×5 rectangular since users will have less probability to locate near the region boundary. When N is large, both the EQLE and the lower bound will be approximately equal to $\frac{1}{2}\sqrt{\frac{S}{N}}$, which validates our results.

2.3.4.3 Asymmetrical Distribution

In this simulation, we create a 1×2 rectangular region, where 1000 users locate in the left half part with probability 0.9 and locate in the other half part with probability 0.1. We use the same collocation pattern for both parts, which indicates that collocation pattern constants in two parts are identical. We allocate measurement points following

$$\mu_i \propto \rho_i^\alpha,$$

where α is the independent variable. Simulation results are shown in Fig. 2.21b, where each data point in the simulation is derived from 1000 experiments. In this figure, the EQLE is minimized when $\alpha = 0.67$, which validates our theoretical analysis in Sect. 2.3.3.

2.4 Scalability with the Number of Users

This section studies the issue from a theoretical perspective, where the upper and lower bounds of the system's localization reliability with respect to the number of users are derived. Our theoretical results can be verified by experiments and thus can provide meaningful guidance for practical system design. The theoretical and experimental results of our work reveal two interesting observations that shed light on the insight into the scalability of the fingerprinting localization system: first, the localization reliability drops dramatically before the number of users increases to a critical point and then decreases smoothly, where the critical point tends to appear when the number of users equals the number of access points (APs) deployed in the service region, and second, even if the number of users approaches to infinity, the fingerprinting localization system still retains a certain level of reliability.

2.4.1 Problem Formulation

2.4.1.1 Interference Region

We consider an indoor space denoted by S, where there are M APs distributed along the boundaries of S and N users within S. As many fingerprinting localization systems [85, 115], we grid S into small square sub-regions, where the center of each square sub-region is set as a reference point (RP). In the offline phase, the fingerprints are measured on RPs; in the online phase, the nearest RP to the user's real location is supposed to be the system's estimated location. If a user performs RSS measurement with respect to an AP, then a transmitter–receiver (T–R) pair is formed. We know that a human body between the T–R pair will impact the RSS measurement; however, due to the limited transmitting power and randomness of radio propagation, the obstacle must be in some region, and then the wireless T–R link can be impacted. We term this region as *interference region*. The concept is implicitly used in studies on passive localization [125, 126], where it is normally believed that the region is in the shape of ellipse. We do not adopt the ellipse assumption because our experimental results show that it is not always true. For users, we have no assumption of the specific probability model describing how likely the user will appear in which location, but we assume the users follow the same model.

2.4.1.2 Localization Reliability Model

In a 2D area S, a location can be identified by a 2D vector $\mathbf{r}$. The RSS observed at $\mathbf{r}$ can be modeled as follows:

$$\mathcal{P}(\mathbf{r}) = \mu(\mathbf{r}) + \sigma \mathcal{Y}, \tag{2.86}$$

where $\mu(\mathbf{r})$ represents how the mean of RSS readings varies with respect to locations. $\mathcal{Y}$ is the normalized Gaussian random variable with $\mathcal{Y} \sim \mathcal{N}(0, 1)$ and σ is a constant representing the variance of the received signal [117]. Modeling RSS as a Gaussian distributed random variable is verified and adopted in a number of work in the literature [32, 36, 39, 77, 85, 101], and the rationale of the modeling is verified with comprehensive experimental results in [116, 117]. Note that Eq. (2.86) is a general model for radio propagation, where the specific expression of $\mu(\mathbf{r})$ is not determined. If we let $\mu(\mathbf{r})$ be a logarithmic function, then the model deteriorates into the LNPL model.

The location estimation process is essentially a mapping from the sample space to the physical space, where it has been found that a user can be localized within δ neighborhood of the real location only if the user's reported fingerprints fall in certain area E of the sample space [116, 117]. We use the simple example as follows to show the modeling approach.

Consider a 1D physical space where the single AP is located at one end, as shown in Fig. 2.23; the corresponding sample space is as shown in the upper part of Fig. 2.23, where each location is distinguished by the observed mean value of the RSS fingerprints μ, and the PDF of the fingerprints observable at the location follows Gaussian distribution based on Eq. (2.86). According to the principle of maximum likelihood estimation (MLE), the user can be localized in the region Q only if the reported RSS fingerprints fall within the region E in the sample space as shown in Fig. 2.23. Corresponding reliability R can be obtained by performing integrating Gaussian PDF over the region E in the sample space. The localization in 2D physical space with high-dimensional sample space is modeled in a similar manner but confronted with more mathematical challenges. It is proved in [116, 117] that the region E in the high-dimensional sample space is a hyper-cylinder with the orthogonal cross-section in the shape of an ellipse.

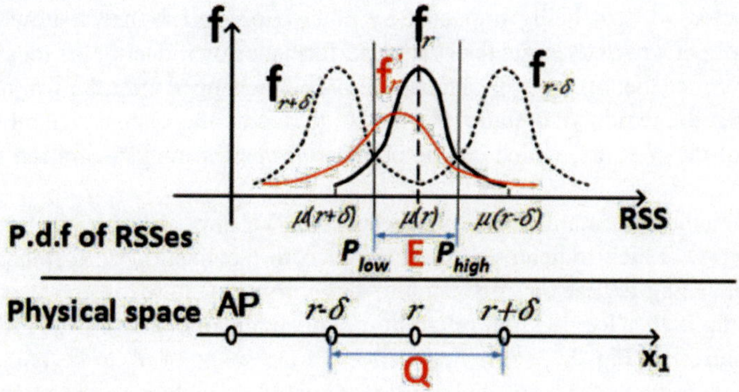

Fig. 2.23 Reliability model

Our work in this book leverages the localization reliability model in [116, 117]; however, our focus is on scalability. Moreover, we consider the practical localization scenario, where fingerprints collection is only performed on RPs [32, 34, 38, 53, 77, 78, 85, 105, 219]. We define the localization reliability as the probability that a user can be correctly localized in a square region surrounded by the other 8 RPs, and the accuracy is the unit length of the grid. This is in contrast to the circle area in [116, 117], where it is implicitly and ideally assumed that each point in the indoor space has to be surveyed.

2.4.1.3 Localization Performance Deterioration by Human Body Blockage

The reliability and accuracy model in [116, 117] as described above implicitly assumes that the probability density function (PDF) representing each location in the offline and online phase is the same. However, the radio propagation environment in the two phases is factually different in practice, where an important reason is the human body blockage effect. Comprehensive studies show that the mean and the variance of the wireless signal can be observed at a location will change, if the signal's propagation path is changed from the line-of-sight path to the non-line-of-sight one [218]. The deviation of the mean and the variance is especially notable with the presence of the human body blockage [122–124]. The deviation makes the shape of the PDF representing the same location **r** change in the two phases. Figure 2.23 shows an example of blockage effect, where the PDF in the offline phase is $f_{\mathbf{r}}$, while it becomes $f'_{\mathbf{r}}$ in the online phase.

2.4.1.4 Strategy of Deriving the Scalability

Consider a fingerprinting localization system, and we assume that there is only one worker performing the site survey in the offline phase. Then the fingerprints database is constructed without being impacted by other people. If N users show up, the blockage effect will deteriorate the system performance by reducing the localization reliability. Since the fingerprints database remains unchanged after the worker's site survey, then the region E is static. We use R' to denote the system reliability after show-up of the N users, which can be obtained by performing integration with $f'_{\mathbf{r}}$ over E.

Although the deformation of $f'_{\mathbf{r}}$ with respect to $f_{\mathbf{r}}$ is unpredictable, we intuitively have $R' \leq R$, which indicates the deterioration in localization performance. The physical meaning is that the mismatch of fingerprints' features in the two phases cannot bring higher localization reliability; mathematically, due to the PDF nature of $f'_{\mathbf{r}}$ as shown in Fig. 2.23, it is impossible for the shape of $f'_{\mathbf{r}}$ to be both higher and wider than that of $f_{\mathbf{r}}$, and the peak of $f'_{\mathbf{r}}$ could deviate from the midpoint of the integration domain, and thus the result of integration for $f'_{\mathbf{r}}$ over E is no greater than that for $f_{\mathbf{r}}$. However, due to the Gaussian property for both of the PDFs and the extremely abstract high dimension of the sample space, finding the expression of R' will be a major challenge. Moreover, the PDF can be changed only if some users are standing in the influence region. The more users are present in the indoor space, the more likely there are some users standing in the influence region of some T–R pairs, and thus more reliability deterioration can occur.

Consequently, *the scalability issue of the fingerprinting indoor localization system is essentially finding how the localization reliability of the system deteriorates with respect to the number of users N.*

We do not discuss the situation where more than one people are present during the site survey process, or the database is updated periodically, because such detailed scenarios can be easily extended from our study.

2.4.2 Localization Reliability Without Blockage

This section presents how to derive R for 2D localization, which lays the foundation of the following studies. The physical space is illustrated in Fig. 2.24, where it is supposed that the user's actual location is at Q_0. In order to make the system accurately estimate the user's location, the user's reported RSSes must be the measurement outcome $\mathbf{o}$, which falls into the region E defined as follows according to the MLE principle:

$$E = \left\{ \mathbf{o} | \prod_{i=1}^{M} f_{Q_0}(P_i) \geq \prod_{i=1}^{M} f_{Q_j}(P_i), j = 1, 2, \ldots, 8 \right\}, \tag{2.87}$$

where P_i is the user's measurement of the RSS with respect to AP_i. Based on the radio propagation model in Eq. (2.86), the outcomes in the event E satisfy the following inequality:

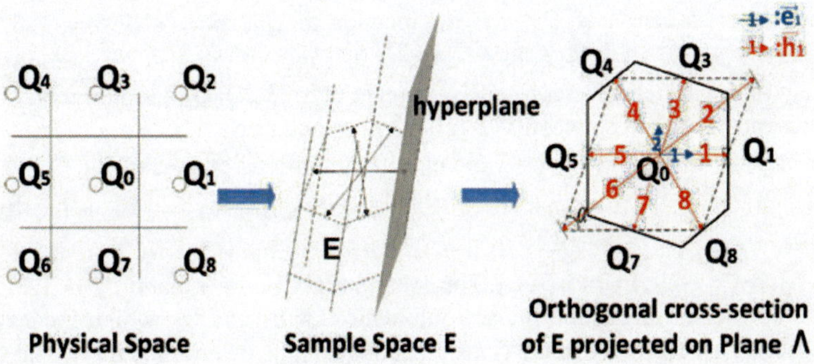

Fig. 2.24 Localization reliability without blockage

$$\prod_{i=1}^{M} \frac{1}{\sigma_i\sqrt{2\pi}} e^{-\frac{(P_i-\mu_i(Q_0))^2}{2\sigma_i^2}} \geq \prod_{i=1}^{M} \frac{1}{\sigma_i\sqrt{2\pi}} e^{-\frac{(P_i-\mu_i(Q_j))^2}{2\sigma_i^2}}, \tag{2.88}$$

which is equivalent to

$$\sum_{i=1}^{M} \frac{\mu_i(Q_j)-\mu_i(Q_0)}{\sigma_i}\left(Y_i - \frac{\mu_i(Q_j)-\mu_i(Q_0)}{2\sigma_i}\right) \leq 0, \tag{2.89}$$

where $\mu_i(Q_j)$ is the mean of RSSes with respect to AP_i at the point Q_j and $Y_i = \frac{P_i-\mu_i}{\sigma_i} = \frac{P_i-\mu_i(Q_0)}{\sigma_i}$.

The process of finding E could be understood with the help of Fig. 2.24. We now find the specific expression of E through mathematical transformation. Let $\mathbf{h_j} = (\frac{\mu_1(Q_j)-\mu_1(Q_0)}{2\sigma_1}, \frac{\mu_2(Q_j)-\mu_2(Q_0)}{2\sigma_2}, \ldots, \frac{\mu_M(Q_j)-\mu_M(Q_0)}{2\sigma_M})^T$, with $j = 1, 2, \ldots, 8$. Then Eq. (2.89) can be presented in the vector form:

$$\begin{cases} 2\mathbf{h_1}(\mathbf{o}-\mathbf{h_1}) \leq 0, \\ 2\mathbf{h_2}(\mathbf{o}-\mathbf{h_2}) \leq 0, \\ \cdots \\ 2\mathbf{h_8}(\mathbf{o}-\mathbf{h_8}) \leq 0, \end{cases} \tag{2.90}$$

where $\mathbf{o} = (Y_1, Y_2, \ldots, Y_M)^T$.

In the square region centered by Q_0 as shown in Fig. 2.24, the mean of the observable RSS fingerprints is unlikely to change dynamically in a very small neighborhood, which is verified by a number of work in the literature [32, 34, 38, 53, 77, 85, 105, 117]; therefore, we have

$$\mu_i(\mathbf{r_j}) - \mu_i(\mathbf{r}) \approx \delta_j|\nabla\mu_i(\mathbf{r})|\cos(\theta_j - \phi_i), \tag{2.91}$$

where $\mathbf{r_j}$ is the location of Q_j, $\mathbf{r}$ is the location of Q_0, $\delta_j = \varepsilon$ if $j = 1, 3, 5, 7$ and $\delta_j = \sqrt{2}\varepsilon$ if $j = 2, 4, 6, 8$, $\varepsilon =$ the unit length of the grid representing the accuracy of the system, θ_j represents the angle between $\mathbf{Q_0Q_j}$ and horizontal axis, ϕ_i represents the angle between $\nabla\mu_i(\mathbf{r})$ and horizontal axis.

We can see that $\theta_j = \frac{(j-1)\pi}{4}$, along with definitions of $\mathbf{h_j}$ and Eq. (2.91), we have $\mathbf{h_1} = (\frac{|\nabla\mu_1(\mathbf{r})|}{2\sigma_1}\varepsilon cos(\phi_1), \ldots, \frac{|\nabla\mu_M(\mathbf{r})|}{2\sigma_M}\varepsilon cos(\phi_M))^T$; $\mathbf{h_2} = \mathbf{h_1} + \mathbf{h_3}$; $\mathbf{h_3} = (\frac{|\nabla\mu_1(\mathbf{r})|}{2\sigma_1}\varepsilon sin(\phi_1), \ldots, \frac{|\nabla\mu_M(\mathbf{r})|}{2\sigma_M}\varepsilon sin(\phi_M))^T$; $\mathbf{h_4} = -\mathbf{h_1} + \mathbf{h_3}$; $\mathbf{h_5} = -\mathbf{h_1}$; $\mathbf{h_6} = -\mathbf{h_1} - \mathbf{h_3}$; $\mathbf{h_7} = -\mathbf{h_3}$; $\mathbf{h_8} = \mathbf{h_1} - \mathbf{h_3}$. The equations above show that $\mathbf{h_j}$, $j = 1, \ldots, 8$, can be represented using the linear combination of $\mathbf{h_1}$ and $\mathbf{h_3}$, which means that those vectors are factually in the same plane that can be determined by $\mathbf{h_1}$ and $\mathbf{h_3}$.

The plane determined by $\mathbf{h_1}$ and $\mathbf{h_3}$ is illustrated in the rightmost sub-figure of Fig. 2.24, which however is not necessarily to be a Cartesian plane. For the convenience of deriving the reliability, we transform the original coordinates system $\{Y_i\}$ to another system $\{\mathbf{e_i}\}$, $i = 1, 2, \ldots, M$, as shown in the rightmost sub-figure of Fig. 2.24. For simplicity of demonstration, we let $\mathbf{e_1}$ be parallel to $\mathbf{h_1}$ and $\mathbf{e_2} \in$ the plane Λ which is determined by $\mathbf{h_1}$ and $\mathbf{h_3}$. We use α to denote the angle between $\mathbf{h_3}$ and $\mathbf{e_1}$(or $\mathbf{h_1}$), then $\mathbf{h_1} = |\mathbf{h_1}| \cdot \mathbf{e_1}$, $\mathbf{h_3} = |\mathbf{h_3}|cos\alpha \cdot \mathbf{e_1} + |\mathbf{h_3}|sin\alpha \cdot \mathbf{e_2}$. Note that we can denote $\mathbf{o} = \sum_i c_i\mathbf{e_i}$ in the new coordinates system, and then the value of c_1 and c_2 can be determined by Eq. (2.89). In particular, the values of c_1 and c_2 can be determined by the inequality set (2.90). This can be more obvious when observing the rightmost sub-figure of Fig. 2.24. We can see that c_1 is constrained by $\mathbf{h_1}$ and $\mathbf{h_5}$ with $\mathbf{h_1} = -\mathbf{h_5}$, and c_2 is constrained by other $\mathbf{h_j}$. Consequently, we have $-|\mathbf{h_1}| \le c_1 \le |\mathbf{h_1}|$, and $L \le c_2 \le H$, where

$$L = \frac{-|\mathbf{h_3}| - c_1\cos\alpha}{\sin\alpha} + \frac{|\mathbf{h_1}|}{|\mathbf{h_3}|\sin\alpha} \cdot \max\{0, -|\mathbf{h_1}| - 2|\mathbf{h_3}|\cos\alpha - c_1, -|\mathbf{h_1}| + 2|\mathbf{h_3}|\cos\alpha + c_1\}, \tag{2.92}$$

$$H = \frac{|\mathbf{h_3}| - c_1\cos\alpha}{\sin\alpha} + \frac{|\mathbf{h_1}|}{|\mathbf{h_3}|\sin\alpha} \cdot \min\{0, |\mathbf{h_1}| + 2|\mathbf{h_3}|\cos\alpha - c_1, |\mathbf{h_1}| - 2|\mathbf{h_3}|\cos\alpha + c_1\}. \tag{2.93}$$

Since $|\mathbf{o}| = \sum_{i=1}^{M} Y_i{}^2 = \sum_{i=1}^{M} c_i{}^2$, which implies that the norm of $\mathbf{o}$ is unchanged after the rotation of the coordinate system, then the probability that the system correctly estimates the user's location is

$$R = \int_E f(\mathbf{P})d\mathbf{P}$$
$$= \int_E \prod_{i=1}^{M} \frac{1}{\sigma_i\sqrt{2\pi}} e^{-\frac{(P_i-\mu_i(\mathbf{r}))^2}{2\sigma_i^2}} dP_1dP_2\ldots dP_M$$

$$
\begin{aligned}
&= \int_E \prod_{i=1}^{M} \frac{1}{\sqrt{2\pi}} e^{-\frac{Y_i^2}{2}} dY_1 dY_2 \ldots dY_M \\
&= \int_{-|\mathbf{h}_1|}^{|\mathbf{h}_1|} \int_L^H \int_{-\infty}^{+\infty} \ldots \int_{-\infty}^{+\infty} \prod_{i=1}^{M} \frac{1}{\sqrt{2\pi}} e^{-\frac{c_i^2}{2}} dc_1 dc_2 \ldots dc_M \\
&= \int_{-|\mathbf{h}_1|}^{|\mathbf{h}_1|} \frac{1}{\sqrt{2\pi}} e^{-\frac{c_1^2}{2}} dc_1 \cdot \int_L^H \frac{1}{\sqrt{2\pi}} e^{-\frac{c_2^2}{2}} dc_2. \qquad (2.94)
\end{aligned}
$$

2.4.3 *Localization Reliability Deterioration by Blockage*

2.4.3.1 Localization Reliability with Blockage

Considering a T–R pair associated with an influence region, if another person appears in the influence region, the RSS value observed by the user is actually profiled by $f'(\mathbf{P})$. The resulted reliability becomes

$$
R' = \int_E f'(\mathbf{P}) d\mathbf{P} = \int_E \prod_{i=1}^{M} \frac{1}{\sigma_i' \sqrt{2\pi}} e^{-\frac{(P_i - \mu_i'(\mathbf{r}))^2}{2\sigma_i'^2}} dP_1 \ldots dP_M,
$$

where $\mu_i'(\mathbf{r})$ and σ_i' represent the deviated mean and standard variance resulted from the human body blockage effect. Note that E does not change since the system still uses the fingerprints database constructed by the single survey worker. We can see that the integrand contains quadratic terms of $\mu_i'(\mathbf{r})$ in the current expression of reliability. This makes it difficult to obtain an analytical solution for the integral. In order to make the reliability's expression more concise, we apply the following new coordinates system to the sample space: let $Y_i' = \frac{P_i - \mu_i'}{\sigma_i'}$, then $Y_i' = \frac{\sigma_i}{\sigma_i'} Y_i + \frac{\mu_i - \mu_i'}{\sigma_i'} \triangleq A_i Y_i + B_i$, where it is straightforward that $A_i > 1$ and $B_i > 0$. Then we have

$$
R' = \int_E \prod_{i=1}^{M} \frac{1}{\sqrt{2\pi}} e^{-\frac{Y_i'^2}{2}} dY_1' dY_2' \ldots dY_M'. \qquad (2.95)
$$

After the coordinates system transformation, the integrand $f'(\mathbf{P})$ only contains quadratic terms of the integral variable itself, which makes the integral operation easier. However, the expression of E is also changed in the new coordinates system; we must first figure out E in the new system and then find analytical solution of R'.

We denote $\mathbf{o}' = (Y_1', Y_2', \ldots, Y_M')^T = (A_1 Y_1 + B_1, \ldots, A_M Y_M + B_M)^T$, and we need to find the form of restrictions to the event E w.r.t. $\mathbf{o}'$ similarly to the vector form in the inequalities set (2.90). Thus we define the following vector w.r.t. the coefficient A_i and B_i, $i = 1, 2, \ldots, M$.

For any $j \in \{1, 2, \ldots, 8\}$, we define $\mathbf{l} = (\frac{B_1}{A_1}, \frac{B_2}{A_2}, \ldots, \frac{B_M}{A_M})^T$, $\mathbf{H_j} = (\frac{\mu_1(\mathbf{r_j})-\mu_1(\mathbf{r})}{2\sigma_1 A_1}, \ldots, \frac{\mu_M(\mathbf{r_j})-\mu_M(\mathbf{r})}{2\sigma_M A_M})^T$ and $\mathbf{h'_j} = k_j \mathbf{H_j}$, where k_j is a constant, $k_j = \frac{|\mathbf{h_j}|^2+\mathbf{l}\mathbf{h_j}}{|\mathbf{H_j}|^2}$.

Similarly to the method as mentioned in Sect. 2.4.2, we have

$\mathbf{H_1} = (\frac{|\nabla\mu_1(\mathbf{r})|}{2\sigma_1 A_1}\varepsilon cos(\phi_1), \ldots, \frac{|\nabla\mu_M(\mathbf{r})|}{2\sigma_M A_M}\varepsilon cos(\phi_M))^T$, $\mathbf{H_2} = \mathbf{H_1} + \mathbf{H_3}$,
$\mathbf{H_3} = (\frac{|\nabla\mu_1(\mathbf{r})|}{2\sigma_1 A_1}\varepsilon sin(\phi_1), \ldots, \frac{|\nabla\mu_M(\mathbf{r})|}{2\sigma_M A_M}\varepsilon sin(\phi_M))^T$, $\mathbf{H_4} = -\mathbf{H_1} + \mathbf{H_3}$,
$\mathbf{H_5} = \mathbf{H_3}$, $\mathbf{H_6} = -\mathbf{H_1} - \mathbf{H_3}$, $\mathbf{H_7} = -\mathbf{H_3}$, $\mathbf{H_8} = \mathbf{H_1} - \mathbf{H_3}$ and $\mathbf{h'_j} = k_j \mathbf{H_j}$, $j = 1, 2, \ldots, 8$.

We then have $k_j + k_{j+4} = \frac{2|\mathbf{h_j}|^2}{|\mathbf{H_j}|^2}$ for $j = 1, 2, 3, 4$ and $0 < k_j + k_{j+4} \le 2$, according to the definitions of $\mathbf{h_j}$ and $\mathbf{H_j}$.

Based on the definitions and analysis above, we have

$$\begin{aligned}\mathbf{h'_j}(\mathbf{o'} - \mathbf{h'_j}) &= k_j\left(\mathbf{H_j}\mathbf{o'} - k_j\mathbf{H_j^2}\right)\\ &= k_j\left[(\mathbf{h_j}\mathbf{o} + \mathbf{l}\mathbf{h_j}) - \frac{|\mathbf{h_j}|^2 + \mathbf{l}\mathbf{h_j}}{|\mathbf{H_j}|^2}\cdot\mathbf{H_j^2}\right]\\ &= k_j\left(\mathbf{h_j}\mathbf{o} - |\mathbf{h_j}|^2\right) = k_j\mathbf{h_j}(\mathbf{o} - \mathbf{h_j}),\end{aligned} \tag{2.96}$$

which establishes the connection between E in the two coordinates systems. Then we have $2k_1 \cdot \mathbf{h'_1}(\mathbf{o'} - \mathbf{h'_1}) \le 0$, $2k_2 \cdot \mathbf{h'_2}(\mathbf{o'} - \mathbf{h'_2}) \le 0, \ldots, 2k_8 \cdot \mathbf{h'_8}(\mathbf{o'} - \mathbf{h'_8}) \le 0$.

Comparing Eqs. (2.94) and (2.95), it can be seen that $f(\mathbf{P})$ and $f'(\mathbf{P})$ have the same form although they are in different coordinates systems. And what we need to do is to analyze the event E in the new coordinates system $\{Y'_i\}$. Since $\mathbf{h'_j} = k_j\mathbf{H_j}$, $j = 1, 2, \ldots, 8$, the coefficient k_j factually reflects how the blockage effect will impact the event E, which results in R'. Figure 2.25a–c shows three possible scenarios how the event E can be impacted by users: (a) if there is a single user in the system just like the scenario of site survey, then the fingerprints are not impacted and $k_1 = k_2 = \ldots = k_8 = 1$, which means that the shape and location of E in sample space are the same as shown in Fig. 2.24, (b) if the RSS readings are insignificantly influenced, then $k_j > 0$, $j = 1, 2, \ldots, 8$, and the shape and position of E will change but in an insignificant manner as shown in Fig. 2.25, which is due to the relationship among E's boundaries, and (c) if $k_5 < 0$, both the shape and location of E will change significantly.

Then we rotate the coordinates system $\{Y'_i\}$ to another orthogonal basis $\{\mathbf{g_i}\}$, $i = 1, 2, \ldots, n$, where $\mathbf{g_1}$ is parallel to $\mathbf{H_1}$ and $\mathbf{g_2} \in$ the plane Λ which is determined by $\mathbf{H_1}$ and $\mathbf{H_3}$. We use β to denote the angle between $\mathbf{H_1}$ and $\mathbf{H_3}$, then $\mathbf{H_1} = |\mathbf{H_1}| \cdot \mathbf{g_1}$, $\mathbf{H_3} = |\mathbf{H_3}|cos\beta \cdot \mathbf{g_1} + |\mathbf{H_3}|sin\beta \cdot \mathbf{g_2}$. Suppose $\mathbf{o'} = \sum_i a_i\mathbf{g_i}$, and then by Eq. (2.89), we have $-k_5|\mathbf{H_1}| \le a_1 \le k_1|\mathbf{H_1}|$ and $L' \le a_2 \le H'$, where

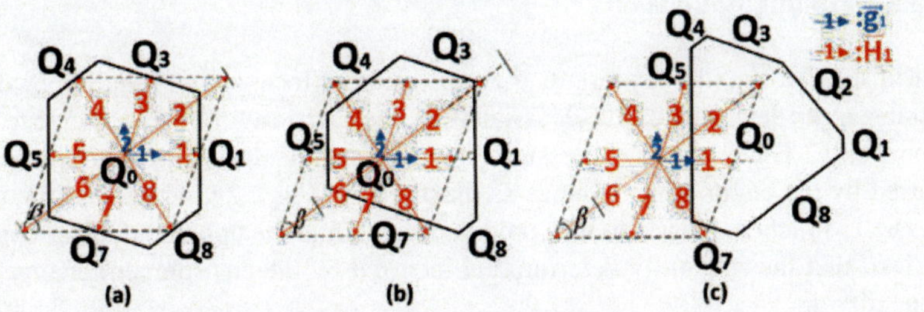

Fig. 2.25 Influence of blockage effect on E in the sample space

$$
\begin{aligned}
L' = -a_1 \cot\beta + \frac{1}{\sin\beta} \cdot \max\{&-k_7|\mathbf{H_3}|, \\
&-k_6|\mathbf{H_3}| - \frac{|\mathbf{H_1}|}{|\mathbf{H_3}|}[k_6(|\mathbf{H_1}| + 2|\mathbf{H_3}|\cos\beta) + a_1], \\
&-k_8|\mathbf{H_3}| - \frac{|\mathbf{H_1}|}{|\mathbf{H_3}|}[k_8(|\mathbf{H_1}| - 2|\mathbf{H_3}|\cos\beta) - a_1]\Big\},
\end{aligned} \tag{2.97}
$$

$$
\begin{aligned}
H' = -a_1 \cot\beta + \frac{1}{\sin\beta} \cdot \min\{&k_3|\mathbf{H_3}|, \\
&k_2|\mathbf{H_3}| + \frac{|\mathbf{H_1}|}{|\mathbf{H_3}|}[k_2(|\mathbf{H_1}| + 2|\mathbf{H_3}|\cos\beta) - a_1], \\
&k_4|\mathbf{H_3}| + \frac{|\mathbf{H_1}|}{|\mathbf{H_3}|}[k_4(|\mathbf{H_1}| - 2|\mathbf{H_3}|\cos\beta) + a_1]\Big\}.
\end{aligned} \tag{2.98}
$$

Since $|\mathbf{o}'| = \sum_{i=1}^{n} Y_i'^2 = \sum_{i=1}^{n} {a_i}^2$, which implies that the norm of $\mathbf{o}'$ is unchanged after the rotation of the coordinate system, then the probability that the system correctly estimates the user's location is

$$
\begin{aligned}
R' &= \int_E \prod_{i=1}^{M} \frac{1}{\sqrt{2\pi}} e^{-\frac{Y_i'^2}{2}} dY_1' dY_2' \ldots dY_M' \\
&= \int_{-k_5|\mathbf{H_1}|}^{k_1|\mathbf{H_1}|} \int_{L'}^{H'} \int_{-\infty}^{+\infty} \cdots \int_{-\infty}^{+\infty} \prod_{i=1}^{M} \frac{1}{\sqrt{2\pi}} e^{-\frac{{a_i}^2}{2}} da_1 \ldots da_M \\
&= \int_{-k_5|\mathbf{H_1}|}^{k_1|\mathbf{H_1}|} \frac{1}{\sqrt{2\pi}} e^{-\frac{{a_1}^2}{2}} da_1 \cdot \int_{L'}^{H'} \frac{1}{\sqrt{2\pi}} e^{-\frac{{a_2}^2}{2}} da_2.
\end{aligned} \tag{2.99}
$$

2.4.3.2 Finding Bounds of R'

We can see that Eq. (2.99) presents the expression of localization reliability, where we transfer the deviation of the Gaussian PDF due to the environment change to that of the area E. The concrete expression of R' requires the exact amount of deviation incurred by the environment change as illustrated in Fig. 2.25, which is difficult to measure in practice. However, we could manage to find the upper and lower bounds of R', so that the reliability deterioration incurred by the environment change can be quantified.

We here use a simple example in the 1D sample space to explain our basic idea for deriving the bounds, which is as shown in Fig. 2.26. We know that the event E is determined in the offline phase, but the event E represented by the blue line segment in the figure is actually the event in the transformed coordinate systems. The coordinate system transformation operations described in the previous subsection could change both the size and the position of E, which is to have a neat expression of f' as shown in the figure. The reliability R' can be obtained by performing integration of f' over E in the figure, where it is relatively easy to find the size but difficult to find the displacement of E in mathematical derivations. The difficulty is particular significant when performing high-dimensional integration, where the specific derivation process to be presented in the following subsections can provide a better understanding.

Although it is difficult to find the displacement of E with respect to the origin of the coordinate system, it is straightforward to see that the upper bound of R' can be obtained if we move E to the position of the line segment in red. For the lower bound, it is obvious that 0 is an option, which however provides negligible information. We want to find the largest lower bound so that the reliability can be tightly bounded. In the process of mathematical derivation, we keep moving the event E toward the right until we find the nearest position from the origin, based on which we can obtain some meaningful results about R'. This is the basic idea how we derive the bounds of R', and detailed derivations can be found in the following subsections.

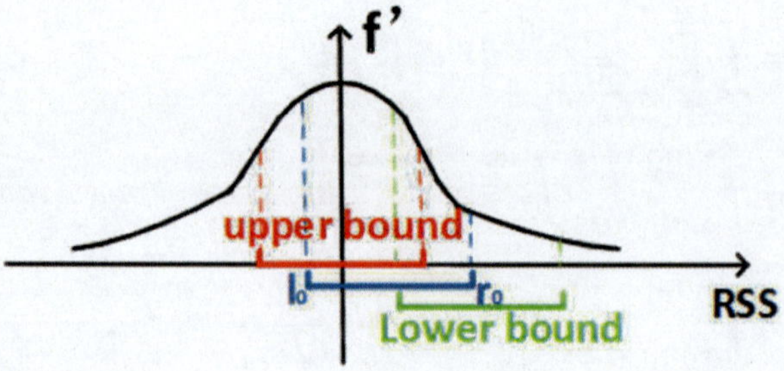

Fig. 2.26 Finding bounds of R'

2.4.3.3 Upper Bound of R'

Due to Eqs. (2.97) and (2.98), $\forall -k_5|\mathbf{H_1}| \le a_1 \le k_1|\mathbf{H_1}|$, $L' \ge -\frac{K_1|\mathbf{H_3}|}{sin\beta} - a_1 cot\beta$ and $H' \le \frac{K_1|\mathbf{H_3}|}{sin\beta} - a_1 cot\beta$, where $K_1 = \max\{k_i, i = 1, 2, \ldots, 8\}$. Consequently,

$$
\begin{aligned}
R' &\le \int_{-k_5|\mathbf{H_1}|}^{k_1|\mathbf{H_1}|} \frac{e^{-\frac{a_1{}^2}{2}}}{\sqrt{2\pi}} da_1 \cdot \int_{-\frac{K_1|\mathbf{H_3}|}{\sin\beta}-a_1\cot\beta}^{\frac{K_1|\mathbf{H_3}|}{\sin\beta}-a_1\cot\beta} \frac{e^{-\frac{a_2{}^2}{2}}}{\sqrt{2\pi}} da_2 \\
&\le \int_{-\frac{k_1+k_5}{2}|\mathbf{H_1}|}^{\frac{k_1+k_5}{2}|\mathbf{H_1}|} \frac{e^{-\frac{a_1{}^2}{2}}}{\sqrt{2\pi}} da_1 \cdot \int_{-\frac{K_1|\mathbf{H_3}|}{\sin\beta}}^{\frac{K_1|\mathbf{H_3}|}{\sin\beta}} \frac{e^{-\frac{a_2{}^2}{2}}}{\sqrt{2\pi}} da_2 \\
&= erf\left(\frac{k_1+k_5}{2}\frac{|\mathbf{H_1}|}{\sqrt{2}}\right) \cdot erf\left(\frac{K_1|\mathbf{H_3}|}{\sqrt{2}\sin\beta}\right) \\
&\triangleq \overline{R'(E)},
\end{aligned}
\tag{2.100}
$$

where $erf(\cdot)$ is the error function defined as $erf(x) = \frac{2}{\sqrt{\pi}} \int_0^x e^{-t^2} dt$.

2.4.3.4 Lower Bound of R'

For any $j \in \{1, 2, \ldots, 8\}$,$k_j \le |k_j| = |\frac{|\mathbf{h_j}|^2 + \mathbf{l}\mathbf{h_j}}{|\mathbf{H_j}|^2}| \le \frac{|\mathbf{h_j}|^2}{|\mathbf{H_j}|^2} + \frac{|\mathbf{l}||\mathbf{h_j}|}{|\mathbf{H_j}|^2}$. Then we have $|\mathbf{l}| = \sqrt{\sum_{i=1}^M (\frac{(\mu_i(\mathbf{r}) - \mu_i'(\mathbf{r}))/\sigma_i'}{\sigma_i/\sigma_i'})^2} \le \sqrt{\sum_{i=1}^M (\frac{\mu_i(\mathbf{r})}{\sigma_i})^2}$, and $\forall j \in \{1, 2, \ldots, 8\}$, $\sqrt{\sum_{i=1}^M (\frac{\mu_i(\mathbf{r})}{\sigma_i})^2} > \sqrt{\sum_{i=1}^M (\frac{\mu_i(\mathbf{r_j}) - \mu_i(\mathbf{r})}{\sigma_i})^2} = |\mathbf{h_j}|$.

Define $K_{2,1} = \max_{j\in\{1,2,\ldots,8\}} \frac{\sqrt{\sum_{i=1}^M (\frac{\mu_i(\mathbf{r})}{\sigma_i})^2}|\mathbf{h_j}|}{|\mathbf{H_j}|^2}$, $K_{2,2} = \max_{j\in\{1,2,\ldots,8\}} \frac{|\mathbf{h_j}|^2}{|\mathbf{H_j}|^2}$, $K_{2,3} = \min_{j\in\{1,2,\ldots,8\}} \frac{|\mathbf{h_j}|^2}{|\mathbf{H_j}|^2}$, and $K_2 = K_{2,1} + K_{2,2}$, and then the following holds: (1) $K_{2,2} > K_{2,3}$ and $K_{2,1}$,$K_{2,2}$,$K_{2,3}$,$K_2 > 0$, (2) $K_{2,1} > K_{2,2}$, (3) $\forall j \in \{1, 2, \ldots, 8\}$, $k_j \le K_2$, and (4) $K_2 - 2K_{2,3} > K_2 - 2K_{2,2} = K_{2,1} - k_{2,2} > 0$.

We here first estimate the lower bound of $\int_{-k_5|\mathbf{H_1}|}^{k_1|\mathbf{H_1}|} \frac{1}{\sqrt{2\pi}} e^{-\frac{a_1{}^2}{2}} da_1$. Since $k_1|\mathbf{H_1}| - (-k_5|\mathbf{H_1}|) = (k_1+k_5)|H_1| = \frac{2|\mathbf{h_1}|^2}{|\mathbf{H_1}|}$ and by the property of the *Gauss error function*,

$$
\int_{-k_5|\mathbf{H_1}|}^{k_1|\mathbf{H_1}|} \frac{1}{\sqrt{2\pi}} e^{-\frac{a_1{}^2}{2}} da_1 \ge \int_{K_2|\mathbf{H_1}| - \frac{2|\mathbf{h_1}|^2}{|\mathbf{H_1}|}}^{K_2|\mathbf{H_1}|} e^{-\frac{a_1{}^2}{2}} da_1.
$$

Since

$$K_2|\mathbf{H_1}| - \frac{2|\mathbf{h_1}|^2}{|\mathbf{H_1}|} \le \left(K_2 - 2 \min_{j\in\{1,2,\ldots,8\}} \frac{|\mathbf{h_j}|^2}{|\mathbf{H_j}|^2}\right)|\mathbf{H_1}|$$
$$= (K_2 - 2K_{2,3})|\mathbf{H_1}|,$$

then

$$\int_{-k_5|\mathbf{H_1}|}^{k_1|\mathbf{H_1}|} \frac{1}{\sqrt{2\pi}} e^{-\frac{a_1{}^2}{2}} da_1 \ge \int_{(K_2-2K_{2,3})|\mathbf{H_1}|}^{K_2|\mathbf{H_1}|} e^{-\frac{a_1{}^2}{2}} da_1. \tag{2.101}$$

Then we estimate the lower bound of $\int_{L'}^{H'} \frac{1}{\sqrt{2\pi}} e^{-\frac{a_2{}^2}{2}} da_2$ in a similar way. If the interval $[L', H']$ moves right, keeping the length of the interval constant, the integral becomes smaller. Since $k_j \le K_2$, $j = 2, 3, 4$, and by Eq. (2.98), we know

$$H' \le -a_1 \cot\beta + \frac{K_2|\mathbf{H_3}|}{\sin\beta} + \frac{|\mathbf{H_1}|}{|\mathbf{H_3}|\sin\beta} \cdot \min\{0, K_2(|\mathbf{H_1}|$$
$$+2|\mathbf{H_3}|\cos\beta) - a_1, K_2(|\mathbf{H_1}| - 2|\mathbf{H_3}|\cos\beta) + a_1\}$$
$$= \frac{K_2|\mathbf{H_3}|}{\sin\beta} + \frac{|\mathbf{H_1}|}{|\mathbf{H_3}|\sin\beta} \cdot \min\left\{-\frac{|\mathbf{H_3}|}{|\mathbf{H_1}|} a_1 \cos\beta,\right.$$
$$K_2(|\mathbf{H_1}| + 2|\mathbf{H_3}|\cos\beta) - \left(1 + \frac{|\mathbf{H_3}|\cos\beta}{|\mathbf{H_1}|}\right) a_1,$$
$$\left. K_2(|\mathbf{H_1}| - 2|\mathbf{H_3}|\cos\beta) + \left(1 - \frac{|\mathbf{H_3}|\cos\beta}{|\mathbf{H_1}|}\right) a_1 \right\}$$
$$\le \frac{K_2|\mathbf{H_3}|}{\sin\beta} + \frac{|\mathbf{H_1}|}{|\mathbf{H_3}|\sin\beta} \cdot [K_2(|\mathbf{H_1}| + 2|\mathbf{H_3}||\cos\beta|)$$
$$\left. - \left(1 + \frac{|\mathbf{H_3}||\cos\beta|}{|\mathbf{H_1}|}\right) a_1 \right] \overset{\text{denoted as}}{=} g_0(a_1).$$

As $g_0(a_1)$ is a linear function and the coefficient of $a_1 < 0$, then $\forall a_1 \in [-k_5|\mathbf{H_1}|, k_1|\mathbf{H_1}|] \subset [-K_2|\mathbf{H_1}|, K_2|\mathbf{H_1}|]$,

$$g_0(a_1) \le g_0(-K_2|\mathbf{H_1}|) = K_2(2|\mathbf{H_1}| + 3|\mathbf{H_3}||\cos\beta|),$$

thus

$$H' \le \frac{K_2|\mathbf{H_3}|}{\sin\beta} + \frac{K_2|\mathbf{H_1}|}{|\mathbf{H_3}|\sin\beta}(2|\mathbf{H_1}| + 3|\mathbf{H_3}||\cos\beta|) \triangleq H''. \tag{2.102}$$

We let the interval $[L', H']$ move right to the interval $[L'', H'']$, keeping the length of the interval constant, which means $H'' - L'' = H' - L'$, then $L'' = H'' - (H' - L')$, and the interval on $[L'', H'']$ is smaller than the interval on $[L', H']$.

By Eqs. (2.97) and (2.98), $H' - L'$ is dependent on the value of a_1. Next we let $L''' = L'' = H'' - \min_{a_1 \in [-k_5|\mathbf{H_1}|, k_1|\mathbf{H_1}|]}(H' - L') > L''$. Then we find that the interval on $[L''', H'']$ is smaller than the interval on $[L', H']$, with L''' and H'' independent of a_1, as is shown below:

$$\begin{aligned}\int_{L'}^{H'} \frac{1}{\sqrt{2\pi}} e^{-\frac{a_2{}^2}{2}} da_2 &\geq \int_{L''}^{H''} \frac{1}{\sqrt{2\pi}} e^{-\frac{a_2{}^2}{2}} da_2 \\ &\geq \int_{L'''}^{H''} \frac{1}{\sqrt{2\pi}} e^{-\frac{a_2{}^2}{2}} da_2. \end{aligned} \tag{2.103}$$

According to Eqs. (2.101) and (2.103), we know

$$R' \geq \int_{(K_2 - 2K_{2,3})|\mathbf{H_1}|}^{K_2|\mathbf{H_1}|} e^{-\frac{a_1{}^2}{2}} da_1 \cdot \int_{L'''}^{H''} \frac{1}{\sqrt{2\pi}} e^{-\frac{a_2{}^2}{2}} da_2, \tag{2.104}$$

where

$$H'' = \frac{K_2|\mathbf{H_3}|}{\sin\beta} + \frac{K_2|\mathbf{H_1}|}{|\mathbf{H_3}|\sin\beta}(2|\mathbf{H_1}| + 3|\mathbf{H_3}||\cos\beta|), \tag{2.105}$$

$$L''' = H'' - \min_{a_1 \in [-k_5|\mathbf{H_1}|, k_1|\mathbf{H_1}|]}(H' - L'). \tag{2.106}$$

By Eqs. (2.97) and (2.98), we have $H' - L' \leq \frac{1}{\sin\beta} \cdot (k_3 + k_7)|\mathbf{H_3}| \leq \frac{|\mathbf{H_3}|}{\sin\beta} \cdot \frac{2|\mathbf{h_3}|^2}{|\mathbf{H_3}|^2} \leq \frac{2K_{2,2}|\mathbf{H_3}|}{\sin\beta}$. Thus

$$\begin{aligned} L''' &\geq H'' - \frac{2K_{2,2}|\mathbf{H_3}|}{\sin\beta} \\ &= \frac{(K_2 - 2K_{2,3})|\mathbf{H_3}|}{\sin\beta} + \frac{K_2|\mathbf{H_1}|}{|\mathbf{H_3}|\sin\beta}(2|\mathbf{H_1}| + 3|\mathbf{H_3}||\cos\beta|) \\ &> 0. \end{aligned} \tag{2.107}$$

Lemma 1 *For any positive number $a, b (a < b)$, we have $\int_a^b e^{-x^2} dx \geq \frac{1}{e^{2a(b-a)+a^2}} erf(b-a)$.*

Lemma 2 *For any positive number a, we have $[erf(a)]^2 = 1 - e^{-\Theta(a^2)}$.*

Applying Lemma 1, we have

$$\begin{aligned} R'(E) \geq & e^{-[(K_2^2 - 4K_{2,3}^2)|\mathbf{H_1}|^2 + 2L'''(H'' - L''') + L'''^2]} \\ & \cdot erf(2K_{2,3}|\mathbf{H_1}|) erf(H'' - L''') \\ \triangleq & \underline{R'(E)}. \end{aligned} \tag{2.108}$$

2.4.3.5 Magnitude of R' w.r.t. m

Theorem 4 *For a localization system with the fingerprinting approach, if there are m APs influenced by the human body blockage effect, then*

$$e^{-\Theta(g(m))}(1 - e^{-\Theta(g(m))}) \leq R'(E) \leq 1 - e^{-\Theta(g(m))4}, \tag{2.109}$$

where $g(m) = \frac{M^2}{(1-A^2)\cdot m + A^2 \cdot M}$, *m is the number of APs that are blocked, and A is the greatest ratio of* σ_i *to* σ_i' *with* $A < 1$.

2.4.4 Number of Impacted Access Points w.r.t. Number of Users

This section derives the number of APs to be impacted by a given number of users. The result can be easy to obtain if we know the specific shape of the interference region because we can count the number of impacted APs when an interferer appears in certain RP. Since the user appearing in certain RP follows some probability model, then the average number of APs to be impacted by a number of users can be obtained. Although existing work implicitly models the interference region in the shape of the ellipse [125, 126], such model however is not always true according to our experimental results.

Another method to find the number of impacted APs by different numbers of users is to conduct experiments; however, experimenting all scenarios with different combinations of the number of users and the positions of users can be labor-intensive. We are to first present our experimental results showing the difficulty of finding the specific shape of the interference region and then present our method to find the number of impacted APs by a number of users, where there is no need for knowledge of the specific shape of the interference region or labor-intensive experiments.

2.4.4.1 Shape of Influence Region

It can be seen that the crux of finding the number of influenced APs is to find the shape of the influence region. Although existing work normally consider the region in the shape of the ellipse [125, 126], the experimental verification is still unavailable. We are to show our experimental results for determining the shape of

[4] We abuse $\Theta(\cdot)$ in this work and modify the definition as follows: $f(n) = \Theta(g(n))$ means $f(\cdot)$ is upper and lower bounded by $g(\cdot)$, that is, $\exists k_1^*, k_2^* > 0, \forall n > 0 : k_1^* \cdot g(n) \leq f(n) \leq k_2^* \cdot g(n)$. Note that the bounds are valid for any given value of n. They are not asymptotical bounds, and there is no need for $n \to \infty$, which is in contrast to the traditional definition of Θ.

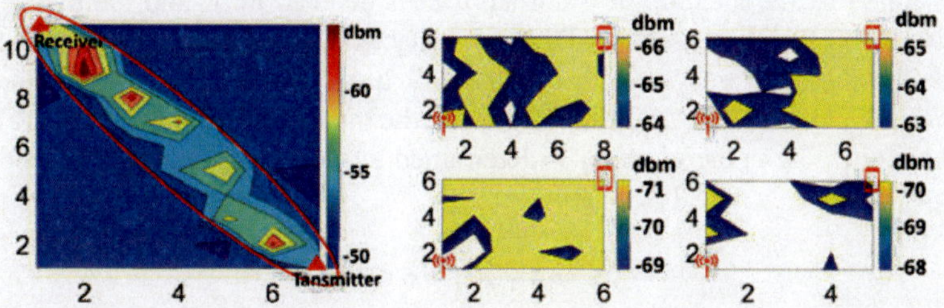

Fig. 2.27 Experimental results of measuring interference region. (**a**) The distribution of the RSS in different cells. (**b**) Experimental results in a larger room

the interference region, which however show that the assumption is not always true. The details of the experiments are as follows.

We set up our testbed in a small room of area 15 m^2, which is grided into 6×10 cells with each cell a 50 cm $\times$ 50 cm small square. We first use ZigBee sensors to do the experiment as in [125, 126]. In particular, we put a ZigBee sensor in the upper left vertex of the room and another one in the lower right vertex and measure the RSS at the receiver side of the T–R pair under the line-of-sight scenario, where the results serve as baseline values. We then let an interferer traversing all cells and record the resulted RSS values at the receiver side. If the remarkable deviation of the observed RSS with respect to the baseline value is observed, we mark the corresponding cell the interferer was at. We use a threshold to specify whether there is a remarkable deviation of the RSS. Figure 2.27a shows the distribution of the RSS in different cells. When we set a smaller threshold, it can be seen that the interference region of the T–R link can be approximately modeled as an ellipse; however, when we set a greater threshold, the interference region is even not a convex region.

We also use the Wi-Fi AP and mobile phone to do the similar experiment in a larger room of area 64 m^2, which is grided into 60 cm $\times$ 60 cm small cells. The results are shown in Fig. 2.27b, which indicates that it is inappropriate to model the interference region into an ellipse area.

We can see from the experimental results that it is unsafe to always model the interference region as an ellipse region, and it is extremely difficult to model the interference region as some regular shape. We have to estimate the number of impacted APs without any prior knowledge of the interference region.

2.4.4.2 Bounding the Number of Impacted APs

As it is difficult to model the shape of the influence region accurately, we try to figure out the number of impacted APs by other means. It is interesting to find that the number of impacted APs in the N-user case can be derived from the simple 2-user case, the result of which can be easily obtained with simple experiments.

Assume that the location of a normal user is denoted by $\mathbf{r}$, and the interferer's location is $\mathbf{r}^*$, which causes $\hat{m}$ APs to be impacted. This means that the interferer enters into the interference regions of the T–R links between the normal user and those $\hat{m}$ APs. As the APs' locations are fixed, the influence regions of the T–R links are determined by $\mathbf{r}$; therefore, $\hat{m}$ is determined by $\mathbf{r}$ and $\mathbf{r}^*$, which can be denoted by

$$\hat{m} = \hat{m}_{\mathbf{r}}(\mathbf{r}^*), \forall \mathbf{r} \in S, \tag{2.110}$$

where $\hat{m}_{\mathbf{r}}(\cdot)$ is determined by the shape of the influence region. It is straightforward that $\hat{m}$ is a random variable with cumulative distribution function (CDF):

$$F_{\mathbf{r}}(x) = P(\hat{m} \leq x) = P(\mathbf{r}^* \in \{\mathbf{r}^* | \hat{m}_{\mathbf{r}}(\mathbf{r}^*) \leq x\}). \tag{2.111}$$

Theorem 5 *For a localization system with N users within the region S and M APs distributed along the region's boundaries, if m out of M APs are impacted by human body blockage effect, then*

$$(a + 1)(1 - b^{N-1}) \leq m \leq \min\{c(N - 1), M\}, \tag{2.112}$$

where $a, b,$ and c are determined by the practical radio propagation environment and can be obtained through the procedures to be presented in Sect. 2.4.4.3.

Note that the concrete form of the bounds for the total number of impacted APs is highly dependent on the aggregated radio propagation environment, which is reflected by the parameters a, b, and c; however, the values of the three parameters are just dependent on $\hat{m}_{\mathbf{r}}(\mathbf{r}^*)$, which can be obtained by merely investigating how one interferer impacts the system, instead of examining all possible scenarios with $N - 1$ interferers.

2.4.4.3 Determine the Environment Dependent Parameters

We here present procedures to obtain values of the three parameters so that our theory can be verified in practice. The basic idea is to first construct $\hat{m}_{\mathbf{r}}(\mathbf{r}^*)$ as a discrete function through practical measurements and then derive values of the three parameters.

There are many possible values of $\mathbf{r}$, but we only select some special locations for measurement in practice. This is based on an important observation: the longer distance between the T–R pair, the larger interference region of the T–R pair will be. This is intuitive since the closer the transmitter is to the receiver, the less likely the AP will get interfered by others, and a larger interference region means that the corresponding AP is more likely to be impacted; therefore, we only need to study the locations that are the nearest to and farthest from all APs. The resulted values of $\hat{m}_{\mathbf{r}}(\mathbf{r}^*)$ for such locations will lead to extreme values of m, which can be used for deriving the scalability bounds of the fingerprinting localization system.

We place k mobile devices at those special locations $\mathbf{r_1}, \mathbf{r_2}, \ldots, \mathbf{r_k}$ and let an interferer standing in different reference points. Then the mobile devices can measure the observed RSS with respect to different APs as the location of the interferer $\mathbf{r}^*$ changes. Given a value of $\mathbf{r}^*$, each mobile device can compare the observed RSS of each AP with that stored in the training phase, based on which the number of impacted APs can be recorded. In this manner, the $\hat{m}_{\mathbf{r}}(\mathbf{r}^*)$ in the form of a discrete function can be obtained. Then we have $a = \max_{\mathbf{r}^*} \hat{m}_{\mathbf{r_j}}(\mathbf{r}^*) - 1$, $b = \frac{||\{\mathbf{r}^* | \hat{m}_{\mathbf{r_j}}(\mathbf{r}^*) = a, \mathbf{r}^* \in S\}||}{||\{\mathbf{r}^* | \hat{m}_{\mathbf{r_j}}(\mathbf{r}^* \in S\}||}$, and $c = E(\hat{m}_{\mathbf{r_i}}(\mathbf{r}^*))$, according to the analysis above.

2.4.5 *Main Results*

Theorem 6 *For the indoor localization system with M APs distributed along boundaries of the region S, which is designed to support N users, the localization reliability of the system R′ satisfies that*

$$e^{-\Theta(\phi(N))}(1 - e^{-\Theta(\phi(N))}) \leq R' \leq 1 - e^{-\Theta(\varphi(N))}, \tag{2.113}$$

where

$$\begin{cases} \phi(N) = \frac{M^2}{(1-A^2)\min\{c(N-1), M\} + A^2 M}, \\ \varphi(N) = \frac{M^2}{(1-A^2)(a+1)(1-b^{N-1}) + A^2 M}, \end{cases} \tag{2.114}$$

and a, b, and c can be obtained by the procedure described above, and A is the ratio of variances mentioned in earlier sections.

2.4.6 *Evaluations of Main Results*

We here verify the main results by examining the trends of both the upper and the lower bound derived. The influence of the number of impacted APs and the number of users on performance bounds is to be examined, which are the major factors in our model. Both numerical and experimental results are to be presented. When conducting experiments, we employ the "bare bone" fingerprinting localization method, which estimates the user's location by only comparing the reported RSS and the pre-constructed radio map. We note that there are variants of fingerprinting localization algorithms, but the scope of this book is to reveal scalability of the fingerprinting localization methodology, instead of proposing new localization algorithms.

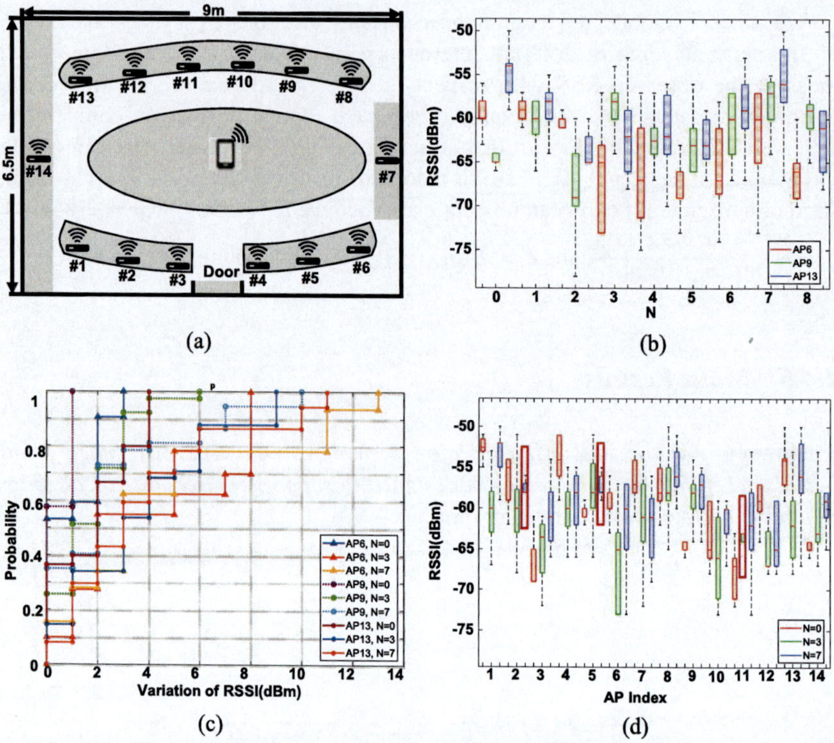

Fig. 2.28 Influence of moving interferers. (**a**) Experiment setup. (**b**) Influence of number of interferers. (**c**) Influence of number of interferers. (**d**) RSSI distribution at different APs

2.4.6.1 Influence of Human Body Blockage

We firstly analyze the influence of the moving interferers on the RSSIs of APs. As shown in Fig. 2.28a, we deploy 14 Wi-Fi APs in a meeting room (6.5 m × 9 m) and put a smart phone at the center of the meeting room to gather the RSSIs of each AP. We measure the RSSIs, while there are a different number of people moving in the meeting room randomly. The influence of the number of interferers is shown in Fig. 2.28b, c. Generally, with the increasing interferers, the fluctuation of APs' RSSIs becomes more severe. However, there are still some APs that are not blocked. Figure 2.28d shows the RSSI distribution of every APs in the meeting room. Though there are 7 interferers in the meeting room, the variation of the RSSI at AP5 and AP2 is less than 2 dBm, which can be utilized for fingerprint based localization.

We now present experimental results for localization, which are to be compared with the theoretical bounds as presented in Sect. 2.4.5. The experimental results are not supposed to be out of the theoretical bounds. We first measure and record RSSes at each cell with respect to all the APs, in order to construct the radio map that associates the ground-truth location with corresponding observed RSSes. We next let one normal user and one interferer enter into the area walking around and then let them stop. Users record the estimated and real locations in the presence of the interferer. We let the user be in 8 different locations and record the corresponding results for 100 times for each case, which yields the reliability of location estimation in the 1-interferer scenario. In this way, we increase the number of interferers one by one to 10 and let interferers be in 8 different kinds of distributions. Then we can find the reliability of location estimation error in different scenarios. To calculate the reliability, we need to specify the tolerance error δ; we consider that one-time localization is successful if the estimated location is in the δ neighborhood of the real location. In our experiments, we calculate the reliability when $\delta = 2\,\text{m}$, 2.5 m, 2.8 m, and 3 m.

2.4.6.2 Numerical Results

To obtain the numerical results of the bounds, we first need to obtain the environment dependent parameters a, b, and c following the procedures presented in Sect. 2.4.4.3.

We set up the testbed in a square $100\,\text{m}^2$ gymnastics room, which is gridded into $1\,\text{m} \times 1\,\text{m}$ cells. Twelve mobile Wi-Fi APs are uniformly deployed along edges of the room, which aims to create a comparatively ideal environment to eliminate unexpected interference from unnecessary details. We choose two special locations to derive the discrete function $\hat{m}_{\mathbf{r_i}}(\mathbf{r}^*)$: the geometric center and a corner of the room denoted by $\mathbf{r_1}$ and $\mathbf{r_2}$ respectively. This is because it is easy to mathematically prove that the center is the nearest location to all APs' and the corner is the farthest. Figure 2.29a, b shows the results when placing the receiving mobile devices in the corner and at the center of the room, respectively, where different colors mean the total numbers of impacted APs observed when the interferer is at different RPs labeled in the room. Based on such setup, we have $E(\hat{m}_{\mathbf{r_1}}(\mathbf{r}^*)) = 5.76$ and $E(\hat{m}_{\mathbf{r_2}}(\mathbf{r}^*)) = 6.92$, thus $a = 8$, $b = 0.96$, and $c = 6.92$.

With the environment dependent parameters, we could profile the trend of upper and lower bounds. Since the expression with respect to $\Theta(\cdot)$ just characterizes the shape of the bounds, we must determine the values of k_1^* and k_2^* of $\Theta(\cdot)$ before obtaining concrete bounding curves. We take the equations from both sides of the inequalities of Theorem 3. For the upper bound, we let $R' = 1 - e^{-k_1^* \varphi(N)}$ equal the highest reliability observed when $N = 1$ in the offline phase, and then we could find the value of k_1^*; for the lower bound, we let $R' = -k_1^* \phi(N)(1 - e^{-k_2^* \phi(N)})$ equal the lowest reliability when $N = 1, 2$ in experiments of obtaining a, b, and c, and then we could find the values of k_1^* and k_2^* as $\phi(N)$ can be obtained once the value of N is fixed. Figure 2.29c, d shows the numerical results of the upper and lower bounds of R'. Figure 2.29e shows the total number of impacted APs as N increases. We can see that as the number of impacted APs increases, the localization reliability decreases, which is realistic.

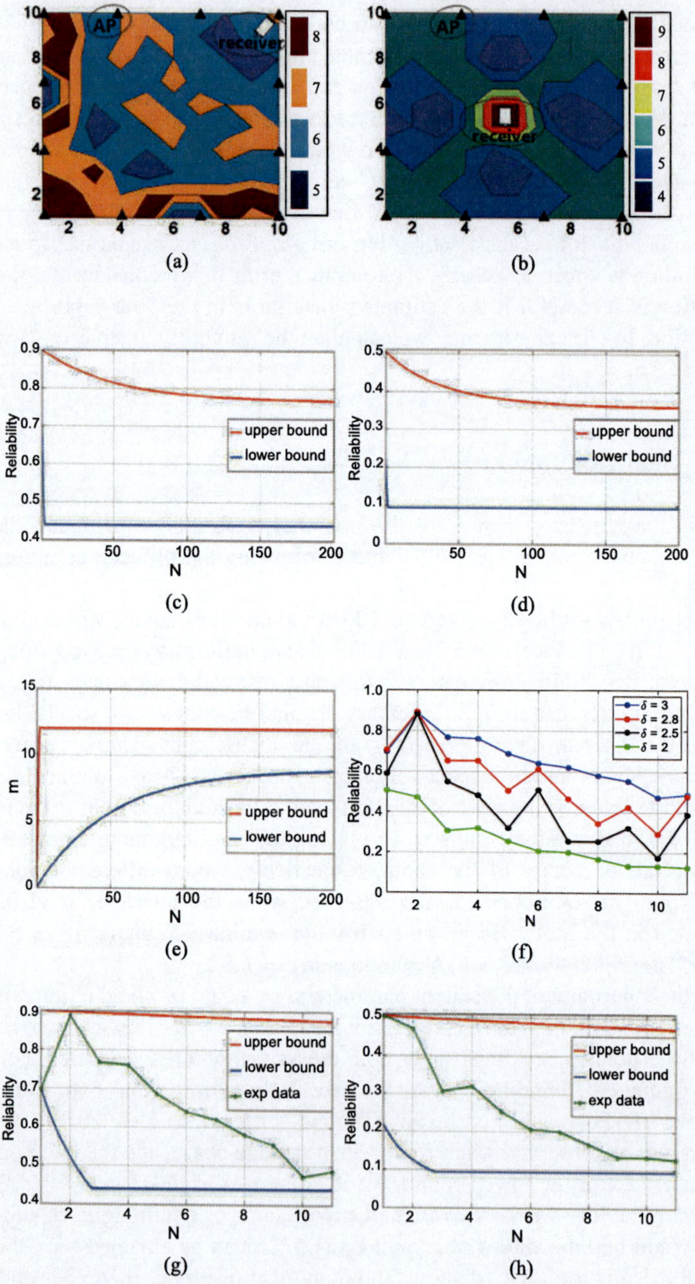

Fig. 2.29 Experimental and numerical results. **(a)** Num. impacted APs when the interferer in different positions. **(b)** Num. impacted APs when the interferer in different positions. **(c)** Numerical results with initial reliability at 90%. **(d)** Numerical results with initial reliability at 50%. **(e)** Num. impacted APs. **(f)** R' w.r.t. N. **(g)** Experimental and numerical results, $\delta = 3$. **(h)** Experimental and numerical results, $\delta = 2$

2.4.6.3 Experimental Results

Figure 2.29f shows the experimental results of localization reliability as N increases under different accuracy standards. Figure 2.29g, h shows the comparative positions of the theoretical bound curves and the experiment curves. We can see that the experimental results are bounded by our theoretical bounds, which validates the main results.

We can see the first-increase-then-decrease part of the reliability curve in those figures. This is primarily caused by the experimental error. The theoretical reliability is in fact the expectation of the corresponding localization reliability of all possible user-distribution scenarios; however, our experiments only record several scenarios. The curve will be smoother if more experiments can be conducted. We also note that the resulted reliability from the experiments almost achieves the lower bound when $N = 11$ because all the 12 APs have been impacted in this case according to Fig. 2.29e. We can expect that reliability curve will be flat if more interferers were added.

2.4.6.4 Important Observations and Analysis

Observation 1 *Localization reliability drops dramatically before the number of users increases to a critical point and then decreases smoothly, where the critical point tends to appear when the number of users equals that of APs deployed in the region.* With Fig. 2.29c, d showing the trend of upper and lower bounds of R', we can see that R' decreases as the number of users N increases, because more people in S means there are users in the interference region between the AP and the normal user with a higher probability; however, it is notable that the overall trends of R' first drop dramatically as N increases to a critical point and then decrease smoothly. This observation can be explained as follows. When the number of users in the system is small, adding a new-coming user can dramatically impact the radio propagation environment; however, if there have been a number of users in the system, it is likely that most of the APs have been impacted, thus adding a new-coming user will make the change of the radio propagation environment not that dramatic.

However, it is difficult to find the exact value of the critical point. Observing the trend shown in Fig. 2.29c, d, the values of the critical point for the upper and lower bounds are different. This is reasonable since the upper bound indicates the best case, where more interferers can be tolerated, but the lower bound indicates the worst case, where it is intolerant to interferers. This is why the critical point for the lower bound appears much earlier than that for the upper bound. The experimental results as shown in Fig. 2.29g, h provide some hint for finding the critical point: it seems that the value of the critical point is around the value of M, the number of APs deployed in the region. This also makes sense to some extent because all the APs could be impacted if $N = M$, with each interferer impacting an AP.

Observation 2 *Even if the number of users $N \to \infty$, the fingerprinting localization system still retains a certain level of reliability.* Our experimental results corroborate our theoretical results; however, what will be the resulted reliability if we keep increasing the number of users N? It is difficult to experiment the scenario of $N \to \infty$, but Theorem 3 has shed some light on the answer.

With Eqs. (2.113)–(2.114), we have $R' \geq e^{-k_1^* \phi(N)}(1 - e^{-k_2^* \phi(N)})$, where $\phi(N) = \frac{M^2}{(1-A^2)\min\{c(N-1), M\}+A^2 M}$ and the parameters c, k_1^*, and k_2^* are all independent of the number of users N in the system but could be related to the number of all APs M. When $N \to \infty$, $\phi(N) = \frac{M^2}{(1-A^2)M+A^2M} = M$, thus $R' \geq e^{-pM}(1 - e^{-qM}) > 0$, meaning that the localization system will still retain certain level of reliability even if there are an infinite number of users.

We now consider the extreme-case lower bound in Theorem 6, which is $e^{-k_1^* M}(1 - e^{-k_2^* M})$ occurring when $N \to \infty$. Mathematically, the trend of the worst-case lower bound can be monotonically increasing or decreasing-first-then-increasing; however, the only possible trend in practice should be monotonically increasing. This can be obtained by using the similar approach of proving Theorem 6. Consequently, an interesting byproduct of the observation is that deploying more APs can improve the extreme-case lower bound.

2.5 Theoretical Guidance on Fingerprints Reporting Strategy

This section investigates how to optimize the fingerprints reporting strategy to improve localization accuracy and how the optimal strategy theory can be utilized to streamline the design of WLAN fingerprinting localization systems. In particular, we first reveal that the fingerprints reporting problem is essentially an NP-Hard size-constrained supermodular maximization problem and then show the inapplicability of the state-of-the-art approximation algorithms to the problem. We then propose a new algorithm and show that if the number of fingerprints measurements is large enough, then the localization accuracy is at most $1 - \varepsilon$ times worse than the optimal value, with ε any given constant close to 0. Moreover, we demonstrate how the optimal strategy theory can be utilized to the improve accuracy of location estimation by resolving the issue of similar fingerprints for both faraway and close-by locations, with an iterative algorithm developed to cross check fingerprints sampled in different locations, in order to derive the best possible result of localization. Furthermore, we reveal the relationship between accuracy of location estimation and coverage of Wi-Fi signals in indoor spaces when planning deployment of APs. Experiment results are presented to validate our theoretical analysis.

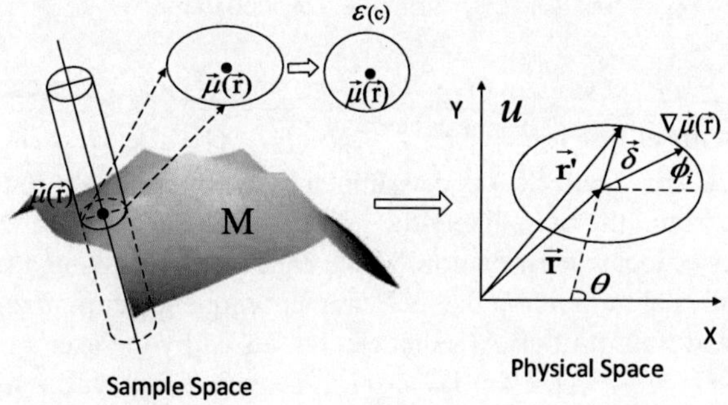

Fig. 2.30 Intersection in sample space

2.5.1 System Model

The theoretical basis of best fingerprints reporting strategy is first presented in [116], where it is shown that reporting RSSes obtained from different APs results in accuracy of location estimations in different levels. The location estimation process of RSS fingerprinting based localization is in fact a mapping from the RSS fingerprints space to the physical space. If we use Q to denote an area in the physical space, which is centered at the user's actual location $\mathbf{r}$ with radius δ, then there must be a corresponding event E in the sample space, which makes the localization system to estimate the user's location in Q. Thus the probability the user is correctly localized in Q is equal to the probability that the event E happens. The event E is a set of outcomes of RSS measurements, the shape of which in the sample space turns out to be a hyper-cylinder [116]. It means that if the reported RSS readings fall into the hyper-cylinder, then the localization system will estimate the user's location in the area Q. The hyper-cylinder intersects with the mean surface of RSSes, and the intersected surface is in the shape of an ellipse, as the example shown in Fig. 2.30.

An interesting finding presented in [116] is that if we use another hyper-cylinder $\mathcal{E}(c)$ to replace the event E, where the intersection between $\mathcal{E}(c)$ and the RSS mean surface is a circle with radius c, then the corresponding area the user will be localized in the physical space is determined by the function $\rho^2(\theta) = \frac{4c^2}{\sum p_i cos^2(\theta-\phi_i)}$, which in fact is the ellipse $\mathcal{U}$ as shown in the right part of Fig. 2.30. In the figure, the location of the user is $\mathbf{r}$ and the location of any point on the boundary of the ellipse $\mathcal{U}$ is $\mathbf{r}'$. Vector $\boldsymbol{\delta} = \mathbf{r}' - \mathbf{r}$ denotes a two-dimensional vector with the direction from the user's actual location to any point on the boundary of $\mathcal{U}$. The symbol θ denotes the angle between $\boldsymbol{\delta}$ and the horizontal axis, and ϕ_i denotes the angle between $\nabla\mu_i(\mathbf{r})$ and the horizontal axis, where $\nabla\mu_i(\mathbf{r})$ is the gradient of the mean of measured RSS with respect to AP_i at location $\mathbf{r}$, and $p_i = (|\nabla\mu_i(\mathbf{r})|/\sigma_i)^2$. The symbol σ_i denotes the observed RSS standard deviation with respect to AP_i.

The ellipse $\mathcal{U}$ can be transformed into the form

$$Q_1\rho^2\cos^2\theta + Q_2\rho^2\sin^2\theta + Q_3\rho^2\cos\theta\sin\theta = 4c^2,$$

where $Q_1 = \sum p_i\cos^2\phi_i$, $Q_2 = \sum p_i\sin^2\phi_i$, and $Q_3 = \sum 2p_i\cos\phi_i\sin\phi_i$, and then the area of the ellipse is $u = 8\pi c^2/\sqrt{4Q_1Q_2 - Q_3^2}$. The event $\mathcal{E}(c)$ determines u in the physical space and $\mathcal{E}(c)$ is determined by reported fingerprints. Hence, the smaller u is in value, the more likely the user can be localized at $\mathbf{r}$, which leads to higher accuracy of location estimation. This means that maximizing the localization accuracy is equivalent to finding the measurement sequence minimizing u.

Specifically, if the set of all APs that can be sensed by the user's mobile device is denoted by $\mathbb{U} = \{AP_i\}, i = 1, \ldots, m$, a sequence of measurements on APs can be denoted by $\mathcal{V}_l = (s_1, \ldots, s_l), s_j \in \mathbb{U}$, and l is the number of times for measurements. The symbol s_j is the index of the measurement in the sequence. Note that s_j does not necessarily mean that the measurement is performed on AP_j, since an AP can be measured more than once in the sequence. The whole set of strategies is denoted as $\mathbb{U}^l$, where the size of the set is m^l.

For the purpose of simplification, the characteristic of AP_i can be described with a complex parameter $Z_i = p_i e^{2i\phi_i}$, where p_i represents the distinctiveness of signal strength influenced by signal gradient and noise, and the corresponding direction is reflected by $2\phi_i$, double of signal gradient direction. With Z_i, minimizing u is equivalent to maximizing

$$\mathcal{F}(\mathcal{V}_l) = \left(\sum_{i\in\mathcal{V}_l} |Z_i|\right)^2 - |\sum_{i\in\mathcal{V}_l} Z_i|^2, \tag{2.115}$$

and thus the optimal fingerprints reporting strategy can be denoted by $\mathcal{V}_l^*$, $\mathcal{V}_l^* \in \mathbb{U}^l$, where

$$\mathcal{V}_l^* = \arg\max_{\mathcal{V}_l\in\mathbb{U}^l} \mathcal{F}(\mathcal{V}_l). \tag{2.116}$$

While interesting insight into location estimation is revealed, the proposed best strategy for fingerprints reporting in [116] is presented in a concise form without details. Our work in this book provides a systematical analysis of the best strategy and exploits the best strategy to streamline design of the localization system.

2.5.2 *Analysis of Optimal Strategy for Fingerprints Reporting*

2.5.2.1 Supermodularity of the Objective Function

Consider the localization system with m APs, each of which can be measured multiple times. Suppose that the mobile device is allowed to perform AP measurements l times, then the objective function of AP selection is equivalent to the function

$g(\cdot) : 2^{[lm]} \to R$, and for any set $S \subseteq [lm]$, we have

$$g(S) = \left(\sum_{a \in S} |s_{\lfloor \frac{a-1}{l}+1 \rfloor}| \right)^2 - |\sum_{a \in S} s_{\lfloor \frac{a-1}{l}+1 \rfloor}|^2, \tag{2.117}$$

where $[lm]$ denotes the set of positive integers $\{1, 2, \ldots, lm\}$.

Intuitively, more times of measurements can lead to more accurate location estimation; moreover, it is straightforward that the first term of $g(S)$ is dominant compared with the second one. Such characteristics indicate that $g(S)$ could be a supermodular function. We provide the rigorous proof in the following.

Theorem 6 *Function* $g(S) = (\sum_{a \in S} |s_{\lfloor \frac{a-1}{l}+1 \rfloor}|)^2 - |\sum_{a \in S} s_{\lfloor \frac{a-1}{l}+1 \rfloor}|^2$ *is a supermodular set function defined on the set* $[lm]$.

Proof 21 We prove the theorem with the definition of supermodular set function [130]. We first perform the following transformation of $g(\cdot)$:

$$\begin{aligned} g(S) &= \left(\sum_{i \in S} |s_{\lfloor \frac{i-1}{l}+1 \rfloor}| \right)^2 - |\sum_{i \in S} s_{\lfloor \frac{i-1}{l}+1 \rfloor}|^2 \\ &= \left(\sum_{i \in S} p_{\sigma(i)} \right)^2 - |\sum_{i \in S} p_{\sigma(i)} e^{2i\phi_{\sigma(i)}}|^2 \\ &= \left(\sum_{i \in S} p_{\sigma(i)}^2 + \sum_{i \neq j \in S} p_{\sigma(i)} p_{\sigma(j)} \right) \\ &\quad - \left(\sum_{i \in S} p_{\sigma(i)}^2 + \sum_{i \neq j \in S} p_{\sigma(i)} p_{\sigma(j)} e^{2i(\phi_{\sigma(i)} - \phi_{\sigma(j)})} \right) \\ &= 2 \sum_{i \neq j \in S} p_{\sigma(i)} p_{\sigma(j)} \sin^2(\phi_{\sigma(i)} - \phi_{\sigma(j)}). \end{aligned}$$

For the convenience of demonstration, we define a new function: $\sigma(i) = \left\lfloor \frac{i-1}{l} + 1 \right\rfloor$ $(1 \leq i \leq lm)$; if we add a new element k to the set S, the value increase of the function introduced by the element is

$$\begin{aligned} &g\,(S \cup \{k\}) - g\,(S) \\ &= 2 \sum_{j \in S} p_{\sigma(k)} p_{\sigma(j)} \sin^2(\phi_{\sigma(k)} - \phi_{\sigma(j)}). \end{aligned}$$

As each term above is non-negative, if adding the same element k into a smaller set, the value of the function must be smaller compared with that if adding it into a larger set S, and thus function $g(\cdot)$ is a supermodular set function.[5]

With the supermodularity of the objective function, the AP selection problem can be intuitively modeled as a size-constrained supermodular maximization problem. In particular,

$$\max_{S}\{g(S) : S \subseteq [N], S = k\}, \tag{2.118}$$

where $g(\cdot) : 2^{[N]} \rightarrow R$ is a supermodular set function on the set $[N] = \{1, 2, \ldots, N\}$. This problem has been proved to be NP-hard [130, 131, 222], and normally the approach for finding the solution is to transform the size-constrained supermodular maximization problem into the corresponding submodular minimization problem [131]. This is because for any supermodular set function $g(\cdot)$, if we define a new function $f(\cdot) = C - g(\cdot)$, and C is a constant that is large enough such as $C = \sup g(\cdot)$, then it is easy to verify that the function $f(\cdot)$ is a submodular set function. With the newly defined function $f(\cdot)$, maximizing $g(\cdot)$ is equivalent to minimizing $f(\cdot)$.

The size-constrained submodular minimization problem is currently with no standardized approximation algorithm to the best of our knowledge [130, 131, 222]. We find two algorithms published recently [130, 222] and examine whether they could be used for resolving our problem with guaranteed performance. In the following, we will show why the two state-of-the-art approximation algorithms are unable to provide the guaranteed approximation ratio, which motivates us to develop our algorithm in Sect. 5.2.

Iyer et al. propose an approximation algorithm for minimizing the submodular set function [222]. Recall the definition of $f(\cdot)$ mentioned above, and the submodularity of $f(\cdot)$ can still hold even without the constant C. The purpose of C is just to ensure that the value of $f(\cdot)$ is non-negative, which is required by [222]. With Iyer's approximation algorithm, the solution can be obtained g_o and the optimal solution g^* satisfies

$$\frac{C - g_o}{C - g^*} \leq 1 + \varepsilon, \tag{2.119}$$

where ε is a constant close to 0. Since C can be a very large constant, the algorithm is unable to ensure a good approximation ratio to the optimal solution for the problem.

Nagano et al. propose a size-constrained submodular minimization algorithm by leveraging the minimum norm base [130]. We here briefly describe the basic

[5] A set function $f(\cdot) : 2^E \rightarrow R$ defined on set E is a supermodular set function if for any set $B \subseteq A \subseteq E$, $f(A \cup \{i\}) - f(A) \geq f(B \cup \{i\}) - f(B)$, where i is an element in set E but not in set A.

idea of the algorithm and then show why the algorithm cannot be applied in the AP selection scenario. The submodular set function can be mapped into an N-dimensional polyhedron, with N being the number of elements in the set. An example of polyhedron corresponding to a submodular set function could be found in our technical report [194].

Theorem 7 *For submodular set function $f(\cdot) = 2g(E) - g(\cdot)$, the minimum norm base $\overrightarrow{b} = (b_1, b_2 \ldots b_N)$ is a vector with the values of all the components being the same.*

Proof 22 According to the definition of minimum norm base, the components $b_1, b_2 \ldots, b_N$ must satisfy the following polyhedron constraints:

$$\sum_{i\in S, S\not\subset E} b_i \le 2\left(\sum_{i\in E}\left|s_{\lfloor \frac{i-1}{T}+1\rfloor}\right|\right)^2 - \left(\sum_{i\in S}\left|s_{\lfloor \frac{i-}{T}+1\rfloor}\right|\right)^2 + \left|\sum_{i\in S} s_{\lfloor \frac{i-1}{T}+1\rfloor}\right|^2 - 2\left|\sum_{i\in E} s_{\lfloor \frac{i-1}{T}+1\rfloor}\right|^2;$$

$$\sum_{i\in E} b_i = \left(\sum_{i\in E}\left|s_{\lfloor \frac{i-1}{T}+1\rfloor}\right|\right)^2 - \left|\sum_{i\in E} s_{\lfloor \frac{i-1}{T}+1\rfloor}\right|^2.$$

Now we first prove that the vector $\overrightarrow{b^*} = (b_i^*)_{1\le i\le N}$ satisfying the following condition is in the polyhedron P_g:

$$b_1^* = b_2^* \ldots = b_N^* = \frac{\left(\sum_{i\in E}\left|s_{\lfloor \frac{i-1}{T}+1\rfloor}\right|\right)^2 - \left|\sum_{i\in E} s_{\lfloor \frac{i-1}{T}+1\rfloor}\right|^2}{N}.$$

It is straightforward to verify that b_i satisfies the second condition of the P_f definition, that is, $\sum_{i=1}^{N} b_i = f(E)$.

Note that

$$\frac{f(S)}{|S|} = \frac{2\left(\sum_{i\in E}\left|s_{\lfloor \frac{i-1}{T}+1\rfloor}\right|\right)^2 - \left(\sum_{i\in S}\left|s_{\lfloor \frac{i-}{T}+1\rfloor}\right|\right)^2}{|S|} + \frac{\left|\sum_{i\in S} s_{\lfloor \frac{i-1}{T}+1\rfloor}\right|^2 - 2\left|\sum_{i\in E} s_{\lfloor \frac{i-1}{T}+1\rfloor}\right|^2}{|S|} \tag{2.120}$$

$$\overset{(a)}{\geq} \frac{\left(\sum\limits_{i \in E} \left| s_{\left\lfloor \frac{i-1}{l}+1 \right\rfloor} \right|\right)^2 - \left| \sum\limits_{i \in E} s_{\left\lfloor \frac{i-1}{l}+1 \right\rfloor} \right|^2}{|S_i|}$$

$$\overset{(b)}{\geq} \frac{\left(\sum\limits_{i \in E} \left| s_{\left\lfloor \frac{i-1}{l}+1 \right\rfloor} \right|\right)^2 - \left| \sum\limits_{i \in E} s_{\left\lfloor \frac{i-1}{l}+1 \right\rfloor} \right|^2}{|E|}$$

$$= b_i.$$

Since the function $g(\cdot)$ is continuously increasing, the inequality (a) holds as $g(S) \leq g(E)$. The inequality (b) holds because $|S| \leq |E|$. For any set S, inequality $\sum\limits_{i \in S} b_i \leq |S| f(S)$ is established, and thus the point represented by vector $\overrightarrow{b^*}$ is in the polyhedron P_f.

According to the Cauchy inequality, we have

$$\left\| \overrightarrow{b} \right\| = \sum_{i=1}^{N} b_i^2 \geq \frac{\left(\sum\limits_{i=1}^{N} b_i\right)^2}{N} = \left\| \overrightarrow{b^*} \right\|;$$

therefore, $\overrightarrow{b^*}$ is the minimum norm base of function $g(\cdot)$.

The theorem above indicates that the algorithm does not fit the AP selection problem under study. The theorem shows that all the components in the minimum norm base of $f(\cdot)$ are the same for the AP selection problem, which makes $|T_i| = N$ in this case. This is to select N elements in an N-norm set, which is meaningless. In the theorem above, we transform the function $g(\cdot)$ into $f(\cdot) = 2g(E) - g(\cdot)$, which is a submodular set function but does not change the essence of the problem.

2.5.2.2 Algorithm for AP Selection

We propose Algorithm 1 to solve the AP selection problem. We are to first present a detailed explanation of the algorithm and then theoretically analyze performance of the algorithm.

The input of the algorithm is an m-dimensional vector, where each component of the vector is a complex number. The complex number z_i of the ith dimension represents the characteristic of AP_i. The output of the algorithm is an l-dimensional vector, where each component of the vector is an integer that is greater than or equal to 1 and less than or equal to l. The output vector shows how many times each AP should be measured in the online phase, e.g., the output vector $(1, 2, 2, 3, 1, 1, 4)$ indicates that if there are totally seven times of measuring opportunities, to measure AP_1 three times, AP_2 twice, AP_3 once, and AP_4 once, respectively, can obtain the

Algorithm 1 Near optimal strategy for fingerprints reporting

Require: complex number vector $(z_1, z_2 \cdots z_m)$ where $z_j = p_j e^{i\phi_j} (0 \le \phi_j \le 2\pi)$ represent the ith AP.
Ensure: A measurement sequence $(n_1, n_2 \ldots n_l)$, where $1 \le n_i \le l$ represent choosing the n_ith AP.

1: $S \leftarrow \emptyset$
2: $w_i = 0 (1 \le i \le lm)$
3: $z^* = 1$
4: **for** $i = 1$ to lm **do**
5: $\quad \sigma(i) \leftarrow \left\lfloor \frac{i-1}{l} + 1 \right\rfloor$
6: **end for**
7: **for** $i = 1$ to l **do**
8: $\quad$ **for** $i \in [lm] \backslash S$ **do**
9: $\quad\quad w_j \leftarrow w_j + p_{\sigma(j)} |z^*| \sin^2(\phi_{\sigma(j)} - \arg(z^*))$
10: $\quad$ **end for**
11: $\quad i^* \leftarrow \arg\max_{i \in [lm] \backslash S} w_i$
12: $\quad z^* \leftarrow s_{\sigma(i^*)}$;
13: $\quad S \leftarrow S \cup \{i^*\}$;
14: **end for**
15: **return** $(\sigma(i))_{i \in S}$

most accurate location estimation. The order of the measurements is not important, since the order does not change the value of the objective function.

In the algorithm, we first initialize the relevant intermediate variables to be used. We use set S to store the set of selected APs; w_i represents the weight of each element in the set $[lm]$. The algorithm will continue to update the weights of the elements in the following steps, which is the basis to decide whether the corresponding AP will be selected. The last variable processed in the initialization stage is z^*, which is also a complex variable and the initial value is 1.

In the fourth row of the algorithm, we map the elements in the set $[lm]$ to a corresponding AP, which is for the rigorousness of narrative. This is because the set function requires each element in the set to be unique, but the fingerprints reporting strategy studied in the work allows an AP to be measured multiple times. After the mapping, each element in the set $[lm]$ can be considered unique, and it is straightforward that choosing element i in the set $[lm]$ is equivalent to select AP with the index $i (\text{mod } m)$.

The 6th and 7th rows of the algorithm represented by the "For" loop is the main part. In the 7th and 8th rows of the algorithm, at the beginning of each loop, the algorithm updates the weight w_i of each element and increases it by $p_{\sigma(j)} |z^*| \sin^2(\phi_j - \arg(z^*))$, and z^* are selected from the last execution of each iteration.

In row 9, the algorithm selects the element with the largest weight in the remaining elements, which is indexed by i^*. Then, the algorithm copies the i^*th corresponding complex number $z_{\sigma(i^*)}$ representing AP_{i^*} to the intermediate variable z^*. Recall that finding the element with the largest weight is factually finding the complex number $z_{\sigma(i^*)}$ that can most effectively decrease the size of the ellipse as described in Fig. 2.1 of Sect. 2.2. Let u_k and u_{k+1} denote the areas of the ellipses that the user will be localized in the physical space when using APs in set $\{1, 2 \ldots k\}$ and $\{1, 2 \ldots k, k+1\}$, respectively, and then we have

$$\begin{aligned} u_k - u_{k+1} &= \frac{1}{2}\int_0^{2\pi} (\rho_k^2(\theta) - \rho_{k+1}^2(\theta))d\theta \\ &= 2c^2 \int_0^{2\pi} \left(\frac{1}{\sum_{i=1}^{k} p_i^2 \cos^2(\theta - \phi_i)} - \frac{1}{\sum_{i=1}^{k+1} p_i^2 \cos^2(\theta - \phi_i)} \right) d\theta \\ &\propto \sum_{i=1}^{k} p_{k+1} p_i \sin^2(\phi_{k+1} - \phi_i). \end{aligned}$$

This is the fundamental reason row 9 of the algorithm is so designed.

In row 11, the algorithm merges the i^*th element into the set S and then enters the next execution of the loop. After looping l times, we can get a set S of l elements, and finally we have the sequence $\sigma(i)$, which indicates how many times which AP should be measured in the online phase.

2.5.2.3 Algorithm Performance Analysis

We first define a new function $h(,) : 2^{[lm]} \times 2^{[lm]} \rightarrow R$, where the arguments of the function consist of two sets, and the value of the function is a real number. For any two sets A and B, the value of the function of the two sets is equal to the sum of the results of the same operation between each corresponding element in the two sets:

$$h(A, B) = \sum_{i \in A, j \in B} p_{\sigma(i)} p_{\sigma(j)} \sin^2(\phi_{\sigma(i)} - -\phi_{\sigma(j)}). \tag{2.121}$$

There is a close relationship between function $h(,)$ and function $g(\cdot)$, as shown in the following lemma.

Lemma 3 *For any two sets $A \in [lm]$ and $B \in [lm]$,*

$$g(A \cup B) = g(A) + g(B) - -g(A \cap B) + h(A \backslash B, B \backslash A). \tag{2.122}$$

Proof 23

$$\begin{aligned} g(A \cup B) &= g(A \backslash B) + g(B \backslash A) + g(A \cap B) + h(A \backslash B, A \cap B) \\ &\quad + h(B \backslash A, A \cap B) + h(A \backslash B, B \backslash A) \\ &= [g(A \backslash B) + g(A \cap B) + h(A \backslash B, A \cap B)] \\ &\quad + h(A \backslash B, B \backslash A) - g(A \cap B) \\ &\quad + [g(B \backslash A) + g(A \cap B) + h(B \backslash A, A \cap B)] \\ &= g(A) + g(B) + h(A \backslash B, B \backslash A) - g(A \cap B). \end{aligned} \tag{2.123}$$

According to the definition of $h(,)$ and $g(\cdot)$, it is straightforward to verify each step of the formula above.

The lemma above provides a more precise description of supermodularity. If we replace the equality with the inequality of the function $h(,)$, it degenerates to the definition of supermodularity. We are to estimate the value of $h(,)$. In fact, estimating function $h(,)$ is to measure how "supermodular" the function $g(\cdot)$ is. Function $g(\cdot)$ is more supermodular with $h(,)$ being larger.

For the convenience of observation, we first transform the function $h(,)$ into the following:

$$\begin{aligned} h(A, B) &= \sum_{i \in A m j \in B} p_{\sigma(i)} p_{\sigma(j)} \sin^2(\phi_{\sigma(i)} - \phi_{\sigma(j)}) \\ &= \sum_{i \in A} \left[p_{\sigma(i)} \sum_{j \in B} p_{\sigma(j)} \sin^2(\phi_{\sigma(i)} - \phi_{\sigma(j)}) \right]. \end{aligned}$$

Note that

$$p_{\sigma(i)} \sum_{j \in B} p_{\sigma(j)} \sin^2(\phi_{\sigma(i)} - \phi_{\sigma(j)}) = \frac{1}{2} \left\{ p_{\sigma(i)} \sum_{j \in B} p_{\sigma(j)} - \text{Re} \left[s_{\sigma(i)} \cdot \sum_{j \in B} s_{\sigma(j)} \right] \right\}, \tag{2.124}$$

where $\text{Re}(\cdot)$ denotes the real part of a complex number. Note that the right part of the equation $\text{Re}[s_{\sigma(i)} * \sum_{j \in B} s_{\sigma(j)}]$ is the real part of the product between complex number $s_{\sigma(i)}$ and the sum of all complex numbers in set B. It is obvious that the equation is less than or equal to the product of the complex number $s_{\sigma(i)}$'s norm and the norm of the sum of all complex numbers in set B; therefore,

$$h(A,B) \geq \sum_{i\in A}\sum_{j\in B} p_{\sigma(i)}p_{\sigma(j)} - \sum_{i\in A}\left(p_{\sigma(i)}|\sum_{j\in B} s_{\sigma(i)}|\right)$$

$$\geq \left(\sum_{i\in A} p_{\sigma(i)}\right)\left(\sum_{j\in B} p_{\sigma(j)} - |\sum_{j\in B} s_{\sigma(i)}|\right).$$

Meanwhile, we note that

$$\sum_{j\in B} p_{\sigma(j)} - |\sum_{j\in B} s_{\sigma(i)}| = \frac{g(B)}{\sum_{j\in B} p_{\sigma(j)} + |\sum_{j\in B} s_{\sigma(i)}|},$$

and thus we have

$$h(A,B) \geq \frac{\sum_{i\in A} p_{\sigma(i)}}{\sum_{j\in B} p_{\sigma(j)} + |\sum_{j\in B} s_{\sigma(i)}|} g(B)$$

$$\overset{(a)}{\geq} \frac{\sum_{i\in A} p_{\sigma(i)}}{2\sum_{j\in B} p_{\sigma(j)}} g(B).$$

The inequality (a) holds because $|\sum_{j\in B} s_{\sigma(i)}| \leq \sum_{j\in B} p_{\sigma(j)}$. Similarly, we have

$$h(A,B) \geq \frac{\sum_{i\in B} p_{\sigma(i)}}{2\sum_{j\in A} p_{\sigma(j)}} g(A). \tag{2.125}$$

We define a new set function $\lambda(\cdot) : 2^{[lm]} \to R$, and then for any set A, $\lambda(A)$ equals the norm sum of all elements in set A, that is, $\lambda(A) = \sum_{j\in A} p_{\sigma(j)}$. Besides, we let $m = \min_{i\in[lm]} p_{\sigma(i)}$, $M = \max_{i\in[lm]} p_{\sigma(i)}$, which will facilitate the following analysis of the algorithm. With the conclusions of the derivations above, we have $h(A,B) \geq \frac{1}{2}\frac{\lambda(B)}{\lambda(A)} g(A)$, and symmetrically we have

$$h(A,B) \geq \frac{1}{2}\frac{\lambda(A)}{\lambda(B)} g(B).$$

Consequently, we have

$$h(A,B) \geq \frac{1}{4}\frac{\lambda(B)}{\lambda(A)} g(A) + \frac{\lambda(A)}{\lambda(B)} g(B)$$

$$= \frac{1}{2}\sqrt{g(A)g(B)}$$
$$\geq \frac{1}{2}\min\{g(A), g(B)\}.$$

We here introduce another concept, *curvature*, to be used in the approximation ratio analysis. Similar to the submodular function, the curvature of supermodular set function is defined as follows.

Definition 4 The curvature of a supermodular function $g(\cdot)$ is

$$\kappa_g = 1 - \min_{j\in E} \frac{g(j)}{g(E\backslash j)}, \tag{2.126}$$

where $g(E\backslash j) = I) - g(E\backslash j)$.

Theorem 8 *We use u_{out} to denote the ellipse area corresponding to the fingerprint reporting strategy yielded by Algorithm 1 and u_{opt} to denote the ellipse area corresponding to the optimal reporting strategy, and then*

$$\frac{u_{out}}{u_{opt}} \leq \frac{1}{\sqrt{1-\delta}} = 1 + \frac{\delta}{2} + O(\delta^2).$$

That is, the approximation ratio of Algorithm 1 is $1 + \frac{\delta}{2} + O(\delta^2)$, and $\delta = 1 - \max\{[(1 + \frac{m}{2M})e^{-(1-\kappa)} - \frac{m}{2M}], (1 + \frac{m}{2Ml})e^{-(1-\kappa)l} - \frac{m}{2Ml}\}$.

Proof 24 Assuming that the optimal solution of the optimization problem is the set S^*. In the second "For" loop of the algorithm, the algorithm will select an element i^*, adding it into set S at the end of each execution of the loop. We assume that the set yielded from step i is S_i, which partially intersects with S^*.

$$\begin{aligned} g(S^*) &= g(S^* \cup S_i) + g(S^* \cap S_i) - g(S_i) - h(S^*\backslash S_i, S_i\backslash S^*) \\ &\overset{(a)}{=} g(S^* \cup S_i) - g(S_i\backslash S^*) - h(S_i \cap S^*, S_i\backslash S^*) \\ &\quad - h(S^*\backslash S_i, S_i\backslash S^*) \\ &\overset{(b)}{=} g(S^* \cup S_i) - g(S_i\backslash S^*) - h(S^*, S_i\backslash S^*) \\ &\overset{(c)}{\leq} g(S_i) + \frac{|S^*\backslash S_i|}{1-\kappa}(g(S_{i+1}) - g(S_i)) - g(S_i\backslash S^*) \\ &\quad - h(S^*, S_i\backslash S^*) \\ &\overset{(d)}{\leq} g(S_i) + \frac{|S^*\backslash S_i|}{1-\kappa}(g(S_{i+1}) - g(S_i)) - \frac{\lambda(S_i\backslash S^*)}{2\lambda(S^*)} g(S^*). \end{aligned} \tag{2.127}$$

Merging $g(S^*)$ from both sides of the inequality, we have the following result:

$$\left(1+\frac{\lambda(S_i\backslash S^*)}{2\lambda(S^*)}\right)g(S^*)\leq\frac{|S^*\backslash S_i|}{1-\kappa}g(S_{i+1})+\left(1-\frac{|S^*\backslash S_i|}{1-\kappa}\right)g(S_i).$$

After appropriate transposition, we can transform the inequality into

$$g(S_{i+1})\geq\frac{1+\frac{\lambda(S_i\backslash S^*)}{2\lambda(S^*)}}{\frac{|S^*\backslash S_i|}{1-\kappa}}g(S^*)-\left(\frac{1-\kappa}{|S^*\backslash S_i|}-1\right)g(S_i).$$

We can regard the inequality above as an iterative inequality of the set S_i, and now we do some transposition to the inequality as follows:

$$\begin{aligned}&\left(1+\frac{\lambda(S_i\backslash S^*)}{2\lambda(S^*)}\right)g(S^*)-g(S_{i+1})\\&\leq\left(1-\frac{1-\kappa}{|S^*\backslash S_i|}\right)\left[\left(1+\frac{\lambda(S_i\backslash S^*)}{2\lambda(S^*)}\right)g(S^*)-g(S_i)\right].\end{aligned}\tag{2.128}$$

Using the formula above, we can perform the computation iteratively until reaching the case of $i=0$. Note that $S_0=\emptyset$, and thus the following inequality holds for any i

$$\begin{aligned}&\left(1+\frac{\lambda(S_i\backslash S^*)}{2\lambda(S^*)}\right)g(S^*)-g(S_i)\\&\leq\left(1+\frac{\lambda(S_i\backslash S^*)}{2\lambda(S^*)}\right)g(S^*)\prod_{j=0}^{i-1}\left(1-\frac{1-\kappa}{|S^*\backslash S_j|}\right).\end{aligned}\tag{2.129}$$

In particular, if $i=l$, we have

$$\begin{aligned}g(S_l)&\geq\left(1-\prod_{j=0}^{l-1}\left(1-\frac{1-\kappa}{|S^*\backslash S_j|}\right)\right)\left(1+\frac{\lambda(S_l\backslash S^*)}{2\lambda(S^*)}\right)g(S^*)\\&\overset{(a)}{\geq}\left(1-\left(1-\frac{1-\kappa}{|S^*\backslash S_l|}\right)^l\right)\left(1+\frac{\lambda(S_l\backslash S^*)}{2\lambda(S^*)}\right)g(S^*)\\&\overset{(b)}{\geq}\left(1-\left(1-\frac{1-\kappa}{|S^*\backslash S_l|}\right)^l\right)\left(1+\frac{|S_l\backslash S^*|m}{2lM}\right)g(S^*).\end{aligned}$$

The inequality (a) can hold because increasing $|S^*\backslash S_j|$ to $|S^*\backslash S_l|$ will make the entire equation smaller in value. According to the definition of function $\lambda(\cdot)$, we have $\lambda(S_l\backslash S^*)\geq|S_l\backslash S^*|m$ and $\lambda(S^*)\leq|S^*|M=lM$, and thus the inequality

(b) also holds. Note that $|S_l| = |S^*| = l$, thus $|S^*\backslash S_l| = |S_l\backslash S^*|$. We use k to represent the number of elements in the two sets. One more step and we will have the following:

$$
\begin{aligned}
g(S_l) &\geq \left(1-\left(1-\frac{1-\kappa}{k}\right)^l\right)\left(1+\frac{km}{2lM}\right)g(S^*) \\
&\overset{(a)}{\geq} \left(1-e^{-\frac{(1-\kappa)l}{k}}\right)\left(1+\frac{km}{2lM}\right)g(S^*) \\
&\overset{(b)}{=} \left(1-e^{-(1-\kappa)t}\right)\left(1+\frac{m}{2Mt}\right)g(S^*).
\end{aligned}
$$

As $(1-\frac{1-\kappa}{k})^l \leq e^{-\frac{(1-\kappa)l}{k}}$, inequality (a) holds. Note that the right hand side of the inequality is actually a function with respect to k. For the convenience of demonstration, we let the variable $t = \frac{l}{k}$, and we have equation (b) with $t \in [1, l]$. Let $q(t) = (1-e^{-(1-\kappa)t})(1+\frac{m}{2Mt})$, and we find the minimum of $q(t)$ by finding t's derivative. Although t is a discrete variable according to the definition of t, we can relax it to a continuous variable when finding the extreme of the function; therefore, the analysis can be facilitated by the derivative. The first order derivative of t in function $q(\cdot)$ is

$$
\frac{dq(t)}{dt} = \frac{e^{-(1-\kappa)t}[(1-\kappa)(t^2+\frac{mt}{2M})+\frac{m}{2M}]-\frac{m}{2M}}{t^2}.
$$

We use $q_1(\cdot)$ to denote the numerator of the formula above and find the derivative of function $q_1(\cdot)$:

$$
\frac{dq_1(t)}{dt} = (1-\kappa)te^{-(1-\kappa)t}\left[2-\ (1-\kappa)\left(t+\frac{m}{2M}\right)\right].
$$

It is straightforward that $\frac{dq_1(t)}{dt} \geq 0$ and $\frac{dq(t)}{dt} \geq 0$, when $0 \leq t \leq \frac{2}{1-\kappa} - \frac{m}{2M}$ and $\frac{dq_1(t)}{dt} \leq 0$, when $t \geq \frac{2}{1-\kappa} - \frac{m}{2M}$. This indicates that $q(t)$ have at most one zero point; therefore,

$$
\begin{aligned}
q(t) &\geq \min\{q(1), q(l)\} \\
&= \min\left\{\left(1-e^{-(1-\kappa)}\right)\left(1+\frac{m}{2M}\right),\right. \\
&\qquad \left.\left(1-e^{-(1-\kappa)l}\right)\left(1+\frac{m}{2Ml}\right)\right\} \\
&= 1-\delta,
\end{aligned}
$$

where $\delta = 1 - \max\{[(1+\frac{m}{2M})e^{-(1-\kappa)} - \frac{m}{2M}], (1+\frac{m}{2Ml})e^{-(1-\kappa)l} - \frac{m}{2Ml}\}$. Based on the results above, we have

$$\frac{u_{\text{out}}}{u_{\text{opt}}} = \frac{\dfrac{8\pi c^2}{\sqrt{\left(\sum\limits_{s_i \in S_l} |s_i|\right)^2 - |\sum\limits_{s_i \in S_l} s_i|^2}}}{\dfrac{8\pi c^2}{\sqrt{\left(\sum\limits_{s_i \in S^*} |s_i|\right)^2 - |\sum\limits_{s_i \in S^*} s_i|^2}}} \leq \frac{1}{\sqrt{q(t)}}.$$

Note that the Taylor expansion of function $\frac{1}{\sqrt{1+x}}$ at $x = 0$ is

$$\frac{1}{\sqrt{1+x}} = 1 + \sum_{i=1}^{\infty} (-1)^i \frac{(2i-1)!!}{(2i)!!} x^i = 1 - \frac{x}{2} + O(x^2);$$

therefore,

$$\frac{u_{\text{out}}}{u_{\text{opt}}} \leq \frac{1}{\sqrt{1-\delta}} = 1 + \frac{\delta}{2} + O(\delta^2).$$

The approximation ratio is $1 + \frac{\delta}{2} + O(\delta^2)$, and $\delta = 1 - \max\{[(1 + \frac{m}{2M})e^{-(1-\kappa)} - \frac{m}{2M}], (1 + \frac{m}{2Ml})e^{-(1-\kappa)l} - \frac{m}{2Ml}\}$.

The algorithm analysis above indicates that every time a new element is picked out from the remaining set, the area of u of the ellipse is constantly decreasing, which means that the accuracy is getting higher and the corresponding difference from the optimal value is decreasing. If the number of selected elements has reached l, when continuing to select the new element to add in according to the selection method of the algorithm, the error will continue to be reduced. It is easy to verify that the inequality (2.129) derived from the analysis above also holds for the situation of $i > l$. We have

$$\begin{aligned} g(S_i) &\geq \left(1 + \frac{\lambda(S_i \backslash S^*)}{2\lambda(S^*)}\right)\left(1 - \prod_{j=0}^{i-1}\left(1 - \frac{1-\kappa}{|S^* \backslash S_j|}\right)\right) g(S^*) \\ &\geq \left(1 + \frac{\lambda(S_i \backslash S^*)}{2\lambda(S^*)}\right)\left(1 - \left(1 - \frac{1-\kappa}{l}\right)^i\right) g(S^*) \\ &\geq \left(1 + \frac{\lambda(S_i \backslash S^*)}{2\lambda(S^*)}\right)\left(1 - e^{-(1-\kappa)\frac{i}{l}}\right) g(S^*). \end{aligned}$$

For any given constant ε close to 0, $g(S_i) \geq (1 + \frac{\lambda(S_i \backslash S^*)}{2\lambda(S^*)})(1 - e^{-(1-\kappa)\frac{i}{l}}) g(S^*) \geq (1-\varepsilon) g(S^*)$, if $i \geq \frac{l \ln \frac{1}{\varepsilon}}{(1-\kappa)}$. That is, if the number of reported signal fingerprints is large enough to make $g(S_i) \geq (1 + \frac{\lambda(S_i \backslash S^*)}{2\lambda(S^*)})(1 - e^{-(1-\kappa)\frac{i}{l}}) g(S^*) \geq (1-\varepsilon) g(S^*)$, then the final localization accuracy is at most $1 - \varepsilon$ times worse than the optimal value.

Now we briefly analyze the time complexity of the algorithm: it is obvious that the algorithm is with polynomial complexity. In rows 4 and 5 of the algorithm, which is the element mapping stage, there are lm loops, with each can be finished in $O(1)$, and thus the complexity is $O(lm)$. In rows 7 to 11 of the algorithm, the dominant factor of the complexity is the "For" loop to update weights, which is nested in another "For" loop. The execution of each loop needs time complexity of $O(lm)$. The complexity of the algorithm's rows 6 to 11 is $O(l^2m)$. Finally, during the algorithm calculating the sequence of returned values, for all elements in set S, the corresponding AP must be found, for which the time complexity is $O(l)$. Consequently, the overall complexity of the algorithm is $O(l^2m)$.

2.5.3 *Applications of the Best Strategy*

2.5.3.1 Location Estimation Leveraging Best Fingerprints Reporting Strategy

The best fingerprints reporting strategy is dependent on the setting of the physical space that needs localization services. Given the indoor space with fixed AP deployment, the best strategy for each location of the space is deterministic and can be derived using the proposed algorithm. Consequently, the best strategy of a location plays a role similar to the local fingerprints stored in the database, which may distinguish one location from another. However, the best strategy is intuitively less sensitive than fingerprints in discriminating one location from another, especially in a small vicinity. Recall the complex vector characterizing an AP, which in essence represents the relative position of the AP with respect to the target location. Since the distance and angle of neighboring locations with respect to surrounding APs are almost the same, best reporting strategies for such locations should be the same.

Such a seemingly frustrating phenomenon factually can also be leveraged to reduce errors in location estimation. Empirical studies show that the estimation errors of pure fingerprinting based localization system could be over 6m [78, 132, 133]. The root cause of such large errors is that physically distant locations may share similar Wi-Fi signal strength, which is due to the dynamic propagation of radio signals. However, the feature of the best strategy described above provides an opportunity to mitigate the impact of such errors. Although multiple faraway locations may have similar fingerprints, their best strategies for fingerprints reporting could differ from each other because their relative positions with respect to surrounding APs are different.

It is worth mentioning that the existing solution to deal with the fingerprints similarity is to utilize the k-nearest neighbors (KNNs) algorithm [2] or the acoustic ranging estimations performed among peer smartphones; however, KNN is a

Algorithm 2 Location determination strategy

Get initial estimated locations $\{r_e\}$, $\{cAP_i\}^0 = \emptyset$.
Initialization(set counter $t = 1$ and calculates $\{cAP_i\}^t \leftarrow \bigcap_{\{r_e\}}\{\mathcal{V}_n^*(r)\}$).
while Radius of $\{r_e\} < r^*$ **do**
  if $|\{cAP_i\}^t - \{cAP_i\}^{t-1}| \neq 1$ **then**
    Set $\{nAP\}$ as the AP that appears most frequently.
  else
    $\{nAP\} \leftarrow \{cAP_i\}^{(t)} \backslash \{cAP_i\}^{(t-1)}$.
  end if
  Report $RSSes$ for AP in $\{nAP\}$.
  Update distance matrix $D_j = |RSSes, \mu(r_j)|$.
  Get new estimated locations $\{r_e\} \leftarrow \{r_j\}, D_j < d^*$.
  Increment counter $t = t + 1$ and recalculate candidate APs $\{cAP_i\}^t \leftarrow \bigcap_{\{r_e\}}\{\mathcal{V}_n^*(r)\}$.
end while
Return estimated location $r_0 = \frac{1}{|\{r_e\}|}\sum\{r_e\}$.

machine learning approach without any theoretical basis for localization, and the acoustic ranging requires collaboration among users [78, 132].

Exploiting the best strategy can reduce large localization errors without consuming extra resources in users' devices. We propose Algorithm 2 to implement a location estimation approach facilitated by the best strategy. We are to provide a walk-through of the algorithm using the example shown in Fig. 2.31. Moreover, our analysis of the algorithm will show that the localization accuracy can be improved not only for reducing large-scale errors but also for discriminating locations in a small neighborhood, where case 1 and case 2 in Fig. 2.31 are used to represent such two scenarios.

The basic idea of the algorithm is to first roughly determine a set of candidate locations. The user measures APs that are included in the best strategies for all candidate locations in each iteration. The server can then reduce the estimation uncertainty according to the reported fingerprints in each iteration and finally localize the user.

As shown in Fig. 2.31, the user reports a fingerprint consisting of RSSes with respect to AP_1 to AP_4. The server calculates the Euclidean distance between the reported fingerprint and the fingerprint associated with the location A, B, and C in the database, respectively, through which the server finds that all the three locations are matching the reported fingerprint. We consider case 1, where the best strategy associated with each location is distinctive because the three locations are far away from each other. The number in the best strategy means the number of times each corresponding AP should be measured. The server finds that AP_3 is included in all three best strategies and then asks the user to measure AP_3. After matching the reported fingerprint with respect to AP_3, the sever finds that the location C's fingerprint with respect to AP_3 in the fingerprints database is the worst match to the reported one, and C can be deleted from the candidates list. Comparing the best strategies of the remaining location candidates A and B, they have AP_2 in common.

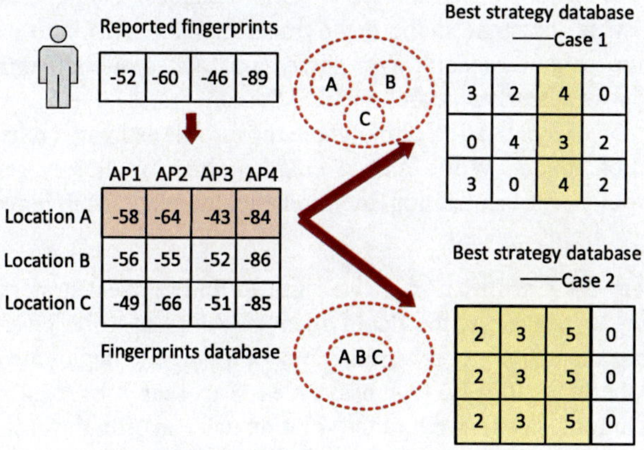

Fig. 2.31 Location determination facilitated by the best strategy

The server asks the user to report another fingerprint with respect to AP_2, and the worst match is also deleted.

Now we consider case 2 in Fig. 2.31, where the best strategy associated with each location is the same because the three locations are in a small neighborhood. In this case, the server asks the user to report a fingerprint with respect to AP_3 because AP_3 appears most frequently in all best strategies. In this way, the server still can cross check fingerprints reported from the user multiple times and find the best estimation. The performance evaluation presented in Sect. 2.6 is to show the effectiveness of such an algorithm.

A key issue of the algorithm is when the iteration should end. The server could execute the iteration until it converges to only one candidate location. It can also end the iteration when the candidate locations are within a certain region denoted by r^*. For example, in case 2 of Fig. 2.31, if r^* is set to be larger than the distance between the three locations, then the iteration will end, and the location that is equally distant to each of the candidate points is returned as the estimated location. The convergence rate of the algorithm is determined by d^*, which shows in each iteration how many candidate points will be accepted to the next iteration. In most cases, d^* can be set with respect to the number of measurement times. If the measurement time is N_0 and there are l candidates, then we can set d^* so that only $\log_{N_0} l$ points will be selected to enter the next iteration, and then after at most N_0 rounds, the iteration will end.

2.5.3.2 Strategy for AP Deployment

Wi-Fi APs have been widely deployed in public indoor spaces, where coverage is an important issue. Bai et al. propose an AP deployment scheme based on evolving diamond pattern, which presents the minimum required number of APs to cover an area [134]. As location based service is gaining increasingly popularity, it could kill two birds with one stone if the deployment of APs takes both coverage and localization into account, especially for newly constructed public buildings. Battiti et al. [103] propose a heuristic search method that integrates coverage requirements with the reduction of localization error, where the error is estimated relied on simple radio propagation model. While AP coverage problem is closely related to AP deployment problem for localization, the fundamental relationship between the two problems has yet been revealed.

AP Deployment for Localization If we want to find the best strategy for fingerprints reporting at location **r**, the deployment of APs has to be given; therefore, it seems to be a paradox to determine the optimal AP deployment based on the theory of the best strategy. Our basic idea is to search over all possible AP deployment solutions, in order to find the solution that can offer the best localization performance. Such AP deployment solution is the best for localization. The crux of the idea is to evaluate the performance of localization based on the best strategy for fingerprints reporting, which is denoted by

$$R(\mathbf{r}) = \left(\sum_{i \in \mathcal{V}_n^*} |Z_i| \right)^2 - \left(|\sum_{i \in \mathcal{V}_n^*} Z_i| \right)^2, \tag{2.130}$$

where $\mathcal{V}_n^*$ is the index of measured APs determined by Eq. (2.116). This is because the larger $R(\mathbf{r})$ the smaller ellipse can be obtained (recall Fig. 2.1), thus the higher localization accuracy can be achieved. Here note the difference between Eq. (2.116) and (2.130). In the discussion of AP deployment problem, we assume that the system always follows the best fingerprints reporting strategy, so that the system accuracy can be maximized.

Equation (2.130) indicates that the change of APs' locations results in the change of $\nabla\mu(\mathbf{r})$. We use $\{x_i, y_i\}$ to denote the location of AP_i and $\{\mathbf{x}, \mathbf{y}\} = \{(x_1, y_1), (x_2, y_2), \ldots, (x_N, y_N)\}$ to denote locations of APs. Consider the localization performance at location **r**, which is denoted by $R(\mathbf{r})$. Suppose the user appears in locations following the probability density function denoted by $f(\mathbf{r})$, the expected performance of localization in such a specific setting is $\int_S R(\mathbf{r}) f(\mathbf{r}) dr^2$. Consequently, the optimal strategy for deploying APs is to fix APs in the following locations:

$$\{\mathbf{x}, \mathbf{y}\}^* = \arg\max_{\{\mathbf{x}, \mathbf{y}\}} \int_S R(\mathbf{r}) f(\mathbf{r}) dr^2, \tag{2.131}$$

where S is the area of the indoor space. This provides fundamental criteria to evaluate different AP deployment strategies for the purpose of localization.

Algorithm 3 AP deployment strategy

```
Initialize {x, y}(Select c locations randomly).
while T_t ≥ t* do
    Generate new strategy by randomly change one location in {x, y}.
    Calculate expected reliability H for the new and old state.
    Accept the new state according to H_new and H_old with probability min{1, e^((H_new − H_new)/T_t)}.
    Update temperature T_t.
end while
{x, y}* ← {x, y}.
```

The challenge for resolving the problem above is that the searching space for optimal $\{\mathbf{x}, \mathbf{y}\}$ is continuous. In practice, we are unable to search the physical space in an inch-by-inch manner, and thus we can limit the locations of APs in discrete positions. Consequently, the problem can be transformed into

$$\{\mathbf{x}, \mathbf{y}\}^* = \underset{\{\mathbf{x},\mathbf{y}\} \subset \{\mathbf{X},\mathbf{Y}\}}{\arg\max} \int_S R(\mathbf{r}) f(\mathbf{r}) dr^2, \tag{2.132}$$

where $\{\mathbf{X}, \mathbf{Y}\}$ denotes the discretized physical space.

The transformed problem can be resolved by a simulated annealing based algorithm as shown in Algorithm 3.

Localization and Coverage In the previous work in the literature, access point deployment problem is considered as a coverage problem, however, for the performance of localization, AP deployment can be further discussed so as to meet the accuracy needs. If a specific AP deployment strategy guarantees that each point in the area can be localized correctly, then each point is surely covered by APs, which means that every point can receive signals higher than the given strength.

Theorem 9 *If an AP deployment plan in a region with area D satisfies the localization accuracy R^*, then every point of the region must be covered by at least one AP within the distance $\frac{\sqrt{N_0}}{\sigma \sqrt[4]{R^*}}$, and there must be at least $\frac{2\sigma\sqrt{3R^*}}{9N_0}(D-2)$ APs need to be deployed; however, if a user is definitely covered by at least one AP at any point of the region, it is still possible that the localization accuracy cannot be satisfied.*

Proof 25 The first part of the theorem is to prove that $\forall R^*, \forall N_0, \exists d$, such that if $\forall \mathbf{P}, \forall AP_i \in \mathbb{U}_l, d(\mathbf{P}, AP_i) > d$, then $R(\mathbf{p}) < \frac{N_0^2}{d^4\sigma^4}$, where $\mathbf{P}$ is a point in the region and $\sigma = \min_i\{\sigma_i\}$.

According to Eq. (2.130), the localization accuracy is determined by the characteristic vector of the AP, thus

$$
\begin{aligned}
R(\mathbf{P}) \leq \left(\sum_{i\in\mathcal{V}_l}|Z_i|\right)^2 = \left(\sum_{i\in\mathcal{V}_l}p_i\right)^2 = \left(\sum_{i\in\mathcal{V}_l}\frac{(\nabla\mu_i)^2}{\sigma_i^2}\right)^2 \\
= \left(\sum_{i\in\mathcal{V}_l}\frac{1}{r_i^2\sigma_i^2}\right)^2 \leq \frac{N_0^2}{d^4\sigma^4}.
\end{aligned}
\tag{2.133}
$$

Equation (2.133) holds when the location is at the center of several equidistant APs with the same RSS variance σ.

If a point can be localized with required accuracy level R^*, then there must be an AP within $\frac{\sqrt{N_0}}{\sigma\sqrt[4]{R^*}}$. To guarantee that every point lies in a coverage range of some APs, the AP deployment strategy must satisfy $\exists AP_i \in \mathbb{U}_l, d(\mathbf{P}, AP_i) < \frac{\sqrt{N_0}}{\sigma\sqrt[4]{R^*}}$.

The second part of the theorem is equivalent to finding the minimum required number of APs to cover the area given the minimum required distance d. Since the space requires indoor localization service can be divided into small square areas, the problem is reducible to circle covering problem which has been proved by Kersher [135].

Based on the Kersher's theory that the least number of circles with radius a denoted by $\psi(a)$, we have satisfies

$$
\pi a^2\psi(a) > \frac{2\pi\sqrt{3}}{9}(D - 2\pi a^2), \tag{2.134}
$$

where D denotes the area of the rectangle. Together with Eq. (2.133), we finally get the minimum required number

$$
\psi > \frac{2\sigma\sqrt{3R^*}}{9N_0}(D - 2). \tag{2.135}
$$

We now give a counter example showing that the localization accuracy cannot be satisfied, while the coverage is satisfied. Assume there are k APs collocated collinearly, every point will be covered if access points have enough power, however, it is impossible to localize the user if the user is standing on any point of the line. This is because the small displacement along the line is with indistinguishable changes in RSSes. If the user moves along the line, it is impossible to localize where the user is.

Equation (2.135) shows the minimum number of APs required is determined by several factors. When the signal channel is clearer, i.e., σ is smaller, the number is smaller. Moreover, the decrease of measurement times N_0 brings the same impact. Moreover, when the required reliability becomes higher, the required number is larger, which increases proportionally to the square root of the reliability.

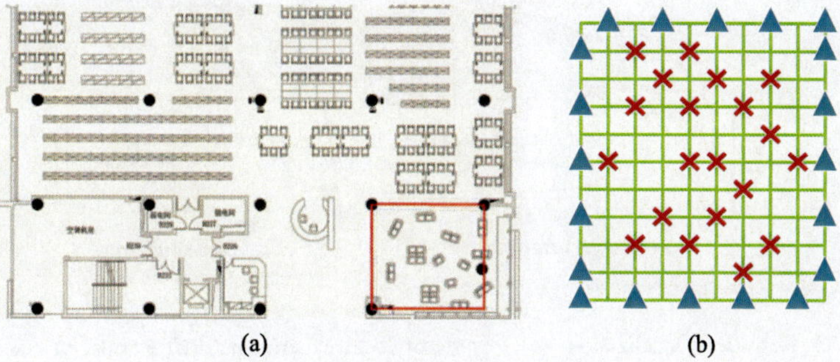

Fig. 2.32 Experiment field. (**a**) Floor plan. (**b**) AP deployment

2.5.4 Performance Evaluation

In this section, we evaluate the proposed algorithms with both local and trace data experiments. We conduct local experiments in an area of our university's library as shown in Fig. 2.32a. The area in the red frame is grided into 10×11 cells and the edge length of each cell is 70 cm. (Note that the black nodes represent pillars.) The APs used in the experiment are uniformly distributed along the edges of the region first, as those blue triangles shown in Fig. 2.32b, and then deployed according to our proposed AP deployment strategy, as those red crosses. We also conduct experiments with the trace data collected by the EVARILOS testbed [193]. The data are collected in an unmanned utility room with many metal objects termed as "Zwijnaarde," where there is almost no outside interference and no persons are present in the environment. Detailed descriptions of the testbed and data could be found in [112–114].

Performance of AP Selection Strategies in CDF We first verify that the proposed best fingerprints reporting strategy indeed can improve the accuracy of localization. We compare the localization results with the best strategy and that with other strategies. If we measure every AP's signal strength at each location of a space, then APs with the strongest average signal strength over different locations can be obtained, and the APs with the strongest signal strength all over the space can also be obtained. *StrongestAvg* strategy is to always measure APs with the strongest average signal strength, and *StrongestMax* is to always measure APs with the strongest signal strength. *SimilarityBased* strategy calculates the similarity matrix of APs and always measures APs with the lowest similarities, where the detailed description of the scheme could be found in [77].

The three strategies mentioned above are to select each AP based on the AP's own importance, where the APs' group effect is not taken into account. We note that a group-discrimination based (GD) AP selection mechanism is proposed recently [129], where the positioning capabilities of a group of APs are investigated. The basic idea of the GD mechanism is to find the best combination of APs and use the

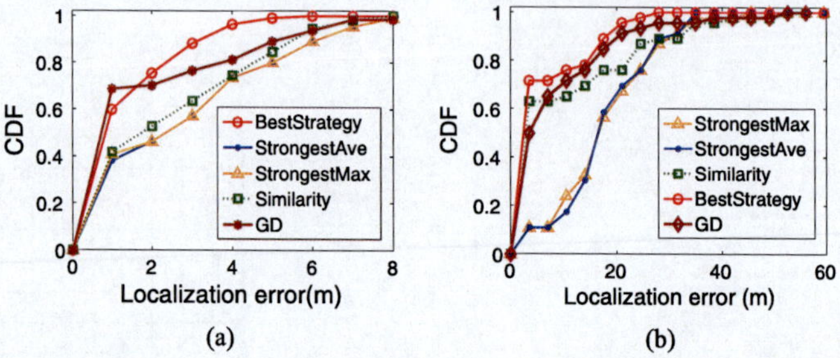

Fig. 2.33 Results of localization. (**a**) Resutls of local experiments. (**b**) Results of trace data experiments

fingerprints generated by those APs as the feature of each location. We also compare our proposed BestStrategy with the GD mechanism. In the experiments, we select the best combination consisting of 6 APs according to the GD algorithm in [129] and use the fingerprints of those 6 APs to distinguish one location from another in the offline phase. We also use the fingerprints from the same 6 APs to perform the online phase to estimate the device location.

Figure 2.33a shows results of the local experiments, where the horizontal axis represents the localization error observed and the vertical axis represents the corresponding cumulative distribution function (CDF) value. In the experiment, we deploy 20 APs in the area with our best deployment strategy as shown in Fig. 2.32. Under such setting, we perform localization in 100 randomly selected points. The results show that the best fingerprints reporting strategy yielded from our algorithm outperforms other strategies. Figure 2.33b shows experimental results with the trace data from the EVARILOS testbed. Since the deployment of APs and landmarks could not be changed, we just evaluate the AP selection strategy. We use the data observed at a part of the landmarks as the training set and that observed at the rest of the landmarks as the test set to perform localization. The results corroborate our local experiment results. Our proposed algorithm outperforms others because other strategies just exploit the statistical properties of the fingerprints, but our strategy is based on the intrinsic mechanism of fingerprinting localization.

It is shown that our proposed mechanism still outperforms the GD mechanism. This is because the process of finding the best combination in [129] is based on empirical metric. According to our theoretical analysis, the number of times for measurements with respect to each AP is also important, which however is not taken into account in GD [129]. It is noted that the performance of the StrongestMax and StrongestAvg is almost the same in both the EVARILOS data experiment and the local experiment. This is because a large proportion of APs selected by the two strategies are the same. If we use the localization accuracy of CDF, 80% as the benchmark to evaluate the algorithms' performance. It shows that at 80% of the time in our local experiments, the localization error is within $2.2m$ and $3.5m$ with the BestStrategy and the second best GD strategy, respectively. In the trace

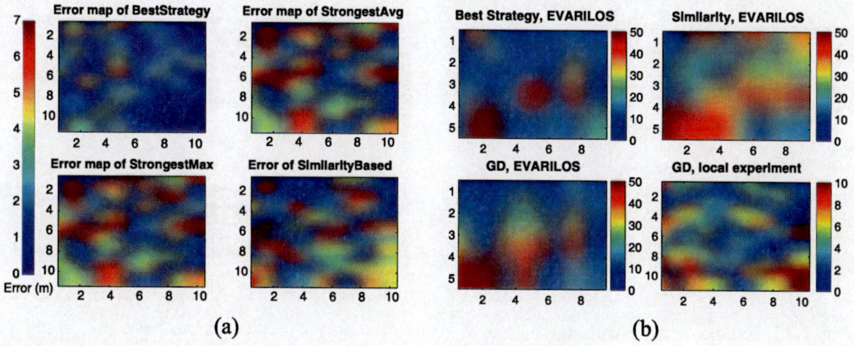

Fig. 2.34 The error map in our experiments. (**a**) Error map of local experiment. (**b**) Error map of selected algorithms

data experiments, those numbers become 13m and 17m. That is, the BestStrategy outperforms the GD strategy by around 37% and 24% in the local and trace data experiments, respectively.

Performance of AP Selection Strategies in Error Map Figure 2.34a illustrates how the localization errors are distributed in the local experiment field. The experiment results with respect to the three strategies all confirm the results in [77]. Moreover, the results also indicate that the best strategy indeed incurs the smallest localization errors. Figure 2.34b illustrates how the localization errors with selected mechanisms are distributed in the EVARILOS experiment field and the local experiment field. We select the BestStrategy, GD, and SimilarityBased strategies. The experiment field of the EVARILOS testbed is a 56 m $\times$ 15 m rectangle space, and each cell in the figure represents a 6 m $\times$ 3.2 m space. All the experimental results show that the proposed BestStrategy outperforms other strategies.

It is worth mentioning that our experimental results are just based on pure fingerprints comparison; other assisting mechanisms such as motion sensors and acoustic ranging as described in Sect. 2.2 are not included. This is because the purpose of the work is to investigate the fingerprints reporting strategy. Moreover, the results of trace data experiments are not as good as that of local experiments, and this is due to the limitation of the trace data. It could be found in the data set that there are only a few RSS readings recorded on some landmarks, and at some landmarks, the RSS readings are exactly the same. This is perhaps because the data are collected by the automatic robot as described in [193].

AP Deployment Strategy We conduct experiments to examine the effect of our proposed AP deployment strategy. In the experiment field as shown in Fig. 2.32, we deploy 20 APs in two ways, where the first is to deploy them uniformly along the four edges, and the second is to follow the proposed AP deployment strategy. We then perform localization in the area for 100 times at 100 randomly selected points. The results of localization are illustrated in Fig. 2.35a, where the curve representing

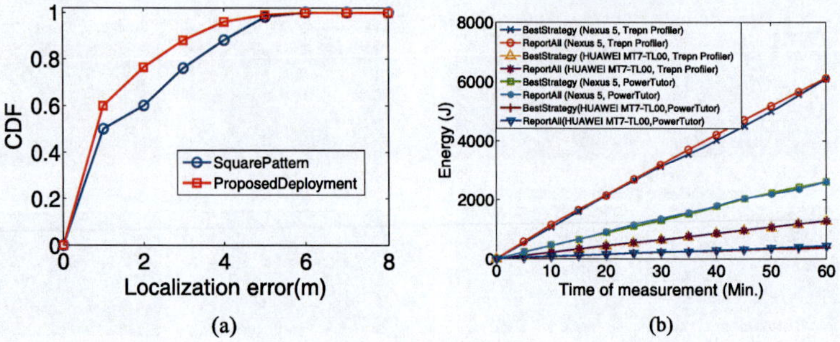

Fig. 2.35 Experimental results and energy consumption. (**a**) Localization error CDF for purposed deployment strategy. (**b**) Energy consumption

uniform deployment is labeled with "SquarePattern." It is straightforward that the proposed deployment strategy can help improving the localization performance.

Energy Consumption We conduct 1-h experiment with a HUAWEI MT7-TL00 and a Nexus 5 smartphone to measure power consumption of RSS fingerprints reporting strategies. We examine two kinds of strategies, where the first one selects 6 APs using our proposed algorithm, and the second one selects all 20 APs. For each strategy, we let the smartphone report the observed RSS fingerprints every 500 ms. Considering that the best strategy may vary in different locations as explained in Sect. 5.1 when the user is moving, we recompute the best strategy every 2 s to make sure that the best strategy for each location is updated in time. This process lasts for 1 h, during which the power consumption results are recorded using both the PowerTutor [195] and the Trepn Power Profiler APP [196] every 5 min. The results are shown in Fig. 2.35b. Since the two kinds of smartphones tailored the Android OS in different ways, their basic energy consumptions are different; moreover, the energy consumption models adopted by the two APPs are different [195], and thus the results by the two different APPs are not the same. However, it is clear that the energy consumption results of the two strategies in different scenarios are almost the same, which means that the energy consumption incurred by the computation of our strategy could be negligible.

2.6 Theoretical Guidance on Optimizing Localization Accuracy

This section theoretically shows that the temporal correlation of the RSS can further improve the accuracy of the fingerprinting localization. In particular, we construct a theoretical framework to evaluate how the temporal correlation of the RSS can influence reliability of location estimation, which is based on a newly proposed radio

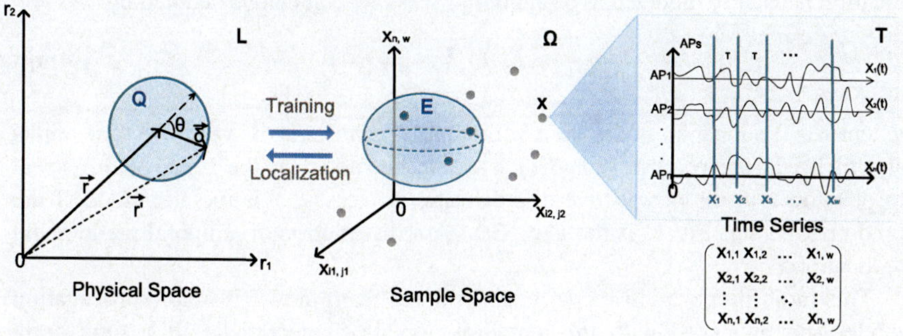

Fig. 2.36 Theoretical localization model

propagation model considering the time-varying property of signals from Wi-Fi APs. The framework is then applied to analyze localization in the one-dimensional physical space, which reveals the fundamental reason why localization performance can be improved by leveraging temporal correlation of the RSS. We extend our analysis to high-dimensional scenarios and mathematically depict the boundaries in the RSS sample space, which distinguish one physical location from another. Moreover, we develop an algorithm to utilize temporal correlation of the RSS to improve the location estimation accuracy, where the process for choosing key design parameters is provided through experiments. Experiment results show that the localization reliability and accuracy can be improved by up to 13% and 30% with appropriately leveraging the RSS temporal correlation information.

2.6.1 Theoretical Model of Location Estimation

Consider an indoor space, which can be modeled as one- or two-dimensional Cartesian space denoted by $L \subset \mathbb{R}$ or $L \subset \mathbb{R}^2$, respectively. Examples of one-dimensional model include hallway and corridor. A user's location in the physical space S can be denoted by $\mathbf{r} = r_1$ or $\mathbf{r} = (r_1, r_2)$ with corresponding dimensions. Based on the localization database constructed in the training phase, a sample space of fingerprints can be induced, which is denoted by Ω^n and n is the number of access points (APs) that can be sensed in the physical space. In the training phase, the site surveyor collects fingerprints of APs in a one-by-one manner at a given location. For an AP, the surveyor samples the observed RSS at certain frequency. Consequently, if there are n APs and each AP is sampled w times, then a point $\mathbf{x}$ in the RSS sample space is as shown in the right part of Fig. 2.36, where $x_{i,j}$ means the RSS observed with respect to AP_i at jth time point. We say this is an n-dimensional sample space and the temporal dimension of sampling is w.

As the radio propagation in the indoor environment is influenced by many factors such as path loss, shadowing, fading, and multipath effect, the signal observed in a location is usually modeled as a random process, which can be denoted by

$$X(\mathbf{r}, \mathbf{t}) = S(\mathbf{r}) + \sigma Y(\mathbf{r}, \mathbf{t}), \tag{2.136}$$

where $\mathbf{r}$ is the location of the observation and $\mathbf{t}$ represents the vector of time points at which RSSes are observed. $S(\mathbf{r})$ is the trend model of the signal with respect to position $\mathbf{r}$ in the perspective of stochastic processes, σ is the amplitude of the randomness, and $Y(\mathbf{r}, \mathbf{t})$ is the joint Gaussian distribution of temporal randomness at location $\mathbf{r}$.

This radio propagation model is generalized from the LDPL radio propagation model and enhanced with the temporal domain characteristic. If a time point t is given, the model degenerated into the radio propagation model in [116], where the RSS can be modeled as a Gaussian random variable with fading and multipath effects integrated into a mean function such as $S(\mathbf{r})$. Such modeling is widely adopted in the literature on indoor localization in the past decade [36, 38, 39, 77, 85, 101, 280]. With the temporal dimension considered in this work, the RSS is modeled as a Gaussian stochastic process, which is also verified by a number of work in the literature [102, 137, 138], where the temporal correlation by the shadowing effect incurred by the fixed building structure is modeled with $\sigma Y(\mathbf{r}, \mathbf{t})$.

According to extensive experimental results and theoretical analysis [77, 280], the mean and variance of the RSS in one location basically remain the same over time and the auto-covariance function of the RSS in one location has the same shape for separate time series, such a random process can be stationary and ergodic, with

$$S(\mathbf{r}') \approx S(\mathbf{r}) + \nabla S(\mathbf{r})(\mathbf{r}' - \mathbf{r}). \tag{2.137}$$

In the localization phase, a user reports observed RSSes to the localization server, which then estimates the corresponding location by matching the reported fingerprints in the fingerprints database. Such a process can be modeled as a mapping from the sample space to the physical space:

$$M : \Omega^n \to L, \quad \mathbf{r}' = M(\mathbf{X}(\mathbf{r}, \mathbf{t})), \tag{2.138}$$

where $\mathbf{r}'$ is the estimated location of the user. This process is illustrated in Fig. 2.36. The user's actual location is at $\mathbf{r}$ and the estimated location is at $\mathbf{r}'$, which incurs the localization error denoted by $\boldsymbol{\delta}$.

Due to estimation errors, the result of the localization is that the user's location is estimated to be in the δ neighborhood of $\mathbf{r}$, which is denoted by Q. To reduce the error of localization is equivalent to mitigating the norm of $\boldsymbol{\delta}$. Since the basis of the estimation is the reported fingerprint by the user, the ideal case is that the user's submitted fingerprints happen to make the system believe that the location of the user is in Q. We use $\mathbb{E}$ to denote such a region in the sample space, so that the user's location can be estimated to be in Q as long as the reported RSSes fall in $\mathbb{E}$.

The probability that the reported RSS fingerprints can fall into the region of $\mathbb{E}$ depends on the model of radio signal propagation, which in fact fundamentally determines the performance of the RSS fingerprinting based approach. The model proposed in [116] considers the observed RSS at one location as a random variable, where temporal correlation of the signal is not taken into account. According to the site survey practice, it is more practical to model the signal as a random process as in this work, where the temporal correlation can be leveraged.

Our investigation in this work focuses on the influence of temporal correlation of the RSS on performance of localization. We are to show that such a seemingly slight change in the radio signal propagation modeling could help reveal interesting findings of the RSS fingerprinting based approach, which have never been presented; the corresponding difficulties in the mathematical analysis are definitely non-trivial. It is worth mentioning that the body-shadowing effect incurred by people's mobility also could influence the radio propagation in the indoor space [122], which is actually an important research field with many efforts dedicated [123, 124, 139]. The body-shadowing effect will be considered in our future work, since taking account of too many factors will make the mathematical analysis untractable and hinders revealing the insight.

2.6.2 *Analysis of 2D Temporal Correlation for 1D Localization*

This section examines a concrete scenario of localization, where both the physical space and the sample space are one-dimensional and the temporal dimension of sampling is two. The purpose of the examination is to find how likely the user can be localized in Q with given δ. It is easier to reveal essence of the fingerprinting approach by analyzing a simple case, where the results could be inspiring for analyzing more complicated scenarios.

2.6.2.1 Finding Region $\mathbb{E}$

Let us first find out what kind of RSSes can be observed at the location $\mathbf{r}$. The one-dimensional physical space can be regarded as a one-dimensional horizontal axis, where the origin of the axis is the location of the AP, and the location of each point can be identified by a scalar r. Based on the radio signal propagation model, the probability density function (PDF) of RSS readings observed follows the Multivariate Gaussian Distribution:

$$f_r(x_1, x_2) = \frac{1}{2\pi\sigma^2\sqrt{1-\rho^2}} e^{-\frac{1}{2}\Delta^2}, \tag{2.139}$$

where x_1 and x_2 are variables representing the RSSes at time points t_1 and t_2 separated by a duration of τ as shown in Fig. 2.37a. Since the random process

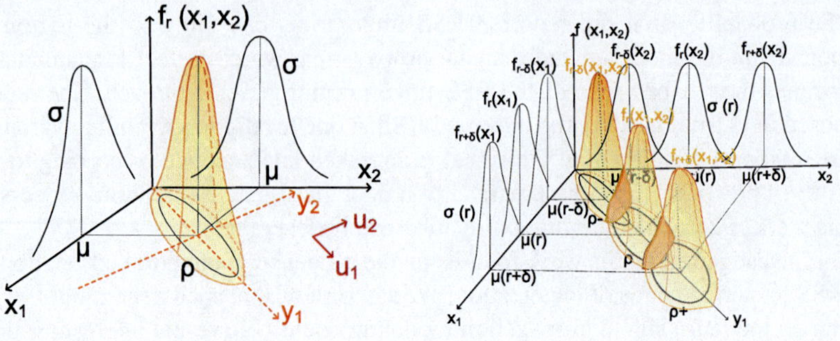

(a) (b)

Fig. 2.37 Joint Gaussian PDF. (**a**) Joint Gaussian PDF of $RSS(t)$ and $RSS(t+\tau)$ at position **r**. (**b**) Joint Gaussian PDFs at different locations

representing the signal is stationary, the following analysis is oblivious to the specific value of t_1 and t_2 as long as they are separated by τ. Symbols μ and σ are the mean and standard variance of the RSS joint distribution at position **r**, respectively; ρ is the autocorrelation coefficient of $f_r(x_1, x_2)$. The Mahalanobis distance is denoted as Δ, where

$$\Delta^2 = \frac{1}{\sigma^2(1-\rho^2)}\Big[(x_1-\mu)^2 + (x_2-\mu)^2 - 2\rho(x_1-\mu)(x_2-\mu)\Big]. \tag{2.140}$$

Since x_1 and x_2 are both observed at r, the corresponding marginal distributions with respect to x_1 and x_2 are the same, according to our signal propagation model; the corresponding means and standard variances of the two marginal distributions are the same as well. This also complies with the conclusion in [116]. Consequently, the covariance matrix of $f_r(x_1, x_2)$ is real, positive, and symmetric, where

$$\mathbf{\Sigma} = \sigma^2 \begin{bmatrix} 1 & \rho \\ \rho & 1 \end{bmatrix}. \tag{2.141}$$

With the same reason, the major axis of the elliptical surface representing $f_r(x_1, x_2)$ should be the angular bisector of the Cartesian coordinates with slope 1.

In order to facilitate our analysis, we put the image of $f_r(x_1, x_2)$ in a new coordinates system with axes y_1 and y_2. We let the major axis of the elliptical surface align to y_1 and the origin of the new coordinates system be $(\mu(r), \mu(r))$ in the old system. Then the PDF in the new system is

$$f_r(y_1, y_2) = \frac{1}{2\pi\sigma^2\sqrt{\lambda_1\lambda_2}} e^{-\frac{1}{2\sigma^2}\left(\frac{y_1^2}{\lambda_1}+\frac{y_2^2}{\lambda_2}\right)}, \tag{2.142}$$

where

$$\lambda_1 = \frac{\sqrt{2}(1+\rho)}{2}, \lambda_2 = \frac{\sqrt{2}(1-\rho)}{2}. \tag{2.143}$$

We now start to find the region $\mathbb{E}$ in this scenario. Refer to Fig. 2.37b, the value of $f_r(y_1, y_2)$ in fact means how likely the user can observe $[y_1, y_2]$ at location r. If the reported RSSes $[y_1, y_2]$ indicate that the user's location is in a small neighborhood of r, then $f_r(y_1, y_2)$ should be higher than $f_{r\pm\delta}(y_1, y_2)$, where $r \pm \delta$ are boundaries of r's neighborhood in the physical space. That is, if the user is localized in the neighborhood of r, the corresponding submitted fingerprints should have fallen into the region

$$\mathbb{E} = \{\mathbf{x} | f_r(\mathbf{y}|\mu(r), \mathbf{\Sigma}(r)) \geq f_{r\pm\delta}(\mathbf{y}|\mu(r \pm \delta), \mathbf{\Sigma}(r \pm \delta))\}. \tag{2.144}$$

The profile of $\mathbb{E}$ is sketched in Fig. 2.37b, which is the space between the two regions in dark color. The two dark-colored regions themselves represent boundaries of intersected neighboring dome-like bodies. Observe marginal PDFs with respect to x_2 for the three locations $r - \delta$, r and $r + \delta$, which are presented by three Gaussian PDF curves on the $x_2 - f(x_1, x_2)$ plane with means $\mu(r - \delta)$, $\mu(r)$ and $\mu(r + \delta)$, respectively. It is worth mentioning that shapes of the three curves are the same, which is determined by the variance of Gaussian noise. This is because Gaussian noise at different locations in a small neighborhood of the physical space is presenting indistinguishable randomness, which have been acknowledged by extensive studies [116, 140]. Due to symmetry of the dome-like bodies, the same thing happens to the marginal PDFs with respect to x_1.

If the temporal correlation of the RSS is not considered, fingerprints observed at different time points with respect to the same AP are independent at each location; therefore, the randomness of the RSS can only be characterized in a 2D curve of the marginal PDF as shown in Fig. 2.37b. Using such randomness to evaluate the performance limit of fingerprinting localization is the basic idea in [116].

Our work in this book characterizes randomness of the RSS with the dome-like bodies as shown in Fig. 2.37b, where the temporal correlation of the signal is taken into account. We can see that our model presents a more accurate description of the randomness of the RSS, where a straightforward observation is the increase of a dimension. Such a model of the RSS provides more distinguishable characteristics of a location compared with that in [116] and thus provides criteria of finer-granularity for localization. This is the fundamental reason why the accuracy performance bound of localization derived in [116] can be further improved if the RSS temporal correlation is taken into account.

2.6.2.2 Analysis on Region $\mathbb{E}$

Since the location estimation is performed based on fingerprints reported by the user, studying properties of $\mathbb{E}$ can help reveal how the system estimates the

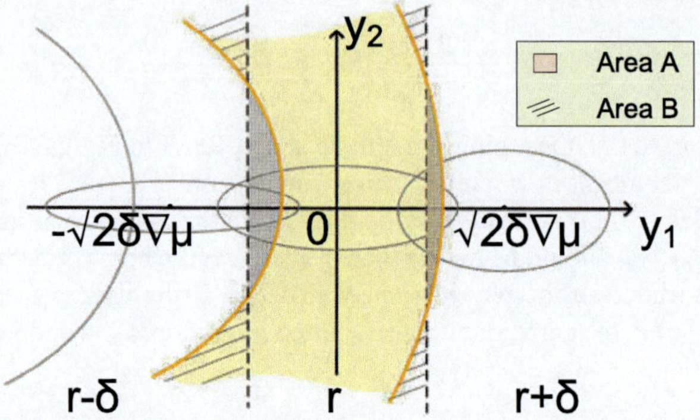

Fig. 2.38 Graphical illustration of region $\mathbb{E}$

user's location. Intuitively, if we project the image in Fig. 2.37b onto the $y_1 - y_2$ coordinates system, the resulted image should be that as shown in Fig. 2.38. The region in yellow should be the projection of the space $\mathbb{E}$, and the two curves in yellow should be boundaries of the region. Consequently, if a user's reported fingerprints fall into the area left to $\mathbb{E}$, the user is more likely at the location $r - \delta$, and if the reported fingerprints fall into the area right to $\mathbb{E}$, the user is more likely at the location $r + \delta$. We are to reveal that the boundaries of $\mathbb{E}$ are in the shape of hyperbolic curve with interesting properties and then reveal challenges for accurately describing the region $\mathbb{E}$ with corresponding analysis provisioned.

Boundaries of Region $\mathbb{E}$ Substituting Eq. (2.142) into Eq. (2.144), we obtain the following inequality:

$$\frac{1}{\sqrt{\lambda_1 \lambda_2}} e^{-\frac{1}{2\sigma^2}\left(\frac{y_1^2}{\lambda_1}+\frac{y_2^2}{\lambda_2}\right)} \geq \frac{1}{\sqrt{\lambda_1^{\pm}\lambda_2^{\pm}}} e^{-\frac{1}{2\sigma^2}\left(\frac{(y_1 \pm \sqrt{2}\delta\nabla\mu)^2}{\lambda_1^{\pm}}+\frac{y_2^2}{\lambda_2^{\pm}}\right)}, \tag{2.145}$$

where λ_1 and λ_2 are scaling factors of ellipse axes for Gaussian PDF at position r, and $\lambda_1^{\pm}$ and $\lambda_2^{\pm}$ are scaling factors at adjacent positions $r \pm \delta$. Specifically,

$$\lambda_1^{\pm} = \frac{\sqrt{2}(1+\rho^{\pm})}{2}, \lambda_2^{\pm} = \frac{\sqrt{2}(1-\rho^{\pm})}{2}. \tag{2.146}$$

Symbols ρ and $\rho^{\pm}$ are the autocorrelation coefficients for the Gaussian distribution at r and $r \pm \delta$, respectively. After simplification, they are equivalent to

$$\begin{cases} \left(\frac{y_1^2}{\lambda_1} + \frac{y_2^2}{\lambda_2}\right) - \left(\frac{(y_1 + \sqrt{2}\delta \nabla \mu)^2}{\lambda_1^+} + \frac{y_2^2}{\lambda_2^+}\right) \leq \ln \frac{\lambda_1 \lambda_2}{\lambda_1^+ \lambda_2^+}; \\ \left(\frac{y_1^2}{\lambda_1} + \frac{y_2^2}{\lambda_2}\right) - \left(\frac{(y_1 - \sqrt{2}\delta \nabla \mu)^2}{\lambda_1^-} + \frac{y_2^2}{\lambda_2^-}\right) \leq \ln \frac{\lambda_1 \lambda_2}{\lambda_1^- \lambda_2^-}, \end{cases} \tag{2.147}$$

which is the specific expression of $\mathbb{E}$ in the sample space. The boundaries of $\mathbb{E}$ can be obtained when the equality holds.

In order to better understand properties of the boundaries, we transform the expressions in inequalities (2.147) into a general form

$$Ay_1^2 + By_1y_2 + Cy_2^2 + Dy_1 + Ey_2 + F = 0, \tag{2.148}$$

where the discriminant Δ equals

$$\Delta = B^2 - 4AC, \tag{2.149}$$

and $A = \frac{1}{\lambda_1} - \frac{1}{\lambda_1^{\pm}}$, $C = \frac{1}{\lambda_2} - \frac{1}{\lambda_2^{\pm}}$. Since $B = 0$, $AC < 0$, then $\Delta > 0$. This means that the two boundaries of $\mathbb{E}$ are in the shape of the hyperbolic curve, where the two foci are on axis y_1.

Note that if $A = C$ and $B = 0$, both of the boundaries are straight lines in parallel. $A = C$ and $B = 0$ also mean that $\lambda_1 = \lambda_2$, $\lambda_1^{\pm} = \lambda_2^{\pm}$, which is to say that measurements with respect to the same AP at different time points are totally independent. This is a degenerated scenario without considering temporal correlation as shown in [116]. The resulted straight-line boundaries are the same as corresponding boundaries of $\mathbb{E}$ in [116]. This is actually corroborating our current result about the shape of boundaries.

Accurate Description of $\mathbb{E}$ Although we have a basic idea about boundaries of $\mathbb{E}$, it is still non-trivial to theoretically prove that the region $\mathbb{E}$ is the same as the intuition as shown in Fig. 2.38. Imagine the detailed scenario that two surfaces representing two joint Gaussian PDFs are intersecting with each other. There are actually two curves of intersection, as the two curves l_1 and l_2 illustrated in Fig. 2.39. This can be mathematically proved through simple derivation by constructing an equation between the two joint Gaussian PDFs.

It is slightly tricky to understand Figs. 2.37b and 2.39. Projections of those domes on planes x_1-$f(x_1, x_2)$ and x_2-$f(x_1, x_2)$ are the same in profile, because this is actually ignoring the temporal correlation of the RSS. Mathematically, the covariance matrix of $f_r(x_1, x_2)$ becomes variance σ^2 as the autocorrelation coefficient $\rho = 0$. However, those joint Gaussian PDFs factually have different autocorrelation coefficients denoted by ρ and $\rho^{\pm}$, as shown in Fig. 2.37b; therefore, if we project those domes on the plane y_1-$f(x_1, x_2)$, the resulted image is just that illustrated in Fig. 2.39.

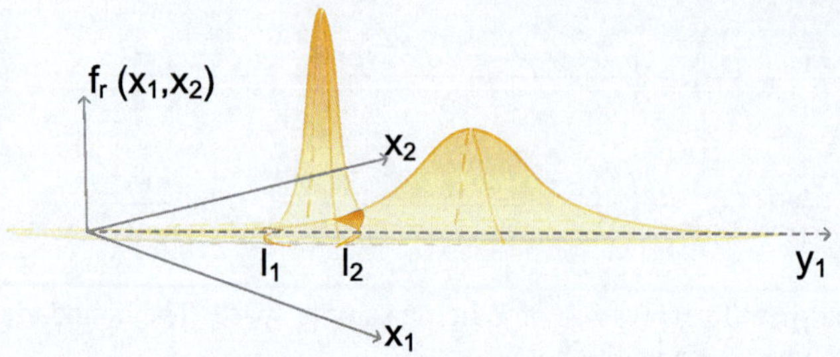

Fig. 2.39 Intersection of two Gaussian PDFs

In the perspective of engineering, the system considers that observing fingerprints around the l_1 is with very low probability if the user is at r, thus it is more meaningful to consider the boundary represented by l_2, in order to ensure an expected localization reliability as high as possible. It is worth mentioning that fingerprints such as those around l_1 indeed can be observed in practice. In this case, the system will estimate that the location of the user is at r', where $f_{r'}(y1, y2)$ has a higher value, although the user is factually at r. Such errors cannot be avoided in the fingerprinting based approach, since small probability events do happen.

We can see that the opening orientation of the boundaries illustrated in Fig. 2.37b is to the left. If $\rho^- < \rho < \rho^+$, the physical meaning of the inequalities (2.137) is that all points with the distance differences between $r-\delta$ to r and r to $r+\delta$ are less than a constant. The opening orientation is to the left, according to the definition of the hyperbola. If $\rho^- > \rho > \rho^+$, the physical meaning of the inequalities (12) is that all points with the distance differences between r to $r-\delta$ and $r+\delta$ to r are less than a constant. The opening orientation is to the right. For convenience of presentation, we here abuse the coordinate in the physical space and use the coordinate to represent the corresponding RSS values in the y_1 axis.

This means that the opening orientation of boundaries are actually determined by the degree of temporal correlation of the RSS at different locations. Moreover, no matter the relationship among ρ and $\rho^{\pm}$, the inequalities of (12) show that $\mathbb{E}$ is in the middle of the two boundaries. As a matter of fact, if we specifically consider the real situation under study, it should be the case $\rho^- < \rho < \rho^+$. Recall our 1D physical model, where the AP is located at the origin of a 1D coordinate axis and $r-\delta$, r and $r+\delta$ are distance to the AP. The farther the location is from the AP, the stronger the temporal correlation of the observed RSS will be; consequently, the orientations of the two boundaries should be to the left as shown in Fig. 2.38.

2.6.2.3 Influence of Temporal Correlation on Accuracy of Localization

We can further verify our theory by examining the expected localization result given special fingerprints. The point $(-\sqrt{2}\delta\nabla\mu, 0)$ in Fig. 2.4 is special, which makes $f_{r-\delta}(-\sqrt{2}\delta\nabla\mu, 0)$ to achieve the maximum value. This means that if a user reports fingerprints $(-\sqrt{2}\delta\nabla\mu, 0)$, the system definitely should estimate the user's location to be at $r - \delta$. Substituting $(-\sqrt{2}\delta\nabla\mu, 0)$ into the first inequality of (12), A natural consequence is supposed to be that the point $(-\sqrt{2}\delta\nabla\mu, 0)$ is definitely to the left of the left boundary of $\mathbb{E}$. However, we are surprised to find that it is possible for the point $(-\sqrt{2}\delta\nabla\mu, 0)$ to be within the region $\mathbb{E}$. That is, the point $(-\sqrt{2}\delta\nabla\mu, 0)$ is to the right of the left boundary of $\mathbb{E}$. This can happen if we set δ to be very small and the difference between ρ^- and ρ to be very large. The gray curve shown in Fig. 2.4 is the resulted boundary if we choose special values of δ and ρ. This event can lead to errors of location estimation, because a user definitely should be localized at $r-\delta$ is in fact localized at r.

The root cause of the phenomenon is that the choice of δ and ρ in a theoretical perspective may not comply with the real situation. In the real world, the temporal correlation in a small neighborhood with respect to the same AP should be varying smoothly. Consequently, if δ is small, the difference between ρ^- and ρ is supposed to be insignificant.

We now compare localization results yielded by considering and ignoring the temporal correlation of the RSS. Recall the study in [116] ignores the temporal correlation of the RSS. The region $\mathbb{E}$ in this case is the region between the two dashed lines. Consider shadowed areas B covered with solid lines. If the user's reported fingerprints fall into such areas, it means that the user supposed to be localized at r is mistakenly localized at $r - \delta$, or the user supposed to be localized at $r + \delta$ is mistakenly localized at r. Similarly, consider the gray areas A. If the user's reported fingerprints fall into such areas, it means that the user supposed to be localized at $r-\delta$ is mistakenly localized at r, or the user supposed to be localized at r is mistakenly localized at $r + \delta$. That is, considering temporal correlation can improve the accuracy of location estimation by providing more accurate criteria for making judgement.

2.6.3 Asymptotic Equivalent Region of $\mathbb{E}$ in High-Dimensional Scenarios

We extend our analysis to scenarios of high-dimensional temporal correlation, sample space and physical space in our conference version [136]; however, it is difficult to obtain the exact information about the shape of $\mathbb{E}$ due to the high dimension and complicated interrelation between the physical space and sample space. In this section, we try to find a closed-form description of $\mathbb{E}'$ that is asymptotically equivalent to $\mathbb{E}$, so that a quantified understanding of $\mathbb{E}$ could be provided.

2.6.3.1 Approximate Matrix

The covariance matrix of the RSS at location $\mathbf{r}$ is denoted by $\Sigma_m(\mathbf{r})$. For the correlation coefficients, it is reasonable that

$$\rho_i(\mathbf{r}') \approx \rho_i(\mathbf{r}) + \nabla\rho_i(\mathbf{r})(\mathbf{r}' - \mathbf{r}), \tag{2.150}$$

because the correlation of the RSS is continuous, which has been verified in our experiments to be presented in [285]. Then we have Eq. (2.151).

$$\Sigma_m(\mathbf{r}') = \Sigma_m(\mathbf{r}) + \delta\cos\theta \begin{bmatrix} |\nabla\rho_0(\mathbf{r})| & |\nabla\rho_1(\mathbf{r})| & |\nabla\rho_2(\mathbf{r})| & \cdots & |\nabla\rho_{m-1}(\mathbf{r})| \\ |\nabla\rho_1(\mathbf{r})| & |\nabla\rho_0(\mathbf{r})| & |\nabla\rho_1(\mathbf{r})| & \ddots & |\nabla\rho_{m-2}(\mathbf{r})| \\ |\nabla\rho_2(\mathbf{r})| & |\nabla\rho_1(\mathbf{r})| & \ddots & \ddots & \vdots \\ \vdots & \ddots & \ddots & \ddots & |\nabla\rho_1(\mathbf{r})| \\ |\nabla\rho_{m-1}(\mathbf{r})| & \cdots & |\nabla\rho_2(\mathbf{r})| & |\nabla\rho_1(\mathbf{r})| & |\nabla\rho_0(\mathbf{r})| \end{bmatrix}. \tag{2.151}$$

We propose to use matrix $\Upsilon_m(\mathbf{r})$ to approximate $\Sigma_m(\mathbf{r})$, as shown in Eq. (2.152).

$$\Upsilon_m(\mathbf{r}') = \begin{bmatrix} \rho_0(\mathbf{r}) & \rho_1(\mathbf{r}) + \rho_{m-1}(\mathbf{r}) & \rho_2(\mathbf{r}) + \rho_{m-2}(\mathbf{r}) & \cdots & \rho_1(\mathbf{r}) + \rho_{m-1}(\mathbf{r}) \\ \rho_1(\mathbf{r}) + \rho_{m-1}(\mathbf{r}) & \rho_0(\mathbf{r}) & \rho_1(\mathbf{r}) + \rho_{m-1}(\mathbf{r}) & \ddots & \rho_2(\mathbf{r}) + \rho_{m-2}(\mathbf{r}) \\ \rho_2(\mathbf{r}) + \rho_{m-2}(\mathbf{r}) & \rho_1(\mathbf{r}) + \rho_{m-1}(\mathbf{r}) & \rho_0(\mathbf{r}) & \ddots & \vdots \\ \vdots & & \ddots & \ddots & \rho_1(\mathbf{r}) + \rho_{m-1}(\mathbf{r}) \\ \rho_1(\mathbf{r}) + \rho_{m-1}(\mathbf{r}) & \rho_2(\mathbf{r}) + \rho_{m-2}(\mathbf{r}) & \cdots & \rho_1(\mathbf{r}) + \rho_{m-1}(\mathbf{r}) & \rho_0(\mathbf{r}) \end{bmatrix}. \tag{2.152}$$

Suppose that the eigenvector of $\Upsilon_m(\mathbf{r})$ is $u = [u_1\ u_2 \cdots u_m]$, which satisfies that $\Upsilon_m(\mathbf{r})u = \tau(\mathbf{r})u$. It is easy to verify that vector

$$u(k) = [1, e^{i2\pi k/m}, e^{i2\pi 2k/m}, \ldots e^{i2\pi(m-1)k/m}](1 \le k \le m)$$

satisfies the equations and the corresponding eigenvalue is

$$\tau_{m,k}(\mathbf{r}) = \rho_0(\mathbf{r}) + \sum_{j=1}^{m-1}(\rho_i(\mathbf{r}) + \rho_{m-i}(\mathbf{r}))e^{i2\pi jk/m}.$$

We note that $\tau_{m,k}(\mathbf{r})$ is a real number since the coefficients of complex conjugate pairs $e^{i2\pi jk/m}$ and $e^{i2\pi j(m-k)/m}$ are identical. For the eigenvalues at location $\mathbf{r}'$, we have

$$
\begin{aligned}
\tau_{m,k}(\mathbf{r}') &= \rho_0(\mathbf{r}') + \sum_{j=1}^{m-1} (\rho_i(\mathbf{r}') + \rho_{m-i}(\mathbf{r}'))e^{i2\pi jk/m} \\
&= \tau_{m,k}(\mathbf{r}) + \delta \cos\theta \sum_{j=1}^{m-1} (|\nabla\rho_i(\mathbf{r})| + |\nabla\rho_{m-i}(\mathbf{r})|)e^{i2\pi kj/m}.
\end{aligned}
$$

To simplify notations, we use $\Delta\tau_{m,k}$ to represent $\sum_{j=1}^{m-1} (|\nabla\rho_i(\mathbf{r})| + |\nabla\rho_{m-i}(\mathbf{r})|)$ $e^{i2\pi kj/m}$, thus $\tau_{m,k}(\mathbf{r}') = \tau_{m,k}(\mathbf{r}) + \delta\cos\theta\,\Delta\tau_{m,k}$.

We are to prove that using $\tau_{m,k}(\mathbf{r})$ to approximate eigenvalues of $\Sigma_m(\mathbf{r})$ could incur the approximation error converging to zero as m goes to infinity.

2.6.3.2 Asymptotical Equivalence Analysis

We take another form of region $\mathbb{E}$ as follows to facilitate understanding:

$$
\sum_{i=1}^{m} \frac{{y_i}^2}{\lambda_{m,i}(\mathbf{r})} - \left[\frac{(y_1 + \sqrt{2}\delta\nabla\cos\theta)^2}{\lambda_{m,1}(\mathbf{r}')} + \sum_{i=1}^{m} \frac{{y_i}^2}{\lambda_{m,i}(\mathbf{r}')}\right] \le \ln\frac{|\Sigma(\mathbf{r})|}{|\Sigma(\mathbf{r}')|},
$$

where $\theta \in [0, \pi]$. If we use eigenvalues of matrix $\Upsilon_m(\mathbf{r}')$ to replace $\lambda_i(\mathbf{r}')$, we could obtain the region $\mathbb{E}'$ characterized by the following equation:

$$
\sum_{i=1}^{m} \frac{{y_i}^2}{\tau_{n,i}(\mathbf{r})} - \left[\frac{(y_1 + \sqrt{2}\delta\nabla\cos\theta)^2}{\tau_{m,1}(\mathbf{r}')} + \sum_{i=1}^{m} \frac{{y_i}^2}{\tau_{m,i}(\mathbf{r}')}\right] \le \ln\frac{|\Upsilon_m(\mathbf{r})|}{|\Upsilon_m(\mathbf{r}')|}.
$$

Lemma 4 $\lim_{m\to\infty} |\Upsilon_m(\mathbf{r}) - \Sigma(\mathbf{r})|^2 = 0$, *where* $|\cdot|^2$ *represents the* Hilbert–Schmidt Norm.

Proof 26

$$
\Upsilon_m(\mathbf{r}) - \Sigma(\mathbf{r}) =
\begin{bmatrix}
0 & \rho_{m-1}(\mathbf{r}) & \rho_{m-2}(\mathbf{r}) & \cdots & \rho_1(\mathbf{r}) \\
\rho_{m-1}(\mathbf{r}) & 0 & \rho_{m-1}(\mathbf{r}) & \ddots & \rho_2(\mathbf{r}) \\
\rho_{m-2}(\mathbf{r}) & \rho_{m-1}(\mathbf{r}) & 0 & \ddots & \vdots \\
\vdots & & \ddots & \ddots & \rho_{m-1}(\mathbf{r}) \\
\rho_1(\mathbf{r}) & \rho_2(\mathbf{r}) & \cdots & \rho_{m-1}(\mathbf{r}) & 0
\end{bmatrix}.
$$

With the definition of *Hilbert–Schmidt Norm*,

$$|\Sigma(\mathbf{r}) - \Upsilon_n(\mathbf{r})|^2 = 2\sum_{i=1}^{n-1} \tfrac{i}{n}\rho_i^2(\mathbf{r}),$$

we will show that

$$\lim_{m\to\infty} \sum_{i=1}^{m-1} \tfrac{i}{m}\rho_i^2(\mathbf{r}) = 0.$$

By applying *Abel Transformation* to the equation above, we obtain that

$$\sum_{i=1}^{m-1} \tfrac{i}{m}\rho_i^2(\mathbf{r}) = \frac{m-1}{m}A_{m-1} - \sum_{i=1}^{m-2} \frac{1}{m}A_i,$$

where $A_i = \sum_{j=1}^{i} \rho_j^2(\mathbf{r})$. Since the covariances are absolutely summable with $\sum_i \rho_i < \infty$, we use A to denote the supremum of $\sum \rho_j^2(\mathbf{r})$, i.e., $A = \sup \sum {\rho_j}^2(\mathbf{r})$. Consequently, we can find N such that $A > A_k > A - \varepsilon$ holds for all the $k \geq N$ for any $\varepsilon > 0$. Hence

$$A > \sum_{i=1}^{m-2} \frac{1}{m}A_i = \frac{1}{m}\left(\sum_{i=1}^{N-1} A_i + \sum_{i=N}^{m-2} A_i\right) > \frac{m-N-1}{m}(A-\varepsilon).$$

Notice that

$$\lim_{m\to\infty} \frac{m-N-1}{m}(A-\varepsilon) = A - \varepsilon$$

and thus have proven that for any $\varepsilon > 0$, $\lim_{m\to\infty} \sum_{i=1}^{m-2} \frac{1}{m}A_i > A - \varepsilon$.

Combined with the fact that $\sum_{i=1}^{m-2} \frac{1}{m}A_i < A$, we can conclude that

$$\lim_{m\to\infty} \sum_{i=1}^{m-2} \frac{1}{m}A_i = A = \lim_{n\to\infty} \frac{m-1}{m}A_{m-1}.$$

The proof is completed.

Lemma 5 *(Wielandt–Hoffman Theorem [142]) Given two Hermitian matrices A and B with eigenvalues α_k and β_k, respectively, and α_k and β_k are ordered such that $\sum_k |\alpha_k - \beta_k|^2$ is minimized, then*

$$\frac{1}{m}\sum_{i=1}^{m}|\alpha_k - \beta_k|^2 \leq |A - B|^2.$$

We present the lemma for the purpose of self-completeness. Without loss of generality, we can assume that $\tau_{m,k}(\mathbf{r})$ and $\lambda_{m,k}(\mathbf{r})$ are in the same order, and otherwise we could rotate the coordinate system to guarantee this property.

Lemma 6 *For any given integer s, we have*

$$\lim_{m\to\infty}\frac{1}{m}\sum_{k=1}^{m}\left(\tau^{s}_{m,k}(\mathbf{r}) - \lambda^{s}_{m,k}(\mathbf{r})\right) = 0.$$

Proof 27 Note that

$$\begin{aligned}
&\frac{1}{m}\sum_{k=0}^{m-1}\left|\tau^{s}_{m,k}(\mathbf{r}) - \lambda^{s}_{m,k}(\mathbf{r})\right| \\
&\frac{1}{m}\sum_{k=0}^{m-1}\left|\tau_{m,k}(\mathbf{r}) - \lambda_{m,k}(\mathbf{r})\right|\left|\sum_{k=0}^{m-1}\tau^{k}_{m,k}(\mathbf{r})\lambda^{s-k-1}_{m,k}(\mathbf{r})\right| \\
&\leq \frac{s\Lambda^{s-1}}{n}\sum_{k=0}^{m-1}\left|\tau_{m,k}(\mathbf{r}) - \lambda_{m,k}(\mathbf{r})\right| \\
&\leq s\Lambda^{s-1}\sqrt{\frac{1}{m}\sum_{k=1}^{m}\left|\tau_{m,k}(\mathbf{r}) - \lambda_{m,k}(\mathbf{r})\right|^2} \\
&\leq s\Lambda^{s-1}\,|A - B|\,,
\end{aligned}$$

where Λ represents the upper bound of the eigenvalues, and obviously s and Λ are constants with respect to m. The penult inequality is based on the Cauchy–Schwarz Inequality and the last inequality is based on Lemma 2. Then the proof is completed.

With Weierstrass' theorem, we know that there exists a sequence of polynomials $P_t(x)$ such that $\lim\limits_{t\to\infty} P_t(x) = \frac{1}{x}$. For every fixed t, we know that $\lim\limits_{m\to\infty}\frac{1}{m}\sum\limits_{i=1}^{m}\left|P_t(\tau_{m,i}(\vec{r})) - P_t(\lambda_{m,i}(\vec{r}))\right| = 0$ according to Lemma 3. Hence combining the two equations above, we can obtain that

$$\lim_{m\to\infty}\frac{1}{m}\sum_{i=1}^{m}\left|\frac{1}{\tau_{m,i}} - \frac{1}{\lambda_{m,i}}\right| = 0.$$

Theorem 7 *The region* $\mathbb{E}'$ *is asymptotical equivalent to region* $\mathbb{E}$*:*

$$(1):\ \lim_{m\to\infty}\frac{1}{m}\left\{\sum_{i=1}^{m}\left[\frac{{y_i}^2}{\lambda_{m,i}(\mathbf{r})}-\frac{{y_i}^2}{\tau_{m,i}(\mathbf{r})}\right]-\sum_{i=1}^{m}\left[\frac{{y_i}^2}{\lambda_{m,i}(\mathbf{r}')}-\frac{{y_i}^2}{\tau_{m,i}(\mathbf{r}')}\right]\right.$$
$$\left.-\left[\frac{(y_1+\sqrt{2}\delta\nabla\cos\theta)^2}{\lambda_{m,1}(\mathbf{r}')}-\frac{(y_1+\sqrt{2}\delta\nabla\cos\theta)^2}{\tau_{m,1}(\mathbf{r}')}\right]\right\}=0;$$
$$(2):\ \lim_{m\to\infty}\left\{\ln\frac{|\Sigma(\mathbf{r})|^{\frac{1}{m}}}{|\Sigma(\mathbf{r}')|^{\frac{1}{m}}}-\ln\frac{|\Upsilon_m(\mathbf{r})|^{\frac{1}{m}}}{|\Upsilon_m(\mathbf{r}')|^{\frac{1}{m}}}\right\}=0.$$

Proof 28 The RSS observed at each location is bounded, hence y_i^2s and $(y_1+\sqrt{2}\delta\nabla\cos\theta)^2$ are bounded. Suppose that their upper bound is M, then

$$\frac{1}{m}\left|\begin{array}{l}\sum_{i=1}^{m}\left[\frac{{y_i}^2}{\lambda_{m,i}(\mathbf{r})}-\frac{{y_i}^2}{\tau_{m,i}(\mathbf{r})}\right]-\sum_{i=2}^{m}\left[\frac{{y_i}^2}{\lambda_{m,i}(\mathbf{r}')}-\frac{{y_i}^2}{\tau_{m,i}(\mathbf{r}')}\right]\\ -\left[\frac{(y_1+\sqrt{2}\delta\nabla\cos\theta)^2}{\lambda_{m,1}(\mathbf{r}')}-\frac{(y_1+\sqrt{2}\delta\nabla\cos\theta)^2}{\tau_{m,1}(\mathbf{r}')}\right]\end{array}\right|$$
$$\le\frac{2M}{m}\sum_{i=1}^{m}\left|\frac{1}{\tau_{m,k}(\mathbf{r})}-\frac{1}{\lambda_{m,k}(\mathbf{r})}\right|.$$

Combined with equation above Theorem 1, we know that the first part of theorem holds. The proof of the second part is similar. By using Weierstrass' theorem again, we know that there exists a sequence of polynomials $Q_t(\tau_{m,k}(\mathbf{r}))$ such that $\lim_{t\to\infty}P_t(\tau_{m,k}(\mathbf{r}))=\ln\tau_{m,k}(\mathbf{r})$. Combined with Lemma 3, we have

$$\lim_{m\to\infty}\frac{1}{m}\sum_{i=1}^{m}\left|\ln\tau_{m,k}(\mathbf{r})-\ln\lambda_{m,k}(\mathbf{r})\right|=0.$$

Notice that

$$\frac{1}{m}\sum_{i=1}^{m}\left|\ln\tau_{m,k}(\mathbf{r})-\ln\lambda_{m,k}(\mathbf{r})\right|\ge\frac{1}{m}\left|\ln\frac{|\Sigma(\mathbf{r})|^{\frac{1}{m}}}{|\Upsilon_m(\mathbf{r})|^{\frac{1}{m}}}\right|,$$

which means that $\lim_{m\to\infty}\frac{1}{m}\left|\ln|\Sigma(\mathbf{r})|^{\frac{1}{m}}-\ln|\Upsilon_m(\mathbf{r})|^{\frac{1}{m}}\right|=0$. With the same virtue, $\lim_{m\to\infty}\frac{1}{m}\left|\ln|\Sigma(\mathbf{r}')|^{\frac{1}{m}}-\ln|\Upsilon_m(\mathbf{r}')|^{\frac{1}{m}}\right|=0$. Combining these two equations, the second part of this theorem is proved.

Figure 2.40 provides the results of a brief numerical analysis of equation (1) and (2) in Theorem 1. It can be seen that the two equations are being satisfied as the value of m increases.

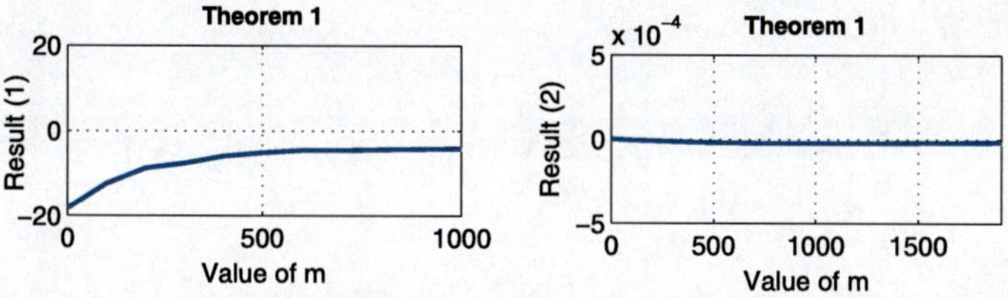

(a) (b)

Fig. 2.40 Convergence rate. (**a**) Numerical analysis of result (1). (**b**) Numerical analysis of result (2)

Theorem 8 *For any* $\varepsilon > 0$*, when* $m > \max\{\frac{4N(\frac{\varepsilon\Delta^2}{4M})\frac{M}{\Delta^2}A}{\varepsilon}, \frac{N(\frac{\varepsilon\Delta}{2})}{\Delta\varepsilon}\}$*, the difference of the expressions of these two regions is at most* ε*, where* $N(\varepsilon)$ *is the convergence rate of the series* $\sum_{i=1}^{\infty}\rho_i^2$*, that is,* $\forall k > \varepsilon$*,* $\sum_{i=1}^{k}\rho_i^2 > A - \frac{\varepsilon}{2}$*.*

Proof 29 Since $\sum_{i=1}^{\infty}\rho_i^2$ converges to A, then $\forall\varepsilon > 0$, there exists an integer $N(\varepsilon) > 0$, such that $\forall k > N(\varepsilon)$, $A \le A_k = \sum_{i=1}^{k}\rho_i^2 > A - \frac{\varepsilon}{2}$. Then we have

$$\sum_{i=1}^{m-1}\frac{i}{m}\rho_i^2(\vec{r}) = A_{m-1} - \sum_{i=1}^{m-1}\frac{A_i}{m} < A - \sum_{i=N(\varepsilon)}^{m-1}\frac{A_i}{m}$$

$$< A - (m - N(\varepsilon))\frac{A - \frac{\varepsilon}{2}}{m} = \frac{\varepsilon}{2} + \frac{N(\varepsilon)}{m}\left(A - \frac{\varepsilon}{2}\right).$$

And if $m > \frac{4N(\frac{\varepsilon\Delta^2}{4M})\frac{M}{\Delta^2}A}{\varepsilon}$, we have

$$\begin{aligned}
&\frac{1}{m}\left|\begin{array}{l}\sum_{i=1}^{m}\left[\frac{{y_i}^2}{\lambda_{m,i}(\mathbf{r})} - \frac{{y_i}^2}{\tau_{m,i}(\mathbf{r})}\right] - \sum_{i=2}^{m}\left[\frac{{y_i}^2}{\lambda_{m,i}(\mathbf{r}')} - \frac{{y_i}^2}{\tau_{m,i}(\mathbf{r}')}\right] \\ -\left[\frac{(y_1+\sqrt{2}\delta\nabla\cos\theta)^2}{\lambda_{m,1}(\mathbf{r}')} - \frac{(y_1+\sqrt{2}\delta\nabla\cos\theta)^2}{\tau_{m,1}(\mathbf{r}')}\right]\end{array}\right| \\
&\le \frac{2M}{m}\sum_{i=1}^{m}\left|\frac{1}{\tau_{m,k}(\mathbf{r})} - \frac{1}{\lambda_{m,k}(\mathbf{r})}\right| \le \frac{2M}{\Delta^2}\sum_{i=1}^{m-1}\frac{i}{m}\rho_i^2(\vec{r}) \\
&< \frac{2M}{\Delta^2}\left(A - \left(m - N\left(\frac{\varepsilon\Delta^2}{4M}\right)\right)\frac{A - \frac{\varepsilon\Delta^2}{4M}}{m}\right) \\
&= \frac{\varepsilon}{2} + \frac{N\left(\frac{\varepsilon\Delta^2}{4M}\right)}{m}\left(\frac{2M}{\Delta^2}A - \frac{\varepsilon}{2}\right) \\
&< \frac{\varepsilon}{2} + \frac{N\left(\frac{\varepsilon\Delta^2}{4M}\right)}{\frac{4N\left(\frac{\varepsilon\Delta^2}{4M}\right)\frac{M}{\Delta^2}A}{\varepsilon}}\left(\frac{2M}{\Delta^2}A - \frac{\varepsilon}{2}\right) \\
&< \frac{\varepsilon}{2} + \frac{\varepsilon}{2} = \varepsilon.
\end{aligned}$$

Moreover, if $m > \frac{N(\frac{\varepsilon\Delta}{2})}{\Delta\varepsilon}$, we have

$$\begin{aligned}\frac{1}{m}\left|\ln\frac{|\Sigma(\mathbf{r})|^{\frac{1}{m}}}{|\Upsilon_m(\mathbf{r})|^{\frac{1}{m}}}\right| &\leq \frac{1}{m}\sum_{i=1}^{m}\left|\ln\tau_{m,k}(\mathbf{r}) - \ln\lambda_{m,k}(\mathbf{r})\right| \\ &\leq \frac{1}{\Delta}\sum_{i=1}^{m-1}\frac{i}{m}\rho_i^2(\vec{r}) \leq \varepsilon.\end{aligned}$$

Consequently, when $m > \max\{\frac{4N(\frac{\varepsilon\Delta^2}{4M})\frac{M}{\Delta^2}A}{\varepsilon}, \frac{N(\frac{\varepsilon\Delta}{2})}{\Delta\varepsilon}\}$, the difference of the expressions of these two regions is at most ε.

2.6.3.3 Boundaries of Region $\mathbb{E}'$

We define the *Fourier Transformation* of the covariance series as

$$g(\omega,\mathbf{r}) = \sum_{j=-\infty}^{\infty}\rho_j(\mathbf{r})e^{i2\pi\omega j}, -\frac{1}{2} < \omega \leq \frac{1}{2}. \tag{2.153}$$

According to *Szego's theorem* [141], for an arbitrary continuous function G, we have

$$\lim_{m\to\infty}\frac{1}{m}\sum_{k=0}^{m-1}G(\tau_{m,k}(\mathbf{r})) = \int_{-1/2}^{1/2}G(g(\omega),\mathbf{r})d\omega. \tag{2.154}$$

Let $G(x) = \ln x$, and we can obtain the approximate expression of the determinant of Σ_m:

$$\ln|\Sigma_m(\mathbf{r})|^{1/m} = \frac{1}{m}\sum\ln\tau_{m,k}(\mathbf{r}) \approx \int_{-1/2}^{1/2}\ln g(\omega,\mathbf{r})d\omega. \tag{2.155}$$

For the matrix $\Sigma_m(\mathbf{r}')$, the corresponding Fourier Transformation is

$$\begin{aligned}g(\omega,\mathbf{r}') &= \sum_{j=-\infty}^{\infty}\rho_j(\mathbf{r}')e^{i2\pi\omega j} \\ &= g(\omega,\mathbf{r}) + \delta\cos\theta\sum_{j=-\infty}^{\infty}|\nabla\rho_j(\mathbf{r})|e^{i2\pi\omega j}, \\ &-\frac{1}{2} < \omega \leq \frac{1}{2}.\end{aligned} \tag{2.156}$$

Similarly, we can use

$$m \int_{-1/2}^{1/2} \ln \left[g(\omega, \mathbf{r}) + \delta \cos\theta \sum_{j=-\infty}^{\infty} |\nabla \rho_j(\mathbf{r})| e^{i2\pi\omega j} \right] d\omega \tag{2.157}$$

to estimate the term $\ln |\Sigma_m(\mathbf{r}')|$.

Notice that

$$\frac{\partial \ln \frac{|\Sigma_m(\mathbf{r})|^{\frac{1}{m}}}{|\Sigma_m(\mathbf{r}')|^{\frac{1}{m}}}}{\partial\theta} = \frac{\delta \sin\theta \sum_{j=-\infty}^{\infty} |\nabla \rho_j(\mathbf{r})| e^{i2\pi\omega j}}{g(\omega, \mathbf{r}) + \delta \cos\theta \sum_{j=-\infty}^{\infty} |\nabla \rho_j(\mathbf{r})| e^{i2\pi\omega j}} = O(\delta \sin\theta). \tag{2.158}$$

Since δ is a bounded real number and $|\sin\theta| \le 1$,

$$\frac{d \ln \frac{|\Sigma_m(\mathbf{r})|^{\frac{1}{m}}}{|\Sigma_m(\mathbf{r}')|^{\frac{1}{m}}}}{d\theta} \approx 0. \tag{2.159}$$

Moreover, $\partial \frac{[\frac{(y_1+\sqrt{2}\delta\nabla\cos\theta)^2}{\lambda_i(\mathbf{r}')} + \sum_{i=1}^{m} \frac{y_i{}^2}{\lambda_i(\mathbf{r}')}]}{\partial\theta} = \sin\theta[h(\cos\theta) - c]$, where $h(\cdot)$ is a monotone function with respect to $\cos\theta$ and c is a positive number. Note that $\theta \in [0, \pi]$, and thus there is at most one root of function $\sin\theta[h(\cos\theta) - c]$ in $[0, \pi]$. We use θ^* to denote this root; therefore, the minimum value of function $\frac{(y_1+\sqrt{2}\delta\nabla\cos\theta)^2}{\lambda_i(\mathbf{r}')} + \sum_{i=1}^{n} \frac{y_i{}^2}{\lambda_i(\mathbf{r}')}$ is achieved under these three cases: $\cos\theta = -1$, $\cos\theta = 1$, and $\cos\theta = \cos\theta^*$.

Consequently, the boundaries of the region $\mathbb{E}'$ can be described by the following three hypersurfaces:

$$F_1 : \sum_{t=1}^{n} \frac{y_i{}^2}{\tau_{n,t}(\mathbf{r})} - \left[\frac{(y_1 - \sqrt{2}\delta\nabla)^2}{\tau_{n,1}(\mathbf{r}) - \delta\Delta\tau_{n,1}} + \sum_{t=2}^{n} \frac{y_i{}^2}{\tau_{n,t}(\mathbf{r}) - \delta\Delta\tau_{n,t}} \right] \le \ln \frac{|\Sigma(\mathbf{r})|}{|\Sigma(\mathbf{r}')|}; \tag{2.160}$$

$$F_2 : \sum_{t=1}^{n} \frac{y_i{}^2}{\tau_{n,t}(\mathbf{r})} - \left[\frac{(y_1 + \sqrt{2}\delta\nabla)^2}{\tau_{n,1}(\mathbf{r}) + \delta\Delta\tau_{n,1}} + \sum_{t=2}^{n} \frac{y_i{}^2}{\tau_{n,t}(\mathbf{r}) + \delta\Delta\tau_{n,t}} \right] \le \ln \frac{|\Sigma(\mathbf{r})|}{|\Sigma(\mathbf{r}')|}; \tag{2.161}$$

$$F_3 : \sum_{t=1}^{n} \frac{y_i{}^2}{\tau_{n,t}(\mathbf{r})} - \left[\frac{(y_1 + \sqrt{2}\delta\cos\theta^*\nabla)^2}{\tau_{n,1}(\mathbf{r}) + \delta\cos\theta^*\Delta\tau_{n,1}} + \sum_{t=2}^{n} \frac{y_i{}^2}{\tau_{n,t}(\mathbf{r}) + \delta\cos\theta^*\Delta\tau_{n,t}} \right] \le \ln \frac{|\Sigma(\mathbf{r})|}{|\Sigma(\mathbf{r}')|}. \tag{2.162}$$

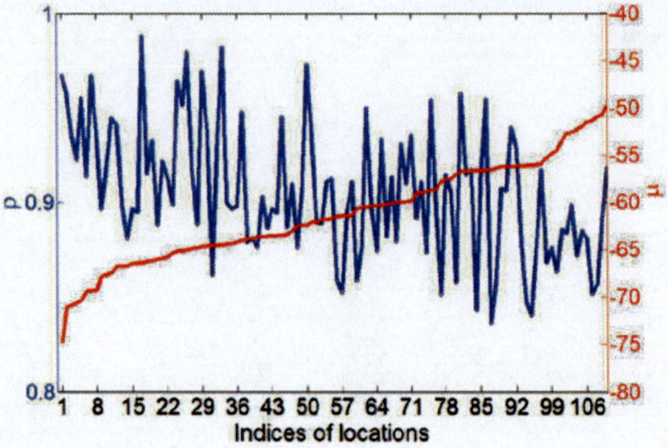

Fig. 2.41 Comparison of ρ and μ

2.6.4 Location Estimation Facilitated by Temporal Correlation

2.6.4.1 Feasibility of Utilizing Temporal Correlation

We first investigate whether location estimation could be facilitated with the information of temporal correlation in practice. We conduct experiments in a square indoor space that is around 40 m^2. The square is divided into grids with the edge length of 60 cm. Figure 2.41 associates the values of μ and ρ with locations, where the horizontal axis represents the indices of locations. We sort the locations according to the locally sampled values of μ, and the corresponding values of ρ are also plotted. It can be seen that the values of ρ present distinct pattern compared with μ in each location. In fact, we calculate the cross correlation between μ and ρ, and the results indicate that the two values are not related to each other. The experiment results illustrated above indicate that the temporal correlation information of the RSS provides a diversified feature compared with the information of μ. The temporal correlation could be used to cross check the result of location estimation. It is worth mentioning that the values of ρ can be derived from the sampled RSSes and thus incur no extra overhead in the training phase. The spatial distribution of μ and ρ is illustrated and analyzed, which could be found in our technical report [194].

Algorithm 4 Temporal correlation for location estimation

Input parameters:
The training data set for each location **r**, x_{ij} $(i = 1 \ldots n; j = 1 \ldots w)$;
The reported RSS sequence t_{ij} $(i = 1 \ldots a; j = 1 \ldots b)$ from a user;
Indoor space L is a set of all the identified locations recorded in the database;
Threshold Th is the critical value of choice for mean vectors.

1. For each location i in L, calculate the mean vector $\boldsymbol{\mu_i} = (\mu_1, \mu_2, \ldots, \mu_m)$, and the correlation vector $\boldsymbol{\rho_i} = (\rho_1, \rho_2, \ldots, \rho_m)$.
2. For the reported data t_{ij} from a user, also calculate the target mean and correlation vectors as μ_t, ρ_t.
3. Find the Euclidean distance between $\boldsymbol{\mu_t}$ and every $\boldsymbol{\mu_i}$. Find all the vectors $\boldsymbol{\mu_k}$ among those $\boldsymbol{\mu_i}$s, and the distance between each $\boldsymbol{\mu_k}$ and $\boldsymbol{\mu_t}$ should be within Th in the sample space, i.e., $|\boldsymbol{\mu_i} - \boldsymbol{\mu_t})| <= Th$. The corresponding locations associated with those $\boldsymbol{\mu_k}$s are denoted using a set $\{\mathbf{l_{kmin}}\}$.
4. Compare the Euclidean distance between ρ_t and ρ_i in $\{\mathbf{l_{kmin}}\}$. Find the vector that is nearest to ρ_t. The nearest distance in correlation sample space is the place we localize the user at.

2.6.5 *Localization Estimation Algorithm*

The basic idea of the algorithm is that we first find a list of candidate locations of the user with the mean value comparison as most of the work in the literature does and then find the most likely location with the temporal correlation comparison. Algorithm 1 illustrates how to utilize temporal correlation for better location estimation, which is in essence a synthetic approach integrating the information of both the mean value and the temporal correlation of the RSS. Detailed methods for finding parameters μ and ρ could be found in our technical report [194].

2.6.5.1 Choice of Design Parameters

Design parameters must be chosen before the algorithm presented above can be appropriately utilized. We conduct experiments to show how parameters such as the number of temporal dimension, the sampling interval and number of samplings could influence the value of ρ in a location. A distinguishable value of ρ in a location is favored in the localization process.

We calculate the temporal correlation coefficient ρ from two-dimension to 280-dimension. The results are shown in Fig. 2.42a. We can find that the value of ρ decreases as the number of dimensions increases and then keeps comparatively flat. The value of ρ is at around 0 when the number of dimensions goes beyond 100, which indicates that it is meaningless to consider over-high dimensions. The current value of RSS becomes almost irrelevant compared with the values of RSS sampled long time ago. We can also find that the number of dimensions 50 is a critical point, after which the trend of ρ becomes flat. Consequently, we recommend to utilize the

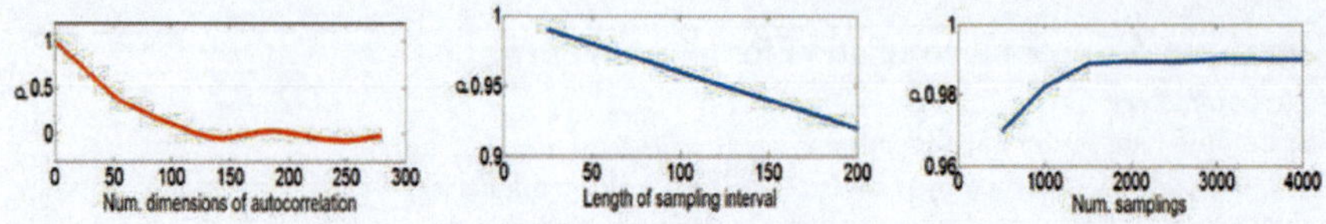

(a) (b) (c)

Fig. 2.42 Experiments for choice of design parameters. (**a**) Number of correlation dimensions. (**b**) Sampling interval. (**c**) Number of samplings

information the temporal correlation with the number of dimensions less than 50 in practice.

The effectiveness of the temporal correlation also depends on the length of the interval that sequential RSS samplings are performed. Figure 2.42b shows the corresponding values of ρ under different sampling intervals varying from 25 ms to 200 ms. It is straightforward that the values of ρ are decreasing as the length of the interval increases, because the longer the interval is the more unrelated the currently sampled RSS value is from the previous ones. However, over-short sampling interval is also unnecessary, since the RSS readings may not change that fast and the observed values of the RSS are basically the same in this case. We choose 100 ms as the length of sampling interval, which is proved to be appropriate in the experiments to be presented later.

After the number of dimensions and the sampling interval are determined, we still need to decide the starting point of the sampling in practice. This is because we could see variance of ρ at the first amount of sampling intervals, as shown in Fig. 2.42c. We need to retrieve those RSSes that yield comparatively stable temporal correlation information. To this end, we sample 2000 RSS values in our system, since the value of ρ tends to be stable after the 1500th sampling as shown in the sub-figure.

2.6.6 Experimental Results

We conduct local experiments and use trace data experiments to verify our model and show the performance of the proposed localization algorithm, where the floor plans are shown as in Fig. 2.43. Detailed descriptions of EVARILOS testbed and the Zwijnaarde room could be found in [112–114, 193]. We first provide experimental results to verify our modeling assumption and then show the influence of the temporal correlation property on the reliability and accuracy of location estimation.

Verification of the Modeling Assumption We did local experiments to verify the modeling assumption in the conference version [136]. We here use the trace data from EVARILOS testbed to further verify the Gaussian assumption. We first filter out those unreliable measurements in the trace data, where there is only 1

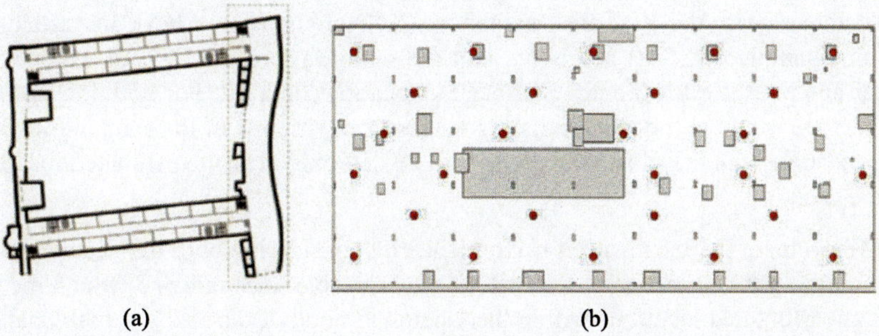

Fig. 2.43 Floor plans of test fields. (a) Local experiment field. (b) Room Zwijnaarde of EVARILOS testbed [113]

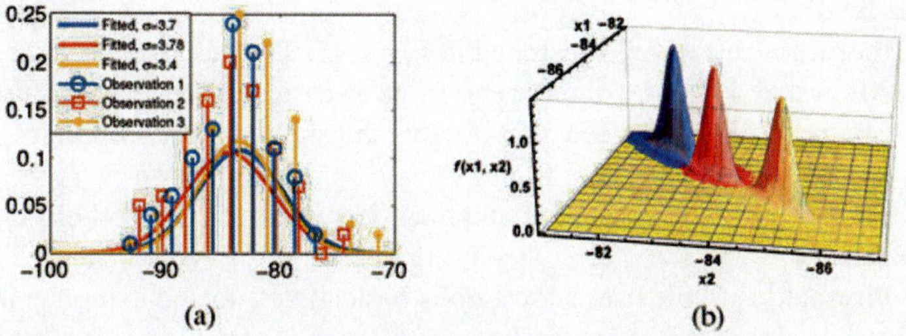

Fig. 2.44 Modeling approach. (a) Gaussian distribution. (b) Gaussian process

or 2 RSS readings recorded or all the RSS readings are exactly the same. We observe the RSS values in three locations (1290, 1980), (1290, 1270), and (1890, 1270), which are with respect to the AP located at (1000, 1712). We calculate the probability each value appears and find that they are approximately to be Gaussian distribution as shown in Fig. 2.44a, with skewness and kurtosis less than |0.5| and |3.3| respectively. The resulted joint PDFs are as shown in Fig. 2.44b. The experimental results support our modeling assumption that the signal could be modeled following Gaussian distribution.

Influence on Localization Reliability We conduct local experiments to examine whether the temporal correlation information of the RSS could improve the reliability of fingerprinting based localization in practice. We choose to use local experiments because the distance among reference points in the experiment can be controlled freely, which is important to examine the reliability improvement when the error tolerance metric is very small. In particular, we use the value of μ and ρ with respect to different APs at each location as the local fingerprints in the localization database. At each sampling location, we collect totally 4000 RSS fingerprints at the sampling interval of 25 ms with respect to each Wi-Fi AP available. In the localization phase, we sample totally 1500 fingerprints at the sampling interval of 25 ms, where the top 5 strongest fingerprints are sampled. We compare the mean value of the sampled fingerprints and those have been stored in the database with the K-nearest neighbors (KNNs) algorithm [105]. In particular, we find sampled RSSes' neighbors that are within a threshold Th in the sample space and put the corresponding associated locations on a list. For each location on the list, we examine the corresponding values of ρ with that of the sampled RSSes. The location with the most matched ρ will be determined as the estimated location of the user.

We evaluate the localization performance by considering both the accuracy and reliability. In the experiment, we first set the error tolerance radius δ, which means that any estimated locations within the δ neighborhood of the user's actual location can be regarded as a correct estimation. We set different radius and randomly pick up 500 estimations to evaluate the performance. With the results of localization, we could find the probability that the user's location is correctly estimated, which is termed as reliability.

The experiment results are illustrated in Fig. 2.45. The radius is set to be 30 cm, 60 cm, 120 cm and 180 cm, respectively. In each case, we increase the threshold Th from 0 to 5. Note that the unit of the threshold is not important, as we consider the normalized distance in the sample space. As shown in Fig. 2.45, the localization reliability increases first and then decreases as the threshold increases in all scenarios.

If the threshold is 0, the user's location is basically estimated using the mean of the RSSes; the information of the temporal correlation is not utilized. If the threshold increases, the system could cross check the candidate locations and find the most matched one; therefore, the reliability is improved. If the threshold is large enough, it means that more candidate locations could be on the list. Since those locations are picked up according to their corresponding fingerprints in the sample space, they may be far away from each other in the physical space. The observed temporal correlation information is unable to effectively tell one location from another. The more the candidate locations are, the higher probability that the location is far from the APs, and the temporal correlation becomes indistinguishable. That is why the reliability becomes worse if the threshold is too large. More detailed experiments data can be found in our technical report [194]. It is shown that the reliability of the fingerprinting localization could be improved by up to 13% if we choose the radius of 0.3 m and the threshold of 1.

Influence on Localization Accuracy With the trace data, we use the location estimation scheme as shown in Algorithm 1 and the regular mean comparison algorithm to perform localization, respectively, in order to evaluate the influence of temporal correlation information usage on the localization accuracy. We compare the two location estimation schemes under different AP selection schemes. With *StrongestAvg*, the user measures APs with the strongest average RSSes can be observed at the to-be-determined location in the online phase. With *StrongestMax*, the user measures APs with the strongest RSSes can be observed. With the *Similarity* strategy, APs are first clustered according to the similarity of their generated RSSes and the representative AP of each cluster is then selected. How to compute the

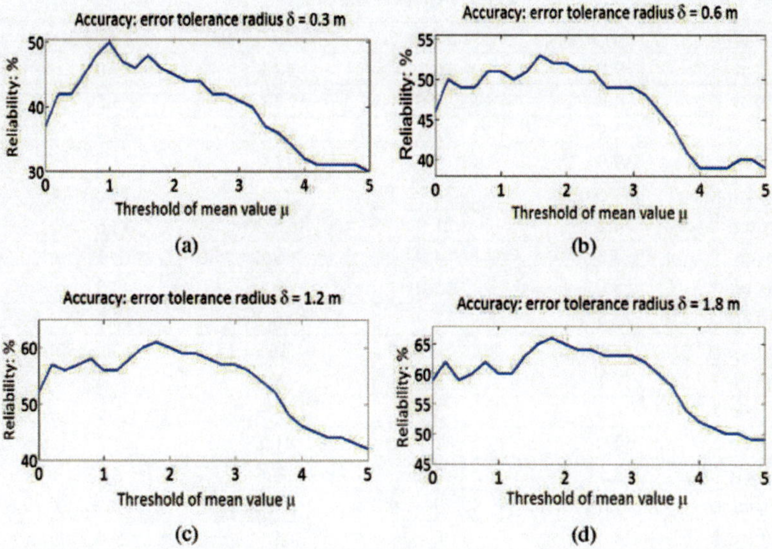

Fig. 2.45 Reliability with different threshold Th and error tolerance radius. (**a**) $\delta = 0.3m$. (**b**) $\delta = 0.6m$. (**c**) $\delta = 1.2m$. (**d**) $\delta = 1.8m$

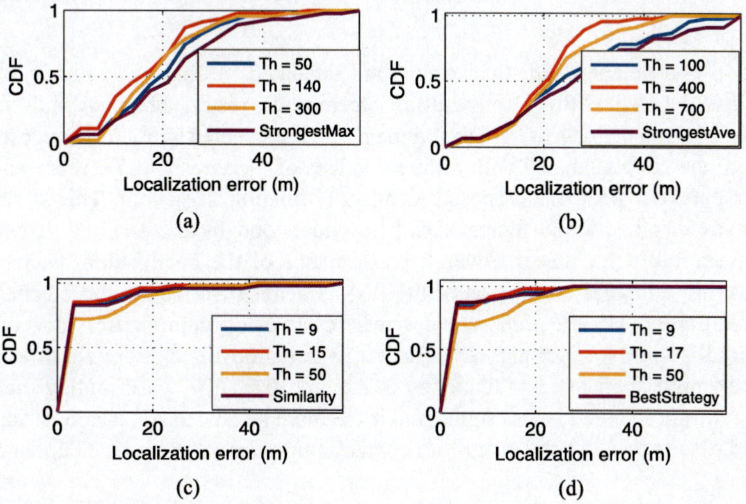

Fig. 2.46 CDF of localization errors with different AP selection schemes. (**a**) StrongestMax. (**b**) StrongestAve. (**c**) Similarity. (**d**) BestStrategy

similarity metric and how to select the representative AP in a cluster are described in [77]. The algorithm for the *BestStrategy* is shown in [111, 116].

After APs are selected, we could run algorithms for localization with and without taking temporal correlation into account. The experiment results are shown in

Table 2.4 Experimental results with EVARILOS trace data

Metric[m]	*StrgstMax*	$Th = 50$	$Th = 140$	$Th = 300$
Avg.err	19.0	16.1	12.8	15.0
Min.err	4.0	3.8	3.6	3.5
Max.err	53.9	43.4	43.4	45.7
Med.err	19.0	15.7	12.2	14.0
Metric[m]	*StrgstAvg*	$Th = 100$	$Th = 400$	$Th = 700$
Avg.err	24.1	21.9	16.3	18.6
Min.err	3.8	4.0	3.8	3.5
Max.err	55.0	54.0	40.2	40.2
Med.err	22.5	18.9	16.2	18.0
Metric[m]	*Similarity*	$Th = 9$	$Th = 15$	$Th = 50$
Avg.err	7.1	6.0	5.7	7.8
Min.err	3.5	3.8	3.6	3.5
Max.err	54.0	41.1	41.1	41.1
Med.err	3.5	3.8	3.6	3.5
Metric[m]	*BestStrategy*	$Th = 9$	$Th = 17$	$Th = 50$
Avg.err	6.4	6.4	5.8	9.5
Min.err	4.0	3.8	3.5	3.6
Max.err	28.1	28.1	28.1	31.5
Med.err	4.0	4.0	3.5	3.5

Fig. 2.46 and detailed statistical results are tabulated in Table 2.4. In Fig. 2.46a, the curve labeled with "StrongestMax" means the results obtained with mean comparison and the APs used in the online phase are selected with the StrongestMax method; the curves labeled with different values of the threshold Th represent the results obtained with the proposed location estimation algorithm. The meanings of curves in other 3 sub-figures could be understood by the analogy. It can be seen from Table 2.4 that the overall performance of the localization mechanism considering temporal correlation of the RSS is better than the regular mechanism under different AP selection schemes, where choosing appropriate value of the threshold is vitally important. In particular, when choosing appropriate threshold, the maximum average error reduction is around 30%. It is worth noting that the accuracy improvement is less significant if the more advanced AP selection scheme is used, this is because the room for accuracy improvement is reduced under such schemes.

Chapter 3
Theoretical Model for CSI Localization

3.1 System Model

3.1.1 Signal Model

We consider an indoor space installed with multiple CSI retrievable anchor points. As in the existing CSI based localization systems [86, 87, 180, 227], those anchor points can retrieve the CSI of wireless signals sent from the user's mobile device. Ideally, the anchor point should have perfect knowledge of the received waveform, based on which the user's location can be estimated. The received waveform can be modeled as

$$r(t) = \sum_{p=1}^{l} \alpha^{(p)} s(t - \tau^{(p)}) + z(t), \tag{3.1}$$

where $s(t)$ denotes the transmitting waveform, l represents the number of paths the signal experienced since transmitted, $\alpha^{(p)}$ is the amplitude of the waveform over the pth path, $\tau^{(p)}$ is the time delay of the signal over the pth path, and $z(t)$ is the additive white noise that can be regarded as a zero-mean Gaussian process with covariance σ^2. Note that $\sigma^2 = N_0/2$ and $N_0/2$ is the two-side power spectral density. The model above is a fundamental description of the received waveform in digital communication, which considers fading, multipath effect, propagation delay, and noise.

However, the Intel 5300 toolkit is unable to provide CSI of the perfect analog waveform. The CSI used for analysis is in fact reflected by the amplitude and phase information of the waveform, which is obtained after the analog signal goes through the ADC module according to basic principles of digital communication. After the ADC module, the analog waveform is sampled and quantized. Since most of the works [86, 87, 180, 227] use the phase information of the waveform

X. Tian et al., *Wireless Localization Techniques*, Wireless Networks,
https://doi.org/10.1007/978-3-031-21178-2_3

to derive the distance between the anchor point and the mobile device, and the quantization process has small influence on the amplitude of the waveform, the waveform modeled in Eq. (3.1) after the sampling process becomes

$$X_m = r(mT) = \sum_{p=1}^{l} \alpha^{(p)} s(mT - \tau^{(p)}) + z(mT), \tag{3.2}$$

with $m = 1, \ldots, L$, where L denotes the total number of samplings and T is the sampling period. Due to the additive Gaussian white noise, the sampled value from the waveform obeys the Gaussian distribution $X_m \sim N(\sum_{p=1}^{l} \alpha^{(p)} s(mT - \tau^{(p)}), \sigma^2)$, where σ^2 is the variance. After sampling, we obtain a series of Gaussian random variables, and the covariance between any two variables is equal to zero with the property of Gaussian process:

$$\begin{cases} \text{cov}(X_i, X_j) = E(z(iT) \cdot z(jT)) = 0 (i \neq j); \\ \text{cov}(X_i, X_j) = D(X_i) = D(X_j) = \frac{N_0}{2} (i = j). \end{cases} \tag{3.3}$$

3.1.2 Problem Formulation

The basic idea of the CSI localization approach is to derive geometric relationship between the target and antenna array equipped on the anchor point. In particular, AoA method tries to find the propagation direction of the received waveform, and ToF method tries to find the flying time of the received waveform incident on each antenna in the antenna array. Such information is basically derived from the phase information of the waveform [86, 87, 180, 227]. Efforts have also been made to utilize the amplitude information to further improve the accuracy [87]. Consequently, the performance of CSI localization depends on how accurately the amplitude ($\alpha^{(p)}$) and phase ($\tau^{(p)}$) information of the received waveform can be measured. Our strategy is to first find the error bound for measuring $\alpha^{(p)}$ and $\tau^{(p)}$ and then explore the corresponding error bound of location estimation.

In the location estimation process, the unknown parameter vector $\boldsymbol{\theta}$ can be formulated as

$$\boldsymbol{\theta} = [\tau^{(1)}, \alpha^{(1)}, \tau^{(2)}, \alpha^{(2)}, \ldots, \tau^{(l)}, \alpha^{(l)}]. \tag{3.4}$$

The basis for estimating the parameter vector is the sampled value of the received waveform, which is in fact a vector of random variables $\mathbf{X}$ as shown in Eq. (3.2). According to the Fisher information theory, the Fisher information matrix (FIM) can be calculated as

$$\mathbf{I}(\boldsymbol{\theta})_{i,j} = -\mathbf{E}\left[\frac{\partial^2}{\partial\theta_i \, \partial\theta_j} \log f(\mathbf{X}; \boldsymbol{\theta}) \middle| \boldsymbol{\theta} \right], \tag{3.5}$$

where $f(\mathbf{X}; \boldsymbol{\theta})$ is the probability density function (PDF) of the observed random variables and θ_i and θ_j are the unknown parameters to be estimated, respectively.

Then, the parameter vector $\boldsymbol{\theta}$ satisfies the information inequality:

$$\mathbf{E}[(\hat{\boldsymbol{\theta}} - \boldsymbol{\theta})(\hat{\boldsymbol{\theta}} - \boldsymbol{\theta})^{\mathrm{T}}] \succeq \mathbf{I}^{-1}(\boldsymbol{\theta}), \tag{3.6}$$

where $\mathbf{A} \succeq \mathbf{B}$ means that $\mathbf{A} - \mathbf{B}$ is positive semidefinite. The inverse of the FIM can be used to evaluate the estimation accuracy of the parameter $\boldsymbol{\theta}$, in particular the variance of the estimated value of the parameter compared with the true value, which is known as the CRB.

Since the PDF $f(\mathbf{X}; \boldsymbol{\theta})$ is following Gaussian distribution according to the signal model, the FIM in this case is for an N-variate Gaussian distributed random variable $\mathbf{X} \sim \mathcal{N}(\boldsymbol{\mu}(\boldsymbol{\theta}), \Sigma(\boldsymbol{\theta}))$, where $\boldsymbol{\theta}$ is a K-dimensional vector of parameters $[\theta_1, \ldots, \theta_K]^{\mathrm{T}}$, $\mathbf{X}$ is the vector of observable random normal variables $[X_1, \ldots, X_N]^{\mathrm{T}}$ with mean $\boldsymbol{\mu}(\boldsymbol{\theta}) = [\mu_1(\boldsymbol{\theta}), \ldots, \mu_N(\boldsymbol{\theta})]^{\mathrm{T}}$, and $\Sigma(\boldsymbol{\theta})$ denotes the covariance matrix of $\mathbf{X}$. The covariance matrix is a constant diagonal matrix because any two random variables are independent of each other. Then the element in the mth row and nth column of the FIM is

$$\mathbf{I}_{m,n} = \frac{\partial \boldsymbol{\mu}^{\mathrm{T}}}{\partial \theta_m} \Sigma^{-1} \frac{\partial \boldsymbol{\mu}}{\partial \theta_n}. \tag{3.7}$$

3.1.3 Challenges

Finding Inverse of FIM Based on the formulation above, we need to find the expression of each element of the FIM to find the CRB; however, the inverse of the FIM with elements as mentioned above cannot be found in practice, because the element of the FIM can be explicitly expressed only if the first-order derivative of the received waveform can be found. This is almost impossible in practice because the Intel 5300 NIC adopts orthogonal frequency-division multiplexing (OFDM) technique, which makes the expression of the signal $s(\cdot)$ unavailable. In particular, the signal is factually an aggregation of the source signal modulated with a bunch of orthogonal subcarriers. According to 802.11n standard [197], there are totally 56 subcarriers at 2.4 GHz and 114 at 5 GHz; however, the CSI can be observed only for 30 subcarriers [87, 143]. This means that although we could know which 30 subcarriers are used based on 802.11n standard [197], the general form of $s(\cdot)$ is unknown for all the subcarriers. Moreover, the expression of $s(\cdot)$ also depends on the source code, which is also unavailable from Intel 5300 toolkit. Consequently, although the expression of the inverse of the FIM and the CRB can be given in certain form, the results make nonsense in terms of revealing the insight of CSI approach or guiding the design of CSI based localization system.

Multipath Effect The fundamentals of the Fisher information theory are that the unknown parameters and the observable random variables are intrinsically connected; however, it is usually difficult to find such intrinsic connection. The radio propagation model such as LNPL indicates the connection between the RSS and distance, but the empirical association hardly reflects the real situation, as the multipath effect is not fully considered in the model. This is why the CRB based on such connection has been proved inaccurate for the RSS-based localization [40]. The multipath effect also impacts the performance of the CSI based approach [86, 87, 227]. In particular, CSI approach is interested in only the amplitude and phase information of the received waveform along the direct path, which can be denoted using $\alpha^{(1)}$ and phase $\tau^{(1)}$, respectively. However, we are unable to just regard $\alpha^{(1)}$ and $\tau^{(1)}$ as the parameter, since in this way the connection between these parameters and the signal model is unable to reflect the real situation, which results in incorrect CRB.

Asynchronization Asynchronization between the transmitter and the receiver is incurred by crystal oscillating circuits implementations, which is inevitable even for devices by the same manufacturer. The asynchronization leads to a sampling frequency offset, which further causes a time shift impacting the phase measurement. Recent works [86, 87, 95, 121, 180] propose methods to remove the asynchronization effect; however, the key issue that whether the asynchronization effect is the dominant factor of localization errors is still unknown.

Antenna Array and Design Guidance The problem formulation with Fisher information theory just considers the received waveform from one antenna; however, the practical CSI approach normally utilizes the antenna array. More antennas will incur the high-rank FIM, which imposes challenges to find the inverse of the FIM. Since antennas are independent of each other in localization, how to aggregate contributions of all antennas to evaluate the overall performance of the localization system is challenging. Moreover, after obtaining the CRB bound, we must reveal the insight into the CSI localization system design, in order to guide improving the system performance.

Our solutions to deal with the challenges mentioned above will be presented in the following sections.

3.2 CRB Analysis in Frequency Domain

This section deals with the first two challenges. In order to perform localization, we are actually interested in parameter $\boldsymbol{\theta}_1 = [\tau^{(1)}, \alpha^{(1)}]$, which is the received signal's property along the direct path from the transmitter to the receiver. This requires revealing the multipath profile so that the direct path component can be picked out. To this end, the Chronos system [227] utilizes the inverse DFT to analyze the multipath effect to obtain the multipath profile. The smart idea of the Chronos system inspires us to perform the CRB analysis [217] in the frequency domain, which makes it possible to find the inverse of FIM.

In the following, we first examine an ideal case that the parameter can be measured with single antenna and then extend our analysis to the practical scenario involving the antennas array.

3.2.1 Parameter Analysis with Single Antenna

Consider the signal along the direct path in the time domain

$$r(nT) = \alpha^{(1)} s(nT - \tau^{(1)}) + z(nT), \tag{3.8}$$

where $n = 1, 2, 3, \ldots, L$. We can apply L-point DFT to the waveform to get the counterpart expression in the frequency domain

$$R(k) = \alpha^{(1)} S(k) e^{-j\frac{2\pi k \tau^{(1)}}{LT}} + \eta(k), \tag{3.9}$$

with $k = 0, 1, 2, \ldots, L - 1$, where the signal spectrum $S(k)$ depends on the transmitting waveform that is usually fixed and $\eta(k)$ is the power spectrum of white Gaussian noise with covariance $L\sigma^2$.

Leveraging the multipath profiling technique mentioned in [227], the previous parameter vector can be cut down to $\boldsymbol{\theta}_1$, and we can establish an accurate intrinsic connection between unknown parameters $\boldsymbol{\theta}_1$ and the observable random variables $S(k)$. We use τ and α to replace $\tau^{(1)}$ and $\alpha^{(1)}$ in rest of the section for simplicity. The observations incorporate all samples of the received signal in the frequency domain. We use vector $\mathbf{X}$ to denote all the random variables:

$$\mathbf{X} = [R(0), R(1), \ldots, R_1(L-1)]^{\mathrm{T}}, \tag{3.10}$$

where the mean vector $\boldsymbol{\mu}$ for the variable vector $\mathbf{X}$ is

$$\boldsymbol{\mu} = \left[\alpha S(0) e^{-j\frac{2\pi 0 \tau}{LT}}, \ldots, \alpha S(L-1) e^{-j\frac{2\pi (L-1)\tau}{LT}}\right]^{\mathrm{T}}. \tag{3.11}$$

With the complex mean vector in the frequency domain, the element of the FIM extended for complex data can be derived [144]

$$\mathbf{I}_{mn} = 2Re\left[\frac{\partial \boldsymbol{\mu}^{\mathrm{H}}}{\partial \theta_m} \sum\nolimits^{-1} \frac{\partial \boldsymbol{\mu}}{\partial \theta_n}\right], \tag{3.12}$$

where $\sum$ is the covariance matrix that is equal to $L\sigma^2$ and H denotes the complex conjugate transpose of the matrix. The FIM for single antenna in frequency domain is

$$\mathbf{I} = \frac{2}{L\sigma^2}\begin{bmatrix} \sum_{k=0}^{L-1} 4\pi^2(\frac{k}{LT})^2\alpha^2|S(k)|^2 & 0 \\ 0 & \sum_{k=0}^{L-1}|S(k)|^2 \end{bmatrix}. \tag{3.13}$$

We here provide the CRB for estimating the delay τ, which is the parameter of interest in both ToF and AoA. The squared error bound with respect to τ is

$$\mathbf{E}\{(\hat{\tau} - \tau)^2\} = \mathbf{Var}(\tau) \geq [\mathbf{I}(\boldsymbol{\theta})^{-1}]_{1\times 1}, \tag{3.14}$$

where $[\cdot]_{n\times n}$ denotes the upper left $n \times n$ sub-matrix of the arguments.

Consequently, the CRB for the time delay with CSI approach termed as time delay error bound (TDEB) in the frequency domain utilizing the FIM in Eq. (3.13) is

$$\mathcal{T}(\tau) = \frac{L^3T^2\sigma^2}{8\pi^2\alpha^2}\frac{1}{\sum_{k=0}^{L-1}k^2|S(k)|^2}. \tag{3.15}$$

We can further simplify the bound with the corresponding channel impulse response (CIR), where $s(t)$ is $\delta(t)$ and $|S(k)| = 1$ for $k = 0, 1, \ldots, L-1$. Thus the bound becomes

$$\mathcal{T}(\tau) = \frac{3L^2T^2\sigma^2}{4\pi^2\alpha^2(L-1)(2L-1)}. \tag{3.16}$$

3.2.2 *Parameter Estimation with Antenna Array*

The analysis above shows the feasibility of performing CRB in the frequency domain with the idea single antenna case; however, the practical mechanism that can obtain the ToF with the single antenna is still unavailable to the best of our knowledge. All existing work in the literature requires using information from multiple antennas at the receiver side, especially the system depending on AoA information to perform location estimation. Consequently, we now try to find the parameter estimation error bound with antennas array.

Suppose there are N antennas at the receiver side; then the received signal of the mth antenna along the direct path at sampling time point n in the time domain can be expressed as

$$r_m(nT) = \alpha_m s(nT - \tau_m) + z(nT), \tag{3.17}$$

with $m = 1, 2, 3, \ldots, N$, $n = 1, 2, 3, \ldots, L$.

The parameters α_m and τ_m denote the amplitude and the time delay of the received waveform along the direct path incident to the mth antenna, and $z(nT)$ denotes the noise effect that is normally following Gaussian distribution, with sampling period T and the total number of samples L. It is reasonable to assume that the total number of samplings for each antenna in the array is the same, since those antennas in the array normally share an ADC module at the receiver in practice. The signal as shown in (3.17) can be transformed into the frequency domain with L-point DFT, which becomes

$$R_m(k) = \alpha_m S(k) e^{-j\frac{2\pi k \tau_m}{LT}} + \eta(k), \tag{3.18}$$

where $k = 0, 1, 2, \ldots, L-1$ and $m = 1, 2, 3, \ldots, N$. The parameter $\eta(k)$ is the zero-mean complex white Gaussian noise with the covariance $L\sigma^2$. The signal power spectrum $S(k)$ depends on the transmitting waveform; therefore, it is reasonable to assume that the signal spectrum $S(k)$ is the same for the receiver waveforms over all antennas.

Since there are N antennas at the receiver, the parameter vector to be estimated is

$$\boldsymbol{\theta} = \underbrace{[\tau_1, \alpha_1, \tau_2, \alpha_2, \ldots, \tau_N, \alpha_N]}_{2N}. \tag{3.19}$$

The observable random variables are the collection of all the received waveforms in the frequency domain over the sampling time. We use vector $\mathbf{X}$ to denote the $NL \times 1$ dimensional random variables:

$$\mathbf{X} = \Bigg[\underbrace{R_1(0), R_1(1), \ldots, R_1(L-1)}_{antenna1}, \ldots\ldots \\ \underbrace{R_N(0), R_N(1), \ldots, R_N(L-1)}_{antennaN} \Bigg]^T. \tag{3.20}$$

Since the parameter $\eta(k)$ is the zero-mean complex white Gaussian noise, the mean vector $\boldsymbol{\mu}$ of $\mathbf{X}$ is

$$\boldsymbol{\mu} = \Bigg[\underbrace{\alpha_1 S(0), \alpha_1 S(1) e^{-j\frac{2\pi\tau_1}{LT}}, \ldots, \alpha_1 S(L) e^{-j\frac{2\pi(L-1)\tau_1}{LT}}}_{antenna_1}, \ldots\ldots \\ \underbrace{\alpha_N S(0), \alpha_N S(1) e^{-j\frac{2\pi\tau_N}{LT}}, \ldots, \alpha_N S(L) e^{-j\frac{2\pi(L-1)\tau_N}{LT}}}_{antenna_N} \Bigg]^T. \tag{3.21}$$

$$\mathbf{H}^H = \begin{bmatrix} \alpha_{\text{ref}} S^*(0) e^{j\frac{2\pi \cdot 0 \|\mathbf{p}-\mathbf{p}_m\|}{LTv}} \left(-\frac{1}{\|\mathbf{p}-\mathbf{p}_1\|^2} + j\frac{2\pi \cdot 0}{LTv\|\mathbf{p}-\mathbf{p}_1\|} \right) \frac{\mathbf{p}^{\mathrm{T}}-\mathbf{p}_1^{\mathrm{T}}}{\|\mathbf{p}-\mathbf{p}_1\|} \\ \vdots \\ \alpha_{\text{ref}} S^*(k) e^{j\frac{2\pi k \|\mathbf{p}-\mathbf{p}_m\|}{LTv}} \left(-\frac{1}{\|\mathbf{p}-\mathbf{p}_m\|^2} + j\frac{2\pi k}{LTv\|\mathbf{p}-\mathbf{p}_m\|} \right) \frac{\mathbf{p}^{\mathrm{T}}-\mathbf{p}_m^{\mathrm{T}}}{\|\mathbf{p}-\mathbf{p}_m\|} \\ \vdots \\ \alpha_{\text{ref}} S^*(L-1) e^{j\frac{2\pi (L-1) \|\mathbf{p}-\mathbf{p}_m\|}{LTv}} \left(-\frac{1}{\|\mathbf{p}-\mathbf{p}_N\|^2} + j\frac{2\pi (L-1)}{LTv\|\mathbf{p}-\mathbf{p}_N\|} \right) \frac{\mathbf{p}^{\mathrm{T}}-\mathbf{p}_N^{\mathrm{T}}}{\|\mathbf{p}-\mathbf{p}_N\|} \end{bmatrix}_{LN\times 1}^{T} \quad (3.22)$$

$$tr\{\mathbf{I}^{-1}\} = \frac{L\sigma^2}{2\alpha_{ref}^2} \frac{\sum\limits_{m=1}^{N} \frac{1}{\beta_m^2}}{(\sum\limits_{m=1}^{N} \frac{1}{\beta_m^2}\cos^2(\theta_m))(\sum\limits_{m=1}^{N} \frac{1}{\beta_m^2}\sin^2(\theta_m)) - (\sum\limits_{m=1}^{N} \frac{1}{\beta_m^2}\cos(\theta_m)\sin(\theta_m))^2}. \quad (3.23)$$

With the analysis above, it seems that we could find the inverse of the FIM and then derive the CRB the same way as for the single antenna case like Eq. (3.12); however, it can be found that the introduction of multiple antennas makes the rank of the corresponding FIM high along this vein. In particular, it is well known that the Intel 5300 NIC has three slots for antennas, which makes the rank of the FIM to be 6 (with the amplitude and the delay). The receiver in the ArrayTrack [86] system has 8 antennas, which makes the rank to be 16. It is tractable to numerically find the inverse of the 6/16-rank FIM; however, it is unlikely to present the corresponding close-form CRB, without which it is impossible to reveal any insight into the CSI approach.

Moreover, even if we could obtain close-form CRB with the high-dimensional FIM, the information-theoretical bound is for estimating the parameters αs and τs, but our ultimate goal is to find the bound for estimating location of the target. Each single antenna plays the unique role in the localization process, and how to incorporate contributions of all antennas in the array to derive the CRB is still unknown in the literature to the best of our knowledge.

3.2.3 *CRB for Location Estimation*

Our strategy to deal with the two issues mentioned above is to associate the parameters αs and τs with location estimation, which is shown in Fig. 3.1. We could observe sampled waveform as a set of random variables $\mathbf{X}$ that represent the effect of parameters τs and αs. We use $\mathbf{p}$ to denote the position of the target and $\mathbf{p_m}$ the position of the mth antenna in the 2D Euclidean space. It is straightforward that

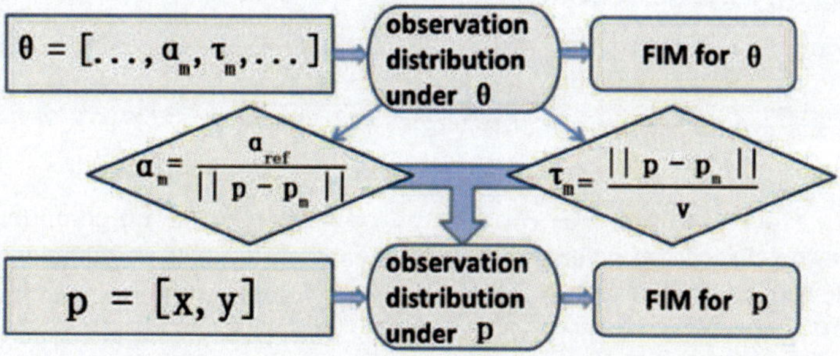

Fig. 3.1 Strategy of CRB for location estimation with antenna array

$$\tau_m = \frac{||\mathbf{p} - \mathbf{p}_m||}{v}, \tag{3.24}$$

where v is the speed of light.

With the multipath profiling technique in [227], we could pick out the CSI over the direct path. Since the influence of small-scale fading caused by multipath effect can be ignored, the amplitude of the received waveform is only related to the distance between the transmitter and the receiver. In most of the localization systems with CSI approach, the transmitter of the target is basically equipped with the omnidirectional antenna [86, 87, 180, 227], which means that the transmitted waveform can be modeled as a spherical wave, and the amplitude of the wave decreases with the distance of propagation. Note that the power of the transmitted waveform is inversely proportional to the square of the propagation distance, but our analysis is in the received waveform level. Thus we have

$$\alpha_m = \frac{\alpha_{ref}}{||\mathbf{p} - \mathbf{p}_m||}, \tag{3.25}$$

where α_{ref} is the reference amplitude at 1 m away from the transmitter.

We now could obtain a new parameter for estimation based on the analysis above. Since CSI based approaches need to utilize the geometric relationship between the transmitter and the anchor points, the precise location information of the anchor points and antenna on them is usually perfectly known. We use $\mathbf{p}_1, \mathbf{p}_2, \ldots, \mathbf{p}_\mathbf{N}$ to denote locations of antennas, based on which we can construct a new parameter vector for estimation $\boldsymbol{\theta} = [\mathbf{p}^\mathrm{T}] = [x, y]$, as shown in Fig. 3.1.

With such a parameter transformation, we could use Fisher information theory to find the CRB for location estimation with CSI approach as shown in Fig. 3.1. The rank of the parameter to be estimated is reduced from $2N$ to 2, and the rank of the FIM is reduced to 2; moreover, we could find the estimation error for localization.

In particular, we first find the FIM $\mathbf{I} = 2Re[\mathbf{H}^H \sum^{-1} \mathbf{H}]$, where $\mathbf{H}$ can be obtained by finding the first derivative of Eq. (3.21) with respect to $\mathbf{p}^\mathrm{T}$ and H denotes the complex conjugate transpose. The result of $\mathbf{H}^H$ is shown in Eq. (3.22).

Consequently, the corresponding FIM is

$$\mathbf{I} = \frac{2\alpha_{ref}^2}{L\sigma^2} \sum_{m=1}^{N} \sum_{k=0}^{L-1} |S(k)|^2 \left(\frac{1}{\rho_m^4} + \frac{4\pi^2 k^2}{L^2 v^2 T^2 \rho_m^2} \right) \mathbf{D}_m, \tag{3.26}$$

where $\mathbf{D}_m$ is a function of $\frac{\mathbf{P}-\mathbf{P}_m}{\|\mathbf{P}-\mathbf{P}_m\|}$. The expression of $\mathbf{D}_m$ is too complicated for us to find the close-form inverse of $\mathbf{I}$; thus we switch to polar coordinates system, where the origin of the system is set at the target's location. Then we can find

$$\mathbf{D}_m = \begin{bmatrix} \cos^2(\theta_m) & \cos(\theta_m)\sin(\theta_m) \\ \cos(\theta_m)\sin(\theta_m) & \sin^2(\theta_m) \end{bmatrix}$$

with θ_m denotes the angle to the mth antenna with respect to the origin.

We let

$$\frac{1}{\beta_m^2} = \sum_{k=0}^{L-1} |S(k)|^2 \left(\frac{1}{\rho_m^4} + \frac{4\pi^2 k^2}{L^2 v^2 T^2 \rho_m^2} \right) \tag{3.27}$$

for convenience of demonstration, and then we have

$$E\{(\hat{\mathbf{p}} - \mathbf{p})(\hat{\mathbf{p}} - \mathbf{p})^{\mathrm{T}}\} \succeq \mathbf{I}^{-1} \tag{3.28}$$

according to the Fisher information theory, based on which we can find the variance of the CSI localization with antenna array:

$$E\{||\hat{\mathbf{p}} - \mathbf{p}||^2\} \geq tr\{\mathbf{I}^{-1}\}, \tag{3.29}$$

where $tr\{\mathbf{I}^{-1}\}$ is the trace of the inverse of FIM as shown in Eq. (3.23).

3.3 Asynchronization Analysis

This section resolves the third challenge mentioned in Sect. 3.1.3. We study another cause of localization error with CSI approach, clock asynchronization between the transmitter and the receiver [78]. The asynchronization is incurred by the crystal oscillating circuits in different hardware, which is inevitable even for devices by the same manufacturer. In particular, the analog received signal goes through the ADC module at the receiver, where the analog waveform is sampled and quantized for the estimation of the CSI. The asynchronization between the transmitter and the receiver incurs a sampling frequency offset, which further causes a time shift impacting the phase measurement [121].

The time shift by clock synchronization is unpredictable but can be regarded as a constant during a short-time measurement as in the localization scenario; thus the time shift is a constant for different subcarriers and antennas. Let τ_0 denote the extra time shift caused by the asynchronization and $r(nT)$ from Eq. (3.8) denote the ideal sampling result of the received signal. The received signal under the influence of asynchronization can be modeled as

$$\widetilde{r}(nT) = r(nT - \tau_0) = \alpha s(nT - \tau - \tau_0) + z(nT - \tau_0), \tag{3.30}$$

where T is the sampling period.

$$tr\{\mathbf{I}_e^{-1}\} = \frac{L\sigma^2}{2\alpha_{ref}^2} \frac{\sum_{m=1}^{N} \frac{1}{\beta_m^2} - \frac{(\sum_{m=1}^{N} \frac{\cos\theta_m}{\gamma_m})^2 + (\sum_{m=1}^{N} \frac{\sin\theta_m}{\gamma_m})^2}{v^2 \sum_{m=1}^{N} \frac{1}{\gamma_m}}}{(\sum_{m=1}^{N} \frac{\cos^2\theta_m}{\beta_m^2} - \frac{(\sum_{m=1}^{N} \frac{\cos\theta_m}{\gamma_m})^2}{v^2 \sum_{m=1}^{N} \frac{1}{\gamma_m}})(\sum_{m=1}^{N} \frac{\sin^2\theta_m}{\beta_m^2} - \frac{(\sum_{m=1}^{N} \frac{\sin\theta_m}{\gamma_m})^2}{v^2 \sum_{m=1}^{N} \frac{1}{\gamma_m}}) - (\sum_{m=1}^{N} \frac{\cos\theta_m \sin\theta_m}{\beta_m^2} - \frac{(\sum_{m=1}^{N} \frac{\cos\theta_m}{\gamma_m})(\sum_{m=1}^{N} \frac{\sin\theta_m}{\gamma_m})}{v^2 \sum_{m=1}^{N} \frac{1}{\gamma_m}})^2}. \tag{3.31}$$

3.3.1 *Parameter Estimation Over Single Antenna*

In this case, the received signal model in the frequency domain is the L-point DFT of Eq. (3.30):

$$\widetilde{R}(k) = \alpha S(k) e^{-j\frac{2\pi k(\tau+\tau_0)}{LT}} + \eta(k) e^{-j\frac{2\pi k\tau_0}{LT}}, \tag{3.32}$$

where $k = 0, 1, 2, \ldots, L-1$ for L-length sampling, τ is the propagation time in the air, and τ_0 is the fixed time shift due to the asynchronization.

It can be seen that the distribution of $\widetilde{R}(k)$ is now dependent on three unknowns τ, α, and τ_0; thus estimation vector $\boldsymbol{\theta}$ becomes $[\tau, \alpha, \tau_0]$. Due to the property of DFT, the mean value of the received signal only contains the signal part, and the mean of $\eta(k)$ is zero. Consequently, the resulted FIM $\mathbf{I}$ of the new parameter vector $\boldsymbol{\theta}$ is

$$\frac{8\pi^2\alpha^2}{L^3T^2\sigma^2} \begin{bmatrix} \sum_{k=0}^{L-1} k^2|S(k)|^2 & 0 & \sum_{k=0}^{L-1} k^2|S(k)|^2 \\ 0 & \frac{L^2T^2}{4\pi^2\alpha^2} \sum_{k=0}^{L-1} |S(k)|^2 & 0 \\ \sum_{k=0}^{L-1} k^2|S(k)|^2 & 0 & \sum_{k=0}^{L-1} k^2|S(k)|^2 \end{bmatrix}. \tag{3.33}$$

Note that FIM **I** is not full rank, since the first row and the third row are the same, which means that we are unable to find inverse of **I** and thus the error of τ cannot be bounded. Based on Eq. (3.32), the phase component of the observable random variables $R(k)$ is determined by time delay τ and time shift τ_0. For the same phase of received signal, there are innumerable solutions for τ as long as the τ_0 changes correspondingly so that $\tau + \tau_0$ remains the same, which is the fundamental reason why the inverse of FIM does not exist. The physical meaning of the result is that we are unable to estimate the parameter τ based on the observable random variables over the single antenna because the effects of τ and τ_0 are aggregated. Without the prior knowledge of τ_0, the propagation delay measured by the single antenna observable results are unreliability.

3.3.2 Parameter Estimation Over Multiple Antennas

Similarly, since different antennas of the receiver share the same ADC module, the received signal model for antenna array has the same asynchronization time shift τ_0. The received signal model considering asynchronization in the frequency domain can be denoted by

$$\widetilde{R}_m(k) = \alpha_m S(k) e^{-j\frac{2\pi k(\tau_m+\tau_0)}{LT}} + \eta(k) e^{-j\frac{2\pi k\tau_0}{LT}}, \tag{3.34}$$

where $k = 0, 1, \ldots, L-1$ for L-length sampling, $m = 1, 2, \ldots, N$ for N antennas, and τ_m denotes the propagation time delay between the receiver's mth antenna and the transmitter.

Similar to the situation in Sect. 3.2.2, now the vector to be estimated $\boldsymbol{\theta}$ becomes

$$\boldsymbol{\theta} = [x, y, \tau_0], \tag{3.35}$$

where x, y are the target's horizontal and vertical localization coordinates. The phase shift of the Gaussian noise component $\eta(k)$ has no influence on the mean vector $\boldsymbol{\mu}$ since the DFT of the Gaussian noise is still a Gaussian noise. Thus compared with Eq. (3.21), the mean vector $\boldsymbol{\mu}$ remains except for a phase shift in the frequency domain due to τ_0 in the time domain. To derive the FIM, we need to take the derivative of the mean vector with respect to the parameter vector. For the convenience of derivation, we use the polar coordinates to express the derivative result. The result of the FIM consists of the sum of the sampling length L and the sum of different antennas. These sums are functioned as a coefficient in the FIM. For simplification, we use the parameter β_m from Eq. (3.27) and let

$$\frac{1}{\gamma_m} = \sum_{k=0}^{L-1} |S(k)|^2 \frac{4\pi^2 k^2}{L^2 T^2 \rho_m^2}, \tag{3.36}$$

where the polar coordinate ρ_m denotes the distance between the mth antenna and the target device. Then the simplified FIM $\mathbf{I}$ of the parameter vector $\boldsymbol{\theta}$ is

$$\frac{2\alpha_{ref}^2}{L\sigma^2}\begin{bmatrix} \sum_{m=1}^{N}\frac{\cos^2\theta_m}{\beta_m^2} & \sum_{m=1}^{N}\frac{\cos\theta_m\sin\theta_m}{\beta_m^2} & \sum_{m=1}^{N}\frac{\cos\theta_m}{v\gamma_m} \\ \sum_{m=1}^{N}\frac{\cos\theta_m\sin\theta_m}{\beta_m^2} & \sum_{m=1}^{N}\frac{\sin^2\theta_m}{\beta_m^2} & \sum_{m=1}^{N}\frac{\sin\theta_m}{v\gamma_m} \\ \sum_{m=1}^{N}\frac{\cos\theta_m}{v\gamma_m} & \sum_{m=1}^{N}\frac{\sin\theta_m}{v\gamma_m} & \sum_{m=1}^{N}\frac{1}{\gamma_m} \end{bmatrix}. \tag{3.37}$$

We then need to find the estimation error bound of the parameter by finding the inverse of the FIM above. However, we are only interested in the first two parameters' estimation error bound, which is to be obtained with the equivalent Fisher information matrix (EFIM) for the convenience of demonstration [145].

The N-length parameter vector $\boldsymbol{\theta}$ can be divided into two parts $\boldsymbol{\theta} = [\boldsymbol{\theta}_1^T, \boldsymbol{\theta}_2^T]^T$, where n-length parameter $\boldsymbol{\theta}_1$ is of the interest. The $N \times N$ FIM $\mathbf{I}_{\boldsymbol{\theta}}$ can be divided corresponding to the parameter vector $\boldsymbol{\theta}$.

$$\mathbf{I}_{\boldsymbol{\theta}} = \begin{bmatrix} \mathbf{A} & \mathbf{B} \\ \mathbf{B}^T & \mathbf{C} \end{bmatrix}, \tag{3.38}$$

where $\mathbf{A} \in R^{n\times n}, \mathbf{B} \in R^{n\times(N-n)}, \mathbf{C} \in R^{(N-n)\times(N-n)}$. The sub-matrix $\mathbf{A}$ is corresponding to the n interested parameters in the parameter vector $\boldsymbol{\theta}$. Then the EFIM is

$$\mathbf{I}_e \triangleq \mathbf{A} - \mathbf{B}\mathbf{C}^{-1}\mathbf{B}^T, \tag{3.39}$$

which has the property that $[\mathbf{I}_{\boldsymbol{\theta}}{}^{-1}]_{n\times n} = \mathbf{I}_e{}^{-1}$. In particular, we are interested in the parameter $\mathbf{p}$; $\mathbf{A}$ is the upper left 2×2 sub-matrix in $\mathbf{I}$ that is corresponding to the first two parameters x and y. We can derive the 2×2 EFIM from Eqs. (3.37) and (3.39). With the EFIM, according to the information inequality, we can get the variance of the CSI localization by calculating the trace of the inverse EFIM.

$$E\{||\hat{\mathbf{p}} - \mathbf{p}||^2\} \geq tr\{\mathbf{I}_{\mathbf{e}}^{-1}\}, \tag{3.40}$$

where $tr\{\mathbf{I}_{\mathbf{e}}^{-1}\}$ is shown in Eq. (3.31) and $tr\{\mathbf{I}_{\mathbf{e}}^{-1}\}$ is the lower bound of the localization estimation error.

Observing Eq. (3.31), it is interesting to find that the expression of the bound is independent of τ_0, but proportional to the variance of the noise σ^2. This does not mean that the location estimation error bound is independent of the degree of asynchronization, because if this is the case, then the bounds in Eqs. (3.23) and (3.31) are supposed to be the same (recall that Eq. (3.23) presents the location estimation error bound without taking asynchronization into account). This result indicates that the asynchronization effect is less effective than the wireless communication noise in impacting the fundamental performance bound of CSI localization.

Since the noise factor is dominating in determining the fundamental CSI localization performance, and the CRB considering asynchronization is over complicated, it is more reasonable to reveal the insight into the design of CSI localization systems based on the CRB shown in Eq. (3.23). We could try to eliminate the time shift by asynchronization as shown in [86, 87, 95, 121, 180]; however, the impact of noise is unable to be eliminated, which is the major factor incurring localization errors in CSI approach.

It is worth mentioning that the introduction of asynchronization does not change the unbiasedness of the estimator. Recall that the estimator function in the synchronized case is $\hat{\mathbf{p}}(\mathbf{X})$, with $\mathbf{E}[\hat{\mathbf{p}}(\mathbf{X})] = \int \hat{\mathbf{p}}(\mathbf{X})d\mathbf{X} = \mathbf{p}$. If we take the asynchronization effect τ_0 into account, the estimator function becomes $\mathbf{E}[\hat{\mathbf{p}}(\tilde{\mathbf{X}})] = \int \hat{\mathbf{p}}(\tilde{\mathbf{X}})d\tilde{\mathbf{X}} = \int \bigcup_{m=1}^{N} \bigcup_{k=0}^{L-1} e^{-j\frac{2\pi k \tau_0}{LT}} \hat{\mathbf{p}}(\mathbf{X})d\mathbf{X} = \int (\bigcup_{m=1}^{N} 1)\hat{\mathbf{p}}(\mathbf{X})d\mathbf{X} = \mathbf{p}$, which is also a unbiased estimator. Although the asynchronization factor τ_0 does not present in Eq. (3.31), it changes the localization estimation error bound $tr\{\mathbf{I}_{\mathbf{e}}^{-1}\}$ compared with Eq. (3.23). This means that the existence of time shift τ_0 actually impacts the lower error bound.

3.4 Insight into CSI Approach

This section reveals insight into the CSI approach based on our theoretical analysis. We first simplify the form of the derived CRB. In practice, the wavelength λ of the Wi-Fi signal is around $9 - 12$ cm, much less than the distance between the target and the antenna array. Given that $\frac{4\pi^2 k^2}{L^2 \upsilon^2 T^2 \rho_m^2} \geq (\frac{2\pi}{\lambda \rho_m})^2 \gg \frac{1}{\rho_m^4}$, Eq. (3.27) can be simplified as $\frac{1}{\beta_m^2} \approx \sum_{k=0}^{L-1} |S(k)|^2 \frac{4\pi^2 k^2}{L^2 \upsilon^2 T^2 \rho_m^2}$.

We let $\mathcal{K} = L\sigma^2 \Big/ (2\alpha_{ref}^2 \sum_{k=0}^{L-1} \frac{4\pi^2 k^2}{L^2 \upsilon^2 T^2})$ to denote the rest part of the CRB. The denominator of Eq. (3.23) can be further simplified using the principles of trigonometric function, which yields

$$G(\mathcal{N}) = \frac{\sum_{m=1}^{N} \frac{1}{\rho_m^2}}{\sum_{m=1}^{N} \sum_{n=1}^{N} \frac{1}{\rho_m^2} \frac{1}{\rho_n^2} \sin^2(\theta_m - \theta_n)}, \tag{3.41}$$

where $\mathcal{N}$ denotes a set of the antennas. Thus the CRB for localization can be rewritten as

$$\mathcal{F}(\mathcal{N}) = tr\{\mathbf{I}^{-1}\} = \mathcal{K} G(\mathcal{N}). \tag{3.42}$$

Theorem 1 *The CRB as shown in Eq. (3.23) must be applicable to CSI localization approaches with Intel 5300 toolkit.*

Proof With the analysis process presented in previous sections, we can see that the CRB can be derived as long as the inverse of the FIM can be found, which is closely related to how the hardware is processing the received waveform. We need to reveal the implicit necessary condition of finding the inverse of FIM in the analysis of previous sections and see whether the Intel 5300 toolkit can satisfy the condition.

Consider a general case where we have an m_1-dimensional observable random variable vector with mean values $\boldsymbol{\mu} = [\mu_1, \mu_2, \ldots, \mu_{m_1}]$ to estimate m_2 parameters. Then the corresponding FIM can be denoted as $\mathbf{I} = \mathbf{U}^T\mathbf{U}$ and

$$\mathbf{U} = \begin{bmatrix} \frac{\partial \mu_1}{\partial \theta_1} & \frac{\partial \mu_2}{\partial \theta_1} & \cdots & \frac{\partial \mu_{m_1}}{\partial \theta_1} \\ \frac{\partial \mu_1}{\partial \theta_2} & \frac{\partial \mu_2}{\partial \theta_2} & \cdots & \frac{\partial \mu_{m_1}}{\partial \theta_2} \\ \vdots & \vdots & \ddots & \vdots \\ \frac{\partial \mu_1}{\partial \theta_{m_2}} & \frac{\partial \mu_2}{\partial \theta_{m_2}} & \cdots & \frac{\partial \mu_{m_1}}{\partial \theta_{m_2}} \end{bmatrix}_{m_1 \times m_2} . \tag{3.43}$$

Suppose $\mathbf{U}$ is a full row or column rank matrix and $\mathbf{x}$ is the nullspace in $\mathbf{U}$; then we have $\mathbf{Ux} = \mathbf{0}$, and

$$\mathbf{Ix} = \mathbf{U}^{\mathrm{T}}\mathbf{Ux} = \mathbf{0},$$

which means that $\mathbf{x}$ is also the nullspace in $\mathbf{U}^{\mathrm{T}}\mathbf{U}$. Similarly, if $\mathbf{x}$ is the nullspace in $\mathbf{I}$, then we have

$$(\mathbf{x}^{\mathrm{T}})\mathbf{U}^{\mathrm{T}}\mathbf{Ux} = (\mathbf{Ux})^{\mathrm{T}}\mathbf{Ux} = ||\mathbf{Ux}||^2 = 0.$$

We thus have $\mathbf{Ux} = \mathbf{0}$. The homogeneous linear equation $\mathbf{Ix} = \mathbf{U}^{\mathrm{T}}\mathbf{Ux} = \mathbf{0}$ has the same basic system of solutions as $\mathbf{Ux} = \mathbf{0}$, which means that $rank(\mathbf{I}) = rank(\mathbf{U}^{\mathrm{T}}\mathbf{U}) = rank(\mathbf{U}) = min\{m_1, m_2\}$. Since $\mathbf{I}$ is an $m_2 \times m_2$ matrix that must be full rank to find the inverse, we have $m_1 > m_2$ so that the inverse of $\mathbf{I}$ can be found.

This means that the sufficient condition to find the inverse of the FIM is that the number of observable random variables must be greater than the number of parameters to be estimated. In our analysis in previous sections, the number of observable random variables is related to the sampling frequency, and the number of parameters to be estimated depends on how the model is constructed, which in the worst case is related to the number of paths the signal components experienced. Intel 5300 NIC's working bandwidth is 20 MHz or 40 MHz, which means that the number of observable random variables is definitely larger than the possible number of parameters to be estimated according to the Nyquist sampling theorem. That is, Intel 5300 toolkit must enable finding the inverse of the FIM, and once the inverse of the FIM can be found, we can derive the CRB as shown in previous sections. ■

Theorem 2 *With the CSI approach using Intel 5300 toolkit, the non-uniformly spaced antenna array outperforms the uniformly spaced one in terms of localizing*

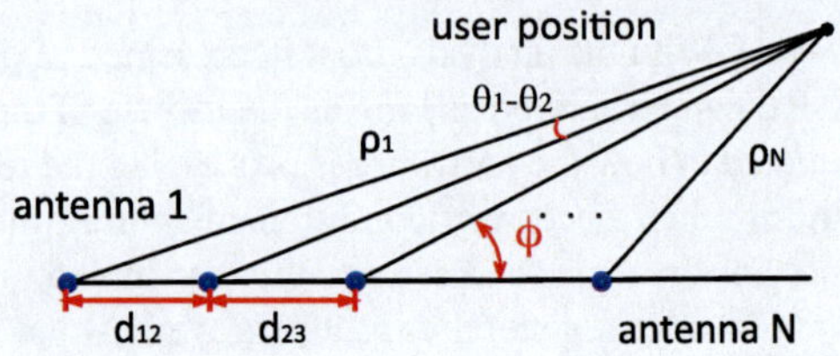

Fig. 3.2 Antenna array in far field

the target that is faraway from the array with $\rho \gg d_{13}$*, where* d_{13} *is the fixed distance between the 1st and 3rd antennas in the array.*

Proof We first consider a general case with N-antenna array; the target is faraway from the array with $\rho \gg d_{1N}$. In this case, the distance between the user and each single antenna is almost the same with $|\rho_m - \rho_n| \approx d_{mn} \cos\varphi$ as shown in Fig. 3.2. The parameter d_{mn} represents the distance between $antenna_m$ and $antenna_n$, and φ denotes the angle between the user position and the antenna array.

Then we have

$$\begin{aligned}\sin^2(\theta_m - \theta_n) &= 1 - \left(\frac{\rho_m^2 + \rho_n^2 - d_{mn}^2}{2\rho_m\rho_n}\right)^2 \\ &\approx 1 - \left(\frac{\rho_m^2 + \rho_n^2}{2\rho_m\rho_n}\right)^2 \left(1 - \frac{2d_{mn}^2}{\rho_m^2 + \rho_n^2}\right) \\ &= \frac{2d_{mn}^2}{\rho_m^2 + \rho_n^2} - \frac{(\rho_m^2 + \rho_n^2)^2(\rho_m^2 - \rho_n^2)^2}{(2\rho_m\rho_n)^2} \\ &= \frac{d_{mn}^2}{\rho^2} - \frac{d_{mn}^2 \cos^2\varphi}{\rho^2} \approx \frac{d_{mn}^2 \sin^2\varphi}{\rho^2}\end{aligned} \tag{3.44}$$

and

$$\begin{aligned}G(N) &\approx \frac{\sum_{m=1}^{N} \frac{1}{\rho^2}}{\sum_{m=1}^{N}\sum_{n=1}^{N} \frac{d_{mn}^2 \sin^2\varphi}{\rho^2} \frac{1}{\rho^2} \frac{1}{\rho^2}} \\ &= \frac{N\rho^4}{\sum_{m=1}^{N}\sum_{n=1}^{N} d_{mn}^2 sin^2\varphi}\end{aligned} \tag{3.45}$$

In the triple-antenna system such as the Intel 5300 NIC that only contains three slots for antennas, according to the average value inequality,

$$
\begin{aligned}
\sum_{m=1}^{3}\sum_{n=1}^{3} d_{mn}^2 &= 2\left(d_{12}^2 + d_{23}^2 + (d_{12}+d_{23})^2\right)\\
&\geq 2\left(\frac{1}{2}(d_{12}+d_{23})^2 + (d_{12}+d_{23})^2\right)\\
&= 3{d_{13}}^2.
\end{aligned}
$$

Only the non-uniformly spaced antenna array ($d_{12} = d_{23}$) can achieve the minimum in the inequality above. This leads to the maximum of $G(\mathcal{N})$ meaning the variance of the estimated location is the greatest. ■

Corollary 1 *The CRB for localizing the target that is faraway from the array with* $\rho \gg d_{1N}$ *is* $\mathcal{K}\frac{6\rho^4}{N(N^2-1)d^2\sin^2\varphi}$, *if the non-uniformly spaced antenna array is used.*

Theorem 3 *For a given target and an antenna array in the polar coordinates system with the target's location as the origin, the utility of adding a new antenna to the array for improving the localization accuracy depends on the polar angle of the new antenna.*

Proof In the polar coordinates system, the target's position is set as the origin. The original antenna array can be expressed as $\mathcal{N}$ with $N-1$ antennas. After adding antenna N, denoted by $\mathcal{A}_N$, the array becomes $\mathcal{N} \cup \mathcal{A}_N$. The utility of the addition is evaluated by the gain that can be brought to accurate location estimation: $\mathcal{F}(\mathcal{N} \cup \mathcal{A}_N) - \mathcal{F}(\mathcal{N})$.

Note that the denominator of $G(\mathcal{N})$ satisfies the Cauchy–Schwarz inequality, which is a positive value, and $\mathcal{K}$ is also a positive constant without affecting the positiveness or negativeness of the gain. To simplify the expression of $G(\mathcal{N})$, we let $A = \sum_{m=1}^{N-1} 1/\rho_m^2$, $B = \sum_{m=1}^{N-1}\sum_{n=1}^{N-1} \frac{1}{\rho_m^2}\frac{1}{\rho_n^2}\sin^2(\theta_m - \theta_n)$, and C and D denote the increment of numerator and denominator of $G(\mathcal{N})$, respectively. Thus $C = 1/\rho_N^2$ and

$$
D = \sum_{m=1}^{N-1} \frac{1}{\rho_m^2\rho_N^2}\sin^2(\theta_m - \theta_N) = C\sum_{m=1}^{N-1}\frac{1}{\rho_m^2}\sin^2(\theta_m - \theta_N),
$$

where the A, B, C, D are all positive values. Then the gain

$$
\begin{aligned}
\mathcal{F}(\mathcal{N} \cup \mathcal{A}_N) - \mathcal{F}(\mathcal{N}) &= \mathcal{K}\frac{BC - AD}{B(B+D)}\\
&= \mathcal{K}\frac{C\left(B - A\sum_{m=1}^{N-1}\frac{1}{\rho_m^2}\sin^2(\theta_m - \theta_N)\right)}{B(B+D)}.
\end{aligned}
\tag{3.46}
$$

According to Eq. (3.46), whether the gain is positive or negative only depends on the polar angle of the newly added antenna θ_N. ■

This theorem can be used in practical deployment of CSI based localization systems. New antenna should be arranged to an appropriate position, which is for the best of most frequently visited locations.

Theorem 4 *The lowest CRB of dual-antenna array localization system with the spacing distance d can be achieved if the target is in the midperpendicular of the two antennas and is $\frac{\sqrt{2}}{4}d$ from the midpoint of the two antennas.*

Proof When there are only two antennas, the corresponding CRB is

$$G(\mathcal{N}) = \frac{\frac{1}{\rho_1{}^2} + \frac{1}{\rho_2{}^2}}{\frac{1}{\rho_1{}^2}\frac{1}{\rho_2{}^2}\sin^2(\theta_1 - \theta_2)}. \tag{3.47}$$

According to the law of cosines, $G(\mathcal{N})$ can also be written as

$$G(\mathcal{N}) = \frac{\rho_1{}^2 + \rho_2{}^2}{1 - \frac{(\rho_1{}^2+\rho_2{}^2-d^2)^2}{4\rho_1{}^2\rho_2{}^2}}. \tag{3.48}$$

Let $t = 1/(\rho_1{}^2 + \rho_2{}^2)$; then

$$G(\mathcal{N}) \geq \frac{\rho_1{}^2 + \rho_2{}^2}{1 - \frac{(\rho_1{}^2+\rho_2{}^2-d^2)^2}{(\rho_1{}^2+\rho_2{}^2)^2}} = \frac{1}{2t^2d^2 - t^3d^4} \geq \frac{27}{32}d^2, \tag{3.49}$$

$iff \rho_1 = \rho_2 = \sqrt{\frac{1}{2t}} = \frac{\sqrt{6}}{4}d$. The distance from the center of the two antennas is $\frac{\sqrt{2}}{4}d$. Proved. ∎

We note that efforts have been dedicated to study influence of anchor point placement on localization performance. For example, Spirito develops a general mathematical framework to analyze distance measurements between the mobile station and multiple base transceiver stations (BTSes) in the cellular network, which theoretically interprets how locations of BTSes can impact the localization accuracy [147]; Tekdas et al. show how to find the minimum number and optimal placement of anchor points for localizing a robot in sensor networks [146]. It also shows that the placement of anchor points in the sensor network can influence the result of CRB [106, 148].

In the works mentioned above, the geometry of anchor points and the target is evaluated by geometric dilution of precision (GDOP), which is the ratio of the change in output location and that in measured data, quantifying how errors in the measurement will affect the final state estimation.

Our work in this section focuses on how the single anchor point should be designed to help approach to the CRLB as much as possible. However, such findings could be the basis for the possible future work studying the fundamental performance in the system with multiple CSI retrievable anchor points. It can be found that Eq. (3.41) is similar to the GDOP expression as they are both in the fractional form, and it is more favored that the values of the both are as less as possible. In fact, GDOP and CRB are closely related to each other, and they can be the expression of each other utilizing the variance of the data measurement and location estimation as described in [106].

3.5 Experimental Results

This section presents numerical results of the CRB obtained and the location estimation errors observed in experiments. The numerical results for a specific location estimation scenario are dependent on the values of the CRB parameters in the scene, such as the variance of environment noise, antenna configuration, and relative positions of the antenna array and the target. We compare the numerical results with the experimental ones in the same settings, so that the theories of the paper can be understood intuitively. Since the theoretical results are based on the assumption of ideal CSI measurements and perfect location estimation algorithms, which cannot be realized in practice, there must be discrepancy between the theoretical performance bounds and the experimental results. However, if the trend of the experimental results is in line with that of the theoretical ones under different impacting factors, our theoretical analysis can be corroborated. The basic experiment settings are tabulated in Table 3.1.

3.5.1 Asynchronization Error Bound

We build an AP capable of retrieving CSI using Intel 5300 toolkit and conduct experiments in a hallway as shown in Fig. 3.3. The TP-LINK TL-WR2041N wireless router is the target that sends out signals for the Intel 5300 NIC in the anchor point to catch. We implement the localization algorithm of SpotFi [87] for processing the CSI data. Intel 5300 NIC has three slots for antennas and can

Table 3.1 Experiment settings

Wireless router	TP-LINK TL-WR2041N
Anchor point	XCY-X30 N2830 + Intel 5300 NIC
Antenna	CISCO AIR-ANT5135DG-R
Software	Linux 802.11n CSI Tool
Corridor	SEIEE Building 4th floor corridor, SJTU
NLOS	SEIEE Building 1-434 lab, SJTU
Room	SEIEE Building 1-434 lab, SJTU
Open space	Square between SEIEE Building 1 and 2, SJTU

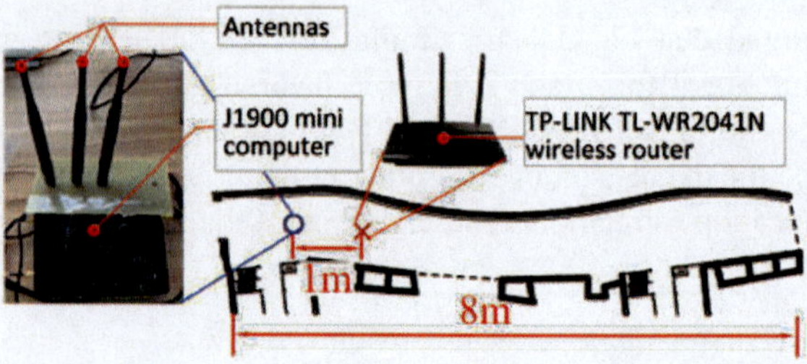

Fig. 3.3 Experiment environment: hallway (NLOS)

retrieve the CSI of 30 subcarriers, which makes 3×30 CSI matrix available. It also provides signal to interference plus noise ratio (SINR), which can be used to estimate properties of environment noise. According to [97, 143], such CSI data are derived from the received waveform after ADC, which is in accordance with the assumptions of our theoretical analysis.

We conduct experiments to verify our asynchronization analysis. In particular, we want to compare the localization performance with and without the asynchronization effect and then find the time delay error bound with the synchronization mechanism.

For the synchronized case, we implement and apply the ToF sanitization algorithm in [87] to eliminate the sampling time offset (STO) in experiments, after which we can construct the smoothed matrix from the 30 subcarrier over each antenna. We further apply the improved MUSIC algorithm in [87] to the smoothed matrix. The steering vector in the MUSIC algorithm is $\mathbf{a}(\tau) = [1, \Omega(\tau), \ldots, \Omega^{14}(\tau)]^T$, where $\Omega(\tau) = e^{-j2\pi\Delta f\tau}$ and Δf is the frequency interval between the subcarriers. Since the sanitization algorithm is unable to eliminate the STO completely, we term the case as quasi-synchronized case. For the asynchronized case, we do not apply the sanitization algorithm, so the errors introduced by the asynchronization effect can be observed. We also plot CRLB of the two cases, where the values of the parameters are assigned based on the experiment observations.

Although not explicitly mentioned in [87], we find that the RF oscillator phase offset between antennas of the AP as observed in ArrayTrack system [86] can also occur in the Intel 5300 based system. Such phase offset is caused during the demodulation procedure, which can significantly impact the localization performance, but can be calibrated with the idea as presented in [86].

The results are shown in Fig. 3.8. We can see that the two curves in the lower part of Fig. 3.4a are CRLBs for the CSI localization system without and with synchronization, respectively. The dots in the upper part of the figure are experimental results in the two cases, where the two fitted curves based on the results are also presented. It can be seen that the theoretical bounds in the two cases are close to each other, which means that perfect precise hardware synchronization can improve the CSI localization performance in an insignificant manner. This corroborates our analysis of CRLB in Eq. (3.31), where it is found that the inevitable noise is the dominant factor impacting the CSI localization performance. It can also be found that the quasi-synchronization approach in [87] can indeed improve the performance in practice.

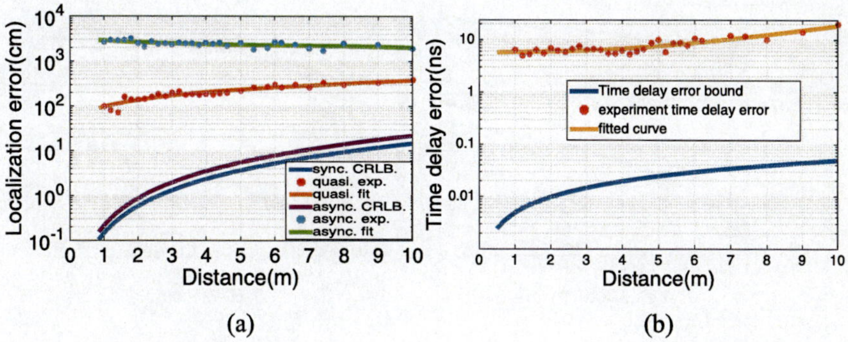

Fig. 3.4 Asynchronization error bound. (**a**) Asynchronization and quasisynchronization comparison. (**b**) Time delay error

We then verify our analysis for time delay errors with the same configuration of the quasi-synchronized case as mentioned above. This is because the perfect synchronization in practice is still unavailable. We set the distance of the T-R pair from 1 m to 10 m and measure the corresponding ToF in each setting for 100 times. Comparing the measured ToF with the ground-truth ToF calculated with the distance divided by the speed of light, we can find the ToF estimation errors from experiments. The experimental results are compared with the theoretical error bound derived, which are illustrated in Fig. 3.4b. The theoretical results are obtained according to Eq. (3.15), where the length of DFT is 512 and the sample rate is set to be 300 MHz. This is equal to the transmitting data rate according to the 802.11n standard. The environment noise power is −95 dBm. The amplitude of the received waveform at the location that is 1 m away from the AP is −40 dBm, and the amplitude measured in locations with different distances from the AP is basically inversely proportional to the distance according to the experiments.

As shown in Fig. 3.4b, we can also find the significant discrepancy between the experimental results and the theoretical error bound. Besides the inherent difference between the theory and practice, the rough ToF sanitization algorithm is another reason, which is unable to estimate accurate ToF [87]. However, we can see that the trend of the experimental results is increasing, because the farther the target is from the AP, the larger the error of estimation can be, which is in accordance with the CRLB trend. This corroborates our theoretical analysis.

3.5.2 Antenna Array Design

We here verify our insight into the antenna array design presented in Sect. 3.4. We first examine whether the localization errors obtained with two antennas will have a greater CRLB than that with three antennas under the same circumstance. Substituting $N = 2$ into the CRLB expression, we can calculate the error bound for 2 antennas. We can construct the smoothed CSI matrix from 2 antennas CSI

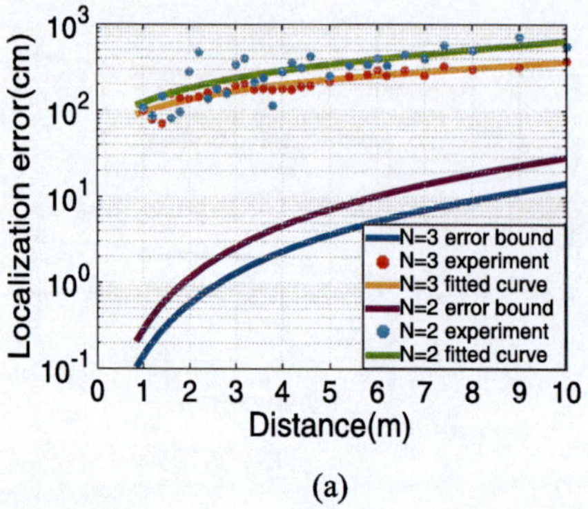

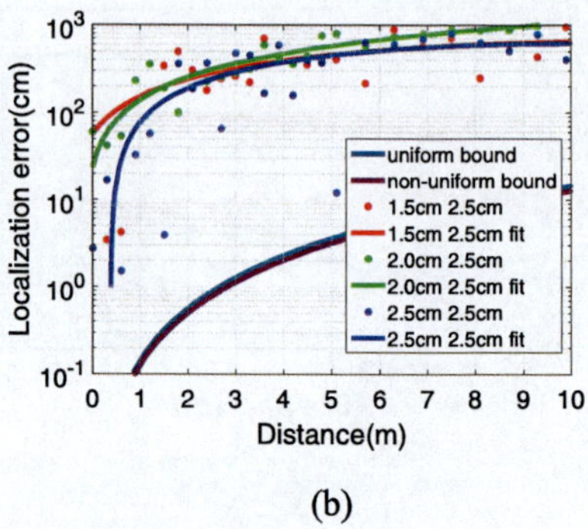

Fig. 3.5 Antenna array design. **(a)** Localization error with two-antenna array at 90 degree. **(b)** Localization error with non-uniform antenna array at 90 degree

and obtain the location estimation with the similar method in [87]. We can see from Fig. 3.5a that the theoretical bound for two antennas is greater than that for three antennas.

We next compare performance of the non-uniformly spaced antenna array with that of the uniformly spaced one. We conduct experiments with the 3−antenna anchor point, and the distances between two adjacent antennas are set as following: (1) $d_{12} = 1.5\,\text{cm}$ and $d_{23} = 2.5\,\text{cm}$; (2) $d_{12} = 2.0\,\text{cm}$ and $d_{23} = 2.5\,\text{cm}$; and (3) $d_{12} = 2.5\,\text{cm}$ and $d_{23} = 2.5\,\text{cm}$. The theoretical error bound is calculated with cases (1) and (3). The target is placed in the midperpendicular of the antenna array, which is moved away from the antenna array in the experiments. We run the SpotFi algorithm as in the previous experiments to estimate the target's position and then calculate the difference from the ground truth. Figure 3.5b illustrates the theoretical and experimental results, from which we can see that the trends of the two kinds of results are basically complying with each other. Recall Theorem 2, the non-uniformly spaced antenna array outperforms the uniformly spaced one if the distance between the antennas and the target is much greater than that of the antenna array's width. We can see from Fig. 3.5b that the non-uniformly spaced antenna array does not outperform the uniformly spaced one when the target is close to the anchor point, but the former outperforms the latter when the distance between the target and the anchor point is around 60 − 80 times of the antenna array's width. When the target is more distant to the anchor point, the distance factor becomes dominant, which corroborates the theoretical results.

3.5.3 *Practical Localization Performance*

The results shown above indicate that it is reluctant to make location estimations with satisfactory accuracy using the antenna array in the single AP. This is because

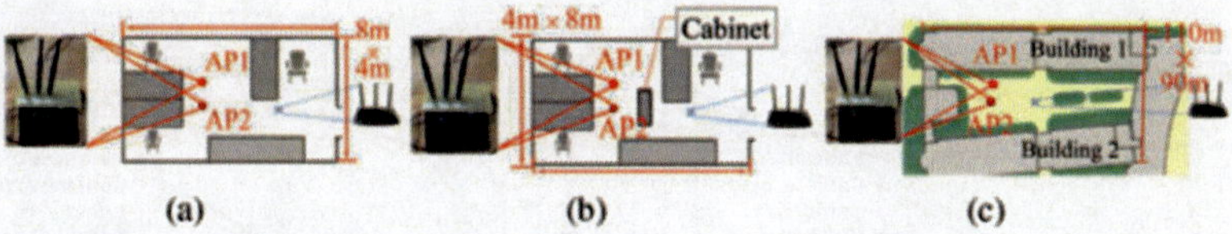

Fig. 3.6 Experiment environment. (**a**) Room. (**b**) NLOS. (**c**) Open Space

the ideally strict synchronization scheme cannot be realized in practice, which makes the estimated ToF unreliable; therefore, the AoA measurement results are the main basis for localization, where the data from multiple APs are more helpful for improving localization accuracy [87]. In the following experiments, we use 2 APs to localize the target in 4 kinds of environment as shown in Figs. 3.3 and 3.6. The 2-AP configuration is the basic AP deployment for examining the practical CSI localization performance.

We investigate the cases that the AP is equipped with 2 and 3 antennas, respectively. We set the midpoint between the 2 APs as the origin point and place the target on the midperpendicular of the line segment joining the 2 APs. We use ρ to denote the distance between the origin point and the target when examining the influence of the distance on the performance. After that, we set the origin at the target and place the 2 APs on the circle centered at the target with a radius of 1.8 m. We let the APs moving along the circle simultaneously and the orientation between the target and the antenna arrays varies. In particular, the angle between the line segment joining the two APs' antenna arrays and the one joining the target and the AP is used to represent the orientation, which is denoted by φ.

Figure 3.7 shows the influence of ρ. It can be seen that the theoretical localization errors are basically in the centimeter level and can achieve millimeter level if ρ is around 1 meter. We perform location estimations in each of the 4 scenes for 100 times, and compare the minimum errors obtained in the experiments with the theoretical CRLBs. We can see that the trends of practical localization errors are in line with the CRLBs in all the 4 kinds of experiment environment. It shows that errors increase as the distance between the target and the origin point increases, which can be multiple meters in practical localization scenarios. We can see that the performance with 3 antennas is better than that with 2, because more antennas are helpful for accurate AoA measurements.

Note that Theorem 4 in the previous section predicts that the CRLB should reach the minimum value when the target is $\frac{\sqrt{2}}{4}d$ from the midpoint of the two antennas; therefore, it is supposed that the localization error achieves the minimum when the target is at the optimal placement and then becomes worse when deviating from the optimal position. However, such trend does not appear in Fig. 3.7. This is because the principle of AoA based localization approach requires the distance between adjacent antennas to be no greater than half of the wireless signal wave length [86, 87], which is less than 3 cm. That means that the target has to be placed around 1 cm from the midpoint of the two antennas, which is impractical to be verified in the experiments.

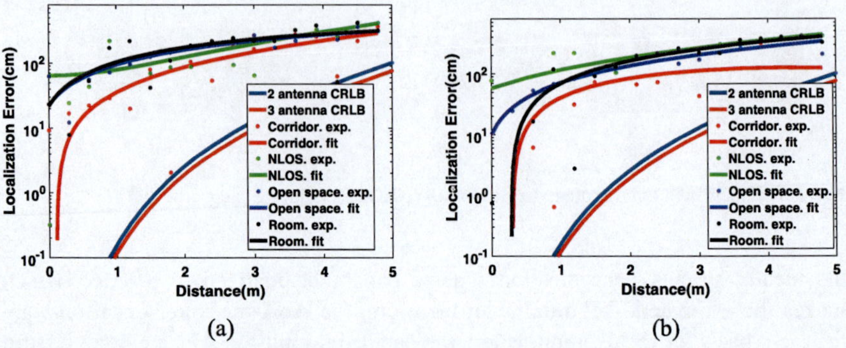

Fig. 3.7 Influence of distance on localization performance. (**a**) Localization error with 2 antennas. (**b**) Localization error with 3 antennas

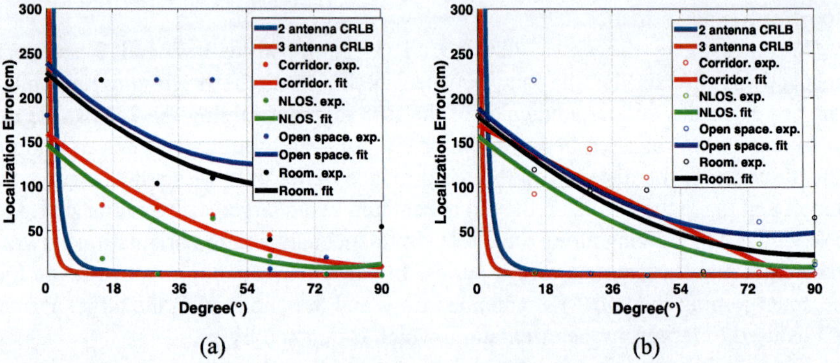

Fig. 3.8 Influence of orientation on localization performance. (**a**) Localization error with 2 antennas. (**b**) Localization error with 3 antennas

Figure 3.8 shows the influence of φ. The trends of the experimental results obtained in all the 4 experiment scenes are also in line with that of the theoretical performance bounds. It can be seen that the localization errors decrease as the value of φ increases.

Figure 3.9 shows more detailed experimental results to examine the influence of distance. It can be seen that the general trend of the experimental results is that the localization performance degrades as the distance between the origin and the target increases. This is in accordance with the trends as shown in Fig. 3.7. Figure 3.10 shows the experimental results to examine the influence of orientation. We can see that the general trend of the experimental results also corroborates that of the theoretical results. The theoretical and practical results can be references for practitioners in their system implementation process. The detailed statistics of the experimental results can be found in [238].

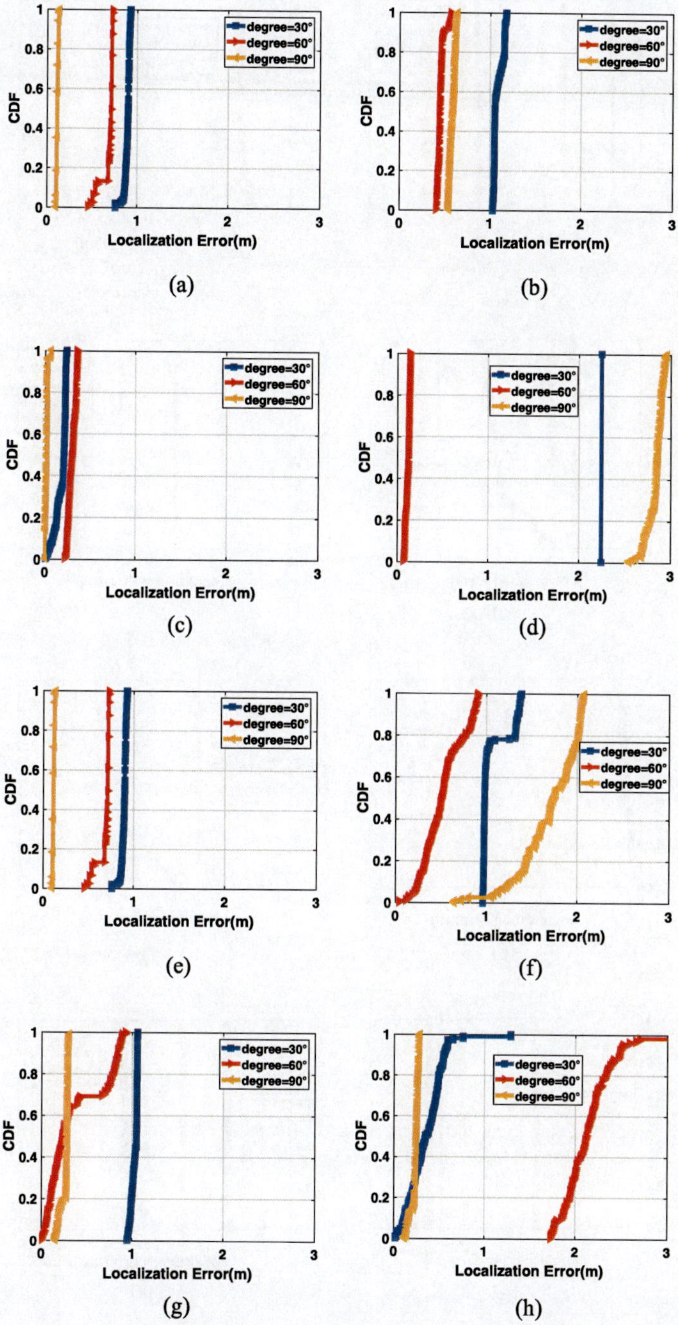

Fig. 3.9 Influence of degree. (**a**) Corridor, 2 antennas. (**b**) Room, 2 antennas. (**c**) NLOS, 2 antennas. (**d**) Open space, 2 antennas. (**e**) Corridor, 3 antennas. (**f**) Room, 3 antennas. (**g**) NLOS, 3 antennas. (**h**) Open space, 3 antennas

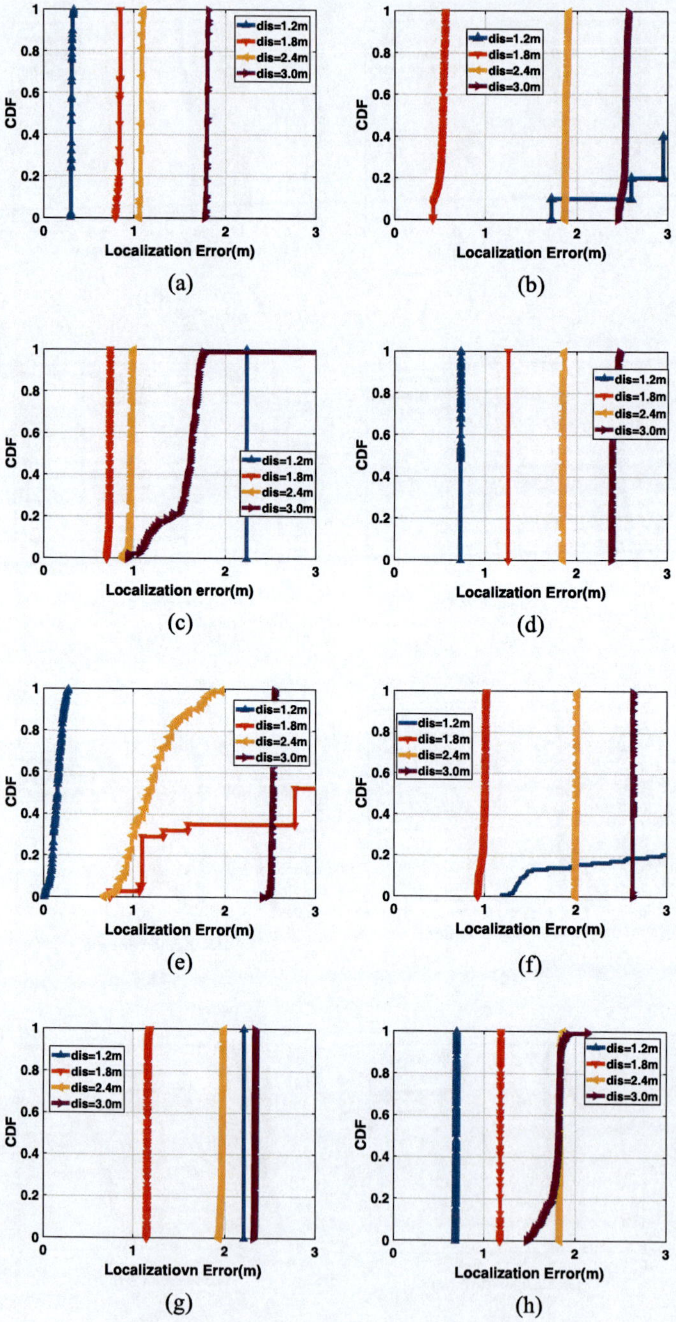

Fig. 3.10 Influence of distance. (**a**) Corridor, 2 antennas. (**b**) Room, 2 antennas. (**c**) NLOS, 2 antennas. (**d**) Open space, 2 antennas. (**e**) Corridor, 3 antennas. (**f**) Room, 3 antennas. (**g**) NLOS, 3 antennas. (**h**) Open space, 3 antennas

Chapter 4
RSS Localization for Large-Scale Deployment

4.1 Online Pricing Mechanism for Crowdsensing Localization Data

This section presents a data pricing scheme dedicated to MCSed fingerprints by enhancing the online learning technique. We first reveal the principle of fingerprints quality assessment for accurate localization. Based on the principle, we design corresponding loss and regret function, which is able to reflect values of the fingerprints with respect to localization accuracy. We then present an online pricing scheme for MCSed data, which results in that the worker's payoff is a random variable following an optimal probability density function (PDF) leading to the minimum expected regret. Further, we extend our scheme to application scenarios with different budget settings, where the pricing strategies for the scenarios of regret minimization with fixed budget and budget minimization for certain fingerprints quality level are investigated. Experimental results are presented to verify our theoretical analysis.

4.1.1 System Model

This section presents the system model in our work. We summarize main notations in the paper in Table 4.1 for the convenience of readers.

MCS System Model Consider a MCS system for fingerprints collection with three kinds of entities, MCS workers, purchase platform, and buyers. The buyer is purchasing fingerprints for a classic fingerprinting based localization system, where the knowledge of the indoor space is perfectly known, including the map, the locations of reference points, and the Service Set Identifiers (SSIDs) of Wi-Fi APs available. The buyer issues the request of RSS fingerprints at the target

X. Tian et al., *Wireless Localization Techniques*, Wireless Networks,
https://doi.org/10.1007/978-3-031-21178-2_4

Table 4.1 Notation table

Section	Symbol	Definition
IV	$\mathbf{r}$	Location in the physical space
	$\mu(\mathbf{r})$	Mean value of RSS fingerprints at $\mathbf{r}$
	S	Two-dimensional Cartesian space
	$\mathcal{Y}$	Normalized Gaussian additive noise
	σ	Amplitude of noise
	$\mathcal{P}(\mathbf{r})$	Observed value of RSS fingerprints
	δ	Maximum estimation error of IMU
	P_{low}, P_{high}	Boundary of E in 1-D sample space
	$F_{error_{\mathbf{r}}}$	Probability of localization error
	N	Size of data sequence
V	P_t	Value of submitted RSS fingerprint at t
	$\mathcal{Z}$	Sampled data space
	h_t	Hypothesis of RSS value at t
	$\mathcal{H}$	Hypothesis space
	$\ell(h_t, P_t), \ell(h_t)$	Loss function
	η	Learning rate
	c_t	Worker's cost
	π_t	Buyer's price
	M	Upper bound of data price
	B	Total budget
	T	Size of online data sequence
	R	Regret function
	$\mathcal{L}$	Risk function
VI	β, λ, μ	Parameters in optimization
	$G_t(c)$	Cumulative distribution of cost
	q_t	Probability of purchasing data at t
	L	Lagrangian function

indoor space to the platform, which then assigns the task to registered MCS workers that are roughly around the area. The chosen workers continuously submitting the current location and corresponding RSS fingerprints observed to the platform in an opportunistic sensing manner [70, 71, 77, 219], where the location of the worker is estimated with dead reckoning by utilizing the inertial measurement unit (IMU) of the worker's mobile device such as accelerometer, magnetometer, and gyroscope [70, 71].

The worker's mobile device installs an agent program such as an APP that is dedicated to perform the opportunistic sensing and interact with the purchasing platform. The agent checks the percentage of resource in the device the MCS task occupies, and calculates the cost of the job. The agent representing the worker also decides if the buyer's offered price is worth of the efforts, where the worker's expectation of taking MCS jobs also can be set to the agent as configuration parameters. If the agent decides not to make the deal, the data submitted to the platform will not be authorized to be exposed to the buyer. We acknowledge the challenge for appropriately designing the agent program, but our work in this paper focuses on how the quality-aware pricing mechanism should be designed for the platform.

It is worth mentioning that the crowd workers are only able to roughly estimate their own locations with the IMU data, and such estimations are very limited not only in accuracy but also in availability [44]. The IMU data based location estimation schemes will incur considerable cumulative errors; some scheme relies on special features in the environment such as the dramatic geomagnetic change near to the server room to calibrate the cumulative errors [128], but such features are not always available anywhere. The existing Wi-Fi signals could also help estimate the user's location information as in [70, 71]; however, such schemes could possibly work well only in certain indoor environments such as hallways, because it requires long-enough space to observe the signal change as the landmark. The work mentioned above partially utilizes the data from IMUs to improve the performance of the indoor localization, or to provide a coarse-grained indoor map. Although not able to provide normal indoor localization services, the roughly estimated location information along with the corresponding fingerprints observed can be a useful source of crowdsensed RSS fingerprints.

Fingerprinting Localization Model The target indoor space S covered by the localization system can be regarded as a two-dimensional Cartesian space with $S \subset \mathbb{R}^2$. We use $\mathbf{r} = (x, y)$ to denote a location in S. We adopt a general model for radio propagation as in [116], where the RSS can be observed at $\mathbf{r}$ with respect to a given AP is

$$\mathcal{P}(\mathbf{r}) = \mu(\mathbf{r}) + \sigma \mathcal{Y}, \tag{4.1}$$

where $\mu(\mathbf{r})$ represents how the mean of RSS readings vary with respect to locations, $\mathcal{Y}$ is the normalized Gaussian additive noise with $\mathcal{Y} \sim \mathcal{N}(0, 1)$, and σ is the amplitude of the noise. This means that the RSS can be observed at $\mathbf{r}$ has a fixed but unknown mean value $\mu(\mathbf{r})$.

The radio propagation model is essentially a generalized model from the LNPL model [77, 85], which itself has been widely adopted by the industrial standardization organizations [109, 110]. Such a modeling approach with the Gaussian distribution assumption is validated by a number of work on indoor localization [38, 39, 77, 280]; moreover, our experimental results with the EVARILOS testbed data [193] also support such modeling [117].

According to the analysis in [116], a point on the mean surface of the RSS sample space $\boldsymbol{\mu}(\mathbf{r})$ corresponds to a location in the physical space $\mathbf{r}$, which is the basic principle of fingerprinting localization. However, the RSS readings observed at a location $\mathbf{r}$ can be impacted by the IMU error, thus the observed $\mu(\mathbf{r})$ could deviate from the true value of the mean of the RSS readings and actually means $\mu(\mathbf{r}')$, which could result in that the user supposed to be localized at $\mathbf{r}$ is estimated at $\mathbf{r}'$.

This is the root cause of inaccurate location estimation of MCS based fingerprints collection approach. Another important information indicated by the fingerprinting localization model is that the closer the reported fingerprint is to the true value of $\mu(\mathbf{r})$, the higher quality the fingerprint has.

Online Learning Model We formulate the quality-aware pricing mechanism design issue under the online learning framework. The reported fingerprints from MCS workers form a data space $\mathcal{Z}$. Since the unavailability of the true value of $\mu(\mathbf{r})$ denoted by $\mu^*(\mathbf{r})$, we could only make the hypothesis of $\mu^*(\mathbf{r})$, and all hypotheses of the value form a hypothesis space $\mathcal{H}$. The fingerprint submitted at time point t for a location is denoted by $P_t \in \mathcal{Z}$. According to P_t and all such data submitted before time point t, we have a hypothesis of $\mu(\mathbf{r})$ at time point t, which is denoted by $h_t \in \mathcal{H}$. In the following, we focus on the situation that the submitted data are for a fixed location, and we use μ^* to denote the true value of the mean of fingerprints for simplicity of presentation. Then the difference between P_t and h_t is termed as the loss function $\ell(h_t, P_t) : \mathcal{H} \times \mathcal{Z} \rightarrow \mathbb{R}$. Our goal is to find the final hypothesis $\overline{h}$ that is the average over all yielded h_t, so that the difference between $\overline{h}$ and μ^* is minimized.

The challenge is that we have no idea about μ^*, and the classic online learning technique as described above is only able to minimize the total loss, which is the sum of $\ell(\overline{h}, P_t)$ over all t. However, even if the total loss is minimized, it is unable to guarantee that the gap between $\overline{h}$ and μ^* is minimized; our work in this paper provides a smart design of the loss function ℓ to this end.

Moreover, a rational buyer has concerns about data purchasing budget, and different buyers have different budget settings. Suppose the worker's cost for collecting the data is c_t at time t ($t = 1, 2, \ldots, T$), and the buyer is willing to pay π_t for the data. If $\pi_t \geq c_t$, then the deal can be made; otherwise, the buyer cannot have access to the data. The buyer may want to obtain the best data possible with a fixed budget B, or obtain a number of data that make the localization system to achieve certain level of accuracy with least possible unfixed budget. The budget settings are also not considered in the classic online learning technique, and we are to deal with the issue in the latter part of the paper.

4.1.2 Quality Evaluation of Fingerprints

This section defines the localization error incurred by the MCSed data, which is used to evaluate the quality of the fingerprints. We are to construct loss function based on such evaluation technique, which is to be proved to have a favored property. That is, the expectation of loss $\ell(\overline{h}, P_t)$ over t achieves the minimum when $\overline{h} = \mu^*$; therefore, we could minimize $\ell(\overline{h}, P_t)$ to obtain a reasonable hypothesis of μ^*.

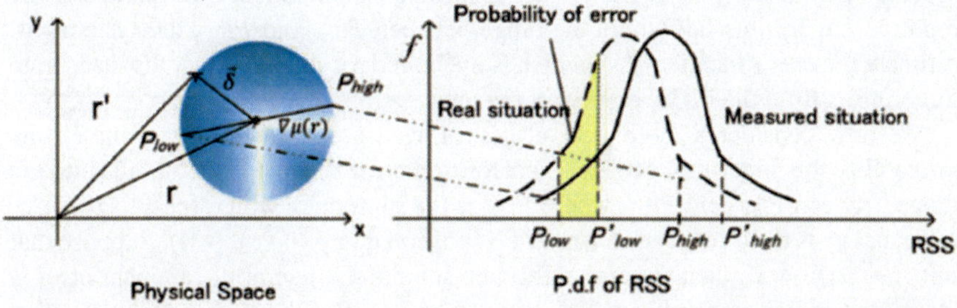

Fig. 4.1 Fingerprinting error

4.1.2.1 Probability Model of Data Error

Assume that the maximum location estimation error with the IMU data is $\boldsymbol{\delta}$, then the worker's actual location is at any point of the $\boldsymbol{\delta}$ neighborhood of $\mathbf{r}$ as shown in left part of Fig. 4.1, when the worker is claiming the submitted fingerprint is observed at $\mathbf{r}$. The boundary of the region is $\mathbf{r}' = \mathbf{r} + \boldsymbol{\delta}$. We consider a simple situation that the worker is reporting the fingerprints with respect to a single AP; although the worker may measure the RSS readings multiple times, only one value will be submitted to the platform in order to save the cost for wireless network access. Thus the submitted data follow the 1-D Gaussian distribution according to the radio propagation model in Eq. (4.1), which is shown in Fig. 4.1.

Suppose the worker is indeed standing at $\mathbf{r}$, the corresponding probability density function (PDF) curve is as the dashed one in the right part of Fig. 4.1. According to the principle of maximum likelihood estimation (MLE), the reported fingerprints in this case should fall within the range between P_{high} and P_{low}, such that $\forall \boldsymbol{\delta}$,

$$\begin{aligned} P_{high} &= sup\{P \mid p_{\mathbf{r}+\boldsymbol{\delta}}(P; \mu(\mathbf{r}+\boldsymbol{\delta})) \geq p_{\mathbf{r}}(P; \mu(\mathbf{r}))\}, \\ P_{low} &= inf\{P \mid p_{\mathbf{r}+\boldsymbol{\delta}}(P; \mu(\mathbf{r}+\boldsymbol{\delta})) \geq p_{\mathbf{r}}(P; \mu(\mathbf{r}))\}, \end{aligned} \tag{4.2}$$

where sup and inf represent the least upper bound (supremum) and the greatest low bound (infimum), respectively; $p_{\mathbf{r}}$ and $p_{\mathbf{r}+\boldsymbol{\delta}}$ denote the probability density function (PDF) describing fingerprints can be observed at location $\mathbf{r}$ and $\mathbf{r} + \boldsymbol{\delta}$ respectively, with $\mu(\mathbf{r})$ and $\mu(\mathbf{r} + \boldsymbol{\delta})$ denote the corresponding mean values of the PDFs.

If the worker's actual location deviates from $\mathbf{r}$, the PDF of corresponding reported fingerprints will deviate from the true value of μ, as the solid curve shown in the right part of Fig. 4.1 with corresponding P'_{high} and P'_{low}. We now examine what will occur if such inaccurate fingerprints were stored in the database and used as if they were accurate. In the localization phase, the user actually at $\mathbf{r}$ will report the observed fingerprint with respect to the same AP. If the reported fingerprints fall within the range between P_{high} and P'_{high}, the system will still estimate the user's location at $\mathbf{r}$, although the information stored in the database is inaccurate. If the reported fingerprints fall within the range between P'_{low} and P_{high}, the system will make the right choice just according to the correct information. If the reported fingerprints fall within the range between P_{low} and P'_{low}, the system will estimate the user's location at $\mathbf{r}'$, which is the boundary condition that the inaccurate fingerprints from the MCS incur errors in localization.

We now consider a more general case. Given a submitted fingerprint P, the probability the fingerprint can incur localization error is equal to that it falls into the range between P_{low} and P'_{low}, where P'_{low} is the supremum with respect to $p_{\mathbf{r}''}(\cdot;\cdot)$. Note that $\mathbf{r}''$ is the true location where P is observed in $p_{\mathbf{r}''}(P;\mu(\mathbf{r}''))$. Suppose that $\mu(\mathbf{r}'')-\mu(\mathbf{r})=d$, then the probability the submitted fingerprint can incur error is that it falls into the range $[P'_{low}-d,\, P'_{low}]$. We can define the quality of the data that is claimed to be collected at $\mathbf{r}$ by how likely the data will incur localization error:

$$F_{error_{\mathbf{r}}}(P;\mu(\mathbf{r}))=\left|\int_{P'_{low}-d}^{P'_{low}} p_{\mathbf{r}}(x;\mu(\mathbf{r}''))dx\right|. \tag{4.3}$$

4.1.2.2 Data Error Analysis

We define that $||P-\mu(\mathbf{r})||=-k\cdot ln(p(P;\mu(\mathbf{r})))$, where k is a positive real number. This is to quantify the distance between P and $\mu(\mathbf{r})$, where $-k\cdot ln(p(P;\mu(\mathbf{r})))$ decreases as P approaches to $\mu(\mathbf{r})$, achieves the minimum when $P=\mu(\mathbf{r})$. In Gaussian Distribution, the $-k\cdot ln(p(P;\mu(\mathbf{r})))$ is exactly the Euclidean distance d^2. We now rewrite the probability of error as the function of $\mu(r)$

$$Err(||P-\mu)||)=F_{error_{\mathbf{r}}}(P,\mu). \tag{4.4}$$

It is straightforward that μ^* should satisfy that

$$\mu^*=arg\min_{\mu(\mathbf{r})}\mathop{\mathbb{E}}_{P}(Err(||P-\mu(\mathbf{r}))||)). \tag{4.5}$$

The following theorem shows the rationale for our definition of error.

Theorem 1 *The expectation of the error $Err(||P-\mu(\mathbf{r}))||)$ over P achieves the minimum when $\mu(\mathbf{r})$ equals true value of the fingerprints μ^*.*

Proof 30 We use $\mu=\mu(\mathbf{r})$ in short. Our target is to prove that

$$\mathop{\mathbb{E}}_{P}[Err(-k\cdot ln(p(P;\mu)))]-\mathop{\mathbb{E}}_{P}[Err(-k\cdot ln(p(P;\mu^*)))]\geq 0.$$

According to the Lagrange interpolation formula, there exists ξ satisfying

$$\begin{aligned}&\mathop{\mathbb{E}}_{P}[Err(-k\cdot ln(p(P;\mu)))]-\mathop{\mathbb{E}}_{P}[Err(-k\cdot ln(p(P;\mu^*)))]\\&=\mathop{\mathbb{E}}_{P}\left[-Err'(\xi)k\cdot ln\left(\frac{p(P;\mu)}{p(P;\mu^*)}\right)\right].\end{aligned}$$

It is obvious that $Err'(\xi) > 0$ for the monotonicity of Err; considering the continuity of Err, $|Err'|$ is finite and there exists a $m > 0$ satisfying that

$$\mathop{\mathbb{E}}_{P}\left[-Err'(\xi)k\cdot ln\frac{p(P;\mu)}{p(P;\mu^*)}\right]\geq\mathop{\mathbb{E}}_{P}\left[-mk\cdot ln\frac{p(P;\mu)}{p(P;\mu^*)}\right].$$

According to Jensen's inequality.

$$\begin{aligned}&-mk\mathop{\mathbb{E}}_{P}\left[ln\frac{p(P;\mu)}{p(P;\mu^*)}\right]\geq -mk\cdot ln\left(\mathop{\mathbb{E}}_{P}\frac{p(P;\mu)}{p(P;\mu^*)}\right)\\&=-mk\cdot ln\left(\int_P\frac{p(P;\mu)}{p(P;\mu^*)}p(P;\mu^*)dP\right)=-mk\cdot ln1=0.\end{aligned}$$

Assume that we have collected N data $P_1,\ldots,P_N$ from the worker, and we let $\hat{\mu}$ be the value that minimizes the $\frac{1}{N}\sum_{i=1}^{N}Err(P_i;\mu)$. Theorem 1 above shows that when the number of data we have collected is large enough, the $\hat{\mu}$ we obtain from the data will approximate to the real mean value μ^*. We will use Hoeffding inequality to show the performance of the approximation.

Theorem 2 *With probability* $1-2e^{-\frac{2\epsilon^2}{N}}$*, the difference between the expectation of the error of* $\hat{\mu}$ *obtained from collected data and the expectation of the error of* μ^* *is less than* ϵ*:*

$$Pr\left(\frac{1}{N}\sum_{i=1}^{N}Err(P_i;\hat{\mu})-\mathop{\mathbb{E}}_{P}[Err(P_i;\mu^*)]\leq\epsilon\right)\geq 1-2e^{-\frac{2\epsilon^2}{N}}.$$

Proof 31 We use $Err(P;\mu)$ to represent $Err(k\cdot ln(p(P;\mu)))$. Assume that $Err(P;\mu)\in[m,M]$, with $0\leq m\leq M\leq 1$, according to Hoeffding inequality, we have

$$Pr\left(\left|\frac{1}{N}\sum_{i=1}^{N}Err(P_i;\mu)-\mathop{\mathbb{E}}_{P}[Err(P_i;\mu)]\right|\geq\epsilon\right)\leq 2e^{-\frac{2\epsilon^2}{N}}$$

for any μ. Then

$$Pr\Big(\frac{1}{N}\sum_{i=1}^{N} Err(P_i;\hat{\mu}) \le \mathop{\mathbb{E}}_{P}[Err(P_i;\hat{\mu})] + \epsilon\Big) \ge 1 - e^{-\frac{2\epsilon^2}{N}},$$

$$Pr\Big(\frac{1}{N}\sum_{i=1}^{N} Err(P_i;\mu^*) \ge \mathop{\mathbb{E}}_{P}[Err(P_i;\mu^*)] - \epsilon\Big) \ge 1 - e^{-\frac{2\epsilon^2}{N}}.$$

It is obvious that the two events are independent to each other, thus

$$\begin{aligned}
&Pr\Big(\frac{1}{N}\sum_{i=1}^{N} Err(P_i;\hat{\mu}) \le \mathop{\mathbb{E}}_{P}[Err(P_i;\hat{\mu})] + \epsilon, \\
&\frac{1}{N}\sum_{i=1}^{N} Err(P_i;\mu^*) \ge \mathop{\mathbb{E}}_{P}[Err(P_i;\mu^*)] - \epsilon\Big) \\
&\ge (1 - e^{-\frac{2\epsilon^2}{N}})^2 \ge 1 - 2e^{-\frac{2\epsilon^2}{N}}.
\end{aligned} \tag{4.6}$$

When

$$\frac{1}{N}\sum_{i=1}^{N} Err(P_i;\hat{\mu}) \le -\mathop{\mathbb{E}}_{P}[Err(P_i;\hat{\mu}) + \epsilon],$$

$$\frac{1}{N}\sum_{i=1}^{N} Err(P_i;\mu^*) \le -\mathop{\mathbb{E}}_{P}[Err(P_i;\hat{\mu}) - \epsilon],$$

we have

$$\begin{aligned}
0 &\le \frac{1}{N}\sum_{i=1}^{N} Err(P_i;\hat{\mu}) \le -\mathop{\mathbb{E}}_{P}[Err(P_i;\hat{\mu})] \\
&\le \frac{1}{N}\sum_{i=1}^{N} Err(P_i;\hat{\mu}) - Err(P_i;\mu^*) + \epsilon \\
&\le \epsilon.
\end{aligned}$$

Combining the Eq. (4.6), the theorem is proved.

The result shows that when N is big enough, the $\hat{\mu}$ converges to the real mean value μ^* in probability. Given that δ is far smaller than $|\mathbf{r}|$ and that $\mu(\mathbf{r})$ is continuous over $\mathbf{r}$, we may make an approximation that for any position $\mathbf{r}'$ on the circle centered at $\mathbf{r}$ with an arbitrary radius δ,

$$\mu(\mathbf{r}') = \mu(r) + \nabla\mu(\mathbf{r})\delta\cos(\phi), \tag{4.7}$$

where ϕ is the angle between $\mathbf{r}$ and $\nabla\mu(\mathbf{r})$ in Fig. 4.1.

Now we consider the thresholds of RSS value within which the estimated location will be at $\mathbf{r}$ rather than the boundary of the circle centered at $\mathbf{r}$. According to the MLE principle [104], it is clear that determining the location at $\mathbf{r}$ requires the RSS value fall into an interval with a higher threshold P_{high} and a lower one P_{low}, and this interval ensures that the RSS value of $\mathbf{r}$ in it should be higher than that at the boundary of the circle centered at $\mathbf{r}$. Then we can derive P_{high} and P_{low} as follows:

$$\begin{aligned} P_{high}(\mu) &= \mu(\mathbf{r}) + \frac{\nabla\mu(\mathbf{r})\delta \min\{\cos\phi, \sin\phi\}}{2}, \\ P_{low}(\mu) &= \mu(\mathbf{r}) - \frac{\nabla\mu(\mathbf{r})\delta \min\{\cos\phi, \sin\phi\}}{2}. \end{aligned} \tag{4.8}$$

Thus we have the specific form of the error

$$P_{error_{\mathbf{r}}}(P;\mu) = \int_{P_{high}(\mu)}^{P_{high}(P)} \frac{1}{\sqrt{2\pi}\sigma} e^{-\frac{(x-P)^2}{2\sigma^2}}\, dx. \tag{4.9}$$

4.1.3 *Quality-Aware Online Pricing Mechanism Design*

In this section, we present our online pricing mechanism design, based on the the standard of data quality evaluation presented in the previous section.

4.1.3.1 Loss and Regret Function

Theorem 1 has shown that the localization error incurred by the noisy data can be utilized to evaluate the data quality. Note that the error function is a concave function measuring the distance between P and $\mu(\mathbf{r})$, which is quite similar to the loss function defined in the online learning framework. In order to determine the price for fingerprints reported sequentially, we transform the error function into the loss function, without changing its monotonicity and the favored property that the function achieves the minimum in the real mean value μ^*. In particular, the concrete form of the loss function is

$$\ell_t(h_t, P) = \int_{P_{low}(h_t)}^{P_{low}(P)} \frac{dP}{\frac{\partial}{\partial h_t} F_{error_{\mathbf{r}}}\left(\frac{P-h_t}{P}; P\right)}. \tag{4.10}$$

In our model, the hypothesis h_t is the parameter of the probability distribution function of the RSS data in location $\mathbf{r}$, and P_t is the value of measured RSS data at time point t. Considering the convexity of the problem, we introduce online convex optimization techniques and rewrite the loss function $\ell_t(h_t, P_t) = \ell_t(h_t)$

for simplicity. The regret function can be defined as

$$R = \sum_{t=1}^{T} \ell_t(h_t) - \sum_{t=1}^{T} \ell_t(h^*), \tag{4.11}$$

where h^* is the optimal choice, causing the least loss in our solution space $\mathcal{H}$. The regret function reflects how the data deviate from the desired value, i.e., the real mean of RSS.

The goal of online learning is to obtain the best hypothesis when data are submitted sequentially. We here use *Online Gradient Descent (OGD)* algorithm [151] as the learning rule for h_t. The basic idea of the OGD algorithm is to minimize the loss of the current hypothesis, which is derived utilizing the data in past iterations. Compared with the traditional gradient descent method, the OGD algorithm calculates the gradient in an online manner, where the information of the complete data set is unavailable. It has been proved that the OGD has an upper bound of regret of $O(\sqrt{T})$, which ensures that the average regret tends to zero when T goes to infinite. With OGD, we obtain a h_t in each time point t according to

$$h_t = h_{t-1} - \eta \nabla \ell_t(h_{t-1}). \tag{4.12}$$

The traditional online learning technique assumes that the buyer can access all the data submitted online. Our data purchasing process collects submitted data that have high quality; however, with the online learning framework, even if the quality of a datum is very low, it can still provide certain information to calibrate the hypothesis. Consequently, we need to compensate the information loss incurred by giving up low-quality data, so that the influence on the loss can be neutralized. In particular, the estimation of loss is $E(\sum_{t=0}^{T} \delta_t \ell_t) = \sum_{t=0}^{T} q_t \ell_t$, where δ_t is the function showing whether the data is chosen and where q_t denotes the probability that the data submitted at t is acquired by the mechanism. However, the definition of regret in (4.11) still includes all the loss at each time t whether it has been purchased or not. In order to get an unbiased estimator of the regret, we define

$$\hat{\ell}_t(h) = \begin{cases} \frac{\ell_t(h_t)}{q_t} & \text{for data chosen,} \\ 0 & \text{else.} \end{cases} \tag{4.13}$$

With the unbiased estimator acquired, we can consider the mechanism as an OGD that receives $\hat{\ell}$ in each round t.

4.1.3.2 Quality-Aware Online Pricing Scheme

Assume that the MCSed data come in the sequence of $d_1, \ldots, d_T$, with each contains a cost $c_1, \ldots, c_T$. The pricing mechanism can determine how much the

Algorithm 5 RSS data pricing mechanism

Require: a sequence of data $d_1, \ldots, d_T$ coming in time $1, \ldots, T$ with each data possessing a cost c_t, $c_t \in [0, M]$
Ensure: a final hypothesis $\overline{h} \in H$
1: **for** t=1,...,T **do**
2: The mechanism acquires a hypothesis h_t from OGD
3: The mechanism posts a price π_t according to a distribution G_t over $[0, M]$
4: **if** $\pi_t \geq c_t$ **then**
5: The mechanism sends the loss function $\ell(h_t)/q_t$ back to the OGD and pay for the posted price π_t
6: **else**
7: The platform rejects the price and the mechanism sends 0 to the OGD
8: **end if**
9: **end for**
10: Return the final hypothesis $\overline{h} = \frac{1}{T}\sum_{t=1}^{T} h_t$

buyer should pay for the data. However, we have no means to know either the quality of data is good enough for localization or there will be a better one coming in the future. We formally define our *RSS data Pricing Mechanism (RPM)* in Algorithm 5.

In the algorithm, we could collect the submitted data periodically. Although the algorithm is oblivious to how to configure the period, we make the system to collect the data every 10ms when we do the experiments. Note that T is the size of the collected data sequence, where we set $T = 1000$ in our experiments, meaning we let the system collect 1000 data samples. Equations (4.12) is a part of the OGD algorithm [151]; Eq. (4.13) is involved in the algorithm in rows 4–6, where the data are chosen in the first case and not chosen in the second one. Although necessary revisions have been made to accommodate the indoor localization application scenario, the algorithm in essence is in the framework of the OGD algorithm, where the convergence property has been proved [151]. Our experimental results also show that the proposed algorithm converges in practice.

The mechanism and online learning algorithm produces a sequence of hypothesis $h_1, \ldots, h_T$. The main goal of our algorithm is to get the best hypothesis $\overline{h}$, the mean value of RSS, from the sequence. One simple approach is to average every hypothesis h_t acquired at each time t

$$\overline{h} = \frac{1}{T}\sum_{t=1}^{T} h_t. \tag{4.14}$$

To evaluate how well our final hypothesis $\overline{h}$ approximate to the optimal hypothesis h^*, we define the risk $\mathcal{L}(h) = \frac{1}{T}\sum_t \ell_t(h)$ of a hypothesis h as the average of $\ell_t(h)$ over t.

Lemma 11 *The expectation of the risk of $\overline{h}$ is less than the $\mathcal{L}(h^*)$ plus R/T [152], that is*

$$\mathop{\mathbb{E}}_{\ell_t} \mathcal{L}(\overline{h}) \leq \mathcal{L}(h^*) + \frac{R}{T}.$$

This means that if $O(R(h^*)) < O(T)$, then the $\overline{h}$ will approximate to the optimal hypothesis h^* as T goes to infinite. However, the crux of the algorithm is to find the best distribution G_t used for the mechanism to post its price.

4.1.4 Data Pricing with Budget Constraints

The price distribution G_t is closely related to the budget setting. We here consider two scenarios: First, the buyer wants to buy the best data with a fixed budget; second, the buyer wants to buy data with certain quality with as least budget as possible.

4.1.4.1 Regret Minimization with Fixed Budget

In this scenario, the buyer has a fixed budget, and the target of the mechanism is to minimize the total regret defined in Eq. (4.11). In this section, we will give the exact form of the distribution G_t and the analysis of the regret bound according to the distribution.

Problem Formulation It is well-known that the regret bound of OGD is $\frac{||h^*||^2}{2\eta} + \eta \sum_{t=1}^{T} \nabla \ell_t(h_t)^2$. Substituting Eq. (14) into the expression above, we have the regret bound of RPM as follows

$$R \leq \frac{||h^*||^2}{2\eta} + \mathop{\mathbb{E}}_{\ell_t, q_t} \left(\sum_{t=1}^{T} \frac{\nabla \ell_t(h_t)^2}{q_t} \right). \tag{4.15}$$

At each time point t, the RPM needs to post a price π_t according to G_t in order to get a minimum regret, we thus reduce the problem of designing a mechanism into an optimization problem

$$\begin{aligned} &\min \sum_{i=1}^{n} \frac{\nabla \ell_i{}^2}{1 - G_i(c_i)} \\ &s.t. \quad \sum_{i=1}^{n} \int_{c_i}^{M} x dG_i(x) \leq B, \end{aligned} \tag{4.16}$$

where $\forall c_i, 0 \leq c_i \leq M$, and $G(0) = 0, G(M) = 1$. Note that the solution we need to find for the optimization problem is the CDF G_t, which is in contrast to the traditional optimization problems where the solutions are variables. We are to use calculus of variations technique to solve it.

Theorem 3 *The optimal solution of the optimization problem (4.18) is in the form of*

$$G_t(c) = \begin{cases} 1 - \frac{\nabla \ell_t}{\sqrt{\lambda c - \beta}} & c \in \left(\frac{\nabla \ell_t^2 + \beta}{\lambda}, M\right]; \\ 0 & else. \end{cases} \tag{4.17}$$

Proof 32 We first give our function space $V = \{y | y(0) = 0, y(M) = 1\}$, and we denote our cost function as

$$M(G_1, \ldots, G_T) = \sum_{t=1}^{T} \frac{\nabla \ell_i^2}{1 - G_t(c_t)}.$$

Then the augmented Lagrange function is derived as

$$J(G_1, \ldots, G_T, \lambda) = M(G_1, \ldots, G_T) + \lambda \left(\sum_{t=1}^{T} \int_{c_t}^{M} x dG_t(x) - B \right).$$

According to the Gateaux Derivative, we obtain that $\forall \hat{G} \in V$,

$$\delta J|_{G_t}(\hat{G}_t - G_t) = \int_{c_t}^{M} (-\frac{\nabla \ell_i^2}{(1 - G_t(c_t))^2} + \lambda x) d(\hat{G}_t(x) - G_t(x)),$$

and if $\overline{G_t}$ is the local minimum, then we have $\delta J(\hat{G}_t - \overline{G_t}) \geq 0$. Note that $\int_0^M d(\hat{G_t}(x) - G_t(x)) = 0$ and the arbitrariness of $\hat{G}_t$, we must have

$$-\frac{\nabla \ell_i^2}{(1 - G_t(c_t))^2} + \lambda x = \beta \geq 0$$

hold for any x on $[c_t, M]$, thus proved.

The major challenge now is to determine β and λ. Notice that $G(x)$ is not continuous, using the Stieltjes Integral, we can rewrite the constraint as follows

$$\begin{aligned} \sum_{t=1}^{T} \int_{c_t}^{M} x dG_t(x) &= \sum_{t=1}^{T} \left(\int_{c_t}^{M} x G_t'(x) dx + (1 - G_i(M)) M \right) \\ &\leq \sum_{t=1}^{T} \nabla \ell_t \left(\frac{2}{\lambda} \sqrt{\lambda M - \beta} + \frac{c_t}{\sqrt{\lambda c_t - \beta}} - \frac{2}{\lambda} \sqrt{\lambda c_t - \beta}\right) \leq B. \end{aligned} \tag{4.18}$$

The Stieltjes Integral here has its practical significance, because we assume that the cost lies between $[0, M]$, in other words, the mechanism does not accept any price higher than M, thus for any posted price c that is higher than M, the mechanism will only pay M instead of c.

Now since we get the solution of the G_t, the remaining is to determine the parameters λ and β. Recall our initial optimization problem of minimizing the regret bound. The Lagrangian is thus given as follows

$$\begin{aligned} L(\mu, \beta, \lambda) =& \sum_t \Bigg(\nabla \ell_t \Big(\sqrt{\lambda c_t - \beta} + \mu \Big(\frac{2}{\lambda} \sqrt{\lambda M - \beta} \\ &+ \frac{c_t}{\sqrt{\lambda c_t - \beta}} - \frac{2}{\lambda} \sqrt{\lambda c_t - \beta} \Big) \Big) \Bigg) - \mu B, \end{aligned} \tag{4.19}$$

and its gradient is obtained accordingly.

$$\begin{aligned} \frac{\partial L}{\partial \lambda} =& \sum_t \nabla \ell_t \Big(\Big(\frac{c_t}{2} - \frac{\mu}{\lambda} \Big) \frac{1}{\sqrt{\lambda c_t - \beta}} + \frac{\mu}{\lambda} \frac{1}{\sqrt{\lambda M - \beta}} \\ &- \frac{1}{2} \frac{c_t^2 \mu}{\sqrt{(\lambda c_t - \beta)^3}} \Big) \\ \frac{\partial L}{\partial \beta} =& \sum_t \nabla \ell_t \Big(\Big(\frac{\mu}{\lambda} - \frac{1}{2} \Big) \frac{1}{\sqrt{\lambda c_t - \beta}} - \frac{\mu}{\lambda} \frac{1}{\sqrt{\lambda M - \beta}} \\ &- \frac{\mu}{2} \frac{c_t}{\sqrt{(\lambda c_t - \beta)^3}} \Big) \\ \frac{\partial L}{\partial \mu} =& \sum_t \nabla \ell_t \Big(\frac{2}{\lambda} \sqrt{\lambda M - \beta} + \frac{c_t}{\sqrt{\lambda c_t - \beta}} \\ &- \frac{2}{\lambda} \sqrt{(\lambda c_t - \beta)} \Big) - B. \end{aligned} \tag{4.20}$$

According to the complementary slackness theorem, $\mu \neq 0$, which means that the constraint condition in Eq. (4.16) for the optimal solution is strict. An intuitive explanation of this property is that the mechanism should use up the budget to get the optimal data. It is infeasible to work out the analytic solution of the optimal value of β and λ due to the complex form of the equation itself and the fact that we do not have access to enough prior knowledge of c_t and $\nabla \ell_t$.

We here give the iterative update of these two parameters. We initially set β to a fixed value and λ to a very small value, e.g., 0.01. Then at each time point t, we update the value of $\lambda^{(t)}$ iteratively according to

$$\lambda^{(t)} = \frac{T^2}{B^2 M} \Big(\sum_{i=1}^{t-1} \frac{\nabla \ell_i (h_t)}{t-1} \sqrt{c_t} \Big)^2 + \frac{\beta}{M}. \tag{4.21}$$

The above iteration formula is straightforwardly derived from the constraint Eq. (4.18).

Regret Analysis When the parameter β and λ is fixed, a higher $\nabla \ell_t$ means a higher probability that a data will be purchased, which is in accordance with the definition of the gradient, the fastest direction that the current h_t descends to the ideal one. For the parameter λ, a smaller λ means that the mechanism would tend to pay a high price for the data. The parameter β can be seen as the most dominant factor that determines the minimum price that the mechanism would pay.

Generally, a more sufficient budget B would lead to a higher β and smaller λ. When budget B goes to infinite, the λ and β goes to zero, leading to a CDF of $G(M) = 1$. That is, when the budget is unlimited, the mechanism will try to purchase the data as many as possible. For a fixed β, we give the estimation of the upper bound of the regret of RPM in Theorem 4.

Theorem 4 *For a fixed β, the regret of RPM is bounded with*

$$R < O\left(\max\{\sqrt{T}, \frac{T\theta}{\sqrt{B}}\sqrt{1 - \frac{\beta B^2}{T\theta^2 M}}\}\right), \tag{4.22}$$

where $\theta = \mathbb{E}\frac{1}{T}\sum_t \nabla \ell_t (2\sqrt{M} - \sqrt{c_t})$.

Proof 33 We prove the theorem by firstly setting the $\beta_0 = 0$, and through simple calculation, we can have an estimation of $\lambda_0 = \frac{T^2}{B^2}\theta^2$. Since that $\partial L/\partial \beta > 0$, $\partial L/\partial \lambda > 0$ hold in β_0 and λ_0, we obtain that the optimal solution (β^*, λ^*) satisfies $\beta^* > \beta^0$, $\lambda^* > \lambda_0$. Considering the discontinuity of the G_t, we introduce $C = \{q_t | c_t < \frac{\nabla \ell_t^2}{\beta}\}$. It is obvious that all the elements in C equals to 1. We substitute λ_0 and β into the estimation of the regret bound given in Eq. (4.15)

$$\begin{aligned}
R &\leq \frac{||h^*||^2}{2\eta} + \eta\mathbb{E}\left(\sum_{t\in C} \nabla \ell_t(h_t)^2 + \sum_{t\notin C}\frac{\nabla \ell_t(h_t)^2}{q_t}\right) \\
&\leq \frac{||h^*||^2}{2\eta} + \eta\left(Te + \mathbb{E}\sum_{t\notin C}\nabla \ell_t c_t \sqrt{\frac{T^2\theta^2}{B^2} - \beta}\right) \\
&\leq \frac{||h^*||^2}{2\eta} + \eta\left(Te + \frac{T^2\theta^2}{B}\sqrt{1 - \frac{\beta B^2}{T^2\theta^2 M}}\right) \\
&< O\left(\max\{\sqrt{T}, \frac{T\theta}{\sqrt{B}}\sqrt{1 - \frac{\beta B^2}{T\theta^2 M}}\}\right),
\end{aligned}$$

and thus acquire the desired result.

4.1.4.2 Budget Minimization for Certain Quality Level

This section considers the scenario where the purpose of the mechanism is to achieve a given error between $\mathcal{L}(\bar{h})$ and $\mathcal{L}(h^*)$. We design another mechanism to pursue a minimum expectation of the money the buyer will pay to achieve the given bound of error. Recall that the difference between the $\mathcal{L}(\bar{h})$ and $\mathcal{L}(h^*)$ is at most the average regret bound $R_0 = R(h^*)/T$. We thus only need to consider the constraint on the average bound of regret which makes the problem feasible.

However, we also encounter the challenge that the objective function of this situation is formed as an integral, which is not a simple task to be resolved with classical optimization methods. Unlike the method we used in Theorem 3, in this case, the constraint of the problem that $\frac{1}{T}\sum_t \frac{\nabla \ell_t^2}{q_t} \leq R_0$ is of a relatively simple form and we observe that the expectation of cost can be approximated by $\sum_t q_t c_t$. Thus we first consider a relatively simple version of the problem with objective function $\min_{q_t} \sum_t c_t q_t$, where $q_t \in [0, 1]$ is the probability of the mechanism to purchase the data in round t. Then we generalize the form between q_t and c_t to obtain the form of CDF for our original problem.

Theorem 5 *The optimal q_t for the minimum $\sum_t c_t q_t$ in the budget minimization scenario is in the form of:*

$$q_t = \min\left\{1, \sqrt{\frac{\lambda}{c_t}} \nabla \ell_t\right\}. \tag{4.23}$$

Proof 34 Considering that the objective function is convex, we can derive the corresponding Lagrangian function

$$L = \sum_t c_t q_t - \lambda\left(-\sum_t \frac{\nabla \ell_t^2}{q_t} + R_0 T\right) - \sum_t \mu_t (1 - q_t). \tag{4.24}$$

The optimal K-T condition of the problem is

$$\frac{\partial L}{\partial q_t} = c_t - \lambda\left[\frac{\nabla \ell_t^2}{q_t^2}\right] + \mu_t = 0. \tag{4.25}$$

In light of complementary slackness, it is clear that when $q_t = 1$, we get $\mu_t \neq 0$ and when $q_t \neq 1$, we get $\mu_t = 0$. Thus based on (4.25), prove the theorem.

Since Eq. (4.23) holds for any c_t, and the cost c_t in our model is arbitrarily given, we may make a generalization of the relation between q_t and c_t, that is, for any $c \in (\lambda \nabla \ell_t^2, M]$, the probability of the data to be acquired is $\sqrt{\frac{\lambda}{c}} \nabla \ell_t$. Thus the cumulative distribution function of the budget saving pricing mechanism is of the form

$$G_t(c) = \begin{cases} 1 - \sqrt{\frac{\lambda}{c}}\nabla\ell_t & c \in (\lambda\nabla\ell_t^2, M], \\ 0 & else. \end{cases} \tag{4.26}$$

Similar to the problem in Eq. (4.16), it is difficult for the mechanism to determine the parameter λ in each round t without prior knowledge of ∇f_t and c_t. According to our constraint on regret and that the optimal condition in Eq. (4.25), λ cannot equal zero, we can get an approximation of the $\sqrt{\lambda^{(t)}}$ through simple calculations

$$\sqrt{\lambda^{(t)}} = \frac{\sum_t \sqrt{c_t}\nabla\ell_t}{R_0 T}. \tag{4.27}$$

After acquiring the form of the G_t, we can now make a relatively more precise estimation of the budget B

$$E(B) = \sum_t \int_{c_t}^{M} dG_t(c) = \frac{T^2}{R_0}\theta\phi, \tag{4.28}$$

where $\phi = \sum_t \frac{1}{T}\nabla f_t(\sqrt{M} - \sqrt{c_t})$ and $\theta = \sum_t \frac{1}{T}\nabla f_t\sqrt{c_t}$.

4.1.5 Experimental Results

We conduct experiments to validate our analysis. First, we set up a Wi-Fi transmitter in the indoor space and the devices held by volunteers automatically report the observed RSS values to the server when they reach into the range of Wi-Fi as shown in Fig. 4.2. Empirically, we set the $\nabla\mu(\mathbf{r}) = 1$ and $\eta = 1/\sqrt{2T}$ to minimize the regret bound through experiments.

Fig. 4.2 Experiment location

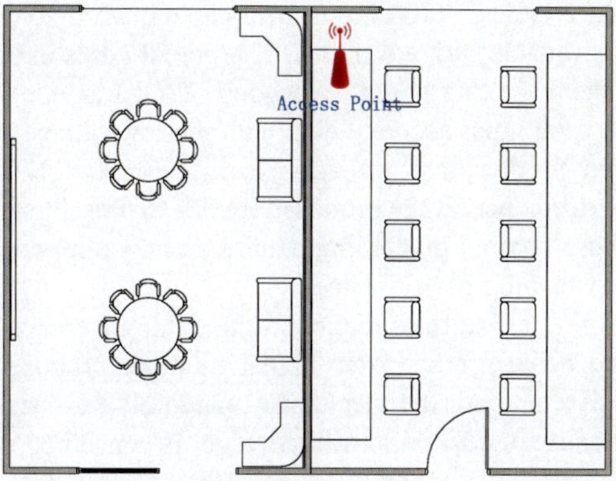

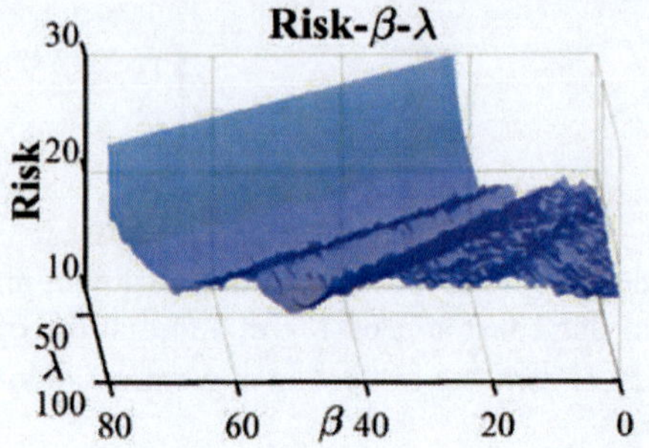

Fig. 4.3 Risk-β-λ

We sample 1000 RSS values at each location **r** and simulate costs of data through a normal distribution with mean value of 0.5 and variance of 1, and the maximum of the cost is bounded by $M = 1$. We first evaluate the impact of the parameter choice on our mechanism. We run the RPM with fixed budget $B = 100$, range λ from 10 to 100 and β from 0 to 100. We use the risk of the final hypothesis $\overline{h}$, $\mathcal{L}(\overline{h})$ defined in Lemma 11 to measure the performance of the mechanism. For each λ and β, we make 100 repeated trails and take the average of the results to diminish the stochastic error. The results shown in Fig. 4.3 indicate that for each λ, there exists a best β that achieves the minimum risk. If the value of β goes greater and that of λ decreases, the performance result of the mechanism tends to be that of the naive mechanism, which is to purchase as many as data with all budget at disposal. This is in accordance with our analysis in Sect. 4.1.4.

We also compare our proposed mechanism RPM with the take-it-or-leave-it (TIOLI) mechanism in [149, 150]. The TIOLI is an incentive model for encouraging participation in the survey, where the value of the private information the subject provides is closely related to the accuracy of the information. Since the subject is always trying to protect their privacy, the provided information's quality varies; however, the TIOLI scheme can guarantee that if the subject provides inaccurate information, then the corresponding reward to the subject is very small. The quality awareness in TIOLI is very similar with the RPM scheme proposed in this paper. Nevertheless, TIOLI must be applied in verifiable scenarios, where the ground truth is available; moreover, TIOLI processes data in batch instead of online as the RPM scheme in this paper. We choose TIOLI as one of the benchmarks because it can buy the most accurate data with the lowest cost. Although the in-batch processing and the verifiable scenario assumption in TIOLI may lead to its advantageous performance, if the proposed RMP can present performance close to TIOLI without those favored processing pattern and assumption, then we can say the RMP has an outstanding performance.

We utilize the samples in the database to evaluate the 3 kinds of mechanisms: the naive method, the TIOLI method, proposed RPM with $\beta = 0$. With the naive method, the mechanism randomly purchases data in an online manner with maximum cost M until the budget is consumed up. With the RPM method, we fix β and adjust λ iteratively according to Eq. (4.21), and use the risk $\mathcal{L}(\overline{h})$ to judge the quality of our final hypothesis. We run each algorithm 100 times and present the average of the results.

The value of RSS fingerprints may vary due to the noise in the indoor environment, which could be even worsened by slightly shaking the sensing devices or blockage of other people. In order to make a reasonable assessment on the mechanism, we analyze collected RSS fingerprint sequences according to how they vary in amplitude. We select two kinds of sequences to conduct the following experiments, where the first kind is the sequence with high variation and the second is with low variation.

We first examine the risk of final hypothesis output by different mechanisms with different budget limits, where the results are shown in Fig. 4.4. It is apparent that the curves of all those mechanisms gradually converge as budget mounts, verifying the online-to-batch conversion we mentioned before. It can be seen that the final risk of RPM outperforms naive method in a large amount while the budget limit is at a median level (100–600 approximately). The difference between RPM and naive method is getting smaller as the budget is increasing. It demonstrates that while the budget is insufficient, RPM works much better than the naive method. There is a remarkable fact that RPM makes prediction according to the data sample received in each round, while TIOLI could obtain all the information from data sequence. The results demonstrate that the performance of RPM catches up with that of TIOLI, indicating RPM is an effective algorithm under the online learning framework.

We then conduct experiments to estimate the mean value of RSS distribution with different budget limits. According to Theorem 1, we approximate the true value of fingerprint by averaging reported data sequence with large amount of RSS samples, and compare it with the prediction generated by different mechanisms. As shown in Fig. 4.5, the difference between RPM and TIOLI is very small and both of them can predict RSS closely to the true value even though the budget is very low, while naive method predicts well only if the budget is relatively high (800 or more). Considering

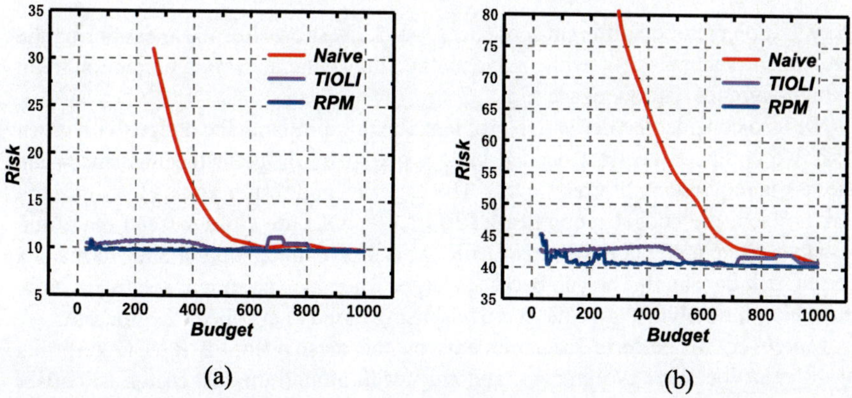

Fig. 4.4 Risk-budget on different sequences. (**a**) Low var. sequence. (**b**) High var. sequence

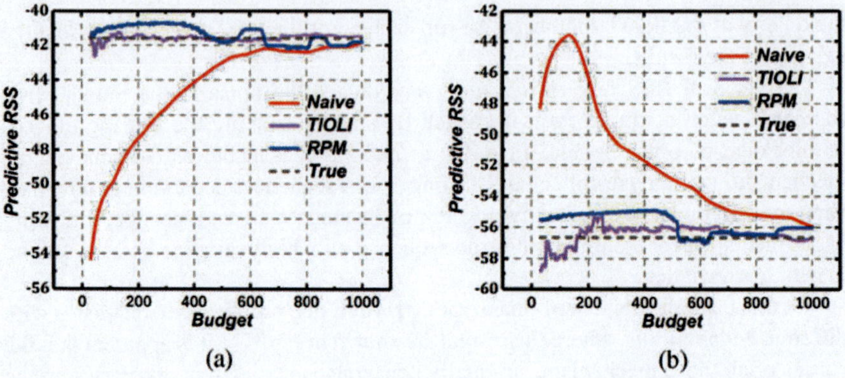

Fig. 4.5 Predictive RSS values on different sequences. (**a**) Low var. sequence. (**b**) High var. sequence

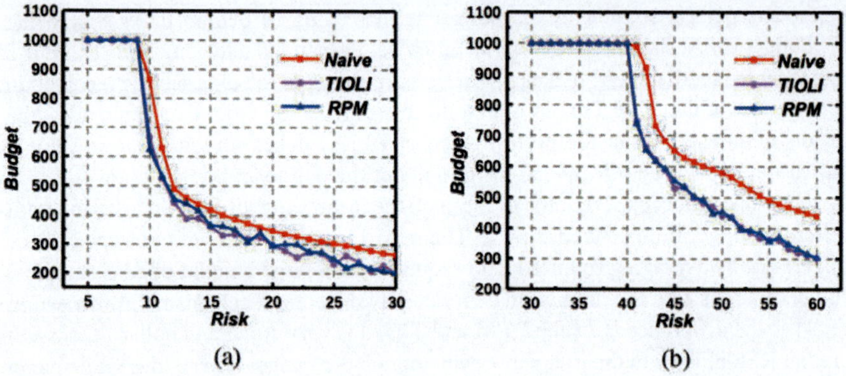

Fig. 4.6 Budget-risk on different sequences. (**a**) Low var. sequence. (**b**) High var. sequence

the variation factor on different sequences, our RPM shows its robustness during the data procurement process, while naive method may generate wrong prediction trend in high-variance data sequence Fig. 4.6.

We also consider the derivative problem about minimizing the budget with given risk. We compare the least needed budget output by different mechanisms when the risk is required at different levels. The experimental results evince that once the risk is fixed, the budget required by RPM and TIOLI are close to each other and less than naive method, indicating that RPM can save much more money to reach a certain risk threshold. The gap becomes larger when the variation is relatively high, indicating the stability of RPM in crowdsensing based data collection process.

Moreover, we perform measurements on the mean value of RSS fingerprints at different locations. Volunteers randomly walk around the test region to collect data sequences at 100 different locations. The predictive error is defined as the difference between true value and the predicted value generated by mechanisms.

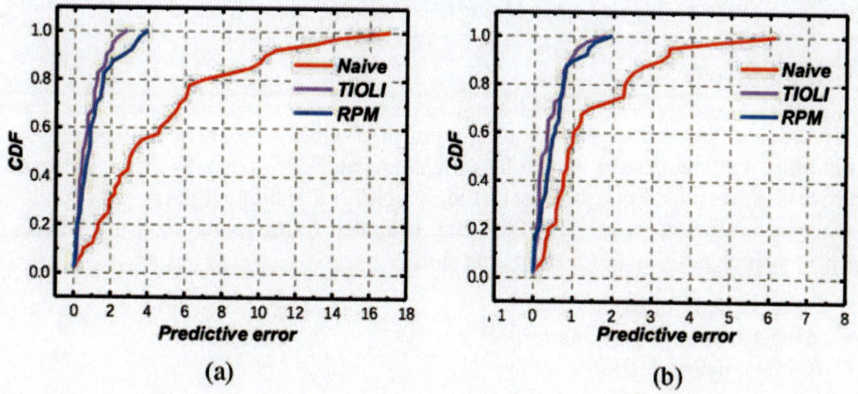

Fig. 4.7 Cumulative distribution function of error. (**a**) Budget = 200. (**b**) Budget = 600

We evaluate different mechanisms by their statistical distribution, which is shown in Fig. 4.7. Although our RPM method has no prior knowledge about the collected data sequence, it still makes accurate estimations on the mean value and achieves a comparative error rate with TIOLI. In practice, the whole fingerprint data set is not always available in advance, and the prediction may vary with the characteristics of arriving data, our online learning based mechanism definitely shows superiority in such practical scenarios. It is also observed that the difference between naive method and RPM is large with low budget($B = 200$) and gets smaller when the budget is getting higher($B = 600$), verifying the effectiveness of our proposed mechanism when the budget is not ample.

4.2 Incentive Mechanism for Mobile Crowd Sensing

These sections propose an incentive mechanism based on a quality-driven auction. The mechanism is specifically for the MCS system, where the worker is paid off based on the quality of sensed data instead of working time as adopted in the literature [247–252]. We theoretically prove that the mechanism is truthful, individual rational, platform profitable, and social welfare optimal. Moreover, we incorporate our incentive mechanism into a Wi-Fi fingerprint-based indoor localization system, in order to incentivize the MCS based fingerprints collection. We present a probabilistic model to evaluate the reliability of the submitted data, which is to resolve the issue that the ground truth for the data reliability is unavailable. We realize and deploy an indoor localization system to evaluate our proposed incentive mechanism and present extensive experimental results.

4.2.1 System Model and Design Challenges

4.2.1.1 System Model

We consider the MCS system consisting of three kinds of players: workers, agent platform, and requesters. The platform aggregates the demands from different requesters, recruits workers, checks the reliability of submitted data, and supplies the selected data to requesters. The time is slotted in the model, and the process below will be performed for each time slot.

- Contract determination
- Winner data set determination
- Payment determination
- Response and update

The main notations used in the paper are tabulated as in Table 4.2.

Table 4.2 Main notations

n	Number of interested workers
N	Set of interested workers
m	Number of data types
M	Compete set of data types
M_i	Subset of M, containing all data types sensed by worker i
F	Set of all submitted data.
x_{ij}	Submitted data of type j measured by worker i
b_{ij}	Worker i's lowest acceptable payment for x_{ij}
C_i	Contract set offered by user i
c_{ij}	Sub-contract offered by worker i for data x_{ij}
c_{-ij}	All sub-contracts except the one for data x_{ij}
N^*	Winner user set, $N^* \subseteq N$
M_i^*	Types of winner data set measured by worker i
W	Winner data set, $W = \{x_{ij} \mid i \in N^*, j \in M_i^*\}$
W^{st}	Winner data set when data x_{st} is not in the winner set
k_{ij}	Cost for x_{ij}
p_{ij}	Payment to user i for data x_{ij}
P_i	Set of p_{ij}, where $j \in M_i$
u_i	Utility of worker i
u_{ij}	Utility of worker i for data x_{ij}
u_p	Utility of the platform
$L(x_{ij})$	The value of data x_{ij}
$R(\cdot)$	Revenue function
$f(W)$	Social welfare function
$f_P(W)$	Social welfare function in platform's perspective

Contract Determination After receiving the demand, a set of workers $N = \{1, 2, \ldots, n\}$ will collect and submit the requested data with each claiming a price b_{ij} for every kind of submitted data x_{ij}, where x_{ij} represents the data of type j collected by user i and b_{ij} is the lowest acceptable payment user i asks for submitting x_{ij}. A worker could collect as many kinds of information as possible. We use $M = \{1, 2, \ldots, m\}$ to denote all types of data need to be sensed in the system, where M_i is a subset of M containing the data types measured by user i. Collecting the data of each type is considered as one task, and there are totally m tasks here. Each pair of data and claimed price is termed as a *sub-contract* denoted as c_{ij}, and all sub-contracts of worker i is termed as a *contract*, i.e., $c_{ij} = (b_{ij}, x_{ij})$, $C_i = \{c_{ij} | j \in M_i\}$. The worker has no need to know other workers' claimed prices and just needs to wait for the response from the platform after uploading the contract.

Winner Data Set Determination The platform needs to determine a winner data set W_j for each data type j after receiving all submitted contracts. We use $F_j = \{x_{ij} | i \in N\}$ to represent the set of all submitted data of type j. The winner data set is the set that can result in the maximum social welfare denoted as $W_j = argmax\{f(W_j) | W_j \in F_j\}$, where $f(W_j)$ denotes the system social welfare of all data in W_j. We use N_j^* to denote the set of the winners who have data of type j being accepted by the requester. Thus the winner data set is $W_j = \{x_{ij} | i \in N_j^*\}$. And we define M_i^* as the set of data types that the worker i collected and accepted by the requester.

Payment Determination After determining the winner data set, the platform needs to calculate the payment p_{ij} the requester should pay for each accepted data x_{ij}. If x_{ij} is not in the winner data set, then $p_{ij} = 0$; otherwise, $p_{ij} > 0$ and p_{ij} should be no less than b_{ij} denoting the claimed price. Note that this is only the payment set given by one requester and p_{ij} may be different for different requesters. In order to incentivize workers to submit high-quality data, we use k_{ij} to denote the cost of user i if submitted data x_{ij} is accepted by a single requester, and use $l \cdot k_{ij}$ to denote the cost if x_{ij} is accepted by l requesters. This is to offer the high-quality data provider high reward.

Response and Update Finally, the platform should pay off all workers for all data they have submitted and accepted by requesters. The platform will respond to user i with a payment set $P_i = \{p_{ij} | j \in M_i\}$. After that, the accepted data will be adopted into the databases of corresponding requesters. The rejected data could be used for the next-round auction and may get accepted by requesters with different requirements on data quality.

Sine workers are paid off according to the quality of submitted data, the proposed model can encourage participants of higher quality who are able to submit higher-quality data. With workers submitting low-quality data get lower or even no reward, the high-quality workers can get higher reward thus the utilization of the rewarding resource can be more effective.

4.2.1.2 Design Challenges

Since workers are individual entities, it is difficult to ask them to negotiate with each other in practice. Moreover, the platform is normally unable to know the private information of workers in advance. The distributed scenario seems to be suitable for the auction model, specifically, the reverse Vickrey auction, the essence of which is simple: First, the platform searches all 2^n possible candidate winner data sets, and calculates the total value and the social welfare each candidate set will bring. Secondly, compare the total value brought by each candidate data set with the platform's required budget; if the total value of the data set is less than the budget, abandon this set. Third, sort all the remaining candidates by the social welfare each can bring, and choose the candidate that can bring the greatest social welfare as final winner data set.

However, the reverse Vickrey auction model has the following drawbacks if it were applied to the MCS system, which may hinder itself from being adopted.

1. All the data (replaceable items) will be regarded as the same in the reverse Vickrey auction; however, the crowdsensed data for a single task in fact vary in their qualities. An effective incentive mechanism is supposed to encourage adoption of high-quality data.
2. The reverse Vickrey auction model will assume the cost of the worker as the quality of data submitted by the worker, which is not always the case in the MCS system. The worker may take many resources to collect some data, but the submitted data can turn out to be with low quality.
3. The platform will have to buy a certain amount of data even if the quality of the data is poor, which incurs inefficiency of funding utilization. This is because the platform could have saved the funding for higher-quality data, instead of buying a group of low-quality data with low value only to consume up the budget.
4. The social welfare in the model will only consider the workers' utilities and the platform's payment, which in together finally equals to the total cost of data in the winner set. It does not take the platform's revenue into account.
5. The model normally will have a high computation complexity. This is because we have to search all possible combinations of submitted data to find the winner data set. The computation complexity is basically $O(2^n)$ if there are n submitted data.

The fundamental reason of these drawbacks is: the reverse Vickrey model has to maintain the truthfulness property, which means that workers' claimed prices are their true costs for sensing the data, but this will lead to the fact that the payment of the platform must be independent of the prices asked by workers. To guarantee the independence, the platform has a fixed required budget that must be spent. The required budget here is the lowest value the platform should obtain from those selected data, which is oblivious to the actual quality of submitted data by workers.

4.2.2 *Quality-Driven Auction*

4.2.2.1 Overview

The idea of *Quality-Driven Auction* (QDA) is as follows. First, calculate a particular value for each sub-contract, which reflects the extent to which the data is worth of buying and sort all sub-contracts by that value. Second, separate the data into three categories and narrow down the searching range so that the candidate winner data are only selected from that range. Third, choose the data set that can maximize the social welfare of the system from the chosen range. The significant difference between the QDA and the reverse Vickrey auction is that we consider the revenue of the platform when calculating the social welfare and we do not need to have a required budget that must be spent, which can avoid buying low-quality data. As we narrow down the searching range, the time spent on the winner data set determination will also sharply decrease. Moreover, QDA has the following favored properties.

Individual Rationality The worker whose submitted data are accepted by requesters will have a utility greater than 0. That is, any worker i's utility for performing tasks:

$$u_i = \sum_{j \in M_i^*} u_{ij} = \sum_{j \in M_i^*} p_{ij} - \sum_{j \in M_i^*} k_{ij} \geq 0. \tag{4.29}$$

Truthfulness No worker can achieve a better utility by submitting a lowest acceptable payment other than its cost. Specifically, for any $i \in N$, $j \in M_i$ and any b_{ij} other than k_{ij}:

$$\begin{aligned} u_{ij}(c_{ij}, c_{-ij}) &= u_{ij}((b_{ij}, x_{ij}), c_{-ij}) \\ &\leq u_{ij}((k_{ij}, x_{ij}), c_{-ij}), \end{aligned} \tag{4.30}$$

where c_{ij} is user i's strategy for data x_{ij} and c_{-ij} is the strategy profile excluding user i's strategy for data x_{ij}.

Platform Profitability The utility of the platform u_p is greater than 0.

$$u_p = R\left(\sum_W L(x_{ij})\right) - \sum_W p_{ij}, \tag{4.31}$$

where $L(x_{ij})$ is an evaluation to the quality of data x_{ij}. We may consider $L(x_{ij})$ as the value of the data to a requester and $R(\cdot)$ is the revenue function with the following properties: $R(0) = 0$, $R'(x) > 0$, $R''(x) < 0$. This is because adding a reliable data ($L(x_{ij}) > 0$) into the winner data set will always bring the platform benefit. With more and more reliable data accepted, the marginal revenue brought by a new data will become less and less. Consequently the platform has a decreasing marginal revenue.

Social Welfare Maximization The total payoffs across all players are maximized. This means that all players including both workers and the platform are taken into account, in contrary to most of the work in the literature, which only focuses on either of them. We use the social welfare function $f(W)$ to quantify the social welfare:

$$\begin{aligned} f(W) &= \sum_{i \in N} u_i + u_p \\ &= \sum_W (p_{ij} - k_{ij}) + R\left(\sum_W L(x_{ij})\right) - \sum_W p_{ij} \\ &= R\left(\sum_W L(x_{ij})\right) - \sum_W k_{ij}. \end{aligned} \tag{4.32}$$

We use W^{st} to represent the winner data set where data x_{st} is definitely rejected:

$$f(W^{st}) = R\left(\sum_{W^{st}} L(x_{ij})\right) - \sum_{W^{st}} k_{ij}. \tag{4.33}$$

Note that the actual cost is only known by the worker himself, thus the platform simply treats the lowest acceptable payment b_{ij} as the cost for sensing data x_{ij}. Consequently, the social welfare in the platform's perspective is:

$$f_p(W) = R\left(\sum_W L(x_{ij})\right) - \sum_W b_{ij}. \tag{4.34}$$

If data $x_{st} \in W$, then its payment will be

$$p_{st} = f_p(W) - f_p(W^{st}) + b_{st}, \tag{4.35}$$

meaning that the incremental contribution data x_{st} does to the whole system. However, if a data is not accepted, then its payment will be 0.

It is worth mentioning that there may exist more than one winner sets, that is, $\exists W_1, \exists W_2$ and $W_1 \neq W_2$, for any other W, $f(W_1) = f(W_2) \geq f(W)$. All these data sets are acceptable to the platform, and none of them violates the rule of payment. It is easy to prove that choosing any one of those winner sets will not hinder the truthfulness and individual rationality of QDA. If the platform chooses W_2 instead of W_1, apparently it will not affect those who are selected in both and those selected

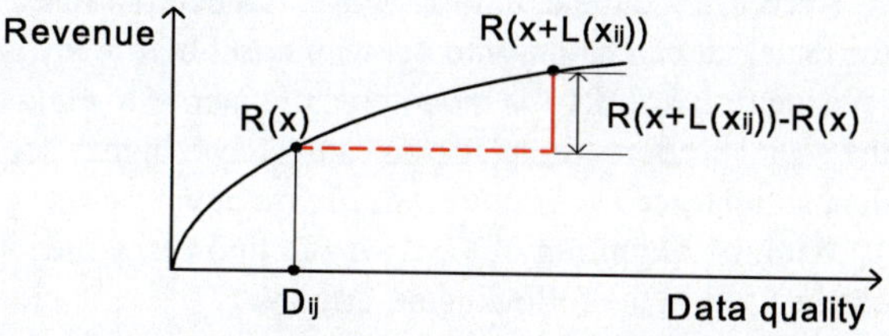

Fig. 4.8 Illustration of D_{ij}

in neither. If $x_{ij} \in W_2, x_{ij} \notin W_1$, $p_{ij} = f(W_1) - f(W^{ij}) + b_{ij}$. Now that $x_{ij} \in W_2, x_{ij} \notin W_1$, $f(W_1) = f(W^{ij})$, $p_{ij} = b_{ij}$. This means that all users will only claim $b_{ij} = k_{ij}$, and the utility for the data x_{ij} is always 0 no matter the data is selected or not.

4.2.2.2 Particular Value of the Sub-contract

The first step of the QDA is to calculate the particular value of each sub-contract mentioned earlier and sort all sub-contracts by the value. This value is a measurement that to what extent the data is worthy to buy, which is influenced by both the data quality and the price. We use D_{ij} to denote the value. Formally, if $R(L(x_{ij})) \geq b_{ij}$, then

$$D_{ij} = \max\{x | R(x + L(x_{ij})) - R(x) - b_{ij} \geq 0\}.$$

We here explain the meaning of D_{ij} with Fig. 4.8. The horizontal axis means the data quality, and the vertical axis means the revenue can be obtained by the platform. The curve stands for $R(\cdot)$, which is the revenue function. For a data x_{ij}, we use the length of a line segment to represent its associated quality $L(x_{ij})$, such as the length of the horizontal dashed line segment in Fig. 4.8. The starting and ending points of the dashed line segment are associated with two values $R(x)$ and $R(x + L(x_{ij}))$ on the curve. The increment of revenue by buying the data x_{ij} can be measured as $R(x+L(x_{ij}))-R(x)$. The utility of buying the data x_{ij} is $R(x+L(x_{ij}))-R(x)-b_{ij}$, where b_{ij} is the price of the data. For a given x_{ij}, the horizontal distance of the two points on the curve is fixed, as well as b_{ij}. If we move the two points from left to the right on the curve while keeping their relative horizontal distance, the value $R(x + L(x_{ij})) - R(x) - b_{ij}$ is decreasing and finally will drop below zero, because $R''(x) < 0$. The value D_{ij} is the largest x that can make the condition $R(x + L(x_{ij})) - R(x) - b_{ij} \geq 0$ still hold. We can see that each data will have an associated D_{ij}, which is only dependent on $R(\cdot)$, $L(x_{ij})$ and b_{ij} and independent of other submitted data.

If $R(L(x_{ij})) < b_{ij}$, $D_{ij} = 0$, which means that the revenue data x_{ij} can bring to the platform is even lower than its own cost, adding data x_{ij} to any set will make the social welfare decrease.

According to the definition above, if a data x_{ij} has a larger D_{ij}, the total value of the data that can be put into the winner set before x_{ij} is selected is larger. Since the revenue function $R(\cdot)$ is monotonically increasing and adding x_{ij} will not attenuate the social welfare, the platform can achieve higher social welfare. Consequently, a data with larger D_{ij} is more worthier to buy.

With the definition of D_{ij}, we can find many attributes of sub-contracts, which can be used in the following description.

Lemma 12 *For* $\forall H \subset F, x_{ij} \in H$, *if* $\sum_H L(x_{ij}) > D_{ij}+L(x_{ij})$, *then* $f(H/x_{ij}) > f(H)$.

Proof 35 According to the definition, $D_{ij} + L(x_{ij})$ is already the largest value after worker i's contribution and it will not decrease the social welfare; however, if there is a set H that has a larger welfare than the former one, that means data x_{ij} actually makes the social welfare lower. ■

Lemma 13 *For* $\forall H \subset F, x_{ij} \notin H$, *if* $\sum_H(L(x_{ij})) < D_{ij}$, *then* $f(H \cup x_{ij}) > f(H)$.

Proof 36 Since $R'()$ is monotonically decreasing, adding data x_{ij} to a set with smaller total value will have a higher marginal revenue while the cost remains the same, which will lead to a higher social welfare. Consequently, adding x_{ij} into a set whose total value is D_{ij} will not decrease the social welfare, and adding it to a set with smaller total value will have an even larger social welfare. ■

We assume that W is a winner data set, and let $L = \sum_W L(x_{ij})$ be the total value of data in the winner data set. $\forall x_{ij} \in F, \forall G \in R$, we divide the data set F into three sets:

$$Q_1(G) = \{x_{ij} | D_{ij} + L(x_{ij}) < G\}$$

$$Q_2(G) = \{x_{ij} | D_{ij} < G < D_{ij} + L(x_{ij})\}$$

$$Q_3(G) = \{x_{ij} | G < D_{ij}\}$$

Note that $Q_1 \cap Q_2 = Q_2 \cap Q_3 = Q_3 \cap Q_1 = \emptyset$, and $Q_1 \cup Q_2 \cup Q_3 = F$.

Theorem 6 $Q_1(L) \cap W = \emptyset$, $Q_3(L) \subseteq W$.

Proof 37 According to Lemma 1, if $D_{ij} + L(x_{ij}) < L$, then $x_{ij} \notin W$ and $Q_1(L) \cap W = \emptyset$; according to Lemma 2, if $D_{ij} > L$, then $x_{ij} \in W$ and $Q_3(L) \subseteq W$. ■

Theorem 7

- *If* $G > \sum_{Q_2(G) \cup Q_3(G)} L(x_{ij})$, *then* $L < G$.
- *If* $G < \sum_{Q_3(G)} L(x_{ij})$, *then* $L > G$.
- *If* $\sum_{Q_3(G)} L(x_{ij}) < G < \sum_{Q_2(G) \cup Q_3(G)} L(x_{ij})$, *then* $min\{D_{ij} | x_{ij} \in Q_2(G)\} < L < \max\{D_{ij} + L(x_{ij}) | x_{ij} \in Q_2(G)\}$.

Proof 38

- If $G > \sum_{Q_2(G)\cup Q_3(G)} L(x_{ij})$ and $L \geq G$, then $L > \sum_{Q_2(G)\cup Q_3(G)} L(x_{ij})$. $W \subseteq Q_2(L)\cup Q_3(L)$, because W is the winner set. $Q_2(L)\cup Q_3(L) \subseteq Q_2(G)\cup Q_3(G)$, because $L > G$. $L = \sum_W L(x_{ij}) \leq \sum_{Q_2(G)\cup Q_3(G)} L(x_{ij}) < G$, which is contradict to $L \geq G$.
- Similar to the proof above.
- We assume that $\sum_{Q_3(G)} L(x_{ij}) < G < \sum_{Q_2(G)\cup Q_3(G)} L(x_{ij})$. If L is larger than G, and $W \cap Q_2(G) = \emptyset$, apparently $W \cap Q_1(G) = \emptyset$, so $W \subseteq Q_3(G)$. Consequently, $\sum_{Q_3(G)} L(x_{ij}) \geq L > G$, a contradiction. Therefore, if L is larger than G, then $W \cap Q_2 \neq \emptyset$. In order to keep at least one element of $Q_2(G)$ in W, there must exist at least one element with $D_{ij} \in Q_2(G)$ with $L < D_{ij} + L(x_{ij})$, thus $L < \max\{D_{ij} + L(x_{ij}) | x_{ij} \in Q_2(G)\}$. The proof for the case when L is smaller than G is likewise. If $L = G$, then $\sum_{Q_3(L)} L(x_{ij}) < L < \sum_{Q_2(L)\cup Q_3(L)} L(x_{ij})$, so $W \cap Q_2(G) \neq \emptyset$.

■

4.2.3 Algorithm of QDA

We present the algorithm of determining the winner data set for a specific type of task for the convenience of presentation. The process of determining the entire submitted data set is similar thus omitted here.

Theorem 10 *The output of Algorithm 1 is the whole set of every* W_j,

$$f(W_j) \geq f(\tilde{W}_j), \forall \tilde{W}_j \subset F_j. \tag{4.36}$$

Proof 39 Algorithm 1 is equivalent to the process illustrated in Fig. 4.9. Each sub-contract can be characterized by two values about the data quality: D_{ij} and $D_{ij} + L(x_{ij})$. We represent each sub-contract using a line segment as shown in Fig. 4.9, with the two ends of the line segment assigned the two values D_{ij} and $D_{ij} + L(x_{ij})$ on the axis of data quality, respectively. If we use a line perpendicular to the data quality axis and cross the axis at L, all those horizontal line segments can be categorized into three classes: class one includes those line segments that

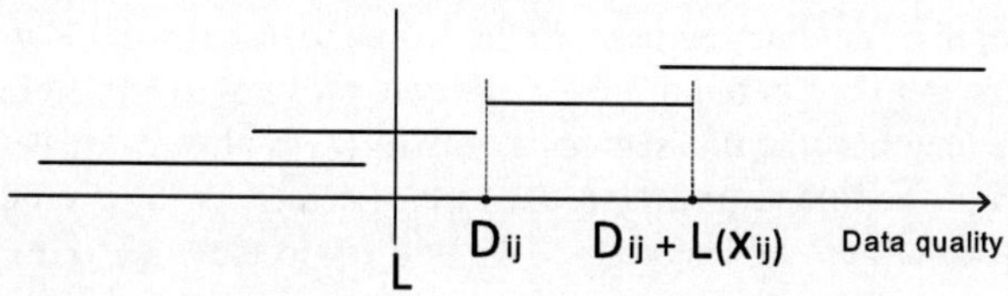

Fig. 4.9 Illustration of Algorithm 6

Algorithm 6 Quality-driven auction

Categorize $\{x_{ij}\}$ into corresponding F_j
$W_j = \emptyset$
$Q_2 = \emptyset$
$G = G_{high} = G_{low} = 0$
$i = 0$
Sort data in F_j according to their values of $D_{ij} + L(x_{ij})$ in the descending order
$G_{high} = \max_{F_j}\{D_{ij} + L(x_{ij})\}$
Sort data in F_j according to their values of D_{ij} in the descending order
$G_{low} = \min_{F_j}\{D_{ij}\}$
while true **do**
  $G = (G_{low} + G_{high})/2$
  for $s = 1$ to n **do**
    if $D_{sj} > G$ **then**
      $W_j \leftarrow W_j \cup x_{sj}$
    else if $D_{sj} + L(x_{sj}) > G$ **then**
      $Q_2 \leftarrow Q_2 \cup x_{sj}$
    end if
  end for
  if $G > \sum_{W_j \cup Q_2} L(x_{ij})$ **then**
    $G_{high} = G$
  else if $G < \sum_W L(x_{ij})$ **then**
    $G_{low} = G$;
  else
    break;
  end if
end while
return $W_j = W_j \cup V$;

completely lie on the left side of L, class two includes those lie completely on the right side of L, and class three includes those intersecting with the vertical line L.

It is straightforward that the three classes are in fact representing the three subcontract sets Q_1, Q_2 and Q_3 mentioned earlier, respectively. As stated in Theorem 1, if we know the exact sum of the quality of the elements in the winner data set, it is safe to say that all the elements in Q_3 are in the winner set while the ones in Q_1 are definitely not. The challenge is to find the value of such "exact sum". Fortunately, we could find the possible range of such value with the facilitation of Theorem 2 and the dichotomy in Algorithm 1.

We first randomly choose a value G, which separates all line segments into three classes. If the sum of the length of the line segments in Q_2 and Q_3 is less than G, the sum of length of the line segments in the winner data set will be on the left side of G by the first item of Theorem 2. We will try to move G to the left in this case. If the sum of the length of the line segments only in Q_3 is already greater than G, the sum of length of the line segments in the winner data set will be on the right side of G by the second item of Theorem 2. We will try to move G to the right in this case. In this way, we are able to narrow down the possible range of L corresponding to the winner data set on the data quality axis by the third item of Theorem 2. All sub-contracts completely on the right side of the range must be in the winner data set, and all those completely on the left side of the range must not. We now only need to search those sub-contracts in Q_2, which can maximize the system's social welfare. Those sub-contracts are denoted to be in the set V in Algorithm 1. Finally, the winner data set should be $W_j \cup V$. ■

It may be noticed that it takes exponential computation complexity finding V; however, the searching range in Algorithm 1 is just Q_2, which is much smaller than it would be in the reverse Vickrey auction (all possible sub-contracts combinations). Thus Algorithm 1 shows much higher computational efficiency in practice as to be shown in the performance evaluation section. It may also be noticed that the computational complexity of Algorithm 1 is also related to $R(\cdot)$ that influences the distribution of D_{ij}. However, it is interesting to find that the computation complexity of Algorithm 1 is in fact more dependent on the characteristics of MCSed data. This is because D_{ij} is dependent on the quality of each data in the first hand, and the quality of submitted data will be normally diversified for the nature of MCS system, where there is no guarantee on the quality of workers. The diversity of workers makes the line segments along the data quality axis sparsely distributed, which reduces the size of Q_2 and thus reduces the computation complexity.

4.2.4 Proving Properties of QDA

In this section, we prove that QDA has the properties of individual rationality, truthfulness, platform profitability and social welfare maximization. To prove the first property, we consider the following two situations: first, the worker claims his true cost as lowest acceptable payment, and second, the worker claims an arbitrary price, where the corresponding winner data sets are W and W^*, respectively.

Lemma 14 *If the data x_{st} is in both W and W^*, then*

$$W = W^*.$$

Proof 40 Since data x_{st} is accepted in both sets and all the other sub-contracts never change, we need to examine if we can find a set of data excluding x_{st}, which can maximize the social welfare. In the platform's perspective, social welfare is $f_P(W) = R(\sum_W L(x_{ij})) - \sum_W b_{ij}$. We can regard $R(\sum_W L(x_{ij}))$ as $R(L(x_{st})) + R_\Delta$, where R_Δ stands for the marginal revenue of all the data except x_{st} in the winner set. Since $R(L(x_{st})) - b_{st}$ is a constant when we know that x_{st} must be in the winner set and its claimed price, no matter what the value of b_{st} is, we need to find a set to maximize $R_\Delta - \sum_{W/\{x_{st}\}}$. Since this expression is independent of x_{st}, the result of finding such set will make no difference, which leads to $W = W^*$. The social welfare in the two cases could be different, but this does not mean that b_{st} can be arbitrary large, or the data may not be accepted, which contradicts the condition of this lemma. ■

Lemma 15 *If the data x_{st} is in both W and W^*, then*

$$f_P(W) = f_P(W^*) + b_{st} - k_{st}.$$

Proof 41 This is equally to prove

$$R\left(\sum_{W} L(x_{ij})\right) - \sum_{W/\{x_{st}\}} b_{ij} = R\left(\sum_{W^*} L(x_{ij})\right) - \sum_{W^*} b_{ij}.$$

According to Lemma 3, $W = W^*$ in this case, the result is straightforward. ■

Theorem 11 *Quality-Driven Auction is truthful.*

Proof 42 We consider the following two situations: first, the worker claims his true cost as lowest acceptable payment; second, the worker claims an arbitrary price. If $x_{ij} \in W$ and $x_{ij} \in W^*$, we prove that the utilities in both cases are the same. If $x_{ij} \notin W$ and $x_{ij} \notin W^*$, the utilities are of course both 0. ■

With Lemma 15, the utility for the data with an arbitrary price is

$$\begin{aligned} u^*_{st} = p^*_{st} - k_{st} &= f_P(W^*) - f_P(W^{st}) + b_{st} - k_{st} \\ &= f_P(W) - f_P(W^{st}) \\ &= f_P(W) - f_P(W^{st}) + k_{st} - k_{st} \\ &= p_{st} - k_{st} \\ &= u_{st}. \end{aligned}$$

In our proof, $f_P(W^{*st}) = f_P(W^{st})$ because x_{st} is in neither W^{st} nor W^{*st}, which means that whatever the contract is will not affect the result of the winner set, thus $W^{*st} = W^{st}$.

If $x_{ij} \in W$ but $x_{ij} \notin W^*$, the user will lose his chance to profit by claiming a price other than true cost. If $x_{ij} \notin W$ but $x_{ij} \in W^*$, for $b_{ij} > k_{ij}$, this will not happen because the lower the asked price is, the greater chance it is accepted. Then, if $b_{ij} < k_{ij}$, we prove that the payment for the data p_{ij} will be even lower than its cost.

$$p_{st}^* = f_p(W^*) - f_p(W^{st}) + b_{st} > k_{st}$$
$$f_p(W^*) - k_{st} + b_{st} > f_p(W^{st})$$
$$R\left(\sum_{W^*} L(x_{ij})\right) - \sum_{W^*} b_{ij} + b_{st} - k_{st} > f_p(W^{st}).$$

In conclusion, if $b_{ij} > k_{ij}$, the data could be accepted or unaccepted, and the corresponding utility is u_{ij} or 0, respectively. If the worker claims the true cost, the data will also have the two results and the utility is the same. Consequently, the worker would rather claim the true cost to get more chance that his data are accepted. If $b_{ij} < k_{ij}$, however, there are three possible utilities for that data, which are u_{ij}, 0 or negative. Therefore, the worker will not claim $b_{ij} < k_{ij}$ to prevent loss.

Lemma 16 *If a data $x_{st} \in W$, then*

$$f_p(W) \geq f_p(W^{st}).$$

Proof 43 Since the winner data set is the set which can maximize the social welfare in the platform's perspective, if $f_p(W^{st})$ is greater than $f_p(W)$, then choosing W^{st} will still be a better choice to maximize the social welfare even if data x_{st} exists. This contradicts the fact that data x_{st} is a winner data, thus $f_p(W) \geq f_p(W^{st})$. ■

Theorem 12 *Quality-Driven Auction is individual rational.*

Proof 44 If data x_{st} is rejected, corresponding payment will be 0, thus its utility is 0. We only need to consider the case when x_{st} gets accepted. In last theorem, we already proved that the user will only claim the true cost. Then, with Lemma 16,

$$\begin{aligned} u_{st} &= p_{st} - k_{st} \\ &= f_p(W) - f_p(W^{st}) + k_{st} - k_{st} \\ &= f_p(W) - f_p(W^{st}) \geq 0. \end{aligned}$$

■

Lemma 17 *If the data x_{st} is in W, then $f_p(W^{st}) \geq f_p(W/\{x_{st}\})$.*

Proof 45 The LHS is the social welfare when data x_{st} is not in the winner set. To obtain W^{st}, the platform may add some other data to the winner set, although the social welfare will not be better than the original case according to Lemma 14. However, $f_p(W^{st})$ will be still larger than $f_p(W/\{x_{st}\})$, which simply deletes x_{st} from the winner set. The process to get W^{st} is to get $W/\{x_{st}\}$ first, meaning to find whether there are other data which can increase the social welfare if included. ■

Theorem 13 *Quality-driven Auction is platform profitable.*

Proof 46

$$
\begin{aligned}
u_p &= R\left[\sum_W L(x_{ij})\right] - \sum_W p_{ij} \\
&= R\left[\sum_W L(x_{ij})\right] - \sum_W \{f_p(W) + b_{ij} - f_p(W^{ij})\} \\
&\geq \sum_{x_{st}\in W}\left\{R\left[\sum_W L(x_{ij})\right] - R\left[\sum_{W/\{x_{st}\}} L(x_{ij})\right]\right\} \\
&\quad - \sum_W \{f_p(W) + b_{ij} - f_p(W^{ij})\} \\
&= \sum_{x_{st}\in W}\left\{R\left[\sum_W L(x_{ij})\right] - R\left[\sum_{W/\{x_{st}\}} L(x_{ij})\right]\right. \\
&\quad \left. - f_p(W) - b_{st} + f_p(W^{st})\right\}.
\end{aligned}
$$

Since $R[\sum_W L(x_{ij})] - f_p(W) = \sum_W b_{ij}$, then

$$
\begin{aligned}
u_p &= \sum_{x_{st}\in W}\left\{f_p(W^{st}) - R\left[\sum_{W/\{x_{st}\}} L(x_{ij})\right] + \sum_{W/\{x_{st}\}} b_{ij}\right\} \\
&= \sum_{x_{st}\in W} \{f_p(W^{st}) - f_p(W/\{x_{st}\})\} \geq 0.
\end{aligned}
$$

■

Theorem 14 *Quality-Driven Auction is social welfare maximal.*

Proof 47 The Quality-Driven Auction is truthful, thus maximizing $f_p(W)$ is equivalent to maximizing the sum of every player's utility in the game, including the platform. We can substitute every $f_p(W)$ with $f(W)$ in all formulas above. The social welfare optimal is important because if we take the users and the platform as a whole sensing system, then the social welfare function can be regarded as the efficiency function of the sensing network, i.e., the revenue of the accepted data, minus the cost spent on sensing. ■

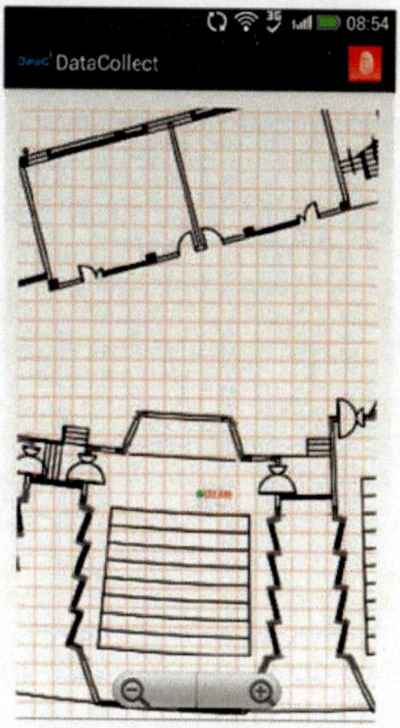

Fig. 4.10 Screenshot of the MCS App

4.2.5 *Applying QDA to the Indoor Localization System*

We apply the QDA based incentive mechanism to an indoor localization system, where the worker needs to report his current location and corresponding Wi-Fi RSS fingerprint. The challenge is how to measure the reliability of the submitted fingerprints. We propose to transform the unreliability of the submitted data to the unreliability of human beings' positioning sense, which can be profiled by experiments performed in advance and once for all.

We develop an App for users who could be enrolled as workers. Figure 4.10 illustrates a screenshot of the App, where the green spot is where a worker thinks he is standing, which is termed as *center* for the rest of the paper. The coordinate (22, 68) denotes the estimated position by the worker, which is measured by counting the number of squares horizontally and vertically. Other 8 cross points surrounding the center are termed as *neighbors*. The task released by the requester is to measure the fingerprint of the center. The worker stands on the place where he believes is the center, presses a button, and the corresponding fingerprint will be sent to a small scale cloud implemented with Cloud Foundry [153]. The measured fingerprint is the data to be submitted, while the corresponding cost of the task is autonomically computed by the App based on the resource of the mobile phone.

However, it is possible that the place the worker stands on is not the exact center the requester is interested in, which incurs error of the submitted data because of human beings' positioning sense error. Most likely, the worker actually chooses a neighbor around the center. Specifically, a worker i actually standing in the area k, thought himself in the area j, will submit the data x_{ij} that actually is x_{ik}.

The requester needs to measure the fingerprints of different centers, which can be modeled as different types of task denoted as $M = \{1, 2, \ldots, m\}$.

We use r_k to denote the probability that the data submitted for center k is indeed measured on center k, and r_{kj} is the probability that the data is actually for center k but the worker thought it is for the center j. Since the two probabilities are closely related to the probability that human beings' positioning sense error occurs, they can be obtained by general purpose experiments performed in advance. In our study, the building where we perform the experiment for our system is divided into grid according to the layout of the ceramic tiles, which are widely used in Chinese buildings. Since the ceramic tile is usually in the shape of 1.2 m×1.2 m-square, it needs reasonable efforts to find the specific center.

After the platform received the data x_{ij}, it actually regards the data as x_{ia}, where a is the area that has the largest $r_k \cdot r_{kj}$. Particularly, each fingerprints requester could hold a probability α as a threshold, in order to benchmark $r_a \cdot r_{aj}$. Then we can define the $L(x_{ij})$ in the indoor localization system as

$$L(x_{ij}) = ln\left(\frac{r_a \cdot r_{aj}}{\alpha}\right),$$

$$a = argmax\{r_k \cdot r_{kj} | k \in M\}.$$

It is straightforward to see that $L(x_{ij}) < 0$ when $r_a \cdot r_{aj}$ is smaller than α. The platform could reject the data since it is not very reliable. With the definition of data reliability, the QDA model can be applied to the indoor localization system.

4.2.6 Performance Evaluation

We perform our experiments in 3 classrooms and 1 corridor of Dongzhong Yuan building in Shanghai Jiao Tong University campus, with 100 m^2 in size. More than 500 fingerprints are collected with 20 mobile phones. The costs of the smart phones for performing the fingerprint collection are configured to be uniformly distributed over $[0, k_{max}]$, which is to model resource levels of large scale crowd. We perform each experiment 100 times and take the average value as the result. We verify if the important properties of our proposed scheme indeed hold in practice, and examine the corresponding cost in terms of the computational complexity.

4.2.6.1 Truthfulness and Individual Rationality

We first verify the individual rationality and truthfulness of the incentive mechanism and show the results in Fig. 4.11. The figure shows that any worker is unable to obtain a higher utility by deviating from the true price, which is the cost incurred to

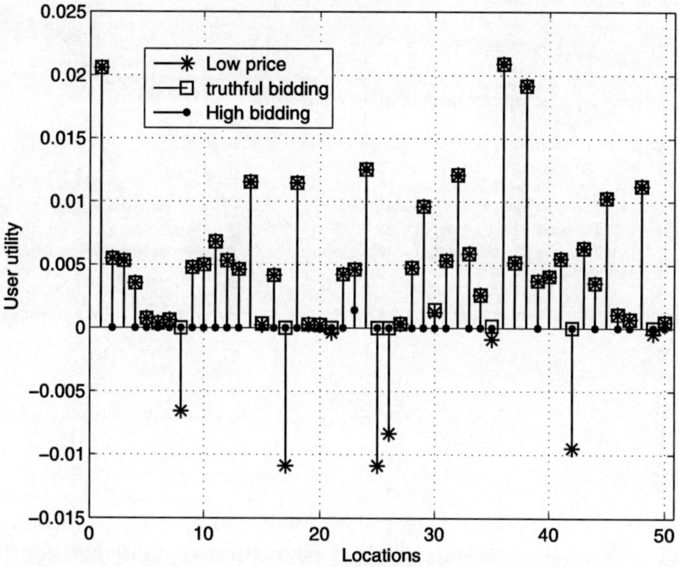

Fig. 4.11 Verification of truthfulness and individual rationality

collect the data. It also shows that any worker will obtain a non-negative utility if the true price is claimed.

In our experiment, the deviation is measured by the ratio of the claimed price to the true price. The claimed price is termed as high price if the ratio of itself to the true price is greater than 1, and it is termed as low price if the corresponding ratio is less than 1. We ask 20 workers to sample 50 locations and let each worker report a random high and low price each for 100 times at each location, respectively. We randomly pick a worker and measure the utility obtained and plot the results in Fig. 4.11. It is obvious that asking a true price is always the best choice, which leads to a non-negative utility. We can see that the worker's utilities are different in different locations, because the reliability of each sampling is different from others. The utilities are almost 0 in all locations when the high price is claimed, this is because the platform will exclude such workers from the candidates set as described in Algorithm 1. The utility could be less than 0 in some cases when the low price is claimed, this is because the payoff cannot even balance off the cost with the low price.

4.2.6.2 Social Welfare

The overall social welfare of the proposed scheme includes workers' utilities plus the platform's utility. The results are illustrated in Fig. 4.12a, where the performance of our mechanism is compared with that of the traditional reverse

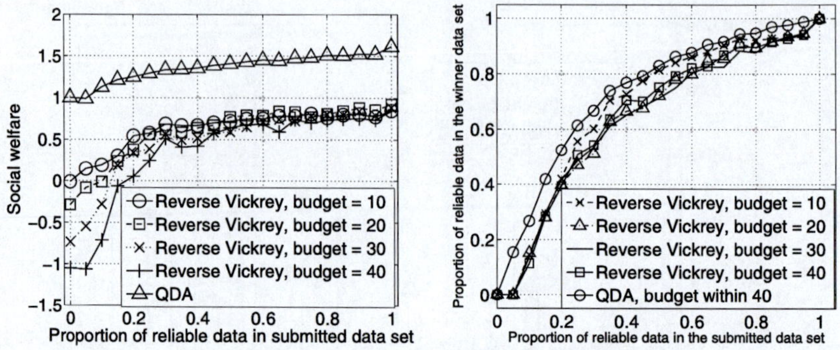

(a) (b)

Fig. 4.12 Social welfare and effectiveness of quality discrimination. (**a**) Social welfare. (**b**) Effectiveness of quality discrimination

Vickrey auction with different budgets. We implement a straightforward extension to the reverse Vickrey auction to let it take the platform's utility into consideration when calculating the social welfare, in order to make a fair comparison. Therefore, the difference between the social welfare is just because of the quality of the accepted data.

Our mechanism is quality driven and the utilization of funding is more effective. In our mechanism, the platform has no need to consume up the budget, and only needs to regard the budget as an upper bound. In the experiments, the budget for our mechanism is set to be within 40, and the platform can spend the funding based on the reliability of the submitted data. It is shown in Fig. 4.12a that the social welfare of our scheme is always higher than that of extended reverse Vickrey auction. With the quality driver, our scheme is always able to select data with high reliability, and refuses to accept the submitted data if all of them are unreliable. Given a data set with different proportions of reliable data, our scheme can always achieve a higher social welfare. In contrast, the extended reverse Vickrey auction accepts all data without checking the corresponding contribution to the social welfare, therefore the performance is lower than that of ours.

4.2.7 *Quality Discrimination*

We examine how effectively our proposed scheme can discriminate data with different levels of reliability. We make different data sets from the entire database of the indoor localization system, and configure the proportion of data with high probability for each set. We want to check if the data with high reliability can be selected by our scheme if the data set were submitted to the platform. The results are shown in Fig. 4.12b, where our mechanism is also compared with the extended reverse Vickrey auction scheme with different budgets. The horizontal axis denotes the proportions of reliable data in the submitted data set, and the vertical axis denotes the proportions of reliable data in the resulted winner data set, which are selected by our proposed scheme. We also set the budget of QDA scheme as 40. It is obvious that the QDA can select more reliable data compared with the reverse Vickrey auction, which indicates the effectiveness of our scheme. It is interesting that the proportion of reliable data in the winner data set selected by the reverse Vickrey auction does not increase with the corresponding budget. This is because the more funding the platform has, the more unreliable data can be selected by the reverse Vickrey auction, since it is not quality driven. The results corroborate the results in the section above. Since more reliable data are selected by our proposed scheme, the corresponding social welfare is higher under the QDA.

4.2.8 Computational Cost

This subsection evaluates the computational cost of the QDA with respect to three factors: the number of workers, the reliability of workers and the cost of workers for performing a task.

The computational cost with regard to the number of recruited workers is closely related to the scalability of any MCS incentive mechanism. Ideally, the computational cost of the platform should be independent of the number of recruited workers. The proposed QDA scheme provides a smart way to avoid searching the entire 2^n contracts, where n is the number of possible contract space. Figure 4.13 presents the computation time it takes for the QDA to find a winner data set in comparison with that for the reverse Vickrey auction. It is obvious that the proposed QDA outperforms the reference scheme. Since the reverse Vickrey auction needs to search all possible contracts to determine the winner data set, it takes more than 1 minute for the platform to finish the calculation when there are only 30 workers. It is easy to see that the proposed QDA is very computational efficient, with computation time negligible.

The computation cost of the platform is also related to the quality of the submitted data and the cost of workers for performing sensing tasks. As described in Sect. 4.2.5, the reliability of the submitted data for the indoor localization system under study is measured by $L(x_{ij})$. Figure 4.14a illustrates how long it takes to obtain the winner data set with submitted data of different levels of reliability. It is easy to see that the proposed QDA takes less time than the reverse Vickrey auction. The reason is the same as above: the searching space is narrowed. It is obvious that the QDA takes longer time to find the winner data set when the reliability of submitted data increases, while the reverse Vickrey auction makes almost no difference. This is because the QDA is sensitive to the data reliability, while the reverse Vickrey auction does not take the reliability into consideration.

The costs of workers to perform tasks will also influence the computation cost. This is because when the price is extremely low, buying any data can bring a

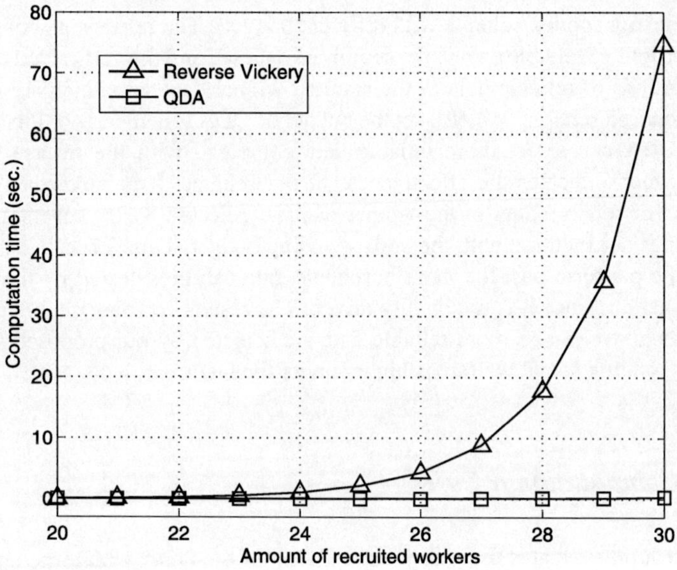

Fig. 4.13 Computation time with respect to the number of recruited workers

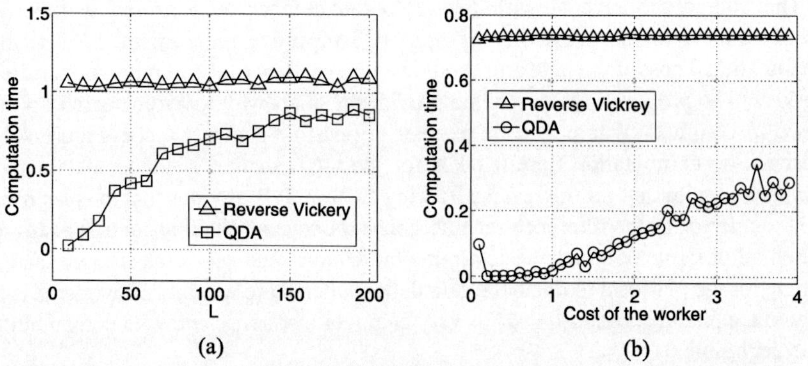

Fig. 4.14 Computation time with respect to the data reliability and worker's cost. (**a**) Computation time with respect to the submitted data reliability. (**b**) Computation time with respect to the worker's cost

marginal utility to the platform, which makes it hard to determine a winner data set. On the other hand, the higher the sensing cost is, the higher payment the platform needs to give. Even if the quality of data is not considered, it is still a challenge to find out a winner data set and keep the platform profitable at the same time. This is why the computation cost increases for both the QDA and the reverse Vickrey auction as shown in Fig. 4.14b. However, the computation cost of the QDA is still less than that of the reverse Vickrey auction.

4.3 Prediction for Fingerprints Data (Quadrotors)

This section presents HiQuadLoc, a RSS fingerprinting based indoor localization system for quadrotors. We propose a series of mechanisms including path estimation, path fitting and location prediction to deal with the negative influence incurred by the high-speed flight; moreover, we develop a 4-D RSS interpolation algorithm to reduce the site survey overhead, where 3-D is for the indoor physical space and 1-D is for the RSS sample space. Experimental results demonstrate that HiQuadLoc reduces the average location error by more than 50% compared with simply applying the RSS fingerprinting based approach for 2-D localization, and the overhead of RSS training data collection is reduced by more than 80%.

4.3.1 Working Process of HiQuadLoc

In this section, we introduce the working process of our system in detail. Like other typical indoor location systems based on RSS fingerprints [255–258], the process of localization in our system consists of *offline data training phase* and *online localization phase*.

4.3.1.1 Offline Data Training Phase

We first divide the space that requires the localization service into cubes with constant size. As a cube theoretically contains unlimited number of points, we just measure the RSS readings of the APs at the center of each cube, which requires great efforts in the case of 3-D localization. In our system, we propose the RSS interpolation algorithm based on 4-D space, which reduces the demand for high fingerprint sampling rate. In practice, we only measure RSS at 1 of each 8 cubes only. All the fingerprints information collected will be uploaded to the localization server and the server then processes the data according to the scheme described in Sect. 4.3.2.2. A detailed description of how to collect fingerprints could be found in Sect. 5.1.

4.3.1.2 Online Localization Phase

In the online phase, we divide the time axis into slots, each of which has a duration of T. During the whole online phase, the quadrotor keeps measuring the RSS from APs in each time slot. The specific process of the continuous-tracking mode consists of the following stages:

Stage 1: When the quadrotor has access to network and intends to localize itself, it first establishes links to the localization server via WLAN or cellular networks, and then sends a message to the server including the length of time slot T set by the quadrotor client. Note that T must be longer than the minimal RSS measurement time according to the quadrotor's hardware performance.

Stage 2: In each time slot, the quadrotor measures the RSS from surrounding APs. The result is included in an RSS-result message, and the client sends the message to the server immediately. Also, the message includes the value of communication delay T_d between the client and the server measured in the last localization process. Moreover, if the Turning Detector finds that the quadrotor is making a turning, the information will also be contained in the message of that time slot (Sect. 4.3.1.3). Note that *Stage* 2 is executed periodically, and the client is allowed to send the RSS-result message even if the localization result for the last time slot has not been returned by the localization server.

Stage 3: Each time when the localization server receives a RSS-result message, it estimates the position of the quadrotor according to the data uploaded as well as the historical localization results. The algorithm structure will be described in Sect. 4.3.1.4. The localization result is sent back to the client and its copy is stored at the server for future localization.

Stage 4: Once the client receives the result, the Delay Tester calculates the time interval between the sending out of the RSS-result message and the return of the result, and sets the interval as the new T_d.

For the single-locating mode, the quadrotor still periodically measures RSS. The difference is that the results are temporarily stored in the client without being uploaded. When the Turning Detector finds that the quadrotor has passed a corner, the data measured before the corner are deleted to save memory space. When the quadrotor needs to locate itself, the Delay Tester first measures the communication delay T_d between the client and the server. Then all the stored data, including T_d, will be uploaded to the server.

4.3.1.3 Turning Detection

To improve the performance of path estimation at the corners, the quadrotor client needs to detect the turning motion of the quadrotor. Our system utilizes the direction sensor installed on the quadrotor to detect turning.

Different from regular UAVs, it is harder for the client to detect the flight direction of quadrotors because a quadrotor can make lateral movements without changing its heading direction. However, we notice that when a quadrotor is moving in a specific direction, its normal vector will have a drift angle for the same direction, as shown in Fig. 4.15a. This is caused by the propulsion principle of the quadrotors. Thus we measure the normal vector instead of the heading direction of the quadrotor.

When fixing on a quadrotor, a typical direction sensor returns three values d_x, d_y and d_z, as shown in Fig. 4.15a. Among them we use d_y and d_z only to calculate

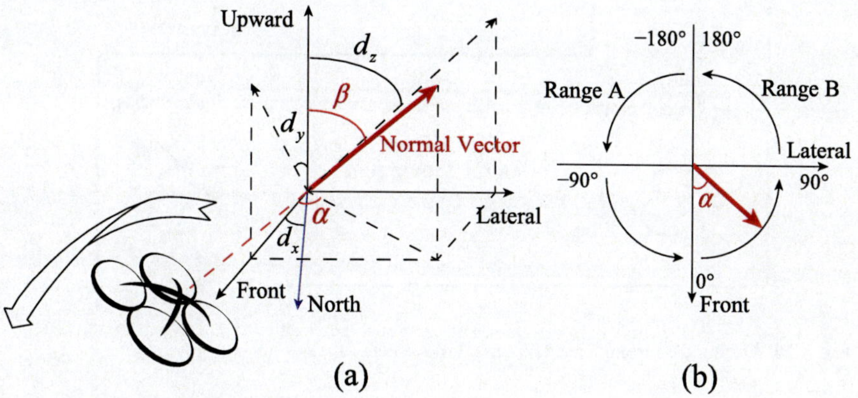

Fig. 4.15 During a turning the direction of the normal vector (determined by α and β) changes. (**a**) Direction vectors. (**b**) Range of α

the drift angle of the normal vector. According to the figure, we get (omit the cases where $d_y = 0$ or $d_z = 0$ for concision):

$$\alpha = \arctan\left(\frac{\tan d_z}{\tan d_y}\right) - 90°\mathrm{sgn}(d_z)[\mathrm{sgn}(d_y) - 1], \tag{4.37}$$

$$\beta = |\arctan\left(\frac{\tan d_y}{\cos\alpha}\right)|, \tag{4.38}$$

where α is the included angle between the projection of the normal vector and the front of the quadrotor. In Eq. (4.37) we project the range of α to $(-180°, 180°]$, as shown in Fig. 4.15b. The angle β is the drift angle of the normal vector. Note that we cannot use the heading direction determined by d_x alone to judge whether the quadrotor is turning, because during a turning based on lateral movement, d_x may not change (Fig. 4.16).

During a flight, the Turning Detector periodically measures d_y, d_z, processes them by low-pass filter, and calculates β for each time interval T_s, where $T_s \ll T$. With constant thresholds α_T and β_T, two different points are simultaneously followed:

- If $\beta < \beta_T$ for continuous duration nT_s, the quadrotor is hovering, during which it may change its direction. A *turning-start signal* will be included in the next RSS-result message. Once $\beta \geq \beta_T$ for another nT_s, a *turning-end signal* will be included in the next RSS-result message. Note that even if the quadrotor does not change its heading direction during the hovering, it is still necessary to notice the Path Corrector adjust the parameters of Kalman filter, otherwise it cannot perform well for the case of hovering.

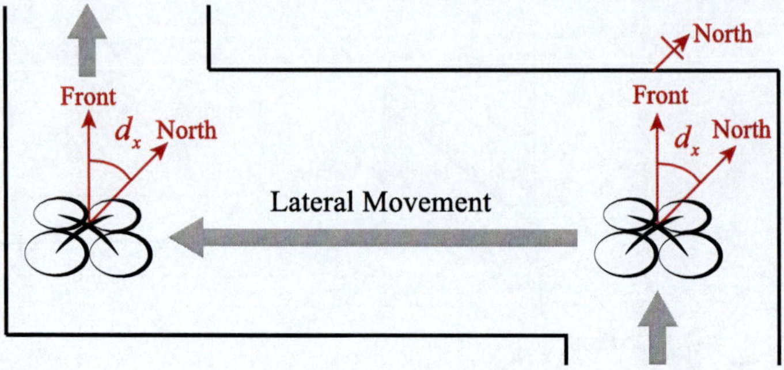

Fig. 4.16 During the turning based on lateral movement, d_x does not change

- When $\beta \geq \beta_T$ for more than nT_s, the Turning Detector starts to calculate α and continuously updates the mean value $\bar{\alpha}$ for its recent N_α values. To avoid the jump between $-180°$ and $180°$, if the recent N_α values simultaneously contain the angles in Range A and B (Fig. 4.15b), we add 360° to the angles in Range A, and add 360° to the newest α if it is also in Range A. Once $|\alpha - \bar{\alpha}| \geq \alpha_T$ for more than mT_s, which means that the quadrotor is turning, the Turning Detector includes the turning-start signal in the next RSS-result message. After that, when $\beta \geq \beta_T$ and $|\alpha - \bar{\alpha}| < \alpha_T$ for more than mT_s, which indicates the end of the turning, the turning-end signal is included in the next RSS-result message. Note that if $\beta < \beta_T$ happens during this time interval, the previous point is still followed. Moreover, the turning based on lateral movement can also be detected by this method.

With the turning-start and turning-end signals, the localization algorithm can take the impact of turning into account, which is to be described in Sect. 4.3.3.2 in detail.

4.3.1.4 Structure of Localization Algorithm

The localization algorithm in our system consists of two subalgorithms: *preliminary localization algorithm* (Sect. 4.3) and *path correction algorithm* (Sect. 4.3.3). The preliminary localization algorithm is based on the statistical properties of RSS fingerprints, and the 4-D RSS interpolation (Sect. 4.2) is applied to reduce efforts for radio map construction. The path correction algorithm contains the methods of path estimation, path fitting and location prediction, which further improves the accuracy of localization based on the result obtained from the preliminary localization algorithm.

In the continuous-tracking mode, when the localization server starts to process a new set of RSS data measured at time slot k, the following steps will be executed:

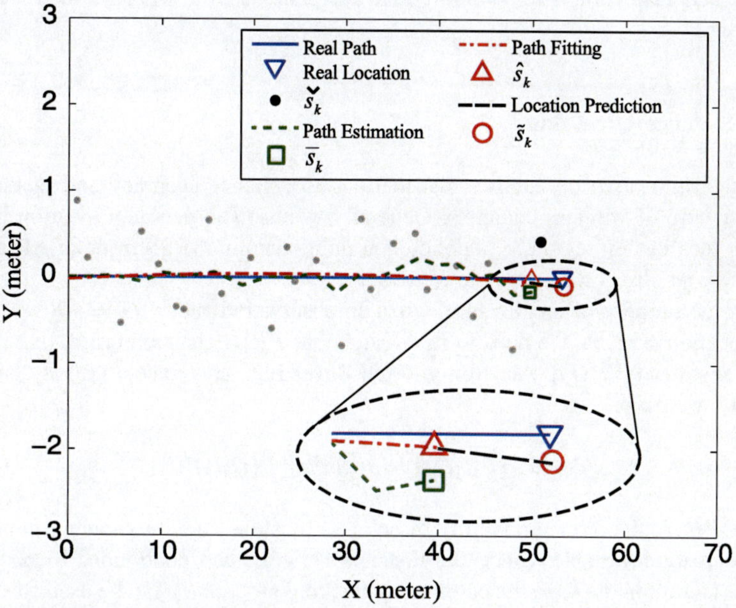

Fig. 4.17 The relationship among $\check{s}_k$, $\bar{s}_k$, $\hat{s}_k$ and $\tilde{s}_k$

Step 1: The preliminary localization algorithm is executed first. The estimated position derived by this algorithm is still preliminary and has only low accuracy. We denote it by $\check{s}_k$.

Step 2: Along with the historical localization results, $\check{s}_k$ is revised by the path estimation module. The new estimated position is denoted by $\bar{s}_k$.

Step 3: Based on the method of path fitting, the motion path is smoothed and $\bar{s}_k$ is replaced by $\hat{s}_k$.

Step 4: According to the communication delay T_d, $\hat{s}_k$ is further processed by the location prediction method, and the final output is $\tilde{s}_k$.

In the single-locating mode, the server continuously executes *Step* 1 and 2 for all the RSS data uploaded, and then execute *Step* 3 and 4 for the last $\bar{s}_k$ only. Figure 4.17 shows an example for the relationship among $\check{s}_k$, $\bar{s}_k$, $\hat{s}_k$ and $\tilde{s}_k$. The algorithms mentioned will be introduced in detail in the following sections.

4.3.2 Preliminary Localization Algorithm

In this section, we present the algorithm for determining the quadrotor location in an 802.11 WLAN environment. The tradition algorithm for data processing in offline data collection is to be modified to accommodate the 4-D RSS interpolation algorithm.

4.3.2.1 Theoretical Basis

Our algorithm uses probability distributions to enhance accuracy and tackle the noisy nature of wireless channels. Generally, we use l to denote a location in the indoor localization region, o to denote a training datum of RSS from an AP, and $\bar{o}$ to denote an observation variable of RSS.

For the purpose of localization, when an observation vector $\bar{O} = \{\bar{o}_1, \ldots, \bar{o}_K\}$ at a location is given, we need to find l such that $P(l|\bar{O})$ is maximized, that is, we want $argmax_l[P(l|\bar{O})]$. According to the Bayes rule, and since $P(\bar{O})$ is constant for all l, we have:

$$argmax_l[P(l|\bar{O})] = argmax_l[P(\bar{O}|l)P(l)], \tag{4.39}$$

where $P(l)$ is the prior probability of being at location l before knowing the value of observation variable. Since the distance between two contiguous localization results is limited by T, in the continuous-tracking mode $P(l)$ can be used to shrink the possible positions of the quadrotor. In our system, only the cubes which are less than $V_m T$ meters from the last localization result have $P(l) = 1$, otherwise $P(l) = 0$, where V_m is the maximum velocity of the quadrotor. Assuming the APs are independent, we have the following method to estimate $P(\bar{O}|l)$:

$$\begin{aligned} P(\bar{O}|l) &= \prod_{r=1}^{K} p(\bar{o}_r|l) \approx \prod_{r=1}^{K} \left[\frac{1}{N} \sum_{q=1}^{N} K(\bar{o}_r; o_l^q) \right] \\ &= \prod_{r=1}^{K} \left[\frac{1}{N} \sum_{q=1}^{N} \frac{1}{\sqrt{2\pi}\sigma} \exp\left(-\frac{(\bar{o}_r - o_l^q)^2}{2\sigma^2} \right) \right], \end{aligned} \tag{4.40}$$

where the function $p(\bar{o}_r|l)$ is the probability distribution of RSS value at the position l. $K(\bar{o}_r; o_l^q)$ is the *kernel function*, and we apply the widely used *Gaussian kernel*. The parameter o_l^q is the q_{th} training observation at the cube l, N is the number of the training data at each cube, and σ is an adjustable parameter that determines the width of the kernel.

4.3.2.2 4-D RSS Interpolation Scheme in Offline Phase

In this section, we introduce the scheme of 4-D RSS interpolation applied in the offline phase. Assume that we have a room with $X \cdot Y \cdot Z$ cubes, where X, Y,

Z denote the units in length, width, height direction respectively. We use G_{ijh} to denote a cube with coordinate (i, j, h), $i = 1, \ldots, X$, $j = 1, \ldots, Y$, $h = 1, \ldots, Z$.

In our localization algorithm, each AP is assigned a fingerprint map denoted by F_s, where $s = 1, \ldots, M$ and M is the number of APs in our system. For each fingerprint map, we use G^s_{ijh} to represent a specific cube in the map. As most of the previous work based on RSS fingerprint, we need to collect training data for each cube in each fingerprint map:

$$O_s = \{O^s_{ijh} | i = 1, 2, \ldots, X; j = 1, 2, \ldots, Y; h = 1, 2, \ldots, Z\}, \tag{4.41}$$

where

$$O^s_{ijh} = \{o^{sq}_{ijh} | q = 1, 2, \ldots, N\}. \tag{4.42}$$

Each O^s_{ijh} here includes N RSS signals measured in G^s_{ijh} and o^{sq}_{ijh} is the q_{th} training observation.

Normally, the cost of such training data collection is reasonable in 2-D localization, because it increases linearly with $X \cdot Y$, i.e., the area of the localization region. However, in the case of 3-D localization, the cost is proportional to $X \cdot Y \cdot Z$, i.e., the volume of the building. Thus the cost cannot be ignored. In our system, we propose the algorithm of 4-D RSS interpolation to accommodate the nature of 3-D localization.

We first let $C \triangleq \{(i, j, h) | i \in [1, X], j \in [1, Y], h \in [1, Z]\}$, $C_o \triangleq \{(i, j, h) | i = 1, 3, 5, \ldots, X; j = 1, 3, 5, \ldots, Y; h = 1, 3, 5, \ldots, Z\}$, and $C_e \triangleq C - C_o$, where we assume X, Y, and Z are odd numbers without losing generality. Different from previous works, we collect training data for each 8 *volume units*:

$$O'_s = \{O^s_{ijh} | (i, j, h) \in C_o\}. \tag{4.43}$$

According to Eq. (4.40), for each cube G^s_{ijh} the resulting density estimate for an observation $\bar{o}$ is a mixture of N equally weighted density functions

$$p(\bar{o} | G^s_{ijh}) = \frac{1}{N} \sum_{q=1}^{N} \frac{1}{\sqrt{2\pi}\sigma} \exp\left(\frac{(\bar{o} - o^{sq}_{ijh})^2}{2\sigma^2} \right), \ (i, j, h) \in C_o. \tag{4.44}$$

Specifically, to get a smoother estimation for RSS distribution, we use $\sigma = 10$. We iterate the observation $\bar{o}$ from its lower bound to upper bound to calculate the density distribution in the G^s_{ijh}. Specifically, we traverse $\bar{o}$ from -150 to 0 dBm and then normalize the density distribution to probability distribution for the G^s_{ijh}.

Let $p_s(\bar{o}, i, j, h) \equiv p(\bar{o} | G^s_{ijh})$, then we could get the values of $p_s(\bar{o}, i, j, h)$ when $\bar{o} \in [-150, 0]$ and $(i, j, h) \in C_o$ according to the training data. To estimate the rest of the values, we could apply the method of interpolation. Note that $p_s(\bar{o}, i, j, h)$ is continuous in the 4-D space

$$\{(\bar{o}, i, j, h)|\bar{o} \in [-150, 0], (i, j, h) \in C\}. \tag{4.45}$$

Thus we can use the cubic spline interpolation in the 4-D space to estimate the $p_s(\bar{o}, i, j, h)$ when $(i, j, h) \in C_e$. Denote these interpolated values by $p_s(\bar{o}, i, j, h)' \equiv p(\bar{o}|G^s_{ijh})'$, and we have the estimated probability for $\bar{o}$ to appear in G^s_{ijh} is

$$\hat{p}(\bar{o}|G^s_{ijh}) = \begin{cases} p(\bar{o}|G^s_{ijh}), & \text{if } (i, j, h) \in C_o \\ p(\bar{o}|G^s_{ijh})', & \text{if } (i, j, h) \in C_e \end{cases}. \tag{4.46}$$

Finally, we train each G^s_{ijh} in each fingerprint map F_s with training data O'_s. Since the value of RSS measured is discrete, we can previously store $\hat{p}(\bar{o}|G^s_{ijh})$ for each integral value of $\bar{o}$ in the database of the localization server. We are to show in Sect. 4.3.4.1 that the 4-D RSS interpolation algorithm can reduce the cost of fingerprint collection effectively, and the reduction in accuracy is acceptable.

4.3.2.3 Preliminary Position Estimation in Online Phase

We assume that the result of each RSS measurement operated by the quadrotor is denoted by

$$\bar{O} = \{\bar{o}^s_r | r = 1, \ldots, K, \ \ s = 1, \ldots, M\}, \tag{4.47}$$

where $\bar{o}^s_r$ is the r_{th} RSS value from the s_{th} AP. The symbol M is the total number of APs, and K is the number of times the quadrotor has measured for the same AP. Note that if the quadrotor is not covered by the s_{th} AP, $\bar{o}^s_r$ will be labeled by NaN and thus does not impact the following calculation.

For each $\bar{o}^s_r$, we can calculate $\hat{p}(\bar{o}^s_r|G^s_{ijh})$ according to Eq. (4.46). Since $\hat{p}(\bar{o}|G^s_{ijh})$ has already been stored, we can get $\hat{p}(\bar{o}^s_r|G^s_{ijh})$ directly by accessing the database in the localization server, which reduces our system's computation complexity.

In this paper, we assume that the signal strengths from the APs are independent. This assumption is justifiable for a well-designed 802.11 network, where each AP runs on a non-overlapping channel; therefore, we could estimate the joint probability using the marginal probability. According to Eqs. (4.39) and (4.40), we get the probability for that the client observes $\bar{O}$ when it is in G_{ijh}:

$$P(G_{ijh}|\bar{O}) = P(G_{ijh}) \cdot \prod_{s=1}^{M} \prod_{r=1}^{K} \hat{p}(\bar{o}^s_r|G^s_{ijh}). \tag{4.48}$$

Note that $P(G_{ijh})$ is retained to shrink the possible location region. Finally, the joint probability matrix can be denoted by

$$P_{joint} = \{P(G_{ijh}|\bar{O})|(i, j, h) \in C\}. \tag{4.49}$$

Based on Eq. (4.39), we then simply find the largest $P(G_{ijh}|\bar{O})$ in P_{joint} and its corresponding G_{ijh} denotes the estimated location for the vehicle. If this is the result for time slot k, we denote it by the preliminary localization result $\check{s}_k = [i, j, h]^T$, which will be further processed by the path correction algorithm. With the help of $P(G_{ijh})$, the algorithm complexity is $O(MKT)$, which are of the constant order.

4.3.3 Path Correction Scheme

In this section, we propose the path correction scheme based on a revised Kalman filter. The information of turning is taken into account to improve the performance of original Kalman filter at indoor corners. Also, we use the method of fitting to reach a higher accuracy, and also consider the negative effect of communication delay.

4.3.3.1 Path Estimation

The basic idea of path estimation is to further optimize the location estimation of the quadrotor, based on the location estimation result obtained from the RSS comparison as described in Sect. 4.3.2.3. We here utilize the concept of Kalman filter [154] for the optimization, where the preliminary location estimation result is regarded as the input of the "Predict" phase of the Kalman filter. The preliminary result is then passed to the "Correct" phase of the Kalman filter, which results in a more accurate estimation.

In the predict phase of Kalman filter, we assume that the quadrotor keeps moving in the 3-D physical space, and the RSS measurement is operated in every time slot. We denote the state of the quadrotor in the k_{th} time slot by

$$s_k = [x_k, y_k, z_k, v_{xk}, v_{yk}, v_{zk}]^T, \tag{4.50}$$

where (x_k, y_k, z_k) is the real position of the quadrotor at the k_{th} time slot, and (v_{xk}, v_{yk}, v_{zk}) is its velocity vector. Thus the motion could be modeled as

$$s_k = \begin{bmatrix} E_3 & TE_3 \\ \mathbf{0} & E_3 \end{bmatrix} s_{k-1} + w_{k-1} = F s_{k-1} + w_{k-1}, \tag{4.51}$$

where E_3 is a three-rank identity matrix, T is the length of a time slot, and w_k is the process noise following Gaussian distribution $w_k \sim N(0, \sigma_w^2)$.

Recall that the preliminary localization result at the kth time slot is denoted by $\check{s}_k = [i, j, h]^T$ in Sect. 4.3.2.3. With the Gaussian noises for localization, the preliminary localization model can be written as

$$\check{s}_k = [E_3, \mathbf{0}]s_k + u_k = Hs_k + u_k, \tag{4.52}$$

where u_k is the preliminary localization noise that follows Gaussian distribution $u_k \sim N(0, \sigma_u^2)$.

Consequently, the details of the "predict" phase in our system are as follows:

- Project the state ahead. This is as shown in Eq. (4.52).
- Project the error covariance ahead.

$$\bar{P}_k = FP_{k-1}F^T + Q_k, \tag{4.53}$$

where $Q_k = \mathbb{E}(w_k w_k^T)$ is the process noise covariance, and initially $P_0 = \mathbf{0}$.

The details of the "correct" phase in our system are as follows:

- Compute the Kalman gain:

$$K_k = \bar{P}_k H^T (H\bar{P}_k H^T + R_k)^{-1}, \tag{4.54}$$

where $R_k = \mathbb{E}(u_k u_k^T)$ is the preliminary localization noise covariance. Note that different from normal Kalman filter, the value of u_k would be changed in each time slot, which is to be shown in Sect. 4.3.3.2.
- Update the estimated state vector with $\check{s}_k$:

$$\bar{s}_k = F\bar{s}_{k-1} + K_k(\check{s}_k - HF\bar{s}_{k-1}), \tag{4.55}$$

where $\bar{s}_0 = [\check{s}_0^T, d_0^T]^T$ initially and d_0 is the initial unit direction vector provided by the direction sensor installed on the quadrotor. Output $\bar{s}_k = [\bar{x}_k, \bar{y}_k, \bar{z}_k, \bar{v}_{xk}, \bar{v}_{yk}, \bar{v}_{zk}]^T$.
- Update the error covariance:

$$P_k = (E_6 - K_k H)\bar{P}_k, \tag{4.56}$$

where E_6 is a six-rank identity matrix.

Since $(\bar{x}_k, \bar{y}_k, \bar{z}_k)$ is the estimated coordinate, we use $\bar{s}_k$ to denote the new estimated position directly in the following sections. Thus for each time slot the path estimation method optimizes the preliminary localization result $\check{s}_k$ and outputs $\bar{s}_k$ as the new estimated position. We are to show that a higher accuracy can be achieved by $\bar{s}_k$ in Sect. 4.3.4.2 with experiment results. As each execution of path estimation is to find the result of the "correct" phase once, the complexity is $O(1)$.

4.3.3.2 Parameter Readjustment During Turning

During the process of path estimation, one disadvantage of Kalman filter is that the localization accuracy at corners decreases obviously, as shown in Fig. 4.18. We can see that the estimated locations of the quadrotor deviate far from the real path of the quadrotor at the corner. This is caused by the unchanged weighting factor for the historical localization results. To overcome this we change the parameters of Kalman filter to adapt it to the turning motions of quadrotors.

As introduced in Sect. 4.3.1.3, once the Turning Detector detects a turning motion during a flight, it uploads turning-start and turning-end signals at the beginning and the end of the turning respectively. Then the following operations are executed:

Step 1: At the beginning of the turning, change σ_u^2 to $\bar{\sigma}_u^2$, where $\bar{\sigma}_u^2$ is a smaller value than σ_u^2.

Step 2: At the end of the turning, change $\bar{\sigma}_u^2$ back to σ_u^2.

The value of R_k should be recomputed after each step, since R_k actually controls the balance for the weighting factors of historical and newest localization results; reducing σ_u^2 leads to a higher weighting factor for the newest localization result, according to Eqs. (4.54) and (4.55). As shown in Fig. 4.18, at corners a high σ_u^2 would cause the overshoots of the path estimated by the Kalman filter. Thus relying more on the newest localization result at the corners can reduce these overshoots. We are to evaluate this revised scheme in Sect. 4.3.4.3. The change of σ_w^2 during turning has similar performance, thus we mainly focus on σ_u^2 in this paper.

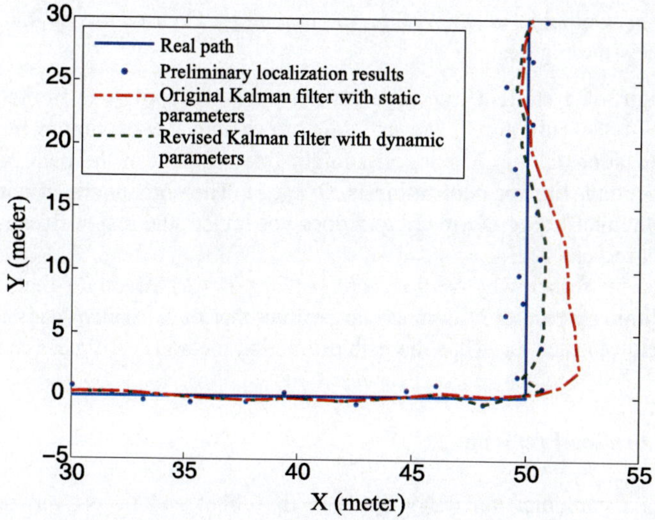

Fig. 4.18 The performance of Kalman filter at a corner (for 2-D case)

4.3.3.3 Path Fitting

Kalman filter works well only when its parameters such as σ_w^2 match with the real situation, which can hardly be satisfied in practice since the indoor environment varies. Thus the jitter of the estimated path as shown in Fig. 4.17 cannot be totally avoided. We apply the method of curve fitting to mitigate this impact. This approach is based on the nature that a quadrotor tends to move in nearly straight lines in indoor environments, e.g., flying along a long corridor. Since the flight path of the quadrotor is actually a 3-D curve, the computation complexity of 3-D curve fitting is large and is severely impacted by the initial parameters of the curvilinear function. To reduce the response time of the localization server, we focus on the projected curve of the flight path on the 2-D ground and the changing curve of $\bar{z}_k$ with time respectively.

Still with the help of the messages about turning motions, the following operations are executed:

Step 1: Once the quadrotor has finished a turning, the current time slot is labeled by T_c.
Step 2: For $\{\bar{s}_k | k = T_c, T_c + 1\}$, output $\hat{s}_k = \bar{s}_k$ directly.
Step 3: For $\{\bar{s}_k | k \geq T_c + 2\}$, we set $\bar{s}'_k = [\bar{x}_k, \bar{y}_k]^T$, and fit the data $\{\bar{s}'_l | T_c \leq l \leq k\}$ with the method of quadratic polynomial fitting. To avoid the case that the flight path is perpendicular to the x axis or the y axis, which leads to the bad performance of fitting, we exchange the two axes and choose the fitted curve with the maximum correlation coefficient. The result is denoted by curve L_k. Approximatively find the nearest point to $\bar{s}'_k$ on L_k, denote it by $[\hat{x}_k, \hat{y}_k]$. Also, we fit the data $\{[\bar{z}_l, l] | T_c \leq l \leq k\}$ with the quadratic polynomial fitting. The value of the result function at time slot k is denoted by $\hat{z}_k$. Output $\hat{s}_k = [\hat{x}_k, \hat{y}_k, \hat{z}_k, \bar{v}_{xk}, \bar{v}_{yk}, \bar{v}_{zk}]^T$. Go back to *Step* 4.3.3.3 once a new turning begins.

The length of a straight line segment between two corners is limited by the dimension of the building. If we use $L_{\max}$ to denote the maximum number of predicted locations along the longest straight line segment in the path predicting phase, the complexity for path fitting is $O(L_{\max})$. The computational complexity of the polynomial fitting is limited and does not impact the responsiveness of the localization server.

Note that the reason why we do not fit $\{\check{s}_l | T_c \leq l \leq k\}$ directly is that, the error of the preliminary results $\check{s}_k$ is much larger than that of $\bar{s}_k$, which leads to severe changes between each L_k. Thus the path estimation method is still necessary.

4.3.3.4 Location Prediction

Due to the communication delay between the client and the server, there still exists non-negligible localization error. Assume that the client sends the RSS-result message at time point t, due to the cost of computation time and transmission delay, when the client receives the localization result returned, the time is $t + T_d$.

Suppose the average velocity of the quadrotor during T_d is v_d, then the additional localization error will be at most $v_d \cdot T_d$. Thus for the high-speed moving quadrotors, it is necessary for our system to take the impact of communication delay into consideration.

In our system, we develop a location prediction scheme to predict the real location of a quadrotor according to the T_d it has uploaded. Moreover, the output of the path fitting scheme, $\hat{s}_k = [\hat{x}_k, \hat{y}_k, \hat{z}_k, \bar{v}_{xk}, \bar{v}_{yk}, \bar{v}_{zk}]^T$ contains the estimated velocity vector of the quadrotor; therefore, we can extend the motion curve of the quadrotor by

$$\tilde{s}_k = [\hat{x}_k + T_d\bar{v}_{xk}, \hat{y}_k + T_d\bar{v}_{yk}, \hat{z}_k + T_d\bar{v}_{zk}, \bar{v}_{xk}, \bar{v}_{yk}, \bar{v}_{zk}]^T. \tag{4.57}$$

Finally, the server responds the client with $\tilde{s}_k$, which is the final result of the localization process at time slot k. The relationship among $\check{s}_k$, $\bar{s}_k$, $\hat{s}_k$ and $\tilde{s}_k$ has been shown in Fig. 4.17, and we are evaluate the performance of the path correction scheme in Sect. 4.3.4.2.

It is straightforward that the complexity of the location prediction is $O(1)$. We can see that the computational complexity in each phase of localization is tractable. As the communication delay is dependent on the hardware and non-changeable, the low computational complexity could help improve the responsiveness of the server, which is important for high-speed quadrotors.

4.3.4 Experiment Results

In this section, we evaluate the performance of HiQuadLoc by conducting experiments with a quadrotor localization testbed. The experiments consist of the following parts: First, we evaluate the performance of the 4-D RSS interpolation algorithm in offline phase by operating an independent test in a lobby. Second, we deploy our system in a corridor and several lobbies, and evaluate the performance of path estimation, path fitting and localization prediction scheme. The benefit of turning detection is to be confirmed, and the impact of velocity on accuracy is to be shown. Third, we build CSI retrievable APs to implement CSI based localization for the quadrotor, and compare the performance of HiQuadLoc scheme with the CSI based scheme. Fourth, we verify that the power consumption of the HiQuadLoc is negligible compared with the basic power consumed by the quadrotor with experiment.

4.3.4.1 Evaluation of 4-D RSS Interpolation Algorithm

As the offline phase could significantly influence the accuracy of localization, we verify that the 4-D RSS interpolation algorithm can reduce the cost of RSS fingerprint collection without causing severe drop in localization accuracy. We conduct the experiment in a lobby with an interior volume of more than $(9 \times 9 \times 5)\mathrm{m}^3$. In the test, we deploy four Wi-Fi APs (HUAWEI E586Bs-2) at the four corners of the lobby.

For the offline data training phase, the lobby is divided into 405 cubes and we collect the RSS fingerprints at the centers of them using 6 Android smartphones. All of the smartphones are Nexus 4 with Android Jelly Bean (4.2.2) as their operation systems. The size of each cube is $1\,\mathrm{m} \times 1\,\mathrm{m} \times 1\,\mathrm{m}$, and we collect 30 sets of RSSes from each AP for each cube. We then generate two different sets of RSS fingerprints. For the *full fingerprints*, we use all the data collected from all the 405 cubes for localization (i.e., O_s in Eq. (4.41)), which takes around 527 minutes. This is the traditional approach widely used by most of works in the literature. For the *interpolated fingerprints*, we use the RSS data of 1 cube for every 8 cubes, i.e., we could use the RSS data collected from only 50 cubes for the training phase. However, we use data collected from 75 cubes (i.e., O_s' in Eq. (4.43)) to conduct the localization experiment due to the irregular shape of the lobby, which takes around 100 minutes. Then we use the 4-D RSS interpolation algorithm to approximate the data in the rest cubes. One example of the RSS probability distribution generated by an AP before and after the interpolation is shown in Fig. 4.19.

For the online localization phase, we use both the full fingerprints and the interpolated fingerprints to perform localization and compare the corresponding performance. To present a fair comparison, we only record preliminary localization results $\check{s}_k$ during each localization process. The localization tests are done by 405 times for each set of fingerprints, and for each test the smartphone will measure RSS 5 times for each AP. Thus we set $M = 4$, $K = 5$ in Eq. (4.48).

The result is shown in Fig. 4.20. It can be seen that, even if the usage of interpolation reduces the accuracy of localization by 0.10–0.17 m when the number of APs varies, the reduced accuracy is still trivial and it will be proved that this decrease is negligible compared with the accuracy gain achieved by the Path Corrector. Moreover, only 19% of the RSS training data are used by the interpolated fingerprints, which means that the workload of fingerprint collection can be reduced by 81%. Thus it is confirmed that our algorithm can significantly reduce the cost of data collection in the offline phase.

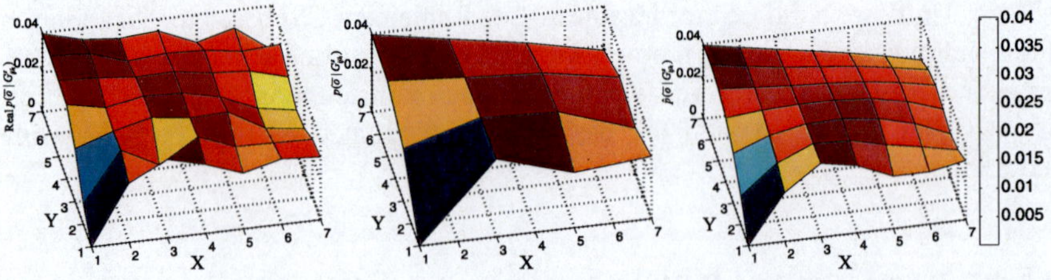

(a) (b) (c)

Fig. 4.19 The probability for RSS= −75 dBm to appear at the cubes which are 1 m from the ground. (**a**) Real data. (**b**) Before interpolation. (**c**) After interpolation

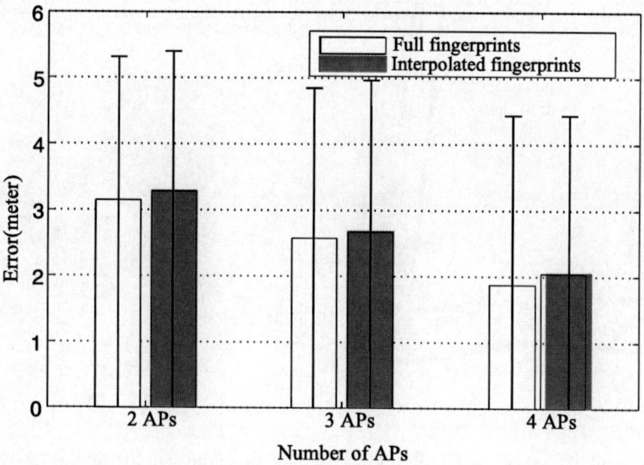

Fig. 4.20 Average localization errors using the two sets of fingerprints

4.3.4.2 Evaluation of Localization Schemes

In this section, we evaluate the performance of the proposed schemes facilitating localization. The 2-D layout of the building where the experiments are conducted is illustrated in Fig. 4.21a. The total area covered is more than 1100 m^2, and the total interior space volume is over 2700 m^3. We install 20 APs (HUAWEI E586Bs-2) along the corridor, and each reference point is covered by at least two APs. The quadrotor used in the experiment is shown in Fig. 4.22. The onboard computer is MSP430F5529, with 128 KB Flash, 8 KB RAM and 25 MHz CPU frequency.

The direction sensor is YAS530B-PZ. The WLAN card HLK-RM04 and MiFi HUAWEI E5776 are installed. The quadrotor can connect to our localization server via cellular networks. Since it is hard for a single AP to cover such a long corridor, M in Eq. (4.48) is set equal to the real number of APs that the quadrotor has discovered. Since the quadrotor keeps moving during the test, $K = 1$ in Eq. (4.48). Moreover, we set $T = 1.5s$, $T_s = 20$ ms, $\alpha_T = 5°$, $\beta_T = 10°$, $m = 25$, $n = 25$, $N_\alpha = 20$, $V_m = 15$ m/s, $\sigma_w^2 = 5$, $\sigma_u^2 = 20$, and $\bar{\sigma}_u^2 = 8$. The communication delay T_d between the quadrotor and the server is 0.335–0.908s during the test. The computing time of each localization process is around 10 ms, which is negligible compared with T_d.

During the offline data training phase, we divide the experiment space into 2700 cubes with size 1 m × 1 m × 1 m, and collect the RSS fingerprints from 485 (18%) of them. For each cube we collect 30 sets of RSS from each AP. The data collected are uploaded to the sever and the 4-D RSS interpolation algorithm is executed.

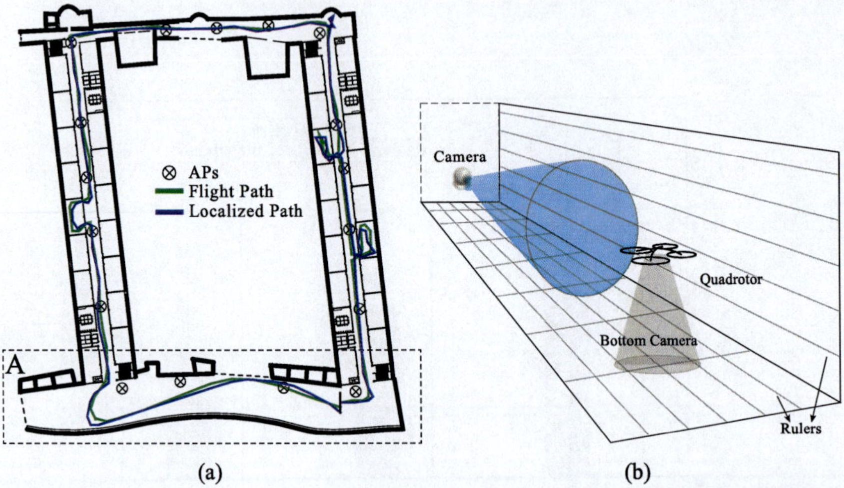

Fig. 4.21 Experiments Settings. The flight path and localized path in (**a**) are the records of a single test (the height is not shown in the figure). (**a**) Experiment field. (**b**) Video recording

Fig. 4.22 The quadrotor experimental platform

During the online localization phase, we use a web camera to record the real location of the quadrotor. As shown in Fig. 4.21b, the bottom of the camera on the quadrotor is used to record its 2D track during the flight. There are rulers drawn on the ground and we thus can get the real position of the quadrotor according to the test time and the video record. For the height of the quadrotor, we use an independent camera deployed at the ends of the corridors, which also serves as the reference point for rulers.

We initialize our client, and control the quadrotor to fly along the corridor. The test is repeated for 10 times, and 1268 localization results are recorded. For instance, according to the video record, the flight path during one test is shown in Fig. 4.21a. We do not strictly constrain the speed of the quadrotor in this case, and the impact of the flight speed is to be evaluated in Sect. 4.3.4.4.

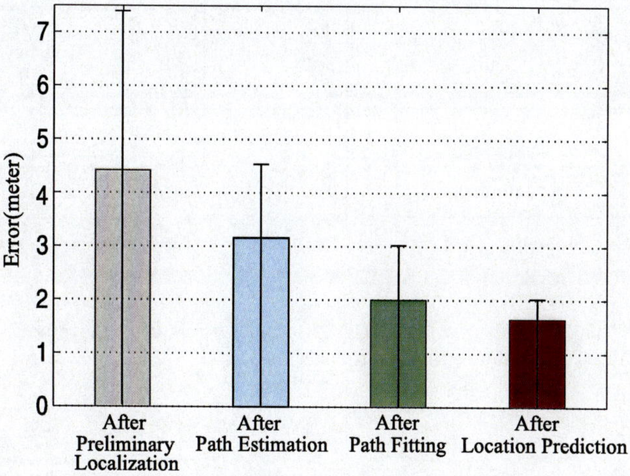

Fig. 4.23 Average location errors after the process of different algorithms

During the flight, we use the continuous-tracking mode only, since the two modes have similar performance. The results of localization are recorded, and we analyze the results output by the methods of preliminary localization, path estimation, path fitting, and location prediction respectively. The average location errors for these results are shown in Fig. 4.23.

For traditional localization systems based on RSS, their performance is the same with that of the preliminary localization algorithm. Thus if they are applied to the case of high-speed quadrotor directly, the average error will be 4.41 m, which can hardly meet with the need of quadrotor localization. If we apply the path estimation method, the average error can be reduced to 3.16 m. Based on the fact that quadrotors move in nearly straight lines for the indoor case most of the time, the method of path fitting further reduces the average error to 2.00 m. Finally, considering the delay of transmission, the location prediction method achieves an average location error of 1.64 m.

To better compare the performance of HiQuadLoc and normal RSS-based localization system, we also plot the cumulative distribution (CDF) of location errors in Fig. 4.24a. As shown in the figure, the location error distribution for the final results of HiQuadLoc (red line) reaches its climax more quickly than the case of normal RSS-based localization system (black line). According to these results, in the field of high-speed quadrotor indoor localization, compared with traditional RSS-based systems, HiQuadLoc has reduced the location error by 62.8%, which is a remarkable improvement in accuracy. The localization errors incurred in each dimension (x, y, and z axes in the physical space) are illustrated in Fig. 4.24b. Note that the location error is the square root of the quadratic sum of the error in each dimension.

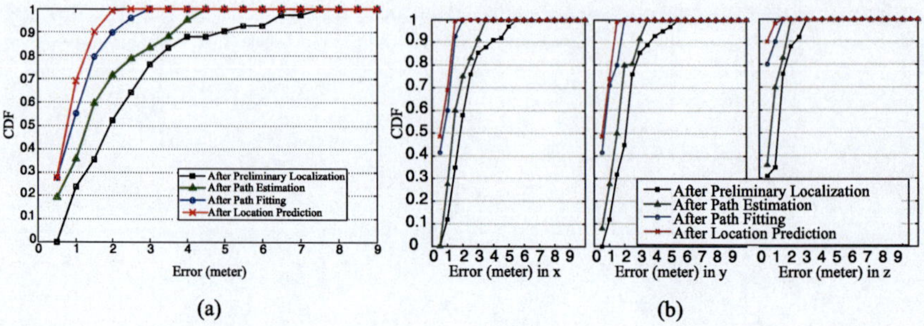

Fig. 4.24 CDF of location errors. (**a**) CDF of location errors after the process of different algorithms. (**b**) CDF of location errors in three axis

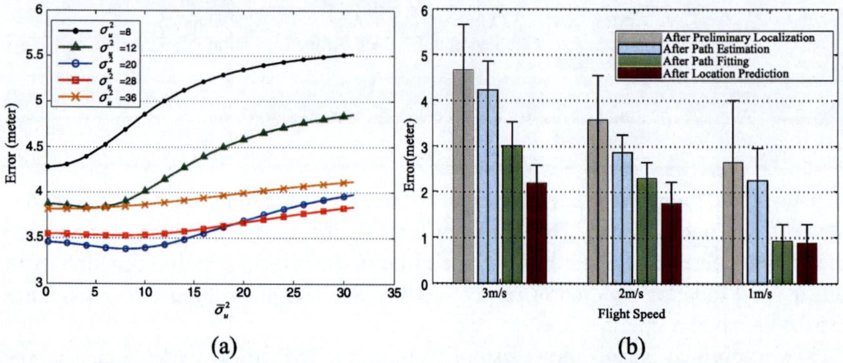

Fig. 4.25 Average location errors. (**a**) Average location errors of $\bar{s}_k$ when σ_u^2 and $\bar{\sigma}_u^2$ vary. (**b**) Average location errors at different speeds

4.3.4.3 Evaluation of Parameter Readjustment During Turning

We mentioned in Sect. 4.3.3.2 that we need to readjust the Kalman filter during the turning of the quadrotor. After the experiment in Sect. 4.3.4.2, we change the values of σ_u^2 and $\bar{\sigma}_u^2$ respectively to confirm that the readjustment of Kalman filter can improve the accuracy of the path estimation. We analyze the output of the path estimation method $\bar{s}_k$ alone to rule out the additional gain of the path fitting and the location prediction method. Since the readjustment mainly happens during turning, we focus on the location error of the ± 5 localization results around each corner of the flight paths.

In Fig. 4.25a, it is shown that when σ_u^2 is fixed, there exists a $\bar{\sigma}_u^2$ that makes the location error minimum. For the case of $\sigma_u^2 = 8$ and $\sigma_u^2 = 36$, the location error is minimum when $\bar{\sigma}_u^2 = 0$, which means that the path generated by Kalman filter should be interrupted during the motion of turning, and a new path is restarted at the first $\check{s}_k$ after the turning. On the other hand, when $\sigma_u^2 = 12$, $\sigma_u^2 = 20$ and $\sigma_u^2 = 28$, the error is minimum when $\bar{\sigma}_u^2 = 5$, $\bar{\sigma}_u^2 = 8$ and $\bar{\sigma}_u^2 = 7$ respectively. Thus we cannot simply set $\bar{\sigma}_u^2 = 0$, because sometimes the historical localization results are still useful during turning. Note that the change of the lines under different σ_u^2 shows that a too large or too small σ_u^2 also leads to the increase of error. Thus we cannot set $\sigma_u^2 = \bar{\sigma}_u^2$ directly.

The experiment results verify that changing the parameter of Kalman filter is necessary at the corners of the flight paths, and the benefit of the proposed revised Kalman filter is validated. In our system, we set $\sigma_u^2 = 20$ and $\bar{\sigma}_u^2 = 8$ according to these results.

4.3.4.4 Evaluation of HiQuadLoc for Different Flight Speeds

Since the quadrotor could flight in different speeds, we also evaluate the impact of speed on the accuracy of localization in our system. To ensure that the quadrotor can fly at approximately constant speed, we limit the flight path in a long lobby in the region A in Fig. 4.21a. The length of the flight path is longer than 80 m. We control the quadrotor to fly in the lobby for 15 times at each speed: 3 m/s, 2 m/s and 1 m/s. All the other settings are same to those in the previous experiments. The highest speed we set here is 3 m/s, which is a reasonable speed for the quadrotor. As a benchmark, a vision-based, autonomous quadrotor equipped with two cameras, an IMU, and an 1.6 GHz Intel Atom processor can fly straight line at the speed of 4 m/s and non-straight line at about 1–2 m/s [198]. We note that some fully loaded quadrotor could fly 20 m/s [199], which, however, is equipped with "high-definition onboard cameras, LIDAR, sonar, inertial measurement units, and other sensors", and our means of localization are basically RSS fingerprints matching and sensors. The purpose of the experiment is to verify the feasibility of the low-cost fingerprinting based approach for localizing the quadrotor in the indoor space, and we do not claim that the high-speed quadrotor indoor localization issue has been completely resolved. The results are shown in Fig. 4.25b. It can be seen that the location error increases as the speed increases. This is caused by the fact that the number of times for RSS measurement is more limited when the quadrotor is moving at higher speed. However, HiQuadLoc still works well in this case. The average location error is 2.19 m, which has been reduced by 53.0% compared with the original error 4.66 m provided by regular RSS-based systems. For the case of a lower speed 2 m/s, that is 1.76 m, reduced by 51.0%. For 1 m/s, that is 0.89 m, reduced by 66.4%. Moreover, we can find that since the location error caused by communication delay is more serious in the higher-speed case, the contribution of the location prediction method in this case is much more obvious than that in the lower-speed case. Figure 4.26a illustrates the CDF of the location errors at different speeds. It can be seen that the path correction algorithm can effectively reduce the location error at all the speeds. Figure 4.26b presents how the localization accuracy decline with speed of the quadrotor after all the localization data processing.

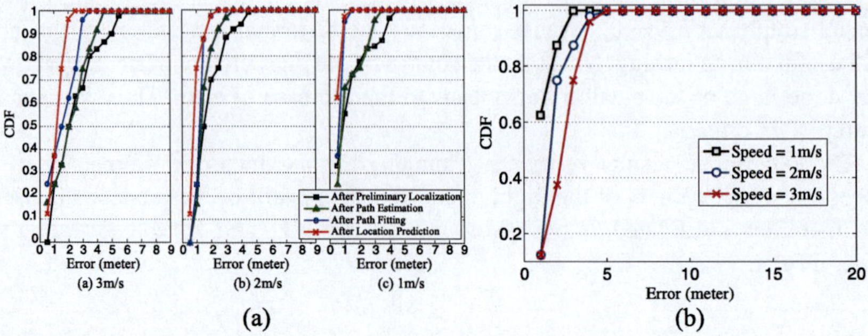

Fig. 4.26 CDF of location errors. (**a**) CDF of location errors at different speeds. (**b**) CDF of location errors after location prediction at different speeds

4.3.5 Comparison with Channel State Information (CSI) Based Scheme

We here compare the HiQuadLoc with channel state information (CSI) based mechanism. In particular, we choose to implement the angle-of-arrival (AoA) method [86, 87] and compare the localization performance with that of HiQuadLoc. The AoA method is chosen because it is the representative scheme utilizing CSI for localization, and it is also the basis of many localization systems such as ArrayTrack [86] and SpotFi [87]. We note that authors of ArrayTrack has built an AP-like hotspot system themselves; however, common deployment of such system is still unavailable to the best of our knowledge. The existing Wi-Fi APs in our experiment building have no interface for retrieving the CSI like most of the off-the-shelf APs.

We build CSI retrievable APs ourselves and the components of the AP are as shown in Fig. 4.27. The experiment setup is configured similar with that in SpotFi. In Fig. 4.27, the Intel 5300 network interface card (NIC) is for collecting the CSI data, Lenovo R400 is acting as the computing platform running the CSI data processing program [97]. Note that the NIC is equipped with 3 antennas, as each Intel 5300 NIC has only three slots for antennas. We built two such APs to cover an area of around 200 m^2 in the corridor as shown in Fig. 4.21a, which has the same deployment density as in [87]. During the experiment, the target to be localized move in different speeds along a track in the shape of "L", and the APs record the corresponding CSI data, which are then used to derive the location information of the target.

We compare the derived locations of the target and the real location of the target, and the CDF of location errors are illustrated in Fig. 4.28, where the performance of both HiQuadLoc and the CSI based scheme are shown. It can be seen that the performance of CSI based scheme is not as good as expected. This could be due to the fact that the moving quadrotor could incur fast changing multipath effects. The multipath effects have been considered in previous CSI based systems such as the ArrayTrack and SpotFi; however, those techniques are basically for the scenario of localizing mobile devices such as smartphones, where the holder of the smartphone

Fig. 4.27 CSI retrievable AP

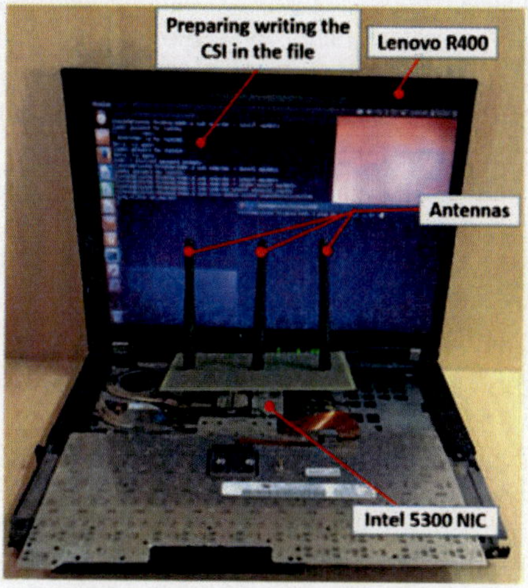

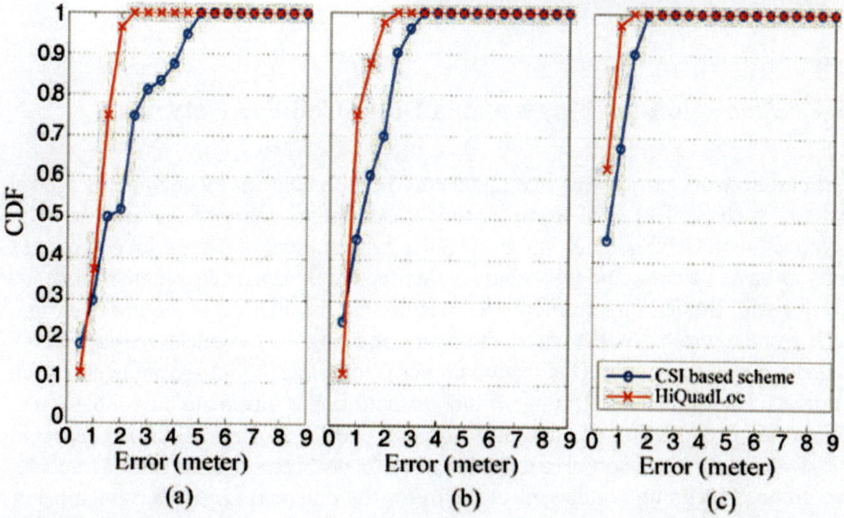

Fig. 4.28 Comparison with CSI based scheme: CDF of location errors at different speeds. (**a**) 3 m/s. (**b**) 2 m/s. (**c**) 1 m/s

has small-scale movement. The quadrotor localization scenario is more challenging as the speed of the quadrotor is faster than the pedestrians; moreover, the flying part of the quadrotor could also incur multipath effect that changes very fast. This is corroborated by the fact that the low speed moving target could be more accurately localized as shown in Fig. 4.28.

Another possible reason of the results is that the CSI based scheme is subtle but not that robust. We find that packet loss could be observed during the experiment, especially when the moving speed of the target is comparatively high and the distance between the target and the AP is far. As the underlying details of Intel 5300 NIC is not revealed, how the packet loss could impact the localization accuracy could not be evaluated. We note that the number of APs could also influence the performance of CSI based localization scheme as described in [87], which also hinders the wide deployment of CSI based localization systems.

In particular, most of APs do not provide the interface for retrieving CSI information. Intel 5300 NIC is a well-known hardware that could help retrieve CSI, which are adopted by a number of work in the literature. However, it is non-trivial to widely deploy APs created with Intel 5300 NIC and the PC in practice. For example, the integration of the NIC into the PC could incur software incompatibility issue. We collect 5 Lenovo R400 laptops that are officially announced supporting Intel 5300 NIC to create the AP [97], among which 3 of them have been found the incompatibility issue. It will be more challenging to widely deploy such systems in practice. In contrast to such systems, there is no need for the HiQuadLoc to retrieve the CSI information. The HiQuadLoc only requires reading RSS information that is naturally supported by all mobile devices, which is very convenient for wide deployment under the current infrastructure.

4.4 Prediction for Fingerprints Data (Cellular Network)

Cellular network positioning is a mandatory requirement for localizing emergency callers, such as E911 in North America. Although smartphones are normally equipped with GPS modules, there are still a large number of users with cell phones only as basic devices, and GPS could be ineffective in urban canyon environments. To this end, the RF fingerprints based positioning mechanism is incorporated into LTE architecture by 3GPP, where the major challenge is to collect geo-tagged RF fingerprints in vast areas. This section proposes to utilize the subspace identification approach for large-scale RF fingerprints prediction. We formulate the problem into the problem of finding the optimal subspace over Stiefel manifold and redesign the Stiefel manifold optimization method with fast convergence rate. Moreover, we propose a sliding window mechanism for the practical large-scale fingerprints prediction scenario, where recorded fingerprints are unevenly distributed in the vast area. Combining the two proposed mechanisms enables an efficient method of large-scale fingerprints prediction in the city level. Further, we validate our theoretical analysis and proposed mechanisms by conducting experiments with real mobile data, which shows that the resulted localization accuracy and reliability with our predicted fingerprints exceed the requirement of E911.

4.4.1 Problem Formulation

4.4.1.1 Fingerprints Prediction: A Subspace Identification Perspective

The fingerprints prediction problem can be formulated into a matrix completion problem [162–164]. The area needs localization service is first divided into grids, and the fingerprints sampled in grids are like elements in a matrix. The purpose of fingerprints prediction is essentially to complete the entire matrix by deriving the unknown elements based on those available ones. A number of mathematical tools for matrix completion are available, such as the singular value thresholding (SVT) [286], singular value partition (SVP) [166], forward-backward algorithm for matrix completion (FBMC) [167] and iterative reweighted least squares (sIRLSp) [168]. Though with different implementation details, those algorithms are generally based on singular value decomposition (SVD).

In particular, given an incomplete $m \times n$ matrix induced from the incomplete radio map, if we assign the values of all unknown elements to be 0s, then we have a complete radio map matrix $A = U \Lambda V^T$ after SVD, where U is an $m \times m$ real unitary matrix, Λ is an $m \times n$ rectangular diagonal matrix with non-negative real numbers on the diagonal and V^T is an $n \times n$ real unitary matrix. It is usually assumed that A is a low-rank matrix, which means that all the column vectors in A are linearly dependent to each other; this is based on the in-practice observation that fingerprints are correlated within a certain area. To exploit the linear dependency, we could keep d greatest singular values lying on the diagonal of Λ making it a $d \times d$ matrix, and make the corresponding parts in U and $V^T m \times d$ and $d \times n$ matrices, respectively. Then multiplying the three parts results in a new $m \times n$ matrix $\hat{A}$, which contains estimations to those unknown elements originally assigned values of 0s in A.

The essence of the SVD method is actually to find a lower-dimensional subspace that contains all the column vector in $\hat{A}$. Consider an m-dimensional space that contains all m-dimensional column vectors in A, if most of those vectors are linearly dependent with each other, then most of them should belong to a lower-dimensional subspace of the m-dimensional space. For example, imagine that there are some points in the 3-D space, if most of the points are linearly dependent with each other, then those points should be in a 2-D plane or a straight line. In SVD, the residual $m \times d$ matrix U is such a lower (d) dimensional subspace induced by the greatest d singular vectors. If the subspace is found, then any vector belong to the subspace are available; this is why the unknown elements can be estimated.

The accuracy of the elements estimation is highly dependent on whether the obtained subspace indeed contains most of the vectors in A. There are infinite number of possible subspaces that can be induced by A, however, the SVD method factually always finds one specific type of the subspace, as it initializes the incomplete elements in A by assigning them with specific values, for example, all 0s, before performing decomposition. Assigning different values to those unknown elements yields different subspaces, but there are infinite number of possible situations, which makes SVD method unable to guarantee that the found subspace is always optimal.

It is worth mentioning that the subspace identified by SVD is optimal in terms of minimizing the Frobenius norm according to the Eckart-Young Theorem [206]. In particular, if we use A_c to denote the matrix obtained after A going through the initialization process of the SVD method, and A_o the matrix obtained after the entire SVD process, then $||A_c - A_o||_{Frobenous}$ is minimized according to Eckart-Young Theorem. However, this does not conflict with our problem formulation in this paper, because our objective is to minimize $||P_\Omega(A) - P_\Omega(A_o)||$ as to be shown in (1). Generally, depending on the method chosen for matrix completion, the unknown elements could be completed to the values that lead to the lower rank matrix during the initialization process, and the initial values will be updated by an iterative process such as the augmented Lagrange multiplier method (ALM) for matrix completion [171]. In contrast, our approach proposed in this paper iteratively finding the optimal subspace directly without initializing assignments of unknown element in A. In the following, we are to show how to find the optimal subspace in the whole set of possible subspaces.

4.4.1.2 Problem Formulation

The fingerprints prediction problem can be formulated into the following matrix completion problem:

$$\begin{aligned} \min_{\hat{A}} \quad & ||P_\Omega(A) - P_\Omega(\hat{A})||, \\ s.t. \quad & \text{rank}(\hat{A}) \leq d, \end{aligned} \tag{4.58}$$

where $||\cdot||$ represents any suitable norm; A is the matrix representing the radio map. Since some elements in A have not been measured thus unavailable, we use $P_\Omega(A)$ to denote those available fingerprints in A. The fingerprints prediction mechanism yields $\hat{A}$; this is a complete estimation of A, which contains estimations to those unmeasured fingerprints in corresponding positions. The constraint rank$(\hat{A}) \leq d$ means that the estimation $\hat{A}$ is under the constraint of low rank, where $d << n$ in practice. The physical meaning of the problem is to find $\hat{A}$ that minimizes the deviation from the available observations denoted by $P_\Omega(A)$, given the low-rank constraint.

The j-th column of A can be regarded as an m-dimensional vector denoted by a_j, and all the column vectors are in an m-dimensional space denoted by $\mathcal{M}$. Since a_js are correlated to each other, then we can assume that the matrix A is a rank-d matrix, which means that all vectors of A belong to a d-dimensional subspace of $\mathcal{M}$. However, as some elements in A are unknown, to obtain a prediction of those elements, we want to find a rank-d matrix A_d based on the known elements in each a_j. The A_d found must be a complete matrix and $A_d = U_d \Lambda_d V_d^T$ after SVD, where the matrix U_d contains dm-dimensional orthogonal vectors, which can span a d-dimensional subspace of $\mathcal{M}$. Finding A_d directly could be challenging, but due to the property of the low-rank matrix, if we can find U_d, then A_d can be derived.

The problem is now transformed into a subspace identification problem in the following form:

$$\min_{\substack{U_d:m\times d\\ w_j:d\times 1}} \sum_{j=1}^{n} ||[U_d w_j]_\Omega - [a_j]_\Omega||_2^2, \tag{4.59}$$

where $U_d w_j$ represents the column vector in A_d corresponding to the column vector in A denoted by a_j. Since a_j could be incomplete, we use $[a_j]_\Omega$ to denote the available elements in a_j; the corresponding elements in A_d is denoted by $[U_d w_j]_\Omega$, as the matrix A_d can be transformed into $A_d = U_d W$ with $W = \Lambda_d V_d^T$, where w_j is the jth column of W.

The problem (4.59) distinguishes itself from other commonly seen optimization problems in that the decision variable U_d is a subspace in the form of a matrix. Recall the m-dimensional space where all vectors of A are in, all d-tuples of orthogonal m-dimensional vectors form a d-dimensional Stiefel manifold; therefore, the problem becomes to find a point in the Stiefel manifold considering the objective function in (4.59). This is a Stiefel manifold optimization problem [165].

It is worth mentioning that multiple points in the Stiefel manifold can possibly form the same subspace, because one subspace may have multiple sets of basis. All those lower-dimensional subspaces in the m-dimensional space form another kind of manifold known as Grassmann-manifold [172–174], which is to be used for convergence proof in the discussion later. We will theoretically prove that the convergence rate of our proposed mechanism based on Stiefel manifold can be higher than performing optimization over the Grassmann-manifold, which especially suits fingerprints prediction in extremely large-scale areas.

4.4.2 *Streamlined Stiefel Manifold Optimization*

4.4.2.1 Algorithm Design

The basic idea of the Stiefel manifold optimization approach is similar to the gradient descent method, which is frequently used in resolving optimization problems. The gradient descent method starts with a given point on a curve representing the objective function, and iteratively takes steps proportional to the negative of the gradient of the function at the current point. The method can be extended to the Stiefel manifold optimization problem [165]; however, the practical scenario of fingerprints prediction problem cannot fit in the general framework presented in [165]. Moreover, it is not mentioned in [165] whether the approach will converge, and how to choose important parameters to guarantee performance, which makes it necessary to streamline the existing Stiefel manifold optimization approach. We are to go through the classic Stiefel manifold optimization approach and show the challenge to be confronted in resolving the fingerprints prediction problem, and then present the streamlined design of the approach for dealing with the challenges.

Challenge 1: Determine the Direction of Iteration The first step in [165] is to determine the direction of iterations, which is realized by finding the Hessian matrix H with respect to the objective function $F = \sum_{j=1}^{n} ||[U_d w_j]_\Omega - [a_j]_\Omega||_2^2$, and then finding the inverse of H. However, we find that H for the fingerprints prediction problem may not have full rank, thus H^{-1} may be unavailable. Specifically, $\frac{\partial F}{\partial U_d} = 2\sum_{j=1}^{n}([U_d w_j]_\Omega - [a_j]_\Omega) w_j^T$ is an $n \times d$ matrix; finding H needs to take the second order derivative of F with respect to U_d, which yields an $n^2 \times d^2$ matrix

$$H = \begin{bmatrix} B_{11} & \cdots & B_{1d} \\ \vdots & \ddots & \vdots \\ B_{n1} & \cdots & B_{nd} \end{bmatrix},$$

where B_{ij} denotes an $n \times d$ sub-matrix. Note that there are unavailable samples in a_j, thus $[U_d w_j]_\Omega$ contains zero-valued elements, which results in that there are all-zero rows in the $n \times d$ matrix after taking the first derivative. This will further make elements from certain B_{i1} to B_{id} in H all zeros after taking the second order derivative. Consequently, it is possible that H contains all-zero rows. To this end, we propose to replace the direction derived from H^{-1} with the gradient of the objective function ∇F, where the intuition is that ∇F can also represent a possible iteration direction.

Challenge 2: Complex Objective Function The second step is to find the common iteration equation:

$$U_{d,t+1} = U_{d,t} M_t + Q N_t,$$

where Q satisfies the QR decomposition of $(I - U_{d,t} U_{d,t}^T)\nabla F$, and M_t and N_t satisfy

$$\begin{bmatrix} M_t \\ N_t \end{bmatrix} = \left(\exp\left(t \begin{bmatrix} U_{d,t}^T \nabla F & -R^T \\ R & 0 \end{bmatrix}\right)\right) \begin{bmatrix} I_d \\ 0 \end{bmatrix}, \quad (4.60)$$

where we replace H^{-1} with ∇F. However, it is noted that U_d and w_j are factually dependent to each other, since $A_d = U_d W$ with $W = \Lambda_d V_d^T$ and w_j is the jth column of W. Moreover, there is a matrix exponential function in the iteration, which we find could hinder finding the iteration equation due to the tedious form involving infinite matrix series.

We propose to replace the original objective function with

$$F(U_d) = \min_{x_j} ||[U_d x_j]_\Omega - [a_j]_\Omega||_2^2, \tag{4.61}$$

which decouples the dependence between U_d and w_j. It is straightforward that $w_j = x_j^* = argmin_{x_j} ||[U_d x_j]_\Omega - [a_j]_\Omega||_2^2$. The new objective function is factually the item of the summation in the objective function in problem (4.59). A natural question is: will the solution with the new objective function be the same as that with the original objective function?

It is straightforward to verify that the second order derivative of $F(U_d)$ with respect to U_d is a semi-definite matrix, which means that the new objective function is convex on U_d, thus definitely can achieve a unique minimum value. However, the solution for optimizing different items to make each item achieve the minimum value may be different. Recall the nature of A that the column vectors a_js of A fall in a d-dimensional subspace due to their correlation. In the process of optimizing each item, the solution must make the corresponding a_j fall in the same d-dimensional subspace. However, even if solutions of optimizing all those items can make those a_js fall in the subspace, they are not necessarily the same, because each solution is factually a set of basis of the d-dimensional subspace according to the definition of Stiefel manifold. It is possible that the solutions of optimizing different items vary, because the subspace can have multiple sets of basis. Nevertheless, the interesting point is that, since each solution is a set of basis of the same subspace, then any element vector in a given set of basis definitely can be represented by any other set of basis. This means that if an item achieves the minimum value, the corresponding solution can also make other items achieve the minimum value; therefore, the solution of the transformed problem is indeed the solution of the original problem.

Challenge 3: Obtain the Common Iteration Equation Based on the revision mentioned above, we now try to obtain the common iteration equation, which requires to perform QR decomposition to the matrix $(I - U_t U_t^T)\nabla F$.[1]

Since $\nabla F = -2r_t w_t^T - U_t(-2r_t w_t^T)^T U_t = -2r_t w_t^T$ with $r_t = P_\Omega(U_d w_j - a_j)$, we have $(I - U_t U_t^T)\nabla F = 2r_t w_t^T$ with $U_t^T r_t = 0$, where $w_t = x_j^*$ for the current vector a_j we are considering. This means that the rank of the matrix is 1 thus it is impossible to perform QR decomposition to the matrix. To deal with this challenge, we propose to replace QR decomposition with SVD, so that we can still obtain an orthonormal matrix Q. We relax the constraint in the classic approach and allow the matrix R to be singular, then we have $Q = [\frac{r_t}{||r_t||} \quad q_2 \quad q_3 \quad \cdots \quad q_n]$ and $R = [2||r_t||w_t \quad 0 \quad 0 \quad \cdots \quad 0]^T$, where $q_2, q_3, \ldots q_n$ are all orthonormal singular vectors and orthogonal to $\frac{r_t}{||r_t||}$.[2] Then we have $M_t = I_d$ and $N_t = \eta_t R$ with the consequent common iteration equation:

[1] Note that we use F to denote $F(U_d)$ and U_t to denote $U_{d,t}$ respectively for the convenience of presentation, and such denotations are also to be adopted in the following discussions.

[2] Hereinafter we use $||\cdot||$ to represent $||\cdot||_2$, the 2-norm of a vector.

$$U_{t+1} = U_t + 2\eta_t \frac{r_t w_t^T}{||r_t|| ||w_t||}, \tag{4.62}$$

where η_t is the step size parameter.

Challenge 4: Determine Step Size of Iteration Before executing the iteration, we must determine the step size first. Imagine the subspace identification process, we can first estimate a subspace U_t and see if the a_j we are considering is in the subspace. If a_j is not in U_t, the angle between the projection of a_j on U_t and a_j itself denoted by θ must be unequal to zero. Then we need to rotate the subspace to decrease θ to find a new estimation of the subspace. The degree of the rotation is factually the step size and it is straightforward that the step size should just make $\theta = 0$. Then we continue to rotate the previously estimated subspace U_t in the same way to make it contain other a_js, and the resulted U_t is the subspace we are finding. More precisely, we are trying to find the set of basis that spans a subspace containing all a_js according to the definition of Stiefel manifold.

Let us define a $d \times d$ matrix W_t in the t-th iteration, where $W_t = [\frac{w_t}{||w_t||} \quad C_t]$. Note that C_t is a $d \times (d-1)$ matrix, whose columns are unit vectors and all orthogonal to $\frac{w_t}{||w_t||}$. According to Gram-Schmidt transformation, C_t can be turned into a new matrix with all columns orthogonal to each other. Here we assume the columns of C_t are orthogonal to each other for the convenience of presentation. Multiply W_t in both sides of Eq. (4.62), we have $U_{t+1}W_t = U_t W_t + 2\eta_t \frac{r_t}{||r_t||} \begin{bmatrix} 1 \ 0 \ 0 \cdots 0 \end{bmatrix}$. The physical meaning of such operations is to perform rotation to the estimated subspace U_t, and the equation above can be further transformed into $U_{t+1}\frac{w_t}{||w_t||} = \frac{p_t}{||w_t||} + 2\eta_t \frac{r_t}{||r_t||}$, where $||p_t|| = ||w_t||$. Note that p_t is the projection of a_j on U_t, $\frac{r_t}{||r_t||}$ is a unit vector that is orthogonal to U_t; therefore, the degree U_t to be rotated is determined by η_t. We need to take an appropriate value of η_t so that a_j can fall in the resulted U_{t+1}. To this end, we should have $U_{t+1}w_t // (p_t + r_t)$, i.e., $(\frac{2\eta_t}{||r_t||}, \frac{1}{||w_t||}) // (1, 1)$, which results in the step size in iteration t:

$$\eta_t = \frac{1}{2} \frac{||r_t||}{||w_t||}. \tag{4.63}$$

After overcoming the challenges above, we present the streamlined Stiefel manifold optimization algorithm (SSOA) as in Algorithm 7, where we use U_Ω to denote the rows of U whose index is in Ω.

Note that in Algorithm 7, after we obtain the optimal subspace U, we need to predict the unavailable fingerprints. Based on the theory of projection, we have the following lemma ensuring the accurate prediction of fingerprints within the subspace, which corresponds Step 12 in Algorithm 7.

Lemma 18 *Assume that A is a rank-d matrix. Denote $e_i, i = 1, 2, \ldots m$ as the unit orthogonal basis vector in which the i-th element is 1 while others are 0. U is the $m \times d$ matrix where each column of A, denoted as a_j, $j = 1, 2, \ldots, n$, satisfies $a_j \in R(U)$, where $R(U)$ means the range space (column space) of U. Then if the*

Algorithm 7 Streamlined Stiefel manifold optimization algorithm (SSOA)

Require:

An initial column-orthonormal $m \times d$ matrix U_0;
sample set Ω, $m \times n$ sample matrix $P_\Omega(A)$;
maximum number of iteration T.

Ensure:

Estimated matrix A_d.

1: $t = 0$;
2: **while** $t < T$ **do**
3: Randomly choose a column index $q \in \{1, 2, \ldots, n\}$, get $[a_q]_\Omega$;
4: $w_t = ([U_t]_\Omega^T [U_t]_\Omega)^{-1} [U_t]_\Omega [a_q]_\Omega$;
5: $p_t = U_t w_t$;
6: $r_t = P_\Omega(a_q - p_t)$;
7: $U_{t+1} = U_t + \frac{r_t w_t^T}{||w_t||^2}$;
8: $t = t + 1$;
9: **end while**
10: $U = U_t$;
11: **for each** $i \in \{1, 2, \ldots, n\}$ **do**
12: $\hat{a}_i = U([U]_\Omega^T [U]_\Omega)^{-1} [U]_\Omega [a_i]_\Omega$;
13: **end for**
14: $A_d = [\hat{a}_1, \hat{a}_2, \ldots, \hat{a}_n]$.

number of sampled elements in a_j, $|\Omega_j|$, is larger than d, and $U_{\Omega_j}^T U_{\Omega_j}$ is invertible, we have our estimation $\hat{a}_j = a_j$ via

$$\hat{a}_j = U(U_{\Omega_j}^T U_{\Omega_j})^{-1} U_{\Omega_j}^T [a_j]_{\Omega_j}.$$

Proof 48 Since $a_j \in R(U)$, then there exists a $d \times 1$ vector c_j, s.t. $a_j = U c_j$. Therefore those indexes of a_j which are sampled satisfy $[a_j]_{\Omega_j} = [U c_j]_{\Omega_j} = [U]_{\Omega_t} c_j$. Due to the fact that $U_{\Omega_j}^T U_{\Omega_j}$ is invertible, hence we have

$$\begin{aligned}\hat{a}_j &= U(U_{\Omega_j}^T U_{\Omega_j})^{-1} U_{\Omega_j}^T [a_j]_{\Omega_j} = U(U_{\Omega_j}^T U_{\Omega_j})^{-1} U_{\Omega_j}^T U_{\Omega_j} c_j \\ &= U c_j = a_j.\end{aligned}$$

In terms of the time complexity of Algorithm 7, note that the main cost lies in Step 3 and Step 11, where a matrix multiplication costs $O(nd^2)$ and a matrix inversion costs $O(d^3)$. So the total time complexity of Algorithm 7 is $O((n + T)nd^2)$.

4.4.2.2 Convergence Analysis

The challenge to prove the convergence of the proposed SSOA in Algorithm 7 is that the elements in each a_j are not completely known. In particular, recall that in problem (4.59) our goal is to find the U_d that minimizes the objective function, we can claim Algorithm 7 converges if we can indeed find the U_d, which spans a subspace containing all vectors a_js in A, but a_j is incomplete itself.

We use $\beta_t(U, U_t) = 1 - \delta_t(U, U_t) = 1 - |U^T U_t|^2$ as the metric of measuring the distance between the estimated subspace in the t-th iteration U_t and its true value U [174], where $|U^T U_t|$ means the determinant of $U^T U_t$. Recall that we assume A is a rank-d matrix, where the subspace U contains all the vectors in A. According to the definition, if U and U_t span the same subspace, then $\beta_t(U, U_t) = 0$; if U and U_t are orthogonal to each other (any column in U is orthogonal to all columns in U_t), then $\beta_t(U, U_t) = 1$ meaning that U and U_t have the largest distance. With the metric, we first prove in Lemma 19 that SSOA converges at least as fast as Grassmann-manifold optimization algorithm with step size η_t. Then we prove in Theorem 15 that SSOA converges faster than Grassmann-manifold optimization algorithm with appropriately choice of η_t. Faster convergence rate is very meaningful especially for large-scale fingerprints prediction.

Lemma 19 *Let* $\eta_t = \frac{||r_t||}{2||w_t||}$, *then SSOA algorithm converges at least as fast as Grassmann-manifold based mechanism.*

Proof 49 Define $\bar{U}_t = U_t W_t = \left[\frac{p_t}{||p_t||}\ C_t\right]$ and $\check{U}_{t+1} = U_t W_t = \left[\frac{q_t}{||w_t||}\ C_t\right]$. Since the first column of $\check{U}_{t+1}$ is not a unit vector, we define $\bar{U}_{t+1} = \left[\frac{q_t}{||q_t||}\ C_t\right]$. With the step size η_t, we have $q_t = p_t + 2\eta_t \frac{||w_t||}{||r_t||} r_t = a_t$. Since W_t is an orthogonal matrix, then $R(\bar{U}_t) = R(U_t)$ and $R(\bar{U}_{t+1}) = R(U_{t+1})$, where $R(U)$ denotes the image of a matrix U, i.e., $R(U) = \{UA | \forall A \in \mathbb{R}^{m \times n}\}$.

Then we define $\bar{U} = UW_t = \left[\frac{a_t}{||a_t||}\ \bar{C}\right]$, where C is a $d \times (d-1)$ matrix with orthonormal columns, and these columns are all orthogonal to $\frac{a_t}{||a_t||}$. Hence $\bar{U}$ is an orthogonal matrix and $R(\bar{U}) = R(U)$. Then there exists an orthogonal matrix Y_t such that $\bar{U} = UY_t$, and $\frac{\delta_{t+1}}{\delta_t} = \frac{|\bar{U}_{t+1}^T \bar{U}|^2}{|\bar{U}_t^T \bar{U}|^2} = \frac{|U_{t+1}^T U|^2}{|U_t^T U|^2} = \frac{|C_t^T C|^2}{|C_t^T C|^2}(\frac{||p_t|| ||a_t||}{p_t^T a_t})^2 = \frac{||a_t||^2}{||p_t||^2}$ which is the convergence rate of the Grassmann-manifold optimization algorithm [174]. Note that p_t is the projection of a_t over the subspace U_t, thus the physical meaning of the equation above means that the distance between U_t and U decreases faster in each iteration.

Theorem 15 *Denote* $\sigma_i(A)$ *as the* i*-th largest singular value of* A *and* $\lambda_i(A)$ *as the* i*-th largest eigenvalue of* A*. If we set the step size* η_t *such that*

$$\frac{\lambda_1(U_t^T U_t)}{\lambda_1(U_t^T U_t) + 4\eta_t^2} \frac{\lambda_d(U_t^T U_t)}{\lambda_2(U_{t+1}^T U_{t+1})} \left(1 + 2\eta_t \frac{||r_t||}{||p_t||}\right)^2 > \frac{||a_t||^2}{||p_t||^2}, \tag{4.64}$$

then the convergence rate of the SSOA is strictly greater than $\frac{||a_t||^2}{||p_t||^2}$*, which is known as the convergence rate of Grassmann-manifold optimization algorithm [174].*

Proof 50 Note that U_{t+1} and U_t are not necessarily with orthonormal columns with the step size η_t now. We first apply the QR decomposition to U_{t+1} and U_t denoted by $U_{t+1} = U^Q_{t+1} R_{t+1}$ and $U_t = U^Q_t R_t$, respectively. Similar to the derivation in Lemma 19, we can derive

$$\frac{\delta_{t+1}}{\delta_t} = \frac{|(U^Q_{t+1})^T U|^2}{|(U^Q_t)^T U|^2} = \frac{|R^{-1}_{t+1}|^2}{|R^{-1}_t|^2} \frac{((p_t + 2\eta_t \frac{||p_t||}{||r_t||} r_t)^T a_t)^2}{(p_t^T a_t)^2}. \tag{4.65}$$

Note that (i) U_{t+1} and R_{t+1} share the same singular values since multiplying an orthogonal matrix does not alter the singular values; (ii) R_{t+1} is a diagonal matrix, thus we have $|R^{-1}_t| = \frac{1}{\prod_{i=1}^d \sigma_i(U_t)}$ and $|R^{-1}_{t+1}| = \frac{1}{\prod_{i=1}^d \sigma_i(U_{t+1})}$ where

$$\sigma_i(U_{t+1}) = \sqrt{\lambda_i(U^T_{t+1} U_{t+1})} = \sqrt{\lambda_i\left(U_t^T U_t + 4\eta_t^2 \frac{w_t w_t^T}{||w_t||^2}\right)},$$

based on the fact that $U_t^T r_t = 0$. Since $w_t w_t^T$ is a rank-1 matrix, with the only non-zero eigenvalue $||w_t||^2$, therefore according to Weyl's inequality [176], we have

$$\lambda_i\left(U_t^T U_t + 4\eta_t^2 \frac{w_t w_t^T}{||w_t||^2}\right) \geq \begin{cases} \lambda_1(U_t^T U_t) + 4\eta_t^2, & i = 1; \\ \lambda_{i-1}(U_t^T U_t), & i \geq 2. \end{cases} \tag{4.66}$$

With Eq. (4.65) and Inequality (4.66), it is easy to derive $\frac{\delta_{t+1}}{\delta_t} \geq \frac{\lambda_1(U_t^T U_t)}{\lambda_1(U_t^T U_t)+4\eta_t^2} \frac{\lambda_d(U_t^T U_t)}{\lambda_2(U^T_{t+1} U_{t+1})} (1+2\eta_t \frac{||r_t||}{||p_t||})^2$. Since RHS of the inequality is greater than $\frac{||a_t||^2}{||p_t||^2}$, which is the convergence rate of Grassmann-manifold optimization algorithm as presented in [174], the convergence rate of Stiefel manifold optimization algorithm is faster. Proved.

Theorem 15 presents the general condition where SSOA outperforms Grassmann-manifold optimization algorithm in convergence rate. In fact, based on experiments on the real big data sets in Sect. 4.4.5.2 we find that (i) $\lambda_1(U_t^T U_t)$ surges rapidly and becomes much greater than $\lambda_2(U^T_{t+1} U_{t+1})$ and $\lambda_d(U_t^T U_t)$; (ii) $\lambda_d(U_t^T U_t)/\lambda_2(U^T_{t+1} U_{t+1}) \in [c, 1]$ with the constant $c > 0$. The verifications are illustrated in Fig. 4.29a and b. Figure 4.29a verifies observation (i) since under different dimension of subspaces ($d = 10, 15, 20$), we find that $\lambda_1(U_t^T U_t)$ is always greater than $\lambda_2(U_t^T U_t)$, and as more iterations are conducted, the gap becomes greater. Figure 4.29b confirms observation (ii) since under $d = 10, 15, 20$, the ratio $\lambda_d(U_t^T U_t)/\lambda_2(U^T_{t+1} U_{t+1})$ is always less than 1 and in our situation we can set $c = 0.45$. Meanwhile, combining both figures we verify $\lambda_1(U_t^T U_t)$ is always greater than $\lambda_d(U_t^T U_t)$.

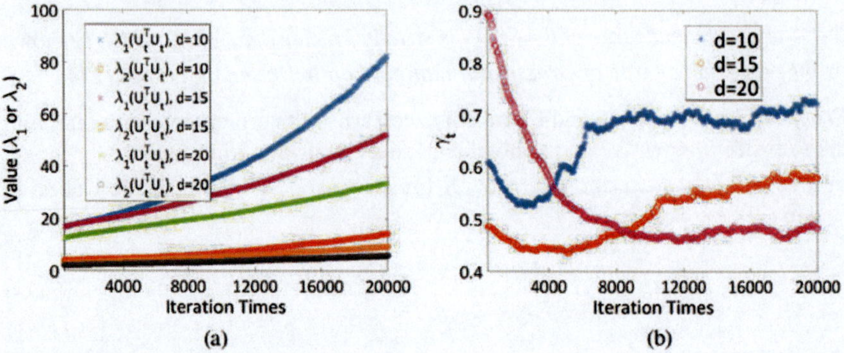

Fig. 4.29 Experimental verifications of observations about Theorem 1. **(a)** $\lambda_1(U_t^T U_t)$ *vs* $\lambda_2(U_t^T U_t)$. **(b)** $\gamma_t = \frac{\sigma_d(U_t^T U_t)}{\sigma_2(U_{t+1}^T U_{t+1})}$

Therefore we present a more practical condition in Proposition 3 that can facilitate determining the step size in practice, which is an approximation to the general Theorem 15.

Proposition 3 *We set* $\gamma_t = \frac{\lambda_d(U_t^T U_t)}{\lambda_2(U_{t+1}^T U_{t+1})} \in [c, 1]$. *If we choose the step size* η_t *such that* $\frac{1}{2}\frac{||a_t||-||p_t||}{\sqrt{c}||r_t||} < \eta_t << \frac{1}{2}\sigma_1(U_t)$, *then the convergence rate of the SSOA is strictly higher than that of Grassmann-manifold optimization algorithm.*

Proof 51 Since $\lambda_1(U_t^T U_t) >> \lambda_2(U_{t+1}^T U_{t+1}) = \gamma_t \lambda_d(U_t^T U_t)$, we can approximate Eq. (4.64) in Theorem 15 as $\gamma_t(1 + 2\eta_t \frac{||r_t||}{||p_t||})^2 > \frac{||a_t||^2}{||p_t||^2}$. Then we can easily obtain

$$\eta_t > \frac{1}{2}\frac{||a_t|| - ||p_t||}{\sqrt{c}||r_t||}.$$

On the other hand, since $\sigma_1(U_t) = \sqrt{\lambda_1(U_t^T U_t)}$we need to ensure that $\eta_t << \frac{1}{2}\sigma_1(U_t)$ to let $\frac{\lambda_1(U_t^T U_t)}{\lambda_1(U_t^T U_t)+4\eta_t^2} \to 1$. Thus when $\frac{1}{2}\frac{||a_t||-||p_t||}{\sqrt{c}||r_t||} < \eta_t << \frac{1}{2}\sigma_1(U_t)$, the convergence rate of SSOA is higher than that of Grassmann-manifold optimization algorithm.

Note that sometimes when $||r_t||$ is small enough, especially when the algorithm is close to convergence, $\frac{||a_t||-||p_t||}{\sqrt{c}||r_t||} > \lambda_1(U_t^T U_t)$. However, in this situation we just use $\eta_t = \frac{1}{2}\frac{||r_t||}{||w_t||}$. Combining the ways above to determine step sizes, we will show that the convergence speed of SSOA outperforms significantly than that of Grassmann-manifold based algorithm under real big data sets in Sect. 4.4.5.2.

4.4.2.3 Discussions

The fundamental reason that the Stiefel manifold optimization approach converges faster than the Grassmann-manifold counterpart is that the physical meaning of a point on the Stiefel manifold is a set of basis of a d-dimensional subspace and that on the Grassmann-manifold is a d-dimensional subspace itself. It is supposed that our optimization algorithm should measure the distance between a_j and the estimated set of basis in each iteration; however, such kind of iteration will definitely incur high computational complexity due to the finer-granularity of distance metric. Since a set of basis can span a subspace, we use the angle between a_j and its projection on the subspace as the distance metric in our algorithm design. In this case, the solution of our algorithm is factually the subspace spanned by the set of basis, instead of the particular set of basis itself the traditional Stiefel manifold optimization mechanism is finding. This is because a subspace can have multiple sets of basis.

However, this means that there may exist multiple solutions can be obtained by our proposed SSOA, and any one of the solutions can satisfy the requirement. Recall that we transform the objective function in the algorithm design (Sect. 3.1), and the new objective function is convex on U_d, then the situation of optimizing over the Stiefel manifold is like that illustrated in the right part of Fig. 4.30, and the left part of the figure shows the situation of optimizing over the Grassmann-manifold. There could be multiple solutions on the Stiefel manifold, but only one solution on the Grassmann-manifold, because a d-dimensional subspace is just regarded as a point on the Grassmann-manifold according to the definition. Consequently, it is easier to find a solution on the Stiefel manifold and the convergence rate is higher.

We present numerical analysis results in Fig. 4.31 showing the average number of iterations for completing 500 randomly generated matrix on the Stiefel and Grassmann-manifold, respectively. The two sub-figures show the number of iterations needed to achieve different levels of prediction accuracy (average error) for the cases that 40 and 60% of the real data are available. It is clear that the proposed SSOA requires less numbers of iterations compared with the Grassmann-manifold optimization algorithm. The experimental results with the real data to be presented in Sect. 4.5.3 also corroborates our analysis.

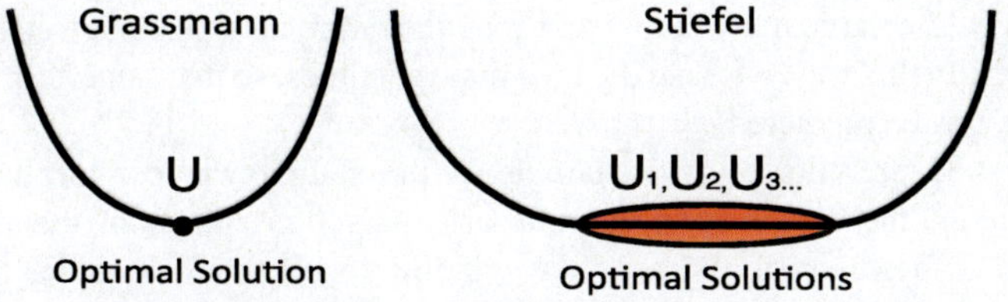

Fig. 4.30 Convergence over Stiefel and Grassmann-manifold

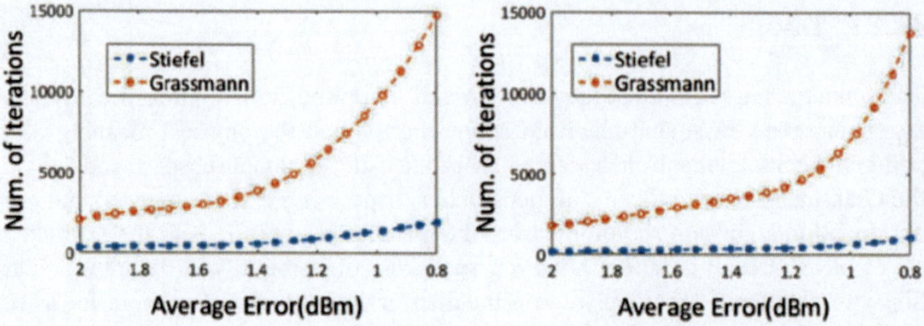

Fig. 4.31 Numerical analysis to the convergence rate (Left: 40%; Right: 60%)

4.4.3 Fingerprints Prediction with Sliding Window

4.4.3.1 Sliding Window Mechanism Design

In practice, the sampled RF fingerprints for cellular networks are unevenly distributed in the extremely vast area, as to be shown in Sect. 4.5. It is impossible to complete the matrix of the entire area in one shot, since the sampled data are sparsely distributed in some subareas. To deal with this issue, we can create a sliding window to scan the entire area in a row-by-row manner, where the crux is to determine the size of the window. In particular, we first grid the entire area into square cells, where the edge length of each cell is dependent on the accuracy requirement (normally in tens of meters). We then let the sliding window cover a number of such cells and move from left to right and top to bottom so that the entire area can be scanned. The sliding window's movement step size in both horizontal and vertical direction is randomly assigned, so that the predicted fingerprints obtained in the window's previous location can be utilized to predict the fingerprints in the current area covered by the window.

Moreover, the window size must be set small enough initially to make sure there are enough amount of data available within the window for prediction. After a round of scanning, we have predicted fingerprints available in some cells, which can be regarded as the training data thus the density of available fingerprints increases. Then we could enlarge the window and scan the entire area again still in a row-by-row manner. In this way, we scan the area multiple times, so that fingerprints in most of the area can be predicted.

Figure 4.32 presents a vivid example of the sliding window mechanism. At the beginning, the sampled fingerprints are sparse, as shown in the upper left of Fig. 4.32, so we set the size of the window to be small, in order to predict the fingerprints just adjacent to the samples at first. During the scanning process, the fingerprints are not always predictable in the window, which depends on the sampling rate inside the window. For example, in the upper left of Fig. 4.32, when the window scans to the red frame, the sampling rate in this frame is high enough, thus the window is predictable. The red elements inside the window are the

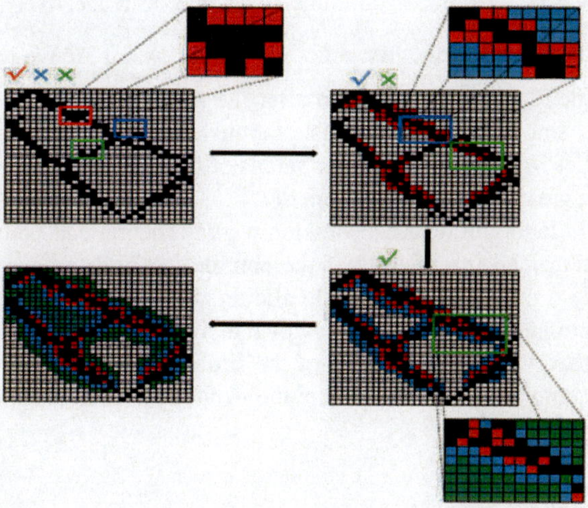

Fig. 4.32 An example of sliding window mechanism

predicted elements. However, when the window scans to the blue and green frames, the sampling rate is very low, thus they are not predicted in this round. With an increasing number of elements predicted, the window expands gradually as shown in Fig. 4.32, and in this way the whole radio map will be recovered finally.

The crux of the sliding window design is to determine the window size, which is essentially to determine the dimension of the subspace d. This is because the length and width of the sliding window must be greater than d, or it is impossible to predict the fingerprints in the unsampled cells. However, the challenge is that the matrix corresponding to the sliding window itself is incomplete. To deal with the challenge, we propose to determine d of the sliding window matrix by sampling a complete sub-matrix within the window. The rationale is that if the vectors in the window matrix are correlated, the correlation should be reflected by any sub-matrix within. The question is how good we can predict the fingerprints in the window matrix if we determine d in this way.

The corner stone assumption of the subspace identification approach is that all the vectors in the window matrix denoted by A_w is in the same subspace denoted by U_d; however, although the fingerprints are correlated, some vectors indeed are not in U_d thus incur prediction errors. Note that a large-valued d can decrease such error since more vectors can be included in U_d, but requires more elements in A_w to be available; in contrast, a small-valued d can increase the error, but it can accommodate more-sparsely sampled A_w.

In particular, suppose $A_w = U \Lambda V^T$ after SVD.[3] We use $\sigma_i = \Lambda_{ii}, i = 1, 2, \ldots, m$ (set $m < n$) to denote the i-th largest singular value of A_w. Suppose we had known d, then we approximate A_w with $A_d = U_d \Lambda_d V_d^T$, where U_d and

[3] Note that we use U, Λ, and V to denote the SVD of window matrix in this section for the convenience of demonstration, which is different from the previous sections where $A = U \Lambda V^T$.

V_d are the first d columns of U and V respectively, and Λ_d is the sub-matrix comprised of the first d columns and rows of Λ. We use $l_d = \sqrt{\frac{\sum_{i=1}^{d}\sigma_i^2}{\sum_{i=1}^{m}\sigma_i^2}}$ to denote the remained information after the approximation which only includes the largest d singular values in Λ. The rationale is that most of the information of the matrix lies in the largest singular values of the matrix after SVD. It is straightforward that a greater d leads to a greater l_d.

In the following discussion we will theoretically prove that the gap between the remained information of the complete sub-matrix we sample within the window, $\tilde{l}_d$, and that of the whole sliding window matrix, l_d, is small enough when the correlation of the columns of the sliding window matrix is strong. This validates that we can determine d of the sliding window matrix by conducting SVD on the complete sub-matrix within the window, with similar remained information.

4.4.4 Remained Information Analysis

We first present some lemmas to prepare for Theorem 16, the remained information analysis.

Lemma 20 *If real numbers $a, b, c, d > 0$, $b << a$, $d << c$, $|a - c| \leq \epsilon_1$ and $|b - d| \leq \epsilon_2$, then we have*

$$|\frac{b}{a} - \frac{d}{c}| \leq \max\left\{\frac{\min\{b, d\}}{ac}\epsilon_2, \frac{\epsilon_1}{\max\{a, c\}}\right\}.$$

Proof 52 This lemma can be proved in a case-by-case manner as follows:

Case 1: If $\frac{b}{a} \geq \frac{d}{c}$ and $a \geq c$ (note that now $b \geq d$), then $|\frac{b}{a} - \frac{d}{c}| = \frac{b}{a} - \frac{d}{c} \leq \frac{|b-d|}{a} \leq \frac{\epsilon_1}{a}$;

Case 2: If $\frac{b}{a} \geq \frac{d}{c}$, $a < c$ and $b < d$, then $|\frac{b}{a} - \frac{d}{c}| = \frac{bc-ad}{ac} \leq \frac{d}{ac}|c - a| \leq \frac{d}{ac}\epsilon_2$;

Case 3: If $\frac{b}{a} \geq \frac{d}{c}$ and $a < c$, $b \geq d$, then $|\frac{b}{a} - \frac{d}{c}| = \frac{bc-ad}{ac} \leq \frac{b}{ac}|c - a| \leq \frac{b}{ac}\epsilon_2$;

Similarly we can conduct the other 3 cases when $\frac{b}{a} < \frac{d}{c}$. Combining all the 6 cases, we finish the proof.

Lemma 21 (The Hoffman-Wielandt Inequality) *Given two $k \times k$ real symmetric matrices X and Y, if the eigenvalues of X are $\mu_1, \mu_2, \ldots, \mu_k$ and that of Y are $\nu_1, \nu_2, \ldots, \nu_k$, with $\mu_1 \geq \mu_2 \geq \ldots \geq \mu_k \geq 0$, $\nu_1 \geq \nu_2 \geq \ldots \geq \nu_k \geq 0$, then we have $\sum_{i=1}^{k}(\mu_i - \nu_i)^2 \leq ||X - Y||_F^2$.*

Lemma 22 ([175]) *If U is an orthonormal matrix and Ω is a random set of row indexes. Then there exists a positive constant C, such that $\mathbf{E}||\frac{m}{|\Omega|}U_\Omega^T U_\Omega - I|| \leq Cm\sqrt{\frac{\log n}{|\Omega|}} \max_{1\leq k\leq m} ||u^k|| \leq 1$, where u^k denotes the k-th row of U.*

Lemma 23 *If* $\hat{A} = \tilde{A}V_p^T$*, where* V_p *is an* $n \times t$ *matrix with orthonormal columns. Then there exists a constant* $C > 0$*, such that* $\mathbf{E}||\frac{m}{|\Omega|}\hat{A}^T\hat{A} - A^TA|| \leq Cm\sqrt{\frac{\log n}{|\Omega|}}\max_{1\leq k\leq m}$
$||u^k||||\Lambda||^2 \leq |||\Lambda||^2$

Proof 53 Note that $A = U\Lambda V^T$ after SVD, and $\hat{A}$ is a sub-matrix of A concatenated by the rows whose indices belonging to Ω. Then we have $\hat{A} = U_\Omega \Lambda V^T$, and

$$\begin{aligned}
&\mathbf{E}||\frac{m}{|\Omega|}\hat{A}^T\hat{A} - A^TA|| \\
&= \mathbf{E}||\frac{m}{|\Omega|}V\Lambda U_\Omega^T U_\Omega \Lambda V^T - \frac{1}{|\Omega|}V\Lambda U^T U\Lambda V^T|| \\
&= \mathbf{E}||V\Lambda\left(\frac{m}{|\Omega|}U_\Omega^T U_\Omega - I_m\right)\Lambda V^T|| \\
&\leq \mathbf{E}||\frac{m}{|\Omega|}U_\Omega^T U_\Omega - I_m||||\Lambda||^2 \\
&\leq Cm\sqrt{\frac{\log n}{|\Omega|}}\max_{1\leq k\leq m}||u^k||||\Lambda||^2.
\end{aligned} \tag{4.67}$$

The last inequality holds based on Lemma 22.

Theorem 16 *We use* l_d *and* $\tilde{l}_d$ *to denote the remained information of the incomplete* $p \times q$ *window matrix* A_w *and a complete* $s \times t$ *sub-matrix* $\tilde{A}_w$ *within, where we set* $p \leq q$ *and* $s \leq t$*, and* d *is the dimension of the subspace obtained by performing SVD to the sub-matrix; if the linear correlation of fingerprints is strong enough in the window matrix* A_w*, then* $|l_d - \tilde{l}_d| \to 0$.

Proof 54 Denote the following SVDs: $A_w = U\Lambda V^T$, $\tilde{A}_w = \tilde{U}\tilde{\Lambda}\tilde{V}^T$, $A_d = U_d\Lambda_d V_d^T$ and $\tilde{A}_d = \tilde{U}_d\tilde{\Lambda}_d\tilde{V}_d^T$. The window matrix A_w is highly linearly correlated, which means that almost all the information is contained within the subspace spanned by several principal axes while the other ones can be reasonably neglected. Then we have $\frac{\sum_{i=d+1}^p \sigma_i^2}{\sum_{i=1}^p \sigma_i^2} \to 0$. Based on Taylor's expansion, we obtain

$$l_d = \sqrt{1 - \frac{\sum_{i=d+1}^p \sigma_i^2}{\sum_{i=1}^p \sigma_i^2}} \approx 1 - \frac{1}{2}\frac{\sum_{i=d+1}^p \sigma_i^2}{\sum_{i=1}^p \sigma_i^2} \approx 1 - \frac{1}{2}\frac{||\Lambda - \Lambda_d||_F^2}{||\Lambda||_F^2}.$$

Similarly, $\tilde{l}_d \approx 1 - \frac{1}{2}\frac{||\tilde{\Lambda} - \tilde{\Lambda}_d||_F^2}{||\tilde{\Lambda}||_F^2}$, thus according to Lemma 20, we obtain

$$
\begin{aligned}
|l_d - \tilde{l_d}| &= \frac{1}{2}|\frac{||\Lambda - \Lambda_d||_F^2}{||\Lambda||_F^2} - \frac{||\tilde{\Lambda} - \tilde{\Lambda}_d||_F^2}{||\tilde{\Lambda}||_F^2}| \\
&= \frac{1}{2}|\frac{||\Lambda - \Lambda_d||_F^2}{||\Lambda||_F^2} - \frac{\frac{p}{s}||\tilde{\Lambda} - \tilde{\Lambda}_d||_F^2}{\frac{p}{s}||\tilde{\Lambda}||_F^2}| \le \frac{1}{2}\max\{S, T\},
\end{aligned}
\tag{4.68}
$$

where

$$
S = \frac{\min\{||\Lambda - \Lambda_d||_F^2, \frac{p}{s}||\tilde{\Lambda} - \tilde{\Lambda}_d||_F^2\}}{\frac{p}{s}||\tilde{\Lambda}||_F^2||\Lambda||_F^2}|||\Lambda||_F^2 - \frac{p}{s}||\tilde{\Lambda}||_F^2|,
$$

$$
T = \frac{1}{\max\{||\Lambda||_F^2, \frac{p}{s}||\tilde{\Lambda}||_F^2\}}|||\Lambda - \Lambda_d||_F^2 - \frac{p}{s}||\tilde{\Lambda} - \tilde{\Lambda}_d||_F^2|.
$$

For S, set $\breve{A}_w = \tilde{A}_w V_p^T$, where V_p is an $q \times t$ matrix with orthonormal columns and $\breve{A}_w$ is an $s \times q$ matrix. Let $Z = \frac{p}{s}\breve{A}_w^T\breve{A}_w - A_w^T A_w$, which is an $q \times q$ matrix. Note that (i) $|||\Lambda||_F^2 - \frac{p}{s}||\tilde{\Lambda}||_F^2| = |\sum_{i=1}^p \sigma_i^2 - \frac{p}{s}\sum_{i=1}^s \tilde{\sigma}_i^2|$; (ii) $\frac{p}{s}\breve{A}_w^T\breve{A}_w$ and $A_w^T A_w$ are both symmetric matrices; (iii) $\breve{A}_w$ and $\tilde{A}_w$ share the same singular values. Thus according to Lemmas 21 and 23,

$$
\begin{aligned}
||Z||_F^2 &= \sum_{i=1}^p \lambda_i^2(Z) \ge \sum_{i=1}^p (\lambda_i(A_w^T A_w) - \frac{p}{s}\lambda_i(\breve{A}_w^T\breve{A}_w))^2 \\
&\ge \frac{1}{p}\left(\sum_{i=1}^p \sigma_i^2 - \frac{p}{s}\sum_{i=1}^s \tilde{\sigma}_i^2\right)^2 = \frac{1}{p}|||\Lambda||_F^2 - \frac{p}{s}||\tilde{\Lambda}||_F^2|^2.
\end{aligned}
$$

Then

$$
S \le \frac{\min\{||\Lambda - \Lambda_d||_F^2, \frac{p}{s}||\tilde{\Lambda} - \tilde{\Lambda}_d||_F^2\}}{\frac{p}{s}||\tilde{\Lambda}||_F^2}\sqrt{n}. \tag{4.69}
$$

Now we focus on T. If $||\Lambda||_F^2 \ge \frac{p}{s}||\tilde{\Lambda}||_F^2$, then

$$
T \le \frac{||\Lambda - \Lambda_d||_F^2}{||\Lambda||_F^2} + \frac{p}{s}\frac{||\tilde{\Lambda} - \tilde{\Lambda}_d||_F^2}{||\Lambda||_F^2} \le \frac{||\Lambda - \Lambda_d||_F^2}{||\Lambda||_F^2} + \frac{||\tilde{\Lambda} - \tilde{\Lambda}_d||_F^2}{||\tilde{\Lambda}||_F^2}. \tag{4.70}
$$

Similarly, if $||\Lambda||_F^2 < \frac{p}{s}||\tilde{\Lambda}||_F^2$, we also have Inequality (4.70). Then if the linear correlation of fingerprints in A_w is strong, $||\Lambda - \Lambda_d||_F^2$ and $||\tilde{\Lambda} - \tilde{\Lambda}_d||_F^2$ approach zero, which makes both S and T approach zero. According to Eq. (4.68), we prove that $|l_d - \tilde{l_d}| \to 0$, which means that using $\hat{A}_w$ to estimate the subspace dimension d for A_w does not incur much deviation in remained information.

4.4.4.1 Sliding Window Algorithm

Now we present our sliding window algorithm as Algorithm 8. Its input is the sampled fingerprint data set while its output is the fully predicted fingerprint database of our target region. Algorithm 8 mainly consists of 3 loops: (i) Each round of the outermost loop (from Line 1 to Line 20) is a complete scanning of the whole target region with the same A_w; (ii) Each round of the intermediate loop (from Line 4 to Line 18) represents the scanning of different rows of A; (iii) The innermost loop (from Line 5 to Line 15) completes the prediction of every column in the row specified by the intermediate loop.

There are several important details required for further explanation: First, in Line 6 we need to check whether the scanning window A_w can be predicted. We give the following Lemma 24 to show the conditions under which prediction by SSOA in A_w will be more likely to be reasonable. Secondly, in Line 13–14 and Line 16–17, we set two random real numbers τ_c and τ_r. The goal of this setting is to guarantee the sliding window scanning the whole target region while not to lead to much higher time complexity: Setting τ_c and τ_r larger than 1 will leave some columns and rows unscanned, while adding *col* and *row* one by one cost too much time.

Now we give Lemma 24 and its proof in detail.

Algorithm 8 Sliding window based fingerprint prediction algorithm

---

Require:
  Sampled matrix $P_\Omega(A)$; Initial subspace dimension d_0.
Ensure:
  Estimated matrix $\hat{A}$.
1: **while** $t < T$ **do**
2:   Construct the scanning matrix A_w, with a_t rows and b_t columns. (i.e. $A_w \in \mathbf{R}^{a_t \times b_t}$)
3:   Set row and *col* as the current position of A_w (represented by an element in A at (row,*col*)).
4:   **while** A_w has not scanned all the rows of A **do**
5:     **while** A_w has not scanned all the columns in current row **do**
6:       Scan a row, check whether the current scanning matrix satisfies conditions in Lemma 24.
7:       **if** conditions in Lemma 24 hold **then**
8:         Conduct SSOA on A_w, and obtain the estimation $\tilde{A}_w$.
9:       **end if**
10:       **if** Some of the elements in $\tilde{A}_w$ are not in reasonable range **then**
11:         Set these elements as 0.
12:       **end if**
13:       Set τ_c as a random real number in (0, 1).
14:       Update $col \leftarrow col + \tau_c b_t$.
15:     **end while**
16:     Set τ_r as a random real number in (0, 1).
17:     Update $row \leftarrow row + \tau_r a_t$.
18:   **end while**
19:   Update a_{t+1}, b_{t+1} and d_{t+1} from a_t, b_t and d_t.
20: **end while**

Lemma 24 *Assume that A_w is of rank d. Set A_w with a_t rows and b_t columns (here we set $a_t \leq b_t$), and in the i-th column of A_w there are m_i unsampled fingerprints. Set d as the dimension of the subspace. Then the prediction has finite solutions under the following necessary conditions:*

- $d \leq a_t$ *and* $d \leq m_i, \forall i$.
- $(a_t - d)(b_t - d) \geq \sum_{i=1}^{b_t} m_i$.

Proof 55 Obviously, if $d > b_t$, then there are no restrictions of these b_t columns since the rank of A_w is less than subspace dimension d. So there are infinite solutions to complete missing parts inside A_w. The same reason for the impossibility of $d > a_t$. Thus $d \leq a_t$.

Then since A_w is of rank d, columns come from the same d-dimensional subspace. Therefore we have the following expressions based on linear algebra

$$\begin{cases} x_{d+1} = k_{d+1,1}x_1 + k_{d+1,2}x_2 + \ldots + k_{d+1,d}x_d \\ x_{d+2} = k_{d+2,1}x_1 + k_{d+2,2}x_2 + \ldots + k_{d+2,d}x_d \\ \vdots \\ x_n = k_{n,1}x_1 + k_{n,2}x_2 + \ldots + k_{n,d}x_d, \end{cases} \tag{4.71}$$

where x_i is the i-th column of A_w, and $k_{i,j}$ is an unknown coefficient.

Before going further, we should note that one can easily set another equation, for example, $x_{d+2} = k'_{d+2,1}x_1 + k'_{d+2,2}x_2 + \ldots + k'_{d+2,d-1}x_{d-1} + k'_{d+2,d+1}x_{d+1}$. However, notice that x_{d+1} are also linear combinations of x_1 to x_d, thus this newly added equation is not an independent one, thus it brings no extra information.

Focus on the first equation. Since there are m_i unknown elements in x_i, therefore in the first equation there are $\sum_{i=1}^{d+1} m_i + d$ unknown variables. With regard to the second equation, it contains d new coefficients and m_{d+2} unknown elements in x_{d+2}. Hence the second equation adds $d + m_{d+2}$ new unknown variables. Likewise, if we add in the j-th new equation, there are $m_{d+j} + d$ new unknown variables. Then totally we have $\sum_{i=1}^{b_t} m_i + (b_t - d)d$ unknown variables.

Meanwhile, note that for each vector equation in Eq. (4.71), it contains a_t element-wise equations. Thus there are $a_t(b_t - d)$ equations in sum.

So if $a_t(b_t - d) \geq d(b_t - d) + \sum_{i=1}^{b_t} m_i$, then the equations have finite solutions. Here it also implies that the sampling rate inside A_w should be at least $1 - \frac{(a_t-d)(b_t-d)}{a_t b_t} = \frac{(a_t+b_t-d)d}{a_t b_t}$.

Remark In Lemma 24, note that we only give two *necessary* conditions for Eq. (4.71) having finite solutions. It can ensure finite solutions for the prediction, but not a single determined accurate solution. Therefore, these two conditions cannot guarantee the absolute accuracy. However, they serve as a promotion of accuracy since they largely eliminate unreasonable outcomes and limit the reasonable prediction inside a finite set.

Table 4.3 Algorithms in comparison

Algorithm	Source
Cell-ID (CID)	[205]
Gaussian mixture model (GMM)	Chakraborty et al. [62]
Tensor completion (TC)	Liu et al. [69]
Sparsity rank SVD (SRSVD)	Gu et al. [169]
Bayesian sparse learning (BSL)	Nikitaki et al. [163]
Alternate minimization (AM)	Jain et al. [170]

4.4.5 Experimental Results

We do experiments with real data sampled by a network operator in two cities, where the data sets are sampled within 48 hours, covering 2.2 and 69.8 km^2 areas in the two cities and containing around 60,000 and 8,820,000 data records, respectively. Each data record contains the GPS location information in terms of latitude and longitude, the corresponding RSRP, the time the measurement is performed, and some other irrelevant parameters.

We will first show the results with the smaller data set, in order to verify that the proposed subspace identification approach outperforms other frequently used matrix completion algorithms [166–168, 286]. Then we will show the results with the larger data set to examine the performance of our proposed SSOA and dynamic sliding window mechanism in fingerprints prediction in the large-scale scenario. To verify our prediction mechanism, we perform localizations with the predicted fingerprints and see if the accuracy and reliability can meet the requirement of E911; the results are compared with a series of algorithms in prior arts as listed in Table 4.3. When evaluating the localization accuracy, we use the GPS location information as the ground truth and the localization errors are deviations of the estimated location from the ground truth. Moreover, we validate our theoretical result that the convergence rate of our proposed SSOA outstrips the Grassmann approach with similar methodology by the larger data set.

4.4.5.1 Experiments on Small Data Set

Overview of Fingerprints Prediction Results We first illustrate the fingerprints prediction results obtained by our proposed mechanism and then compare our mechanism with others in terms of different metrics. The top sub-figure in the first column of Fig. 4.33 shows the data sampled on the main road of the target area. This data set also contains fingerprints obtained from branches of those main roads. We use the proposed mechanisms to predict fingerprints on those branches, and compare the predictions with the real data value. The distribution of RSRP fingerprints is shown in the bottom sub-figure in the first column. The rest of the sub-figures illustrate the process of matrix completion with our proposed SSOA combined with the dynamic sliding window scanning. The top sub-figure in the

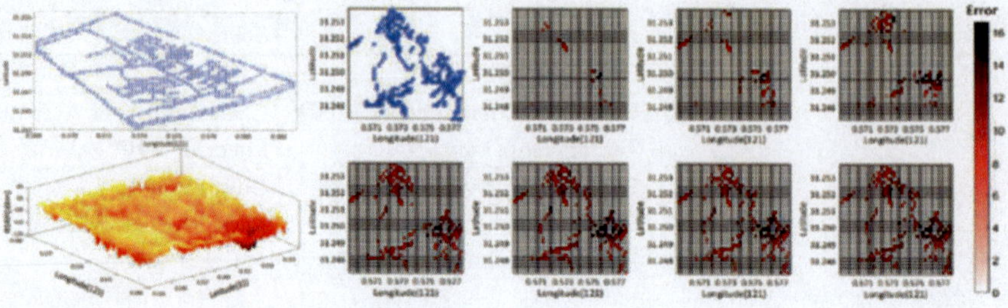

Fig. 4.33 Completion process with small data set

second column shows the locations on those branches of the main roads where real data are sampled. The following sub-figures show the distribution of the completion errors in dBm after each round of scanning with the sliding window. It can be seen from the results that the fingerprints can be predicted, and the average error and standard deviation are 5.1 dBm and 3.5 dBm respectively.

Algorithms in Comparison We compare the performance of the SSOA with that of the following matrix completion methods: *Singular value thresholding (SVT)* is a variant of SVD, where there is a threshold τ to determine the rank of the rectangular diagonal matrix; *Singular value partition (SVP)* is another variant of SVD, which partitions available observations into subsets and perform completion on each subset; *Forward-backward algorithm for matrix completion(FBMC)* scheme formulates the matrix completion problem into a convex optimization problem, with minimizing the objective function that is a combination of completion error and the rank of the estimated matrix; *Iterative reweighted least squares(sIRLSp)* is a family of algorithms, which conducts a least square minimization problem with the decision variable as a matrix.

Performance Metrics To evaluate the performance of each mechanism, we use a subset of the whole data set to form the training set, based on which the fingerprints in other areas will be predicted, and the rest of the data form a test set denoted by Ω_{test}, where the data in the set are the ground truth for testing the predicted fingerprints. All the predicted fingerprints form a set denoted by Ω^c, but note that Ω_{test} and Ω^c are not necessarily the same, since we may predict fingerprints of some area that is not surveyed by the technicians. The performance of the mechanisms mentioned above is evaluated with metrics including $ErExp(A)$, $ErStd(A)$, $NSE(A)$ and $ITR(A)$, where detailed definitions are available in [207].

Prediction Performance 1 Figure 4.34 shows the performance under different proportions of real data in the test set Ω_{test}, which is termed as the sampling rate. If we just use the data obtained from the main road to perform prediction, the sampling rate is 0, if we take 50% of the data in Ω_{test} out also as the training set then the sampling rate is 50%. It is straightforward that the subspace identification approach

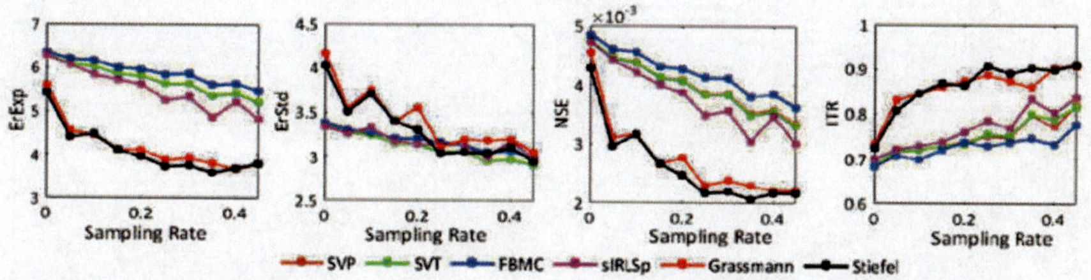

Fig. 4.34 Metrics with different sampling rates on branch roads

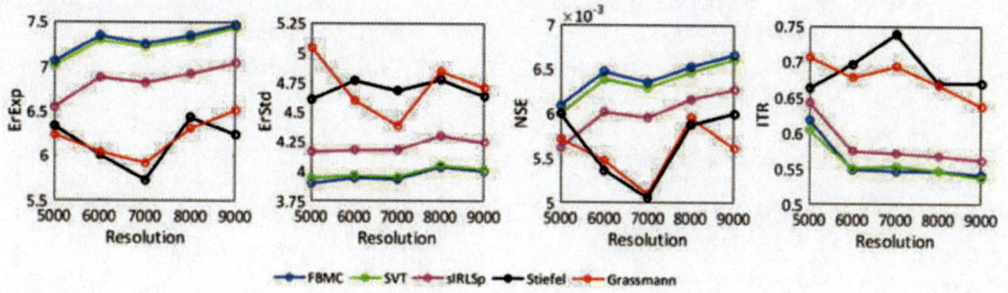

Fig. 4.35 Metrics with different resolutions of the area

no matter over the Stiefel manifold or the Grassmann-manifold outperforms other mechanisms, except for the $ErStd$ metric. Although the subspace identification approach results in a higher standard deviation, the expectation of the errors *ErExp(A)* is smaller than that of others. This is because the real data set itself fluctuates dramatically, but the results by other mechanisms are unable to reflect such variation.

Prediction Performance 2 Figure 4.35 shows the performance under different ways of griding. We divide the region into cells with different edge length, which is termed as resolution. Based on the operator's localization accuracy requirement, the target area may be divided into grids with different resolutions, which in essence is to change the size of the matrix A. In Fig. 4.35, the resolution 5000 means that the target area is divided into 5000 equally sized small cells over each edge. We only use the main-road data as the training set to predict fingerprints in the branch roads. The results in Fig. 4.35 shows that our proposed mechanisms work well under different ways of griding.

Remark Note that both Stiefel manifold and Grassmann-manifold based mechanism perform well, with faint difference in performance metrics above. However, we will show the significantly higher efficiency of Stiefel manifold based mechanism in larger data set, which coincides with its faster convergence rate shown in Theorem 16 and promises its higher practicality in large-scale fingerprints prediction than Grassmann-manifold based mechanism.

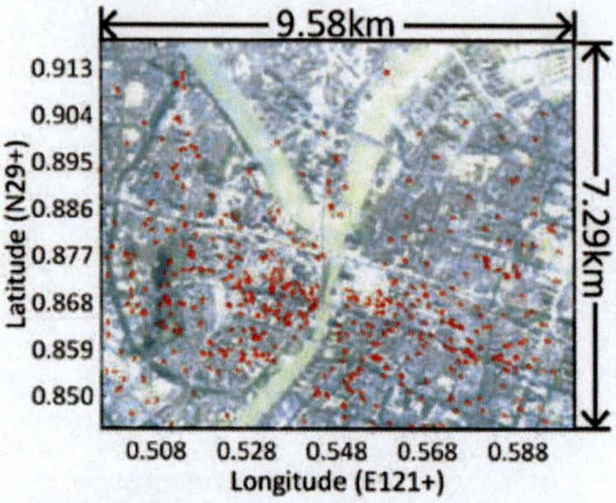

(a) Target region and BSs distribution

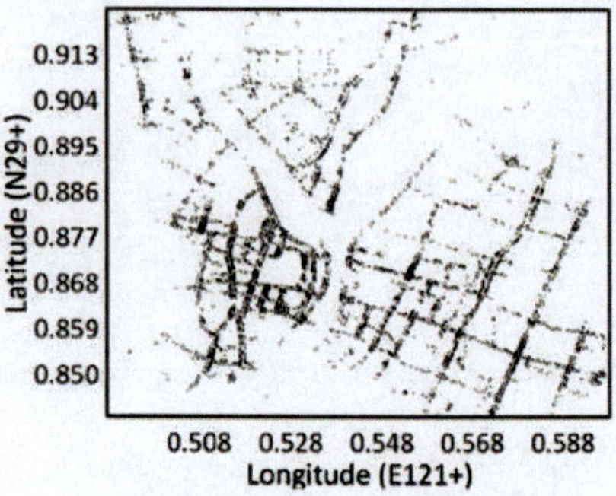

(b) Fingerprints on main roads

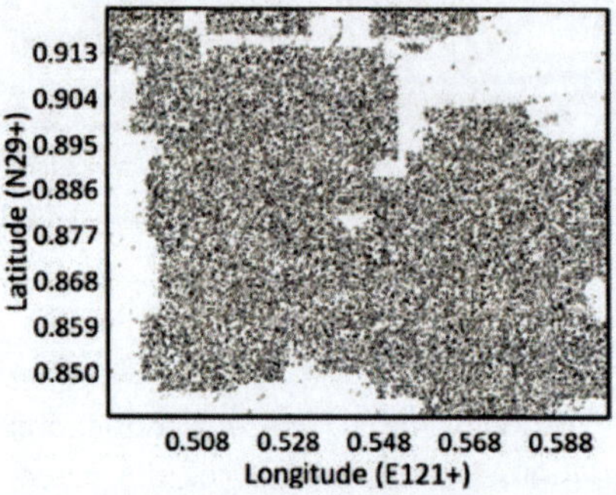

(c) Predicted fingerprints

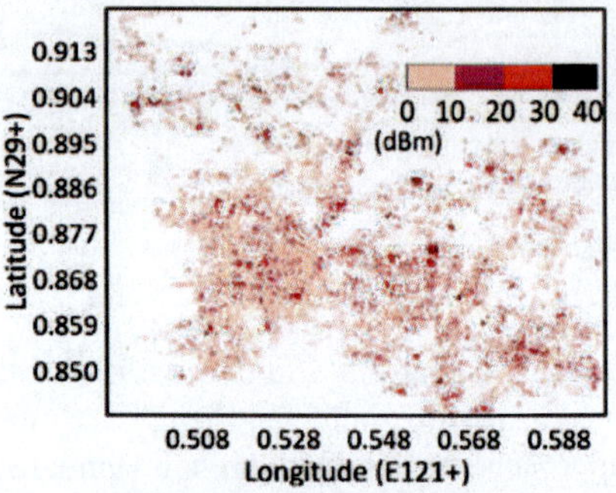

(d) Prediction errors distribution

4.4.5.2 Experiments on Large Data Set

Overview of Fingerprints Prediction Results The map of the city where the data were sampled is shown in Fig. 4.36a; the red dots on the map represent the location of the BSs. We use fingerprints collected along the main roads of the city to predict the fingerprints on those branching roads. The spatial distribution of fingerprints on main road are shown in Fig. 4.36b, which accounts for only 6.7% of the whole region. After 7 iterations of sliding window based prediction mechanism with SSOA, we obtain Fig. 4.36c shown the prediction result. The predicted region accounts for 73.2% of the whole region, having fingerprints in most of the locations predicted. To examine the prediction accuracy, we compare the predicted results with the ground true, and show corresponding error of each prediction in Fig. 4.36d, where different colors represent different levels of errors in dBm. We find that the average and median predicting errors are 8.46 and 7.09 respectively.

Local Performance of Fingerprints Prediction We here show the local performance of fingerprints prediction in Fig. 4.36d. Recall that the sliding window mechanism can be viewed as a filtering process to some extent, and the error could

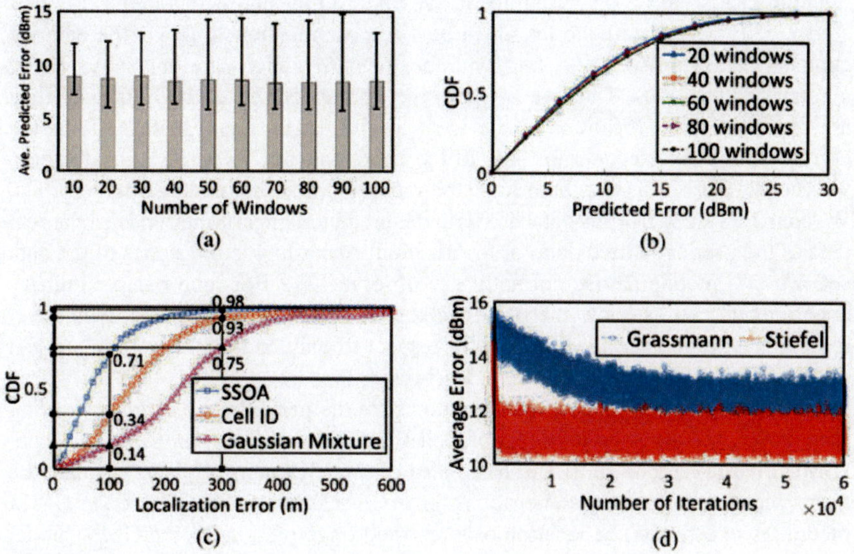

Fig. 4.36 Experimental results on 69.8km^2 data set. (**a**) Average prediction errors. (**b**) CDF of prediction errors. (**c**) CDF of localization errors. (**d**) Convergence comparision

increase, which is influenced by estimations obtained by the previous iteration. This is actually reflected in Fig. 4.36d, where it can be seen that dots with deep color usually exist by clusters. The phenomenon can be incurred by multiple factors when executing the sliding window algorithm, such as the initial state of the sliding window, the spatial distribution of samples, and the noise of samples. In the experiments, we randomly sample a 250 m × 300 m sub-area over the city region as shown in Fig. 4.36d multiple times, and examine the prediction performance within the window each time. Then the average and distribution of the errors can be obtained.

Figure 4.36a shows the average, maximum and minimum prediction errors when we select different numbers of sub-areas to examine. It can be found that the average predicting error (the bars) fluctuates slightly around 8.5 dBm in different number of windows varying from 10 to 100, with the standard deviation 0.17 dBm. This indicate the stability of average predicting performance in different sub-areas. We can also see that the minimum value is generally much closer to the average error than the maximum value, indicating that the good predictions are more than the bad ones. Figure 4.36b shows the cumulative distribution function (CDF) of prediction errors. It can be seen that the CDFs under different numbers of samplings approximately overlap with each other, indicating that the prediction performance in each sub-area is stable.

Positioning Results with Predicted Fingerprints We here validate that the predicted fingerprints can be utilized for location estimation with the accuracy and reliability satisfying E911 requirement. We first grid the entire area into 871 × 663 square cells with each edge length to be 11 m. As mentioned above, the data set contains data from 611 BSs, but a number of base stations are only observed at a couple of locations. Thus we first sort the BSs according to the frequency they are observed at all locations of the area, and select the top 135 BSs. Then the corresponding data account for 96% of the entire data set. To perform localization, we choose those cells that both have the measured and the predicted fingerprints. We construct a fingerprints database with the predicted fingerprints, and use the real data as the user's reported data for localization. According to the statics of the data set, a user's mobile device normally can observe 1–12 BSs, and our preliminary experimental results show that the localization accuracy will be unacceptable if the user just report the fingerprint with respect to only one BS; therefore, we just consider the cells that can observe at least two BSs.

With our predicted fingerprints, we compare the performance of fingerprinting localization performance with that of Cell ID (CID) and Gaussian Mixture Model (GMM) based method [62]. The basic idea of the CID approach is to estimate the user's location to be the geometric center of all BSs the user can observe; GMM method is to estimate the location of a reported fingerprint using the GMM model constructed based on the Gaussian radio propagation model, which also can be regarded as a method to predict a given fingerprint's location.

We perform localizations for around 4500 test samples by three methods respectively. For SSOA and Gaussian mixture method, we use the predicted fingerprints and part of the sampled fingerprints to form a fingerprints database, and use the rest of the sampled fingerprints as the test set. Given some fingerprints in the test set, we run the regular localization algorithm to estimate the corresponding location, and the localization errors are deviations of the estimated location from the ground truth. We draw the CDF of localization errors for each method, as shown in Fig. 4.36c. The localization error is the Euclidean distance between the user's estimated location and the ground truth. We use the E911's localization requirement benchmark to evaluate the three localization methods, which is "within 100 m for 67% and within 300 m for 90%". We can see that the fingerprinting method using our predicted fingerprints by SSOA achieves "within 100 m for 71% and within 300 m for 98%", CID method achieves "within 100 m for 34% and within 300 m for 93%", and the GMM method achieves "within 100 m for 14% and within 300 m for 75%". This is because CID's performance is impacted by the unbalanced distribution of BSs, and GMM's assumption that the received signal strength at a given location is a multivariate Gaussian distributed random variable [62] is not always realistic especially in urban environment with more serious shadowing and multipath effects.

Convergence Rate Our convergence analysis reveals that the proposed SSOA mechanism converges faster than the Grassmann-manifold optimization algorithm, and we now provide experimental results to validate this claim. We consider the entire area as a giant matrix, and use 40% of the data as the training set to predict the rest of the data. We let the SSOA and the Grassmann-manifold optimization algorithm iterate 60,000 times and examine the prediction error after each iteration.

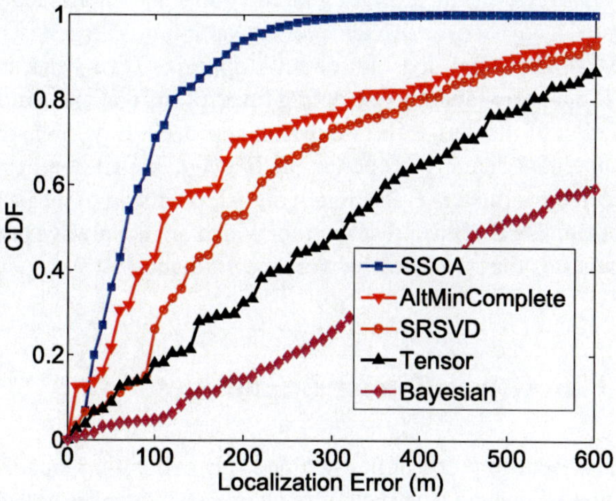

Fig. 4.37 Comparison of the SSOA with other matrix completion methods

The prediction error is found by comparing the predicted data and the real data in the other 60% of the data set, and each error is represented as a point in Fig. 4.36d. It shows that the average prediction error using SSOA reaches around 11 dBm within 1000 times, while the error using Grassmann method only reaches around 12.5 dBm after 30,000 times. It takes 75 min for the Grassmann method to reach 12.5 dBm error, while our proposed SSOA just consumes around 2 min to achieve 11 dBm error.

Comparison to other Matrix Completion Approaches We note that efforts have been dedicated to utilize matrix completion approach for predicting RF fingerprints in the indoor environment. Although considering different localization environment, in essence such mechanisms share some similarities with the proposed one. We here compare performance of the following 4 such algorithms with that of the proposed approach SSOA in the outdoor scenario: AltMinComplete minimizes the fingerprints estimation error by optimizing the two unitary matrices in the SVD process in an alternating manner [170]; Tensor combines the signal values with respect to different BSs together to form a tensor, which can be regarded as a 3-D matrix, in order to complete those unavailable fingerprints [69]; SRSVD reduces the possible interference in Wi-Fi signal which may incur negative impact on fingerprints completion [169]; Bayesian sparse learning approach conducts localization by exploiting the low-rank and sparsity properties of the sampled signals [163].

The localization experiments are conducted in the same way as mentioned before, and the results are shown in Fig. 4.37. It can be seen that the proposed SSOA mechanism outperforms all the other ones. AltMinComplete presents the second best performance, closest to SSOA, because it directly optimizes the unitary matrices in SVD; however, there is no guarantee that the alternating optimization approach adopted can necessarily achieve the optimum. Tensor is realized in a similar manner as SSOA but does not directly optimize U, instead, it predicts the subspace that U spans by randomly selecting fingerprints that are sampled. SRSVD is a slight variation of the classical SVD-based approach. Bayesian sparse learning considers the potential sparsity properties of RF signals, but it relies on setting a slew of pre-defined parameters (variation of signal, precision of measuring) and the assumptions of signal's Gaussian distribution, which are not always practical in real localization scenarios thus presents the worst performance.

4.5 Floor Plan Generation for Localization

These sections present FineLoc, a fine-grained self-calibrating localization system based on the freely deployed Bluetooth low-energy (BLE) nodes and crowdsourced data, which can profile more detailed layout information of the indoor space. We first reveal that existing systems can only generate inaccurate floor plans owning to the coarse-grained Wi-Fi reference information. Then we utilize the increasingly popular BLE beacon nodes as the source of reference information, with which a series of dead-reckoning optimization and new schemes particularly for finer-grained indoor map construction are presented. We implement a prototype FineLoc system, which is deployed in around 11,000 m^2 areas. Our experimental results with the prototype show that FineLoc can achieve 80% localization errors within 1.6 m, 1.4 m, and 1.1 m in the library, classroom building, and office building respectively, with an average density of deployed BLE nodes less than 2.6/100 m^2.

4.5.1 Motivation

Constructing indoor maps requires reference information that can reflect the physical layout of buildings [54]; moreover, the reference information must be uniquely identifiable, and functioning with no need to deploy special infrastructures, considering the scalability and cost. The natural signatures such as the fluctuation of magnetic fluctuation in certain part of the building as in UnLoc [128] need no dedicated infrastructure but cannot be uniquely identified in multiple buildings. The spatial or timing feature of Wi-Fi signals can be uniquely identified with the help of the Wi-Fi AP's MAC address [70, 71]; however, such reference information has limited capability to reflect the building's physical layout. This is why the floor plan constructed with Wi-Fi based reference information is with coarse granularity. This section first reveals the fundamental reason why the Wi-Fi based reference information leads to coarse-grained indoor map, and then presents our choice of reference information source for finer-grained indoor map.

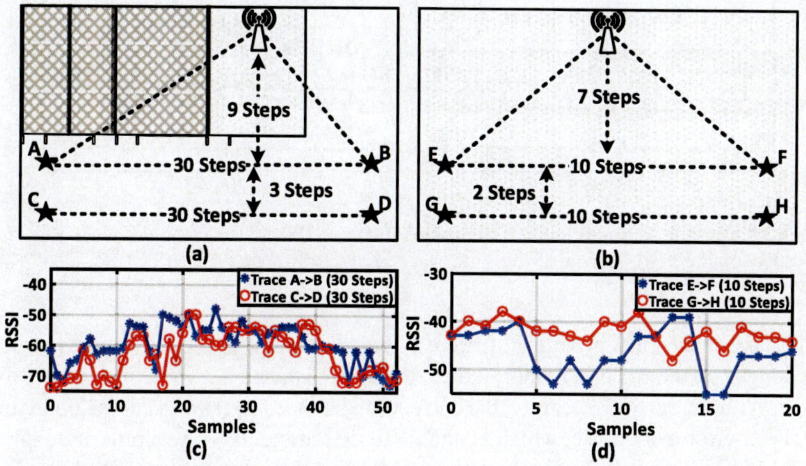

Fig. 4.38 Wi-Fi RTTP shows less scalability. (**a**) Scenario A. (**b**) Scenario B. (**c**) RSSI in scenario A. (**d**) RSSI in scenario B

4.5.1.1 Analysis of Wi-Fi Landmarking

Wi-Fi signals can reflect the building's physical layout to some extent. When a user passes through the covered area of a Wi-Fi AP, the user's mobile device can observe that the RSSI of the AP increase as the user moves closer toward the AP, and decrease as the user moves past the AP. Figure 4.38a shows the scenario, where it shows that the RSSI increases to a peak point and then decreases as the user moves from A to B. The *RSSI tread tipping point* (RTTP) corresponds to a fixed position on the way that is closest to the AP, which can serve as a landmark to provide location reference information in map construction [287]. In reality, the RTTP can be obviously identified only if the length of trace AB (30 steps) is notably greater than the distance between the AP and the trace (9 steps). In contrast, once the length of trace EF (10 steps) is comparable to the distance between the AP and the trace (7 steps) in Fig. 4.38b, we can see from Fig. 4.38d that the RTTP is difficult to identify. The hidden reason is that Wi-Fi RSSI changes slightly when the distance between the user and the AP varies a little. The observation indicates that the Wi-Fi landmarks only appear when the user is walking along a long path, which explains why Walkie-Markie can only construct maps for pathways; because the size of room is usually limited, where the users cannot walk along a long straight path for RTTP identification.

In PiLoc [71], the user's mobile device records the observed Wi-Fi RSSI while walking; the recorded time series of the RSSI from different users are compared for correlation, which is a critical reference information to determine whether the RSSI series are observed from the same location in the building. Although claiming independent of landmarks, PiLoc factually leverages the RTTP feature to determine the correlation of two traces in an implicit manner. The similar RTTP observed in

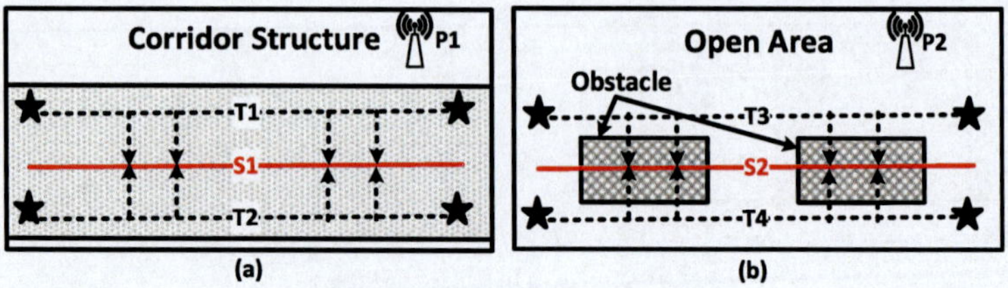

Fig. 4.39 Existing schemes cannot recognize obstacles due to coarse-grained merging. (**a**) Corridor structure. (**b**) Open area

two traces is the most prominent feature of the two traces, which only appears when the length of the trace is notably larger than the distance between the AP and the trace. Without the RTTP feature, the correlation between the two traces is dominated by the environment noise, which is unable to determine the trace similarity. Figure 4.38d shows the scenario, in which the correlation of the two traces is 0.143.

The experimental results presented in Walkie-Markie and PiLoc factually corroborate our analysis. In particular, Figure 4 in [287] and Fig. 7–9 in [71] show that the RTTP appears when the length of the trace is at least 30 steps. This is far beyond the dimension of regular rooms in the building; therefore, both Walkie-Markie and PiLoc focus on constructing pathway maps.

If the RTTP only appears in the long trace, will the reference information based on Wi-Fi signals be suitable for wide open areas such as libraries or museums? Unfortunately, the answer is negative. If the two paths are labeled with the RTTP from the same AP, it will be difficult to determine distances between the paths and the AP respectively. As shown in Fig. 4.38c, paralleled trace AB and CD can obtain almost the same RTTP. Thus it is impossible to recognize the relative distance between AB and CD (3 steps). Figure 4.39a illustrates the two traces T_1 and T_2 in real situation, where the space between the two traces can be observed. Due to coarse-grained RTTP recognition, the current solutions in [71, 287] just merge the two traces to form a combined trace as shown in S_1, which makes the space between T_1 and T_2 unable to be reflected in the resulted floor plan. Regular pathways may be modeled with line segments; however, the trace merging schemes can prevent the constructed floor plan from reflecting the obstacles such as the bookshelf, showcase, and booth between two traces, as shown in Fig. 4.39b, which results in coarse-grained map with little usefulness.

The trace merging method is adopted in both Walkie-Markie and PiLoc, because the nature of the Wi-Fi signal can provide only coarse-grained reference information. The infrastructure Wi-Fi AP is designed to cover a hundred-meter-radius area; APs are normally sparsely deployed to save the cost and avoid MAC layer collisions. Even if some traces can observe strong RSSI with respect to an AP, it is still hard to determine the distance between the trace and the reference point such as RTTP generated by the Wi-Fi AP. Consequently, it is reluctant for the coarse-grained relative location information of the user's trace with respect to instable reference information to provide basis to construct a fine-grained indoor map. It is necessary to find finer-grained reference information.

4.5.1.2 BLE Landmarking

We find that BLE nodes such as iBeacons can be a promising source of reference information for indoor map construction, which can be uniquely identified and have been deployed in many public places. For example, major league baseball (MLB) has been using iBeacons since the start of the 2014 baseball season to track MLB app users and send relevant messages to enhance the ballpark experience [200]; airline companies and retailers have been using iBeacons to send flight information and coupons [201, 202]; the mobile social app WeChat owning hundreds of millions of active users has been providing users with the mini program interface to discover iBeacons [203]; there have been iBeacon based localization systems deployed leveraging the proximity approach [202, 203]. Research efforts also have been dedicated to building BLE based localization systems [155, 157, 158], where it has been verified that BLE has some favored characteristics in localization compared with Wi-Fi; however, how to leverage BLE nodes for indoor map construction is still an open issue.

The fundamental reason that BLE nodes can provide finer-grained reference information is that the nodes can be close to the user's mobile device, which is rooted in the BLE design principles to provide smaller coverage for power saving. Such a favored feature leads to that the relative location of the user's trace with respect to the BLE node can be more accurate and reliable. This provides opportunities to identify spaces between traces and thus can reflect more detailed information of the building's physical layout.

However, to enable the fine-grained self-calibrating localization system, we are confronted with the following new challenges:

Challenge 1: Fine-Grained Map Construction As shown in Table 5.2, FineLoc is the only fine-grained self-calibrating indoor localization system. In contrast to previous floor plan construction methods, integrating users' traces with more detailed information instead of just illustrating the skeleton increases the difficulty to realize map construction. Since there are many factors such as coarse-grained RTTP recognition and IMU error affecting the construction performance, it is challenging to construct fine-grained floor plan.

Challenge 2: Dynamic Landmarks Environment Since BLE nodes can be in reach of people, it may happen that the BLE nodes are moved accidentally, which results in an environment with dynamic landmarks. As the most important information to label different traces, dynamic landmarks might lead to false floor plan construction. Therefore, it is necessary to design a new trace merging method to recognize landmarks movement and update the map dynamically.

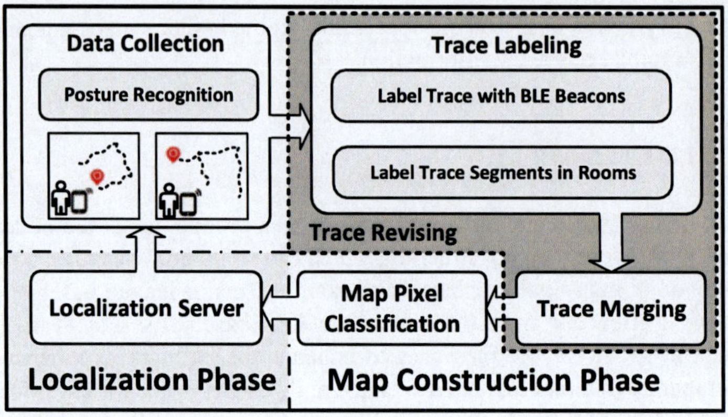

Fig. 4.40 System architecture

4.5.2 *System Overview*

FineLoc constructs indoor maps in a crowdsourcing manner, where the BLE reference information associated with the crowd worker's traces are used to integrate different traces into a map. There is no need to have prior knowledge of the BLE nodes' locations, and FineLoc tolerates that BLE nodes can be moved or running out of power. We develop mechanisms to categorize traces submitted by different users, from which we derive detailed information reflecting the building's physical layout. FineLoc supports map construction in wide open areas such as the library, where we can recognize the position of obstacles.

FineLoc is shown in Fig. 4.40, which shows that the system consists of two phases. FineLoc first collects trace data from mobile users and processes the data with several sequential mechanisms to construct the indoor map, which is termed as the *map construction phase*; the constructed map is then used for positioning in the *localization phase*. According to our experiments to be presented in Sect. 4.5.8, we can construct finer-grained floor plan in contrast to previous works [70–72]. Our work in this paper uses BLE beacon based approach in the localization phase, which presents 80% localization errors of 1.6 m, 1.4 m and 1.1 m in the library, classroom, and office building respectively. The average density of the BLE nodes is less than 2.6/100 m^2.

4.5.3 Data Collection

4.5.3.1 Trace Data Format

FineLoc supports crowdsourcing in a non-participatory manner, where crowd workers can collect data when walking around within the indoor space in daily life. The trace data set is denoted by $T = \{\tau_i, i = 1, 2, \ldots, m\}$, where $\tau_i =< id, p, f >$ is a specific trace. The element id is factually a tuple $< userid, traceid >$ meaning the trace data is collected by which user. Note that a user can submit multiple traces labeled by different $traceid$s, where each labels the segment of the user's trace when the user is in a specific posture, and the $traceid$ will change if the user switches the posture during walking. This is because the same IMU trace data indicate different walking distances when the user is in different postures. The element $p = \{p_t, t = 1, 2, \ldots, l\}$ records the user's position in each user's relative coordinate system in each sampling time slot with $p_t =< x, y >$. We set the initial position and the direction of a newly generated trace as (0, 0) and along x-axis in the relative coordinate system, respectively. In contrast to the previous work [71, 128, 160] assuming the availability of heading direction and stride length of each trace, which is not always true in practice, our trace data are in different relative coordinate systems and will be merged into the same coordinate system in the trace merging process to be presented in Sect. 4.5.4.3. The element $f =< mac, rssi >$ records detected BLE MAC address and RSSI series during the trace, which is to be utilized for clustering and merging traces later.

4.5.3.2 Posture Recognition in Dead Reckoning

Posture Recognition The trace information is derived from the IMU data [57], where the orientation and step count are fundamental for determining the direction and length of the trace. How to recognize change of the walking direction has been well studied in [128], which is also adopted in our work. For the step count determination, detecting acceleration period is a widely adopted approach [53, 71, 128, 156], the basis of which is the observation that the acceleration readings in the three axes of the accelerator present periodicity.

However, such an approach pays limited attention to the practical scenario that users may switch their postures of holding the mobile device during walking. Figure 4.41a illustrates periods of accelerator readings when the user is using different postures, where the acceleration period can be identified by observing the accelerator's peak value. However, when the user holds the smartphone horizontally, one period of the accelerator's readings indicates that the user moves one step, but when swinging the smartphone, one period means two steps.

This is because the readings in X, Y, and Z axes of the accelerator are dependent on the orientation of the accelerator's coordinate system with respect to the direction of gravity. When holding the smartphone horizontally, each step can result in a slight

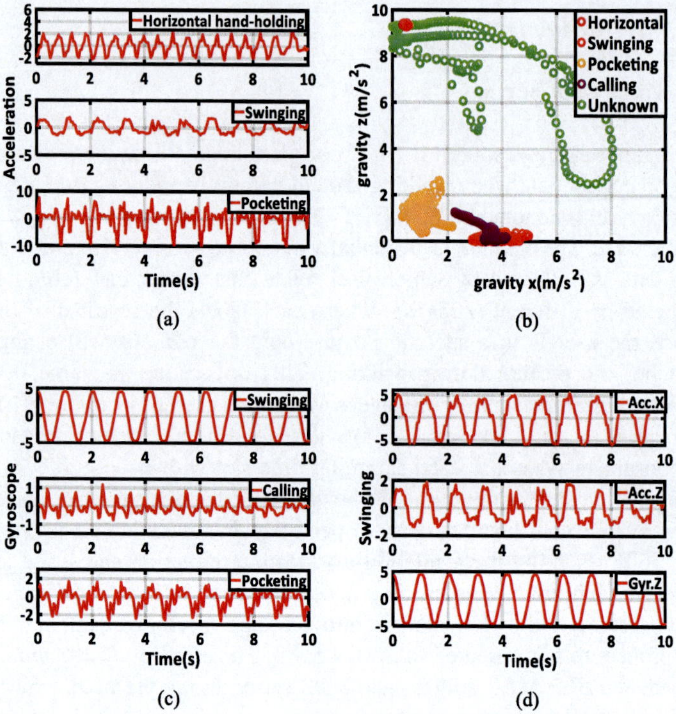

Fig. 4.41 Posture recognition. (**a**) Acceleration periods. (**b**) Gravity distribution. (**c**) Gyroscope readings. (**d**) Periodicity

vibration in the gravity direction, which indicates a period of walking. The gravity direction happens to be in accordance with the direction of Z axis in this case, thus each period of the Z-axis readings represents one step. However, when the user is swinging the smartphone during walking, the gravity direction is in accordance with the direction that is perpendicular to the gravity direction. It can be found that the readings in Z axis in this case show the following pattern: peak (right foot) $\sim$ zero (one step) $\sim$ valley (left foot) $\sim$ zero (two steps) $\sim$ peak (the next period). This period takes twice time compared with the period in the horizontal hand-holding case, so a period represents two steps.

We propose to improve the accuracy of the step counter by recognizing the user's posture of holding the mobile device. Normally, the user's postures can be categorized into the following classes: horizontal hand-holding, swinging, pocketing, and calling, among which the horizontal hand-holding posture is usually used when the user is performing localization. Since the user may have unpredictable postures other than the 4 classes, we categorize the rest of possible postures into the unknown state. Our posture recognition scheme is based on the observation that the gravity accelerator readings are different under different postures of the

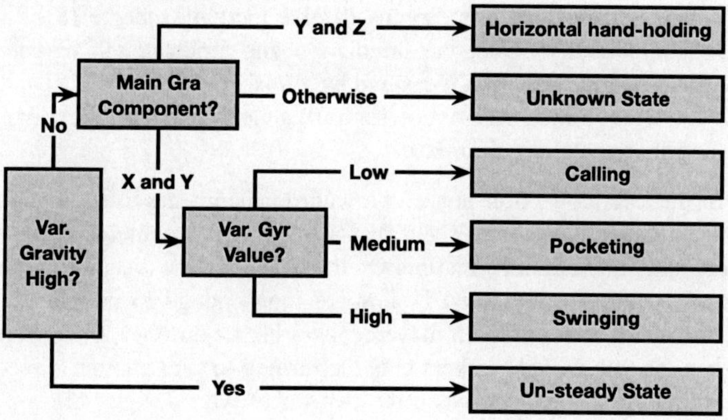

Fig. 4.42 Posture recognition classifier

user's, as shown in Fig. 4.41b. It can be seen that most of the postures can be well recognized, but some of them have very similar features. For example, mobile device is usually placed vertically when calling, swinging, and pocketing. To address this issue, we find that referring to the gyroscope data can help cross check the user's posture, as shown in Fig. 4.41c. It is found that the amplitude and pattern of the gyroscope readings vary under different postures. Based on this observation, we have designed a classifier to recognize different postures as shown in Fig. 4.42. To separate disturbed states (for example, irregular shaking) from regular states, we first calculate the variance of gravity to evaluate the stability of smartphone placement. This is because smartphone placement is relatively steady under regular states. Then we obtain smartphone placement relative to absolute gravity direction of the earth by analyzing main gravity component, where Y denotes the vertical placement and Z denotes the horizontal one. Finally, with gyroscope readings, we can further determine smartphone postures.

Step Counter With the user's posture determined, we find that the readings in X, Y, and Z axes of both accelerator and gyroscope may all present periodicity in Fig. 4.41d. Besides, for a certain posture, different readings suffer from noise pollution in varying degrees. We could choose the readings with more obvious periodicity to obtain a more accurate step counts, which is in contrast to existing schemes considering sum of square of readings in all the 3 axes of the accelerator [53, 128, 156], which could amplify the noise by aggregating the noise in each axis.

Heading Angle and Stride Length Though many researches have been conducted to improve heading angle recognition and stride length evaluation[70, 71, 159], these two problems remain serious. For heading angle recognition, compass readings would be interrupted by magnetic field and gyroscope can hardly provide absolute direction under different postures; For stride length evaluation, since a certain person might cover varied stride length, a fixed or even trained stride length is still unreliable. In this paper, we propose an adaptive merging scheme (Sect. 4.5.4.3), which can automatically adjust the orientation and stride length of each trace. Therefore, only relative direction measured by gyroscope readings[128] and a fixed stride length are required in this section for trace generation. Then these two factors would be considered and synchronized.

Remark In the data collection phase, we want to record the user's trace as long as possible, because it is more likely that a longer trace contains more position information and landmark information. On the one hand, the distance between two parallel traces could be identified if they are long enough to include turns. For example, if the two traces contain movements of taking turnings, the different step counts when the two crowd workers take the turning toward another direction can help determine the distance between the two traces. Our design philosophy is in contrast to the previous work [71] preferring the short traces along one direction, since our work tries to reflect the physical layout of the indoor space as detailed as possible, but the previous work is implicitly designed to construct the pathway layout, where it is convenient for the 1-D short traces in parallel to be integrated into line segments. On the other hand, a longer trace may go through more landmarks while their relative positions could be determined to reflect the physical layout. Such information could be utilized to recognize dynamic landmarks in Sect. 4.5.4.3.

4.5.4 Trace Labeling

4.5.4.1 Labeling Traces with BLE Beacons

With the collected trace data, we now need to associate each trace with BLE beacon nodes using the beacons' MAC addresses. We associate the trace with a beacon if a part of the trace is within 1 m of the beacon, where "associate" means that the beacon is considered in the way of the crowd worker and can be regarded as a label of the trace. This is because the BLE node's coverage is very limited, the mobile device can observe strong RSSI in the 1 m neighborhood of the beacon. We could leverage the labeled traces to derive the skeleton of the floor plan. However, the challenge is that the corresponding RSSI when the mobile device is in the 1 m neighborhood of the beacon varies due to the random wireless signal propagation and the heterogeneity of beacons and the mobile devices. It is impractical to set a fixed threshold of the RSSI to determine if the mobile device and the beacon are close enough to each other.

To address the issue, we propose a dynamic threshold updating scheme, where the basic idea is to update the maximum RSSI observed by crowd workers, and adjust the information of the traces that are collected before updating accordingly. In particular, consider a BLE landmark i, for which the maximum observed RSSI of the landmark is currently recorded in the server as $r^i_{\max}$. If we have a trace with a series of RSSI $f^i = \{r_1^i, r_2^i, \ldots, r_m^i\}$ observed by a crowd worker along the trace $p = \{p_1^i, p_2^i, \ldots, p_m^i\}$ with respect to the landmark i, we first define $w(r_k^i, r_{\max}^i)$ to represent the weight r_k^i with respect to $r_{\max}^i$:

$$w\big(r_k^i, r_{\max}^i\big) = \frac{(r_k^i - r_{\max}^i - \gamma)^2 I(r_k^i > r_{\max}^i - \gamma)}{\gamma^2}, \tag{4.72}$$

where I is an indicator function, and γ is a protecting threshold to determine the transmission range of the landmark with an empirical value $\gamma = 15$. Then we sum the weight of each value and assign a weight to the trace defined by

$$\mathcal{W}(l^i) = \sum_{k=1}^{m} w\big(r_k^i, r_{\max}^i\big), \tag{4.73}$$

where l^i is the BLE landmark i. The trace has a weight with respect to l^i only if it contains a RSSI observation with respect to l^i such that $r_k^i > r_{\max}^i - \gamma$. The greater the RSSI observed in the series, the higher weight the trace is with l^i, thus the closer the trace is to l^i. Note that the square of both the numerator and denominator is to better distinguish weights of two traces. Our empirical study shows that if the weight is greater than 0.8, the trace should be associated with the landmark.

As more crowd workers pass by the landmark l^i, the FineLoc server may get updated by greater $r_{\max}^i$, which means that the weight of the trace with respect to l^i should also be updated. A formal description of the update is

$$\mathcal{W}(l^i | r'_{\max}) = \mathcal{W}(l^i | r_{\max}) w(r_{\max}, r'_{\max}), \tag{4.74}$$

where $\mathcal{W}(l^i | r_{\max})$ can be updated to $\mathcal{W}(l^i | r'_{\max})$ and $w(r_{\max}, r'_{\max})$ could be calculated through Eq. (4.72).

After determining weight of the landmark, we could acquire the landmark's position in the current coordinate system. With acquired weights w_k and positions $p = \{p_1^i, p_2^i, \ldots, p_m^i\}$, we can obtain the position $l^i = [x_i, y_i]$ by

$$l^i = \frac{\sum_{k=1}^{m} p_k w_k}{\mathcal{W}(l^i)}. \tag{4.75}$$

4.5.4.2 Labeling Trace Segments in Rooms

A fine-grained indoor map should be able to reflect room information, which requires the trace data to be categorized in terms of sub-areas. The solution in WILL system [56] calculates Euclidean distance between two fingerprints to determine whether the fingerprints are observed from the same room, which is based on the observation that the RSSI changes due to the wall blockage. In contrast to WILL utilizing Wi-Fi signals, FineLoc uses BLE signals, which provide an opportunity to derive more accurate trace segments classification. However, since RSSIs change rapidly when the user is near to the node, tradition Euclidean distance could not be employed directly. Here, we use a discrete function $\mathcal{F}_V(\cdot)$ to map the RSSI into corresponding level. In particular, the RSSI falling into the following ranges {>-65, -65∼-75, -75, -75∼ −85, −85, −85∼ −95, otherwise} will be mapped into the following levels $\{1, 2, 3, 4, 5, 6, 7\}$. We calculate dissimilarities among the RSSI series along the trace by

$$\mathcal{V}_{k,k+1} = (\mathcal{F}_V(f_k) - \mathcal{F}_V(f_{k+1}))^2, \tag{4.76}$$

where f_k denotes RSSIs with respect to all observable BLE nodes in position p_k. If $\mathcal{V}_{k,k+1} > 30$, then a new cluster will be generated. In FineLoc, trace segments represent the traces with respect to different rooms. Since there is usually one door between adjacent rooms, we can define the position satisfying $\mathcal{V}_{k,k+1} > 30$ as the *segmented position* for door recognition. Detail methods to recognize the door would be presented in Sect. 4.5.6.

In this way, each step in the trace will be categorized into a cluster, and steps with RSSIs near to each other are put in the same cluster. All clusters reflect the spatial characteristics of the RSSI in the building, which lay the foundation of recognizing rooms. We now could obtain a set $S_l = \{l^1, l^2, \ldots, l^n\}$, which includes all landmarks observed along the trace segment. Based on our empirical study, a landmark l^i can be included into subset S_l if one of the following event occurs: First, the crowd worker passes by the landmark and observed RSSI makes the indictor function $I(\cdot) = 1$ (Eq. (4.72)); second, more than 60% of the RSSI observed along the trace segment satisfies $r_k^i \leq -75$.

Remark Instead of labeling remote traces (about 5 ∼ 10 m) based on Wi-Fi RTTP [70] or RSSI correlation [71, 72] to reveal the rough skeletal structure, our system constructs the accurate floor plan with precise information of traces and landmarks. We associate the trace with a beacon if a part of the trace is near to the beacon (about 0 ∼ 2 m) and calculate the beacon's position. We develop a weighted labeling method based on dynamic threshold and then design corresponding methods to update derived labeling weight. Proposed methods make FineLoc more adaptive to the random wireless signal propagation and the heterogeneity of devices.

4.5.4.3 Trace Merging

We now integrate the clustered trace segments obtained from the previous processes into a skeleton of the floor plan. This is factually to merge the traces in different relative coordinate systems to a unified coordinate system. Our basic idea is to merge two traces first, and then merge the resulted trace with another trace and so on.

Figure 4.43 illustrates an example scenario of the merging process. We need to first translate the entire trace so that their geometric center happens to be the origin

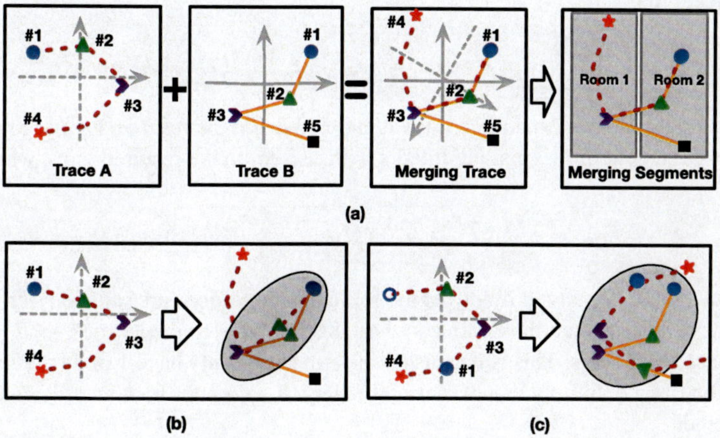

Fig. 4.43 Process of trace merging. **(a)** Trace merging algorithm. **(b)** IMU error. **(c)** Landmarks movement

of the relative coordinate system, and then rotate the coordinate systems so that the landmarks in the two systems can exactly overlap. Although the BLE landmarks can provide references to combine the two traces, the relative distances among those landmarks in the two systems can be different, thus we need to calculate the misalignment of the landmarks. Such operations can be tedious to realize in practice especially when there are a large number of traces needed to be merged. To address this issue, we present the following matrix transformation based method.

We use $\overline{P}^A = (\overline{x}^A, \overline{y}^A)$ and $\overline{P}^B = (\overline{x}^B, \overline{y}^B)$ to denote the coordinates of the geometric centers of landmarks of trace A and B, respectively, which can be obtained by averaging the horizontal and the vertical coordinates of those landmarks. The two traces have some landmarks in common, and we use $\mathcal{L}^A$ and $\mathcal{L}^B$ to denote the coordinates vectors of those landmarks in the two traces, respectively. Then we have the normalized coordinates vectors of those common landmarks with respect to $\overline{P}^A$ and $\overline{P}^B$, which are denoted by $\hat{\mathcal{L}}^A$ and $\hat{\mathcal{L}}^B$, respectively. We use K to denote the scaling coefficient for translation, and $\mathcal{R}$ the rotation matrix:

$$\mathcal{R} = \begin{bmatrix} \cos\theta & \sin\theta \\ -\sin\theta & \cos\theta \end{bmatrix}, \tag{4.77}$$

where θ is the angle trace A needs to rotate counterclockwise to comply with trace B. Trace merging is actually finding appropriate K and $\mathcal{R}$ so that

$$\hat{\mathcal{L}}^A K\mathcal{R} = \hat{\mathcal{L}}^B. \tag{4.78}$$

To this end, we first construct $\mathcal{L}_k^M = [x_k^A + y_k^A, x_k^A - y_k^A]$ and $\mathcal{L}_k^N = x_k^B + y_k^B$, Then we can transform Eq. (4.78) to the following: $\mathcal{L}^M[Kcos\theta\ Ksin\theta]^T = \mathcal{L}^N$, and obtain that

$$[K cos\theta \ K sin\theta]^T = ((\mathcal{L}^M)^T \mathcal{L}^M)^{-1} (\mathcal{L}^M)^T \mathcal{L}^N. \tag{4.79}$$

With such results, we can merge trace A and B. We use the *residual error* e to denote the misalignment of original landmarks with respect to the resulted ones, where

$$e = \sqrt{|(\hat{\mathcal{L}}^B - \hat{\mathcal{L}}^A K\mathcal{R})^T (\hat{\mathcal{L}}^B - \hat{\mathcal{L}}^A K\mathcal{R})|}. \tag{4.80}$$

Two traces can be merged if both of the following conditions are satisfied: First, they must have at least 3 landmarks in common; Second, the resulted error $e < T_e$, where T_e is set to be 8 steps. This is because if the two traces only have 1 or 2 landmarks in common, they definitely can be perfectly merged no matter how they deviate from each other.

In FineLoc, e might result from the malfunctioning step counter or displaced landmarks. In reality, since step counter error is slowly cumulative and evenly distributed, it usually leads to a smaller e compared with displaced landmarks as shown in Fig. 4.43b. In contrast, once a landmark has been moved to another position as shown in Fig. 4.43c, we can hardly merge the traces collected, because landmark movement leads to a large residual error e. Therefore, by observing the abnormal residual errors larger than T_e in the trace merging process, FineLoc is able to detect unreliable landmarks, which are to be updated. This will ensure the robustness of the system in dynamic environment.

We have merged the two traces as of now, the next step is to cluster the segments of the traces, so that the trace segments in different rooms can be recognized. Recall that each segment of a trace is associated with a set of landmarks S_l. We define the similarity of S_l and S_l' as

$$\mathcal{F}_s(S_l, S_l') = \frac{S_l \cap S_l'}{S_l \cup S_l'}. \tag{4.81}$$

If $\mathcal{F}_s(S_l, S_l') > 0.6$, the two segments of traces can be categorized into the same group in our system according to our empirical study.

Remark In contrast to previous works slicing the trace into segments such as turns or straight lines and then merging the similar segments [70–72], we merge complete traces directly. The advantage of our method is that we can realize the overall layout of beacons as shown in Fig. 4.43. In this way, we present a method to recognize the movement of beacons and adapt FineLoc to dynamic landmarks environment. Note that "segment" in this paper represents the traces relative to different rooms, in contrast to turns or straight lines in previous works [71, 72].

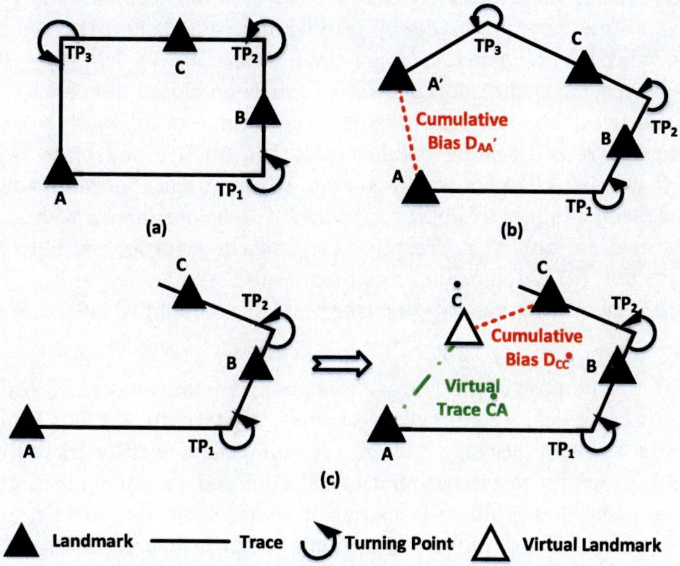

Fig. 4.44 Process of trace revising. (**a**) Ground truth. (**b**) Trace self-revising. (**c**) Trace co-revising

4.5.5 Trace Revising

In trace labeling and trace merging stages, the system performance highly relies on the accuracy of IMU sensors. Though posture recognition algorithm has been proposed and step counter with 98% accuracy has been realized, floor plan construction also suffers from direction error because the error in direction will be gradually accumulated as the user moves on. This problem might lead to serious error, which, however, has not been resolved by previous floor plan construction systems[70, 71, 74] yet. In this section, we propose a novel scheme to revise user's direction where there are two stages: trace self-revising and trace co-revising.

In the trace self-revising stage as shown in Fig. 4.44a, there are three BLE landmarks A, B, C and three turning points TP_1, TP_2, TP_3. When a user goes through a closed-loop trace $A - TP_1 - B - TP_2 - C - TP_3 - A$, we might obtain a measured trace as shown in Fig. 4.44b due to IMU errors. We here define cumulative bias $\mathcal{D}_{AA'}$ as an indicator of trace accuracy. Intuitively, the system should modify the values of TP_1, TP_2, TP_3 to minimize $\mathcal{D}_{AA'}$ which can be formulated into the following optimization problem: $E_{TP} = arg \min_{E_{TP}} \mathcal{D}_{AA'}$, where $E_{TP} = \{TP_1, TP_2, \ldots, TP_n\}$ denotes the sequence of turning points. To address this issue, our system utilizes particle filter to select TP_i to generate new particles and finally determine E_{TP}.

In practice, we may not be able to obtain a closed-loop path all the time. To extend our scheme to non-closed-loop scenarios, we combine trace merging stage and trace revising stage, which yields a trace co-revising algorithm. The basic idea is to construct virtual closed-loop path with merging landmarks. As shown in Fig. 4.44c, when the user generates a non-closed-loop trace $A-TP_1-B-TP_2-C$, self-revising scheme cannot work; however, if this non-closed-loop trace is merged into a global trace, we can perform co-revising scheme as follows: (1) Determine rotation matrix $\mathcal{R}$ and scaling coefficient K (Eq. (4.79)), and place landmarks $A, C, \dot{A}, \dot{C}$ into the same coordinate system, where $\dot{A}$ and $\dot{C}$ mean landmarks in global trace with a higher reliability; (2) Move $\dot{A}$ to be overlapped with A, and we obtain a virtual position of landmark $\dot{C}$; (3) Connect A (or $\dot{A}$) and $\dot{C}$ to generate virtual trace $\dot{C}A$; (4) Based on the closed-loop path $\dot{C} - A - TP_1 - B - TP_2 - C$, we can define cumulative bias $\mathcal{D}_{C\dot{C}}$ and the problem is simplified into a self-revising problem.

Remark For the purpose of IMU error elimination, previous works [71, 72] propose inter- and intra- trajectory correcting schemes, which correct traces by merging multiple similar traces into one trace, thus cannot recognize obstacles as illustrated in Fig. 4.39. To address this issue, our data collection and trace merging stages prefer merging complete traces directly instead of slicing them into smaller segments (turns and straight lines). However, merging complete traces suffers more from direction error because the error in direction will be gradually accumulated as the user moves on. Since previous trace revising method is course-grained, it is necessary to propose our trace revising method to eliminate the IMU error.

4.5.6 Map Pixel Classification

The skeleton of the indoor floor plan can be obtained with the process as shown in Fig. 4.45. We can divide the map into small square sub-regions, each of which can be regarded as a pixel. Besides pixels covered by those trace segments, there are still a large number of pixels that need to be classified as walls, doors, pathways, and so on. We here present our map pixel classification method.

Suppose we have a map skeleton such as the one shown in Fig. 4.45, for each pixel, we first calculate the distance between the pixel and each kind of trace segment. Give a trace segment of cluster ID i, the distance d_i is the sum of the distances between the pixel and the k nearest points in the trace. Suppose that there are m clusters of trace segments, then we have $\{d_i, i = 1, 2, \ldots, m\}$, among which $d_{1\min}$ and $d_{2\min}$ denote the shortest and the second shortest distances respectively. We define $d_\Delta = d_{2\min} - d_{1\min}$. Then we can classify the pixel according to the following rules:

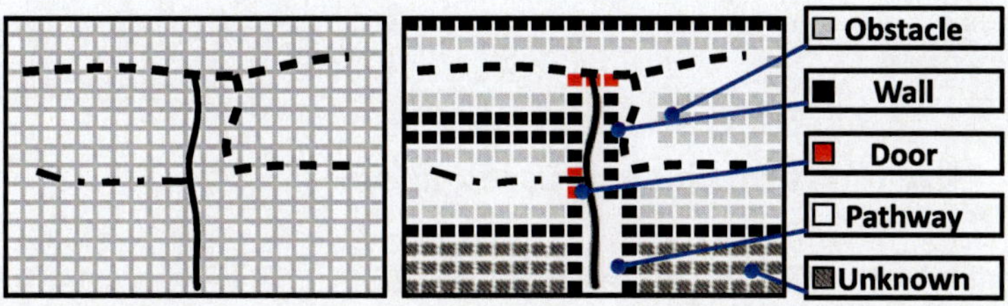

Fig. 4.45 Map pixel clustering

$$Pixel\ role = \begin{cases} Pathway, & d_{1\,\min} < 0.5k; \\ Obstacle, & 0.5k < d_{1\,\min} < 2k; \\ Unknown, & d_{1\,\min} > 2k; \\ Wall, & d_{\Delta} < 0.5k\ d_{2\,\min} < 2k. \end{cases}$$

Note that the distance between the pixel and a single point in the trace segment can be erroneous due to the unpredictable abrupt malfunction of the step counter or the sudden change of BLE RSSes, that is why we use the k nearest points in the trace to determine the shortest distance, where k is set to be 4 in our system; moreover, since the absolute distance between two pixels is indistinguishable in a fine-grained map, k also can amplify the distance differences, so that the pixels can be classified more accurately.

The rationales of the rules presented above are as follows. The pixel that is 0.5 m from a trace is likely to be a pathway pixel, since the width of the space a walking human body needs to occupy is around 0.5 m. The pixel satisfies the second rule can be an obstacle, since normally the dimension of an obstacle in the indoor space is around 2 m such as desks and bookshelves. This rule can not 100% guarantee identifying the obstacles, since there indeed are some small obstacles such as sofa and chairs; however, if we have sufficient trace data, such small obstacles still can be recognized. If the pixel is comparatively far away (2 m) from the nearest trace, we temporarily regard it as an unknown point; if no other traces are near to it, probably nobody ever passes by, thus it is indeed unknown. The wall is probably adjacent to or separate two traces, and the wall is actually one kind of obstacle thus we let $d_{2\,\min} < 2k$. Moreover, in Sect. 4.5.4.2, we define segmented position when we observe a new cluster. Since the segmented position represents the border between different rooms, we regard one pixel as the door when the distance between the pixel and the segmented position is lower than $0.5k$.

Remark Compared with previous works which can only sketch the skeleton of floor plan as shown in the left sub-figure of Fig. 4.45, we for the first time develop the map pixel classification method to realize fine-grained map construction. With proposed methods, we can construct floor plan with limited traces, where sub-rooms and obstacles could be recognized.

4.5.7 Map Construction and Localization

After introducing the frame and methodology of FineLoc, we here show how to integrate our proposed schemes (Sec. 4.5.3 ~ Sect. 4.5.6) into algorithms to perform map construction and localization.

Trace Generation Algorithm In Algorithm 9, we would collect data, recognize smartphone postures and generate user trace. The system generates a new trace when the user's posture switching occurs since posture switching would introduce error in dead reckoning and damage floor plan construction.

Algorithm 9 Trace generation

Require: Sensor readings
Ensure: Labeled trace

```
for each time slot k do
    Parse acc, gra, gyr from sensor readings;
    Section 4.5.3.2 - Posture recognition
    Recognize smartphone posture Pos_t;
    if Pos_k ≠ Pos_{k-1} then
        Generate a new traceid;
    else
        Update position p_k;
        Section 4.5.4.1 - Trace Labeling
        Parse RSSI series f^i = {r^i_1, r^i_2, ..., r^i_m};
        if I(·) = 1 (Eq. (4.72)) then
            if r^i_k <= r^i_max then
                Calculate w(r^i_k, r^i_max) and W(l^i) (Eq. (4.73));
            else
                Update r'_max = r^i_k and W(l^i|r'_max) (Eq. (4.74));
            end if
            Attach landmark l^i to trace (Eq. (4.75));
            Section 4.5.5 - Trace self-revising
            if l^i is historical landmark of this trace then
                Revise trace and obtain E_TP;
            end if
        end if
        Section 4.5.4.2 - Label trace segments
        Calculate dissimilarities V_{k,k+1} (Eq. (4.76));
        if V_{k,k+1} > 30 then
            Generate new cluster and obtain segment set S_l;
        end if
    end if
end for
```

Floor Plan Generation Algorithm The purpose of Algorithm 10 is to merge, revise trace, and finally finish map pixel classification. Here we define collected trace space E_T^{Local} and merged trace space E_T^{Global}, where E_T^{Local} denotes the traces collected by users and E_T^{Global} represents the traces after merging. With more traces collected and merged, we can enlarge the scale and reliability of E_T^{Global}. Finally, detailed floor plan would be constructed through map pixel classification scheme.

Algorithm 10 Floor plan generation

Require: Collected trace space E_T^{Local}, merged trace space E_T^{Global}
Ensure: Finer-grained floor plan
  for all trace $T_A, T_B \in E_T^{Local} \cup E_T^{Global}$ **do**
    Calculate overlapping landmarks L^A and L^B
    if $size(L^A)$ and $size(L^B)$ >= 3 **then**
      Section 4.5.4.3 - Trace megring
      Determine $\mathcal{R}$, K and e (Eqs. (4.79), (4.80));
      if $e > T_e$ **then**
        Determine dynamics of landmarks;
      else
        Section 4.5.5 - Trace co-revising
        if T_A or $T_B \in E_T^{Global}$ **then**
          Perform trace co-revising
        end if
        Merge trace T_A,T_B and obtain global trace T_G;
        Add global trace T_G into trace space E_T^{Global}
        Merge the same segment; (Eq. (4.81));
        Section 4.5.6 - Map pixel classification
        Perform map pixel classification for trace T_G;
      end if
    end if
  end for

We use Fig. 4.46 to illustrate how the floor generation algorithm works. As shown in Fig. 4.46a, we collect trace A, C, B sequentially. Since trace A and C have no intersection, we first consider that they belong to different indoor maps. Once trace B is collected, we can merge trace A and B via trace merging module and derive trace AB. Meanwhile, since there exists intersection between trace AB and C, we finally merge all traces and construct more completed floor plan. Proposed methods can be directly employed in multi-floor or multi-building scenarios without any modification. As shown in Fig. 4.46b, there are four traces A, B, C, D coming from different buildings. In our system, we derive BLE's MAC address to uniquely identify different landmarks. Differing from Wi-Fi which covers a hundred-meter-radius area, BLE can only cover limited area, which means no interface among multiple floors or multiple buildings. Since there can never be landmark intersection between traces coming from different indoor maps, we thus independently construct respective indoor map (AC and BD).

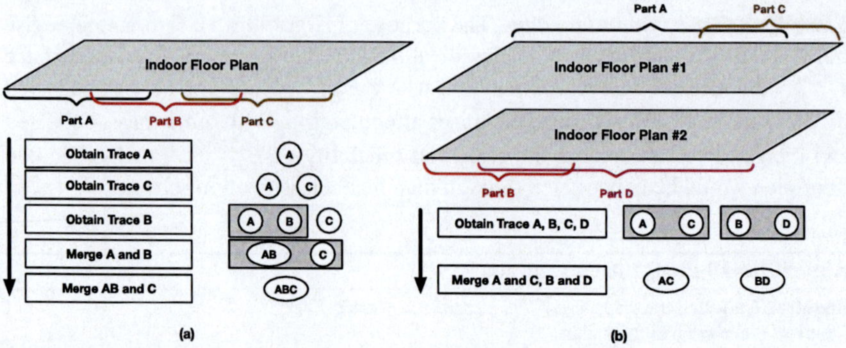

Fig. 4.46 Process of floor plan generation. (**a**) Single indoor map construction. (**b**) Multiple indoor maps construction

Localization Phase In the mapping phase, our system utilizes the crowdsourced data to construct finer-grained floor plan. Meanwhile, FineLoc attempts to localize the user with collected data. In previous work, dead reckoning [128] and fingerprinting based [53, 70, 71] localization schemes have been employed for localization. With the improvement of IMU accuracy, dead reckoning shows higher reliability than fingerprinting within a small region. To further eliminate cumulative error of dead reckoning, landmarks could be utilized.

In this paper, localization is based on dead reckoning and BLE landmarks. Compared with traditional landmarks (for example, water dispenser and turns[74]), BLE landmarks show higher reliability and uniqueness. Specifically, the system first generates user's trace and then determines which map the trace belongs to. Due to uniqueness and limited coverage of BLE nodes, it is simple to find corresponding map. It should be noted that when new landmarks are detected, we would perform mapping phase to construct a new region map. Secondly, to localize the user, we just need to merge his/her trace into existing floor plan, where rotation matrix $\mathcal{R}$ and scaling coefficient K are required. Therefore, according to the merging rule in Eq. (4.78), we can obtain $P = (P^A - \overline{P}^A)K\mathcal{R} + \overline{P}^B$, where P^A denotes the position in current trace's coordinate system, P represents the position in global floor plan, $\overline{P}^A = (\overline{x}^A, \overline{y}^A)$ and $\overline{P}^B = (\overline{x}^B, \overline{y}^B)$ denote the coordinates of the geometric centers of landmarks of trace A and B.

4.5.8 *Performance Evaluation*

4.5.8.1 Experimental Setups

We implement FineLoc on different Android phones (HUAWEI Mate 7, InFocus M512, Nexus 5) and conduct experiments in three buildings to evaluate our

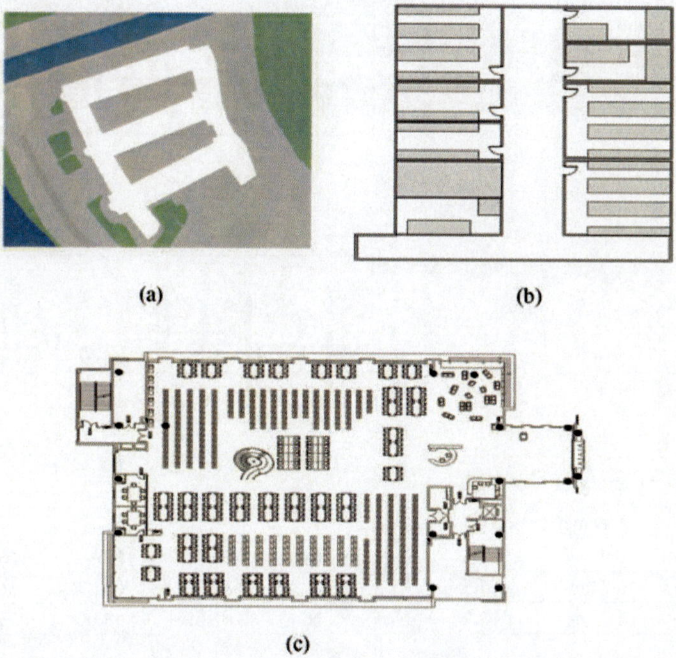

Fig. 4.47 The ground truth of three scenarios. (a) Classroom building (pathway). (b) Office building (subroom). (c) Library (open area)

proposed mechanisms, where the first testing environment is a 10,000 m^2 pathway area in a classroom building, the second is a 500 m^2 area in an office building with rooms, and the third is a 1200 m^2 wide area, where the floor plans are shown in Fig. 4.47.

4.5.8.2 Accuracy of Posture Recognition

We enroll 5 volunteers to test the accuracy of the proposed posture recognition scheme. Each volunteer is asked to walk 400 steps twice, during which freely switching postures is encouraged. We record the time consumed by each walk, the duration and the number of steps for each posture. Figure 4.48a shows the experimental results in time, where the horizontal axis is the index of the experiment and the vertical axis is the proportion of time each posture is correctly recognized; we can see that the scheme can correctly identify different postures in more than 99% of the time, where most of the can-not-tell situations occur during the posture switching. Figure 4.48b shows the results in the number of steps, which corroborates the results in Fig. 4.48a. We can see that only situations with less than 4 steps may be incorrectly identified for each posture.

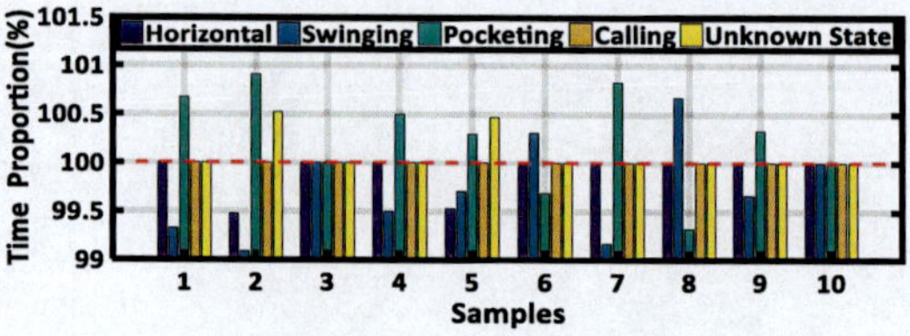

(a)

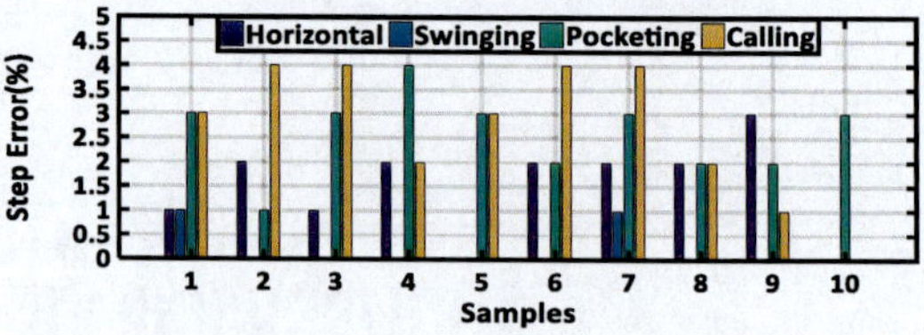

(b)

Fig. 4.48 Accuracy of posture recognition. (**a**) Accuracy in time. (**b**) Accuracy in num. of steps

Table 4.4 Accuracy of floor plan

Scenario	Mean error	$\mathcal{P}_{room}$	$\mathcal{R}_{room}$	$\mathcal{F}_{room}$	$\mathcal{P}_{obstacle}$	$\mathcal{R}_{obstacle}$	$\mathcal{F}_{obstacle}$
Classroom	0.97 m	N/A	N/A	N/A	N/A	N/A	N/A
Office	0.81 m	81.3%	83.8%	82.5%	68.7%	64.8%	66.7%
Library	1.21 m	N/A	N/A	N/A	61.6%	53.0%	57.1%

4.5.8.3 Performance of Map Construction

We construct the indoor floor plans of the three testing areas, and randomly select a number of points, including 420 points from the classroom building, 154 points from the office building and 249 points from the library. Those points are then compared with the corresponding points in the real map to find the deviation. There are 18 and 68 obstacles in the office building and the library respectively; the pathways in the classroom building contain no obstacles. We evaluate the proposed map construction mechanism in terms of the 8 metrics as tabulated in Table 4.4. Note that the first 5 metrics are also used in SensorWit[74] and CrowdMap[161], where $\mathcal{P}_{room} = \frac{|S_{generate} \cap S_{true}|}{|S_{generate}|}$, $\mathcal{R}_{room} = \frac{|S_{generate} \cap S_{true}|}{|S_{true}|}$, $\mathcal{F}_{room} = 2 \times \frac{\mathcal{P} \times \mathcal{R}}{\mathcal{P} + \mathcal{R}}$.

We can see that $\mathcal{P}_{room}$ denotes the ratio of correct sub-room area to generated area while $\mathcal{R}_{room}$ means the radio of correct sub-room area to ground truth. $\mathcal{F}_{room}$ combines $\mathcal{P}_{room}$ and $\mathcal{R}_{room}$. Since there is no sub-room in classroom and library, the second and fourth rows employ N/A in Table 4.4. Meanwhile, since our proposed mechanism is also able to recognize obstacles, we utilize $\mathcal{P}_{obstacle}$, $\mathcal{R}_{obstacle}$ and $\mathcal{F}_{obstacle}$ to verify the performance of obstacle construction. In particular, the definition of $\mathcal{P}_{obstacle}$, $\mathcal{R}_{obstacle}$ and $\mathcal{F}_{obstacle}$ is also similar to that of $\mathcal{P}_{room}$, $\mathcal{R}_{room}$ and $\mathcal{F}_{room}$.

Figure 4.49 shows the layout of the classroom building and the constructed pathway map. We deploy 16 landmarks along the 500 m-long pathway. We only utilize 4 traces to construct the map, which takes less than 20 mins. This outperforms

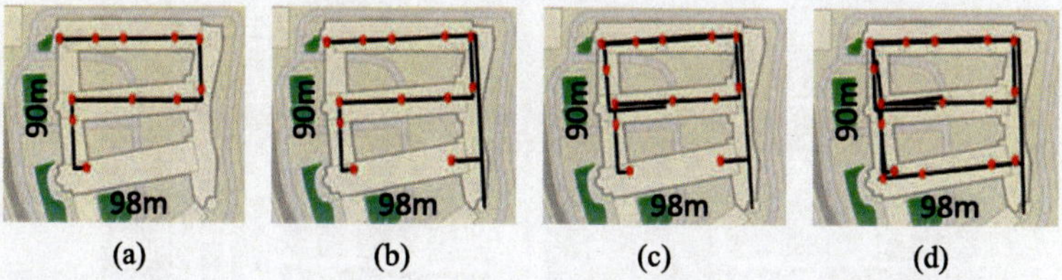

Fig. 4.49 Floor plan of classroom building. (**a**) One trace. (**b**) Two traces. (**c**) Three traces. (**d**) Four traces

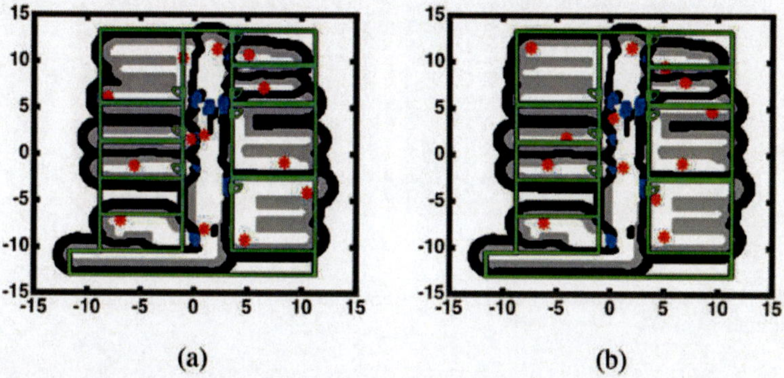

Fig. 4.50 Floor plan of the office building area. (**a**) Before landmarks movement. (**b**) After landmarks movement

the Walkie-Markie and PiLoc, where the performance are 260 m within 50 mins and 150 m within 10 mins respectively. This is because we record long trace segments instead of dividing the trace into very small segments in previous systems[70–72]. This will save the amount of data required and improve the data processing efficiency.

Figure 4.50 shows the layout of testing area in the office building, where there are 8 rooms and 13 landmarks denoted by red stars. In contrast to existing scheme that can only profile pathways, the FineLoc can sketch the outlines of rooms, obstacles, doors, and pathways, which are in the color of black, grey, blue, and white in the figure, respectively. The map shown in Fig. 4.50a is constructed after 20 mins of volunteers' free activities, and that shown in Fig. 4.50b is constructed while 5 out of the 13 landmarks are moved to somewhere else, which shows the capability of automatic landmark updating.

Figure 4.51 shows the layout of the testing area in the library, where there are 68 obstacles and 16 landmarks denoted by red stars. Compared with existing scheme that can only sketch the outline of the whole building, the FineLoc can sketch obstacles in details, which is in the color of grey. The map shown in Fig. 4.51a is constructed after 10 mins of volunteers' free activities, and those shown in Fig. 4.51b and c are constructed after 1 hour and 2 hours of volunteers' free activities. Then we

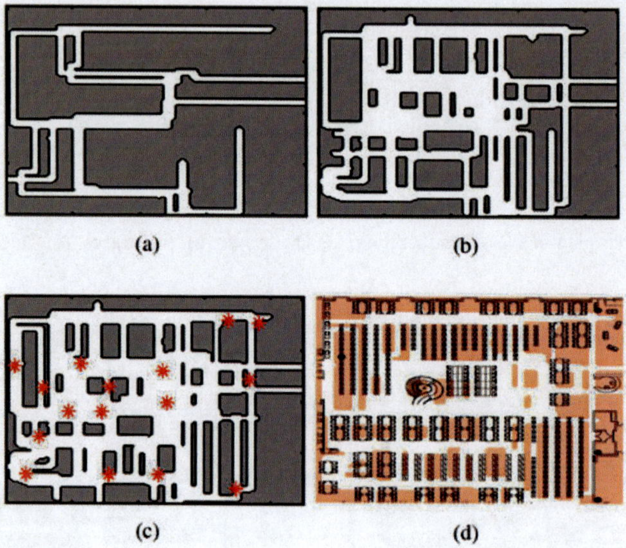

Fig. 4.51 Floor plan construction of university library. (**a**) 10 minutes. (**b**) 1 hour. (**c**) 2 hours. (**d**) Ground truth

have placed the floor plan into a ground truth and show the performance with a 16.2% obstacles missing and 20.6% extra obstacles in Fig. 4.51d.

4.5.8.4 System Comparison

In this section, we compare the FineLoc system with existing methods to verify that FineLoc can realize finer-grained indoor map construction. During experiments, we collect Wi-Fi/BLE signals and construct floor plan based on PiLoc and FineLoc. In particular, existing Wi-Fi floor plan construction methods can only sketch the skeleton of floor plan as shown in Fig. 4.52a. Meanwhile, there is also false trace merging due to failed correlation detection. In contrast, FineLoc can recognize different sub-rooms, obstacles, walls, and doors. Thus we can realize finer-grained floor plan construction as shown in Fig. 4.52b.

To verify the performance of sub-room recognition and obstacles recognition, we then compare our system with SensorWit. As shown in Table 5.2, one important advantage of FineLoc is unique calibration. It means that FineLoc can simultaneously construct multiple indoor maps without interrupting each other. In addition, FineLoc can realize a higher accuracy as shown in Table 4.5.

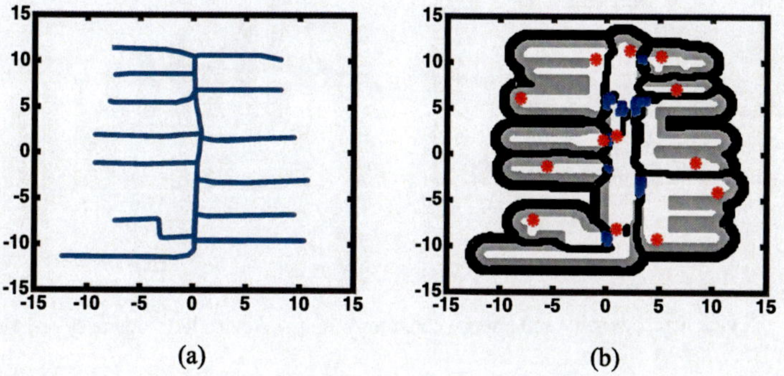

Fig. 4.52 Map construction comparison. (**a**) Floor plan (PiLoc). (**b**) Floor plan (FineLoc)

Table 4.5 System comparison

System	FineLoc	PiLoc	SensorWit
$\mathcal{F}_{room}$	82.5%	Not work	76.0%
$\mathcal{F}_{obstacle}$	57.1% ∼ 66.7%	Not work	Not work

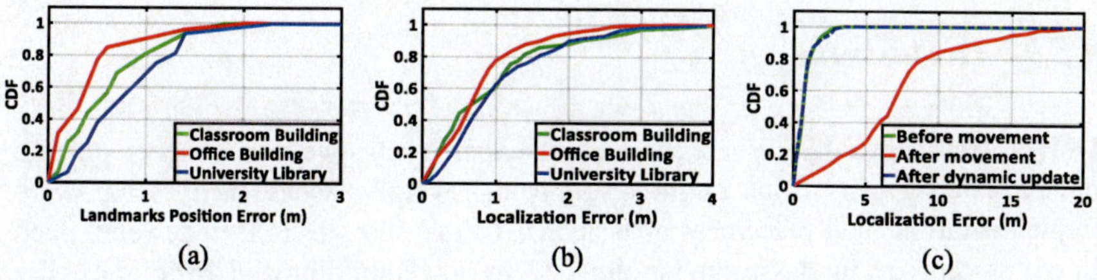

Fig. 4.53 Localization performance. (**a**) Landmarks position error. (**b**) Localization error. (**c**) Dynamic environment

4.5.8.5 Localization Performance

There are 45 landmarks employed in our system. To measure the performance of landmark's position, we show the results in Fig. 4.53a, where 80% errors are less than 1.0 m, 0.5 m, and 1.2 m respectively. Meanwhile, We collect 84 traces from the three buildings, and select 2459 points in buildings to perform location estimation, where there are 475 points from the classroom building, 726 points from the office building and 1258 points from the library. We utilize BLE beacon nodes and the smartphone to do localization experiments. The resulted CDFs are illustrated in Fig. 4.53b, where 80% errors are less than 1.4 m, 1.1 m, and 1.6 m respectively. We can see that our system performs stably in different indoor environments.

Figure 4.53c shows the localization results of dynamic environment, where BLE landmarks can be moved to other places. The experiments are conducted in the office building. We move 5 out of 13 BLE nodes to other randomly selected places and then perform the map updating and location estimation simultaneously. We can see from Fig. 4.53c that the localization error increases dramatically when BLE

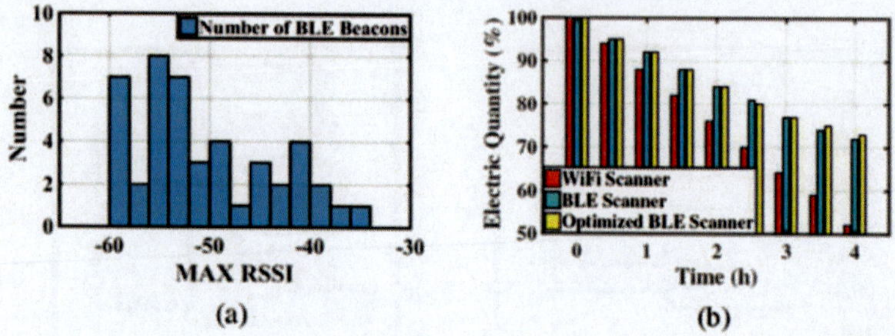

Fig. 4.54 Device heterogeneity and energy consumption. (**a**) Device heterogeneity. (**b**) Nexus 5

nodes are moved, because the smartphone may consider the BLE's current position as the position in the original constructed map. However, with our map updating scheme, the new map accommodating the current locations of the BLE nodes can be generated after around 20 mins of the volunteers' free activities.

4.5.9 Discussions

Device Heterogeneity When conducting experiments, we can observe notable device heterogeneity. This might result from several reasons such as hardware implementations and placement of beacons. To measure 45 BLE nodes employed in our system (16 in classroom building, 13 in office building and 16 in university library), we plot historical maximum RSSIs in Fig. 4.54a. This indicates that it is necessary to perform our trace labeling algorithm instead of a fixed RSSI threshold method.

Deployment of BLE Beacons As shown in Figs. 4.49, 4.50 and 4.51, the densities of BLE nodes are 0.18/100 m^2, 2.6/100 m^2 and 1.3/100 m^2 respectively. This means that the proposed mechanism can work with sparsely populated BLE nodes. In real situation, the sub-room structure such as the office building area shown in Fig. 4.47b requires the most iBeacons to construct the floor plan. It is because that at least one iBeacon is necessary to label each room (Sect. 4.5.4.2). Even so, FineLoc can construct fine-grained floor plan shown in Fig. 4.50 with limited BLE nodes. Moreover, due to proposed trace labeling and merging mechanism, FineLoc prefers the BLE Beacons that are easier to be in reach of people. It is because that the trace labeled with more landmarks can be merged and revised with less error.

System Cost During our experiments, the cost of each iBeacon is about \$6 [204], which leads to \$0.01/m^2, \$0.16/m^2 and \$0.08/m^2 cost under the three scenarios shown in Fig. 4.47. We can thus construct fine-grained floor plan with the cost less than \$0.16/m^2. Such cost can be easily accepted for localization applications.

Scanner of BLE RSSI Compared with Wi-Fi scanner, the activation rate of general BLE scanning on smartphone might be low. Through our experiment we find that BLE scanner on smartphone just works in Android 4.3 or higher version and obtains RSSI each 3–5 seconds sometimes. Fortunately, current Android version is normally newer than Android 4.3 and we can improve the scanning rate to 1 time/second. This can be realized by turning on and off BLE adapter periodically using *adapter.stopLeScan()* and *adapter.startLeScan()*. This would increase scanning rate without incurring much energy consumption (Fig. 4.54b).

Energy Consumption The energy consumption of BLE nodes is low; however, scanning BLE beacon signals will consume energy in users' mobile devices. As we mentioned, we could improve the scanning frequency of the mobile device to detect the BLE beacon signal, but will this incur higher energy consumption in the mobile device? We find the answer to the issue through experiments, for which we set the scanning interval to be 1s for 3 kinds of Android phones (HUAWEI Mate 7, InFocus M512, Nexus 5). We maintain the scanning process for 4 hours and show the energy consumption results in Fig. 4.54b. We can observe that the resulted energy consumption by improving the scanning frequency is slightly higher than the default configuration, but still lower than half of that for scanning Wi-Fi. For the other two kinds of Android phones, we can observe similar results during the experiments.

Chapter 5
CSI Localization for Large-Scale Deployment

5.1 Extended MUSIC Algorithm

OFDMA Wi-Fi backscatter can significantly improve the communication efficiency and meanwhile maintain ultra-low power consumption; however, the ground-up reworking on the core mechanism of traditional Wi-Fi system revolutionizes the basis of many existing Wi-Fi based mechanisms. In this section, we explore how localization can be realized based on OFDMA backscatter, where a batch localization mechanism utilizing concurrent communication in the OFDMA backscatter system is proposed. We present a series of mechanisms to deal with the fundamental change of assumptions brought by the new paradigm. First, we process signals at the receiver in a finer granularity for signal classification. Then we remove phase offsets in real time without interrupting the communication. Finally, we propose an extended MUSIC algorithm to improve accuracy with limited localization information in OFDMA backscatter mechanism. We implement a prototype under the 802.11 g framework in WARP, based on which we conduct comprehensive experiments to evaluate our propose mechanism. Results show that our system can localize 48 tags simultaneously, while achieving average localization errors within 0.49 m. The tag's power consumption is about 55–81.3 μW.

5.1.1 Preliminaries

5.1.1.1 OFDMA Backscatter System

We implement concurrent localization based on Wi-Fi OFDMA backscatter systems [118], thus we first highlight the OFDMA backscatter system, so that the following contents of this paper can be understood.

X. Tian et al., *Wireless Localization Techniques*, Wireless Networks,
https://doi.org/10.1007/978-3-031-21178-2_5

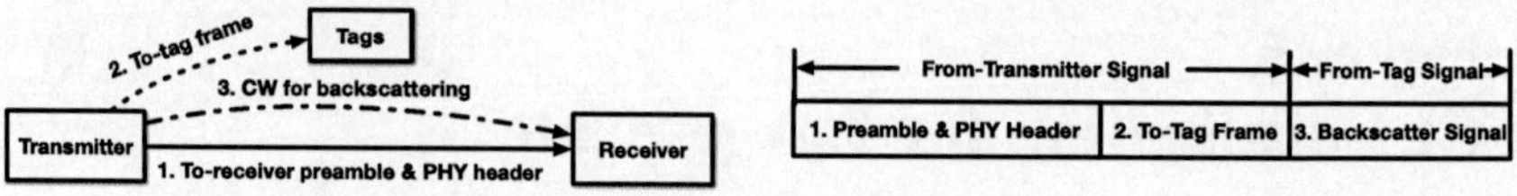

Fig. 5.1 Synchronization Mechanism in Wi-Fi OFDMA Backscatter

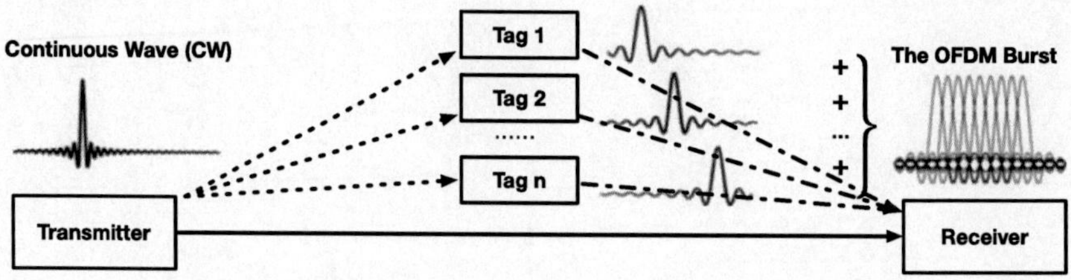

Fig. 5.2 Wi-Fi OFDMA backscatter system

Wi-Fi backscatter provides a more effective power saving solution for low-power IoT devices [276–279]. The first Wi-Fi backscatter system in 802.11b framework is proposed in passive Wi-Fi [288], where the authors realize 11 Mbps transmission with 59.2 μW energy consumption. The key design of Wi-Fi backscatter is the frequency shifting mechanism in tags, where the tag shifts ambient wireless signals from one channel to another channel. Since backscatter systems can realize frequency shifting operation without the energy-consuming RF component, they enable low-power communication.

In contrast to Wi-Fi backscatter in 802.11b framework, Wi-Fi OFDMA backscatter enhances system concurrency and capacity. The crux of any OFDMA system is the effective synchronization mechanism. To synchronize the OFDMA backscatter system, there are three parts in OFDMA backscatter designs as shown in Fig. 5.1. First, the 802.11 g preamble and PHY header are transmitted to synchronize the clock of transmitter and receiver; Second, excitation signal transmitters produce to-tag frame to synchronize tags and the transmitter. Since the transmitter and receiver have been synchronized in the first part, we can synchronize the whole system; Third, the transmitter broadcasts continuous wave (CW) that would be backscattered to different subcarriers by tags. Based on three steps in Fig. 5.1, we can coordinate the frequency shift in multiple tags as shown in Fig. 5.2, which is equivalent to assign different subcarriers to different tags. In this way, backscatter signals coming from multiple tags make up a completed OFDM signal that could be demodulated at the receiver.

5.1.1.2 Challenges

Since the lack of strict synchronization mechanism makes OFDMA backscatter unfeasible, the synchronization designs mentioned above are necessary. However, such a mechanism brings three new challenges to concurrent localization. This section introduces particular challenges in batch localization systems, so that readers could focus on certain sections in the following discussions.

Challenge 1: Collecting Feasible CSI Existing localization techniques [87, 179, 180, 227] localize the target based on CSI collected by Intel 5300 tool kit [97] which derives 30 CSI measurements for 64 subcarriers from each packet. To distinguish this kind of CSI from the CSI collected in our system, we define the CSI collected by Intel 5300 tool kit as ***packet-level CSI***. In the OFDMA backscatter system, as shown in Fig. 5.1, signals observed by the receiver is the combined information from many paths. For example, if 48 tags are employed in the system, the receiver would observe the combined information across 49 different propagation paths which consist of 48 Tags-RX paths and one TX-RX path. The combination of signals can be observed in symbol domain and frequency domain, where symbol domain represents different OFDM symbols in the OFDM burst and frequency domain means different OFDMA subcarriers. In symbol domain, one OFDM burst contains both from-transmitter and from-tag signals as shown in Fig. 5.1. Since Intel 5300 tool kit derives packet-level CSI based on the preamble [97] that exactly comes from the transmitter instead of tags, packet-level CSI cannot be utilized to localize tags. In frequency domain, one OFDM burst contains backscatter signals of multiple tags as shown in Fig. 5.2. Since packet-level CSI can only provide 30 measurements, it cannot be utilized to classify 48 different tags. Therefore, instead of utilizing existing packet-level CSI directly, it is necessary to process received signals in a finer granularity for the path classification. Only in this way, we can derive feasible CSI for target localization.

Challenge 2: Eliminating Phase Offsets As verified in previous experiments [180], phase offsets severely affect localization performance. In particular, uncalibrated phase offsets would lead to an average 60 degrees error of AoA calculation, which is not feasible in localization systems. To address *challenge 1*, we derive fine-grained CSI of each OFDM symbol. However, in contrast to packet-level CSI, our fine-grained CSI suffers even more from phase offsets which consist of continuous dynamic phase offset and down conversion phase offset. ***Continuous dynamic phase offset*** could be addressed via the method introduced in [91], we thus focus on discussing why it is challenging to remove the second type of phase offset without dedicated packet transmission. In particular, ***down conversion phase offset*** is caused by the down conversion step at the receiver. To address this challenge, it is necessary for prior works [86, 94, 178] to perform phase offset calibration which requires to interrupt the communication. In particular, ArrayTrack [86] utilizes RF cables to manually measure this phase offset. Since down conversion phase offset would be reset as a new random value once device initializes, the manual RF cable calibration is not practical. Meanwhile, Phaser and AWL [94, 178] require special calibration communication between transceivers, which inevitably interrupts normal communication of the tag. When there are a large number of tags in OFDMA backscatter systems as shown in Fig. 5.2, such a calibration scheme means massive interruptions of communication. Therefore, it is necessary to design a scheme to calibrate or remove down conversion phase offset without interrupting the communication, which has not been realized by any existing work.

Challenge 3: Localizing with Limited Information Caused by OFDMA backscatter mechanisms, limited subcarriers could be assigned to each tag. It is because in OFDMA backscatter systems, multiple subcarriers are assigned to different tags. Since the amount of subcarriers is constant, with the increase of concurrently localized tags, the number of subcarriers assigned to each tag decreases. In fact, such a design makes prior localization systems depending on the information of redundant subcarriers unfeasible. For example, RF-Echo [289] and WiTag [179] derive ToF based on CSI of multiple subcarriers. When OFDMA backscatter systems assign each tag only a single subcarrier for high concurrency, these systems cannot provide satisfying solution. Another possible scheme against limited subcarriers is sending massive packets for more CSI measurements [87]. However, frequent communication is energy-consuming and cannot be satisfied in practical low-power systems. Since a single CSI measurement is exactly unstable and therefore leads to inaccurate AoA measurements, this challenge is necessary to be addressed in the batch localization system. However, high concurrency mechanisms limit the employment of redundant subcarriers while low-power communication mechanisms make continuous massive packets inoperable. Thus, it is challenging to localize targets accurately with extremely limited information.

5.1.2 System Overview

Our system can localize multiple backscatter tags concurrently, which consists of the following four modules shown in Fig. 5.3 to address the three challenges mentioned above.

OFDM Burst Processing Module As discussed in *challenge 1*, each OFDM symbol contains CSI corresponding to multiple paths. In particular, the signals of different

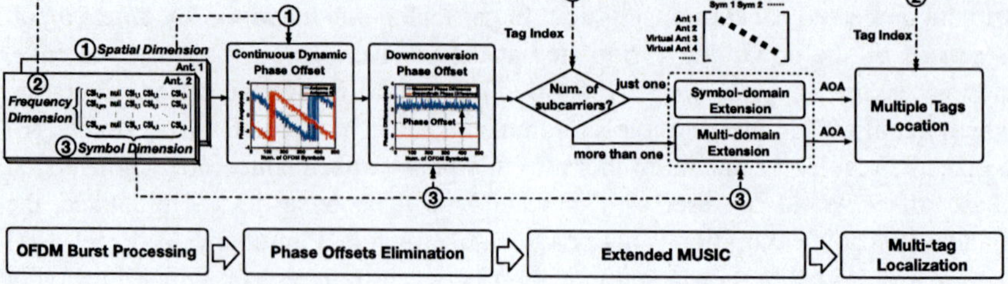

Fig. 5.3 System overview

subcarriers usually come from different tags as shown in Fig. 5.2. In addition, each subcarrier signal consists of the from-transmitter part and the from-tag part as shown in Fig. 5.1. In this module, we process the OFDM burst and derives CSI based on three different dimensions, i.e., spatial dimension, frequency dimension and symbol-dimension. Among them, ***spatial-dimension CSI*** labeled with ① represents CSI at different antennas, and it could be utilized to remove continuous dynamic phase offset. ***Frequency-dimension CSI*** labeled with ② means CSI across different subcarriers, and it could be employed to separate multi-tag signals for concurrent localization. ***Symbol-dimension CSI*** labeled with ③ means CSI of different OFDM symbols, and it could be utilized to remove down conversion phase offset (*challenge 2*) and realize our extended MUSIC scheme (*challenge 3*).

Phase Offsets Elimination Module As described in *challenge 2*, we must remove phase offsets to enable AoA measurement. In this module, we utilize spatial-dimension and symbol-dimension CSI to remove these two phase offsets respectively. We first remove dynamic continuous phase offset based on spatial-dimension CSI as shown in [91]. Secondly, the basic idea of down conversion phase offset elimination is that we can observe the same phase offset across different propagation paths. Since symbol-dimension consists of both from-transmitter and from-tag CSI respectively depicting the channel state of different propagation paths, we can therefore remove down conversion phase offset. In contrast to previous methods, we for the first time remove phase offsets without dedicated packet transmission, which is more important with the rapid rise of connectivity needs from IoT devices. Besides, since our phase offset elimination module works in real time, extra performance evaluation and re-calibration processes proposed in previous work [94] are also not necessary.

Extended MUSIC Scheme Module Since limited subcarriers can be allocated to each tag and there are usually $2 \sim 3$ antennas equipped in COTS devices, we can hardly derive accurate AoA for localization (*challenge 3*). To address this challenge, we proposed extended MUSIC scheme consisting of ***symbol-domain extension*** and ***multi-domain extension***. As shown in Fig. 5.3, we first determine the number of subcarriers assigned to the localized tag and then perform corresponding extended MUSIC scheme. Symbol-domain extension and multi-domain extension share the same basic idea, i.e., improving performance by establishing the virtual antenna array. Mathematically, we extend the dimensions of traditional MUSIC matrix to enable more accurate AoA measurements. In particular, the OFDM burst processing module provides us with the fine-grained channel state depiction and about $500\times$ samples of 500 OFDM symbols compared with packet-level CSI. Therefore, we can utilize symbol-dimension information to extend MUSIC matrix for AoA measurements. Meanwhile, when more than one subcarriers assigned, we can also employ both symbol-dimension and frequency-dimension information to further extend MUSIC matrix for more accurate results.

Multi-tag Localization Module To localize backscatter tags, it is necessary to derive at least two AoAs with respect to different transceivers. In multi-tag localization module, we combine AoAs relative to different transceivers and then localize the tag. Since backscatter tags can always be indexed by corresponding subcarriers, we can derive multi-tag position based on frequency-dimension information.

5.1.3 OFDM Burst Processing

In this section, we describe in detail how we process the OFDM burst for feasible CSI collection to address *challenge 1*. For this purpose, we first analyze advantages of our finer-grained CSI and then show how to derive it in our system. Since the finer-grained CSI consists of three different dimensions, i.e., spatial dimension, frequency dimension and symbol-dimension, we define it as ***three-dimension CSI*** (***3D-CSI***) in the following contents of this paper.

5.1.3.1 3D-CSI Analysis

Compared with previous works [86, 87, 94, 95, 179, 289] which employ multiple antennas, subcarriers, and packets for pure performance improvement, we explore the nature of different kinds of CSI and design a more practical localization system. What we concern in our system are concurrency, energy conservation, communication compatibility and accuracy. *Concurrency* represents that we can simultaneously localize multiple tags. Without concurrent mechanisms, we must schedule multiple tags in sequence to derive feasible CSI [290], which is considered to be inefficient [182]. *Energy conservation* represents that we do not require continuous massive packets for accurate localization due to inconsistency between continuous massive packets and low-power designs. Meanwhile, compared with single-packet localization, massive-packet localization will lead to lower efficiency. *Communication compatibility* means that we cannot interrupt other devices' communication while localizing certain targets or calibrating phase offsets. It is because communication interruption will become more and more insufferable with the increase of connected IoT devices. *Accuracy* means we can derive accurate target position, which is obviously important for any localization system. Now we discuss different CSI dimensions as follows:

Spatial Dimension The spatial dimension represents CSI at different antennas. One function of spatial-dimension CSI is to calculate AoA and then localize multiple tags. As shown in many prior works [86, 87, 94, 178, 227], the antenna array is essential infrastructure for AoA localization. In addition, spatial-dimension CSI could also be employed to ensure accuracy, where signals received at different antennas could be utilized to remove continuous dynamic phase offset.

Frequency Dimension The frequency dimension represents CSI across different OFDM subcarriers. Since our system assigns different OFDMA subcarriers to different tags, the ID of subcarriers could be utilized to index corresponding tags. As we know, there are 48 data subcarriers in Wi-Fi OFDM systems. Therefore, compared with existing localization methods which can only assign all subcarriers to one certain target, we can assign these subcarriers to 48 targets and therefore collect multi-tag CSI with 48× efficiency. In this way, we can improve localization concurrency and maintain at most 48 devices communication compared with existing systems.

Symbol-Dimension The symbol-dimension represents CSI of different OFDM symbols, which could be utilized to ensure system energy conservation, communication compatibility and accuracy. First, as discussed in *challenge 1*, the synchronization mechanism makes previous packet-level CSI not feasible. However, we regard such a challenge as an opportunity to solve down conversion phase offset problem (*challenge 2*). It is because down conversion phase offset is constant for different propagation paths. Based on symbol-dimension information, we can separate from-transmitter CSI and from-tag CSI, which respectively depict channel states of different propagation paths. Since we can simultaneously obtain above information based on one single OFDM burst and then remove down conversion phase offset without extra calibration communication, we realize communication compatibility. Second, for from-tag signals, we can obtain 500 CSI measurements corresponding to 500 OFDM symbols, which provides 500× samples compared with prior works. With such fine-grained information, we design the MUSIC extension scheme to improve localization accuracy without sending continuous massive packets in prior works [87, 179, 227], we thus ensure energy conservation and accuracy.

From above contents, we can realize the advantages of 3D-CSI. Here we discuss how to obtain such a fine-grained CSI in our system.

5.1.3.2 OFDM Burst Processing

As shown in Fig. 5.4, we first discuss how to acquire symbol-dimension and frequency-dimension CSI matrix at one certain antenna. Then the proposed scheme can be directly applied to the other antennas. In particular, we represent CSI matrix in Fig. 5.4 as $\boldsymbol{H}$. According to the basic knowledge of signals and systems, channel response $H(f)$ can be derived using the following equation, $H(f) = S_r(f)/S_t(f)$, where $S_r(f)$ and $S_t(f)$ are respectively frequency-domain descriptions of received signals and transmitted signals. For matrix calculation, we have $\boldsymbol{H} = \boldsymbol{S}_r./\boldsymbol{S}_t$, where $\boldsymbol{S}_r./\boldsymbol{S}_t$ denotes element-by-element division of matrix $\boldsymbol{S}_r$ and $\boldsymbol{S}_t$. Since symbol-dimension CSI contains three different parts, we calculate them independently. Especially, we represent $\boldsymbol{H}$ as a block matrix $[\boldsymbol{H}^{pre} \;\; \boldsymbol{H}^{null} \;\; \boldsymbol{H}^{back}]$, where $\boldsymbol{H}^{pre}$, $\boldsymbol{H}^{null}$ and $\boldsymbol{H}^{back}$ denote CSI of the preamble, to-tag frame and backscatter signals respectively. To calculate this block matrix, we have

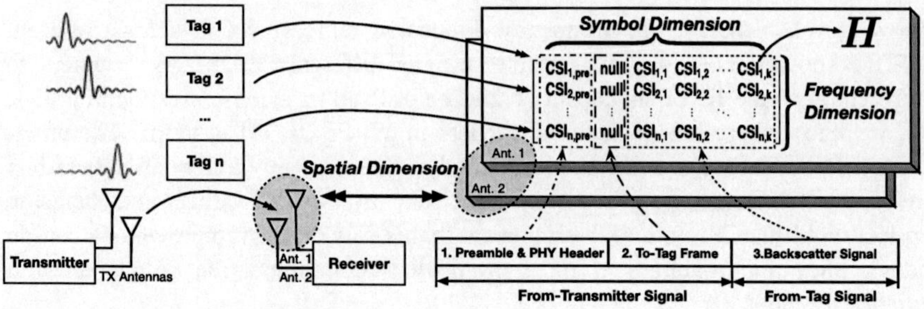

Fig. 5.4 Concurrent CSI collection

$$H = \left[H^{pre} \quad H^{null} \quad H^{back} \right] = \left[S_{rx}^{pre} \quad S_{rx}^{null} \quad S_{rx}^{back} \right] ./ \left[S_{tx}^{pre} \quad S_{tx}^{null} \quad S_{tag}^{back} \right], \tag{5.1}$$

where S_{rx}^{pre}, S_{rx}^{null}, and S_{rx}^{back} respectively represent the preamble, to-tag frame, and backscatter signals at the receiver while S_{tx}^{pre}, S_{tx}^{null}, and S_{tag}^{back} respectively denote these three parts at the transmitter or tags. Among these variables, since the to-tag frame is to coordinate the frequency shift in multiple backscatter tags, they are ignored.

Now we show how to obtain S_{rx}^{pre}, S_{rx}^{back}, S_{tx}^{pre}, and S_{tag}^{back} in our system. First, since the preamble and backscatter signals are captured by the receiver, we can acquire S_{rx}^{pre} and S_{rx}^{back} via baseband processing in WARP boards. Second, in regular OFDM systems, the transmitter preamble S_{tx}^{pre} is known in advance, which is utilized to synchronize the clocks of the transmitter and receiver. Therefore, what we should consider here is how to obtain S_{tag}^{back}, which denotes communication information of backscatter signals and is unpredictable. To decode S_{tag}^{back}, the most challenge is that there is an unknown phase offset of received backscatter signals, which is caused by unknown channel state. Consequently, we can utilize known information called tag-to-receiver preamble to estimate channel state and then decode backscatter signals with estimated results. In particular, tags embed tag-to-receiver preamble before communication symbols. When the receiver captures backscatter signals, it first estimate channel state with the known tag-to-receiver preamble and then decodes communication symbols. In this way, we can acquire all essential matrices to calculate H. *To be stressed that despite the same words utilized, above phase offsets calibration is one necessary communication process and is not equivalent to the phase offsets calibration introduced in challenge 2.* Finally, we process received OFDM signals at different antennas based on the same method and thus derive 3D-CSI. In the following contents of this paper, we employ $C(m, n, k)$ to denote 3D-CSI, where m, n, and k respectively represent the index of antenna, subcarrier, and OFDM symbol. Specifically, $C(m, n, pre)$ denotes CSI of the transmitter preamble.

In contrast, according to the description in CSI tools [291], packet-level CSI comes from the preamble of the whole OFDM burst, i.e., from-transmitter preamble.

In reality, the tag-to-receiver preamble is special in OFDMA backscatter systems and cannot be processed by off-the-shelf CSI tools. Since packet-level CSI only depicts channel state between the transmitter and receiver, it cannot be utilized to localize backscatter tags, which is verified in our experiments in Fig. 5.11a.

Remarks Section 5.1.3 introduces the advantage of our 3D-CSI, where we can ensure concurrency, energy conservation, communication compatibility and accuracy at the same time. Then we process OFDM burst to derive 3D-CSI in OFDMA backscatter systems. The application of OFDMA backscatter in localization systems brings significant promotion to concurrency without nearly performance degradation in other aspects for the following reason. In contrast to OFDM assigning all subcarriers to one single device, OFDMA assigns multiple subcarriers to different devices for concurrent connectivities. Consequently, with multiple devices connected simultaneously, communication capacity of one certain device decreases. However, IoT applications are featured by a large number of devices with short bursts of data, which makes communication capacity of each device less important. Therefore, our system can improve localization concurrency without nearly performance degradation.

5.1.4 Phase Offset Elimination

The OFDM burst processing module proposed in previous section allows us to acquire 3D-CSI. However, in contrast to packet-level CSI, 3D-CSI suffers even more from phase offsets as discussed in *challenge 2*. In Fig. 5.5, there are two kinds of phase offsets consisting of continuous dynamic phase offset and down conversion phase offset. This section reveals the root reason for each phase offset and removes phase offsets in real time without previous calibration.

We first represent the phase angle of $C(m, n, k)$ utilizing $\varphi_{m,n,k} = angle(C(m, n, k))$, which consists of three parts as follows:

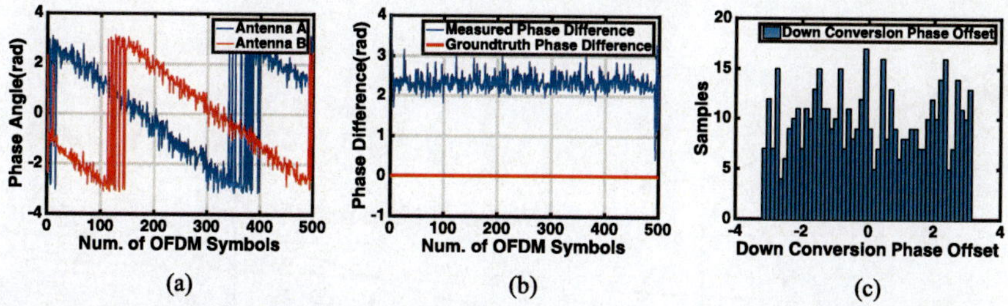

Fig. 5.5 Phase offset elimination. (**a**) Continuous dynamic phase offset. (**b**) Down conversion phase offset. (**c**) Phase offset distribution

$$\varphi_{m,n,k} = \varphi^{ex}_{m,n,k} + \varphi^{co}_{m,n,k} + \varphi^{down}_{m,n}, \tag{5.2}$$

where $\varphi^{ex}_{m,n,k}$ denotes the exact phase angle, $\varphi^{co}_{m,n,k}$ means continuous dynamic phase angle, and $\varphi^{down}_{m,n}$ means down conversion phase angle. Moreover, $\varphi^{down}_{m,n}$ does not change with the time [94, 178] while the other variables are time-varying.

5.1.4.1 Continuous Dynamic Phase Offset

As shown in Fig. 5.5a, we can observe continuous dynamic phase angles at each antenna, which come from the residual frequency offset. In particular, the slopes of phase angles at different antennas are exactly the same i.e., $\varphi^{co}_{m_1,n,k} \equiv \varphi^{co}_{m_2,n,k}, \forall m_1, m_2$. In this way, we can remove continuous dynamic phase offset by calculating phase difference between consecutive antennas and obtain $\varphi_{\Delta m,n,k} = \varphi_{m_2,n,k} - \varphi_{m_1,n,k} = [\varphi^{ex}_{m_2,n,k} - \varphi^{ex}_{m_1,n,k}] + [\varphi^{down}_{m_2,n} - \varphi^{down}_{m_1,n}]$. For simplification, we utilize $\varphi^{ex}_{\Delta m,n,k}$ and $\varphi^{down}_{\Delta m,n}$ to respectively represent exact phase difference and down conversion phase offsets between consecutive antennas, where $\varphi^{ex}_{\Delta m,n,k} = \varphi^{ex}_{m_2,n,k} - \varphi^{ex}_{m_1,n,k}$ and $\varphi^{down}_{\Delta m,n} = \varphi^{down}_{m_2,n} - \varphi^{down}_{m_1,n}$. We thus have

$$\varphi_{\Delta m,n,k} = \varphi^{ex}_{\Delta m,n,k} + \varphi^{down}_{\Delta m,n}. \tag{5.3}$$

Continuous dynamic phase offset can be removed without interrupting regular AoA calculation. Review the process of AoA calculation in Fig. 5.6. The basic idea is when the signals arrive at the antenna array, we can observe a corresponding phase shift at consecutive antennas. Since the distance between the receiver and target is much longer than inter-distance of the antenna array, we can regard propagation paths between target and different antennas as a series of parallel lines. Therefore, in Fig. 5.6, based on geometric knowledge, we can observe an extra flight distance of the signal at antenna 2 labeled with the red line. It could be written as $dsin(\theta)$, where d is the distance between consecutive antennas and θ is AoA. Then, according to principle of communication, we have corresponding phase shift $-2\pi \times dsin(\theta) \times f/c$, where f is the frequency of the signal and c is the speed of

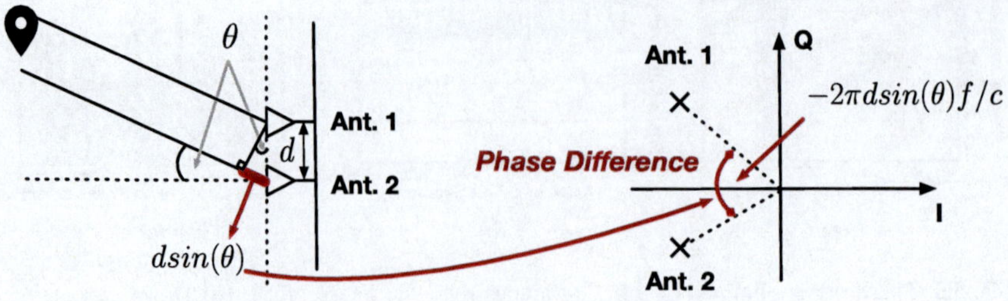

Fig. 5.6 Relation between AoA and phase difference

light. Recall the definition of $\varphi^{ex}_{\Delta m,n,k}$ in Eq. (5.3), which denotes exact phase angle difference between consecutive antennas caused by extra distance $dsin(\theta)$. Since d is known in advance, we can derive AoA as follows:

$$\theta(n,k) = asin\left(-\frac{c \times \varphi^{ex}_{\Delta m,n,k}}{2\pi df}\right), \tag{5.4}$$

where $\varphi^{ex}_{\Delta m,n,k}$ is necessary to derive AoA and is also maintained after continuous dynamic phase offset elimination. However, by subtracting the phase angle of one antenna from the other antenna, we can just obtain $\varphi_{\Delta m,n,k}$ in Eq. (5.3), which consists of two components. Between them, $\varphi^{ex}_{\Delta m,n,k}$ is what we want for AoA calculation while $\varphi^{down}_{\Delta m,n}$ is an unknown phase offset caused by down conversion step. In reality, since we can just obtain $\varphi_{\Delta m,n,k}$ without any other information about its two components, without previous calibration, prior works cannot obtain $\varphi^{ex}_{\Delta m,n,k}$ for AoA calculation. Especially, prior experiments [94] show that down conversion phase offset might be larger than exact phase difference and leads to an average 60 degrees error of AoA calculation. To address this problem, we now discuss how to remove down conversion phase offsets.

5.1.4.2 Down Conversion Phase Offset

Figure 5.5b shows phase difference between consecutive antennas, where the blue line denotes phase difference $\varphi_{\Delta m,n,k}$ before down conversion phase offset elimination and the red line denotes exact phase difference $\varphi^{ex}_{\Delta m,n,k}$. We can observe a phase offset $\varphi^{down}_{\Delta m,n}$ between two lines caused by down conversion steps. Though down conversion phase offset does not change with the time, it could be reset as a new value once device initializes. We conduct 500 experiments to verify this features and Fig. 5.5c shows the distribution of down conversion phase offsets, where we can see down conversion phase offsets are exactly random values.

Since down conversion phase offset is an unpredictable value and makes AoA measurement available, existing systems [94] must first calibrate it with dedicated packet transmission. In this section, we propose how to dynamically remove down conversion phase offset without dedicated packet transmission. As shown in Fig. 5.7, each OFDM burst consists of TX-RX and Tag-RX sub-signals (Part C). For CSI $C(m,n,pre)$ of the transmitter's preamble, we can also utilize $\varphi_{m,n,pre} = angle\ (C(m,n,pre))$ to denote its phase angle. Calculating phase difference of $C(m,n,pre)$ between consecutive antennas (Part B), we have

$$\varphi_{\Delta m,n,pre} = \varphi^{ex}_{\Delta m,n,pre} + \varphi^{down}_{\Delta m,n,pre}, \tag{5.5}$$

where $\varphi^{ex}_{\Delta m,n,pre}$ denotes exact phase difference (Purple area in part D) and $\varphi^{down}_{\Delta m,n,pre}$ denotes down conversion phase offset (Gray area in part D). As discussed above, since $\varphi^{down}_{\Delta m,n,pre}$ is constant for different paths, we have $\varphi^{down}_{\Delta m,n,pre} \equiv \varphi^{down}_{\Delta m,n}$.

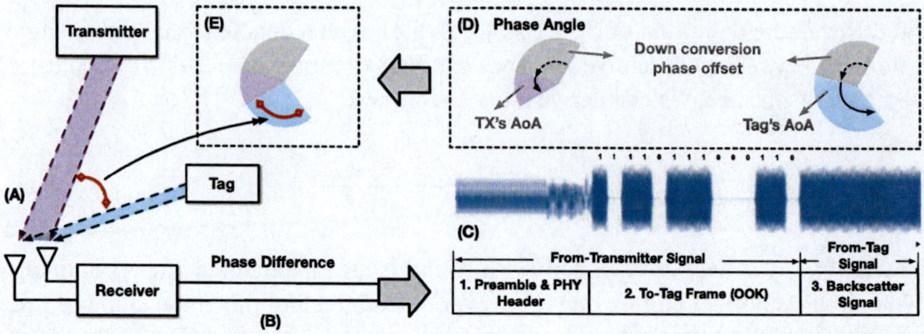

Fig. 5.7 Down conversion phase offset elimination

By subtracting $\varphi_{\Delta m,n,pre}$ in Eq. (5.5) from $\varphi_{\Delta m,n,k}$ in Eq. (5.3), we have $\varphi^{relative}_{\Delta m,n,k} = \varphi_{\Delta m,n,k} - \varphi_{\Delta m,n,pre} = \varphi^{ex}_{\Delta m,n,k} - \varphi^{ex}_{\Delta m,n,pre}$ (Part E), where $\varphi^{relative}_{\Delta m,n,k}$ denotes the relative phase offsets of nth subcarrier and kth symbol. From this equation, we can observe that $\varphi^{relative}_{\Delta m,n,k}$ is composed of $\varphi^{ex}_{\Delta m,n,k}$ (Blue area in part E) and $\varphi^{ex}_{\Delta m,n,pre}$ (Purple area in part E), which are corresponding to AoAs of the tag and transmitter respectively. Since position of the transmitter is usually constant and known during localization, we can acquire AoA of the transmitter in advance to determine $\varphi^{ex}_{\Delta m,n,pre}$ and therefore derive $\varphi^{ex}_{\Delta m,n,k}$ using

$$\varphi^{ex}_{\Delta m,n,k} = \varphi^{relative}_{\Delta m,n,k} + \varphi^{ex}_{\Delta m,n,pre}. \tag{5.6}$$

Remarks Section 5.1.4 clarifies how to utilize the symbol-dimension information to extract the CSI from different targets and realize the down conversion phase offset elimination without the communication interference. In this section, our system shows three advantages compared with prior works [86, 94, 178]. First, since previous works try to calibrate phase offsets instead of removing it directly like our scheme, they might introduce an extra calibrating error. Second, previous works require special calibration communication, for example, the transmitter sends a message to the receiver to calibrate phase offsets. During their calibration process, multi-tag communication and localization are terminated. In contrast, since the OFDM burst contains both from-transmitter and from-tag signals in our system, we can remove phase offsets while maintaining normal communication and localization. Third, interrupted by environment noise, calibrated results might be inaccurate and re-calibration is necessary to address this problem. Since previous calibration methods cannot work in real time, they request extra performance evaluation to judge calibrated accuracy. In contrast, our system removes phase offsets dynamically, which is more practical and robust.

5.1.5 Extended MUSIC Scheme

The phase offset elimination introduced in previous section allows us to remove continuous dynamic phase offset and down conversion phase offset without calibration. To localize targets, recently proposed SpotFi [87] derives accurate AoA based on multiple subcarriers and packets. However, as discussed in *challenge 3*, allocated subcarriers for each target are usually limited in OFDMA backscatter systems and numerous packets would consume much energy. In the following contents, we will first formulate our math model and then reveal why traditional MUSIC schemes and SpotFi cannot provide satisfactory accuracy. Finally, we introduce our extended MUSIC scheme in detail.

5.1.5.1 Derive AoA of Tags with Limited Information

As shown in Fig. 5.3, since we perform different extended MUSIC schemes based on the number of subcarriers assigned, we need to establish mapping relationships between tags and subcarriers. For this purpose, we define the subcarrier set of tag T as $\mathcal{F}_T$, which includes all subcarriers assigned to tag T. For example, supposing that subcarriers #1, #4, #6 are assigned to the tag #2, we have $\mathcal{F}_2 = \{1, 4, 6\}$. Then we count the elements of each $\mathcal{F}_T$ and determine corresponding extended MUSIC scheme.

To derive AoA of tags, we here formulate our math model. Since multiple tags share the same scheme for AoA calculation based on respective subcarrier set, we introduce our extended MUSIC scheme by just discussing a single tag. For each subcarrier, supposing that there are L paths arriving at the receiver, we can observe $x_1, x_2, \ldots, x_m$ at m different antennas, which contains the signals coming from L paths. Meanwhile, as shown in Fig. 5.6, there is a fixed phase difference φ^{ex} between consecutive antennas. Since signals propagating across different paths arrive at the receiver with different AoAs, we utilize φ_l^{ex} to denote this fixed phase difference corresponding to the AoA of lth path. If all signals of L paths arriving at the first antenna could be expressed as $[s_1, s_2, \ldots, s_L]^T$, we can utilize $\left[s_1 e^{\varphi_1^{ex}}, s_2 e^{\varphi_2^{ex}}, \ldots, s_L e^{\varphi_L^{ex}}\right]^T$ to represent the signals at the second antenna. In this way, corresponding signals at the mth antenna could be denoted by $\left[s_1 e^{(m-1)\varphi_1^{ex}}, s_2 e^{(m-1)\varphi_2^{ex}}, \ldots, s_L e^{(m-1)\varphi_L^{ex}}\right]^T$. Since received signals x_m observed at the mth antenna is the combination of multipath signals, we have

$$\begin{bmatrix} x_1 \\ x_2 \\ \vdots \\ x_m \end{bmatrix} = \begin{bmatrix} 1 & 1 & \cdots & 1 \\ e^{\varphi_1^{ex}} & e^{\varphi_2^{ex}} & \cdots & e^{\varphi_L^{ex}} \\ \vdots & \vdots & \ddots & \vdots \\ e^{(m-1)\varphi_1^{ex}} & e^{(m-1)\varphi_2^{ex}} & \cdots & e^{(m-1)\varphi_L^{ex}} \end{bmatrix} \begin{bmatrix} s_1 \\ s_2 \\ \vdots \\ s_L \end{bmatrix} + \begin{bmatrix} n_1 \\ n_2 \\ \vdots \\ n_m \end{bmatrix}, \tag{5.7}$$

where n_m denote noise at the mth antenna while the other variables have been defined. For simplification, we can write Eq. (5.7) as

$$\mathbf{X} = \mathbf{AS} + \mathbf{N}, \tag{5.8}$$

where **X** denotes the received signals matrix, **A** is the steering matrix, **S** represents propagation signals of L paths and **N** is the noise matrix. Especially, each column in the steering matrix **A** is defined as the steering vector that could be written as

$$\mathbf{a}(\theta) = [1, \Phi(\theta), \ldots, \Phi(\theta)^{m-1}]^T, \tag{5.9}$$

where $\Phi(\theta) = e^{\varphi^{ex}} = e^{-j2\pi d sin(\theta) f/c}$ denotes the phase difference between consecutive antennas. Equation (5.8) is the basic expression of MUSIC, which is first proposed in 1986 [185] and widely employed in recent years [86, 87, 94, 227]. The function of MUSIC is to calculate **a** with the only knowledge about **X**. Since MUSIC is a common algorithm, we do not discuss its principle in detail here. Instead, we just reveal why MUSIC cannot provide satisfactory accuracy by analyzing how it works. In particular, *to find steering vectors* **a**, *we first calculate the eigenvectors of* $\mathbf{XX}^*$ *corresponding to the near-zero eigenvalues, and then compute the steering vectors orthogonal to these eigenvectors*. In reality, for $M \times 1$ matrix **X**, we can obtain $M \times M$ matrix $\mathbf{XX}^*$. When M is small, the eigenvectors of $\mathbf{XX}^*$ is also in low dimension and at most $M-1$ eigenvectors could be found corresponding to near-zero eigenvalues. Due to limited number and low dimension, it is difficult to find accurate steering vectors. Since M denotes the number of antennas and limited antennas could be employed in practical systems, we can see why MUSIC cannot provide satisfactory accuracy.

To address this challenge, SpotFi [87] utilizes multiple subcarriers and packets for accurate localization. However, in OFDMA backscatter systems, multiple subcarriers are assigned to different backscatter tags for concurrent localization while massive packets would lead to insufferable energy consumption and lower localization efficiency. Therefore, we here propose extended MUSIC schemes, which consist of symbol-domain extension and multi-domain extension. In particular, symbol-domain extension utilizes symbol-dimension CSI to extend the dimension of **X** while multi-domain extension utilizes both symbol-dimension and frequency-dimension CSI. Since symbol-domain extension could be regarded as a specific multi-domain extension where just one subcarrier is available, we first introduce symbol-domain extension and then multi-domain extension.

5.1.5.2 Symbol-Domain Extension

The basic idea of symbol-domain extension is to extend the matrix dimension utilizing 3D-CSI $C(m, n, k)$. In particular, as shown in Fig. 5.8, symbol-domain extension scheme consists of four phases as follows. Here we discuss each phase in detail.

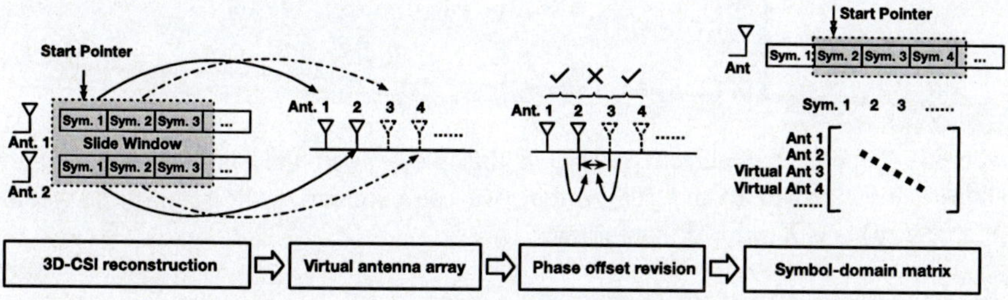

Fig. 5.8 Symbol-domain extension

3D-CSI Reconstruction As discussed in Sect. 5.1.4, $C(m, n, k)$ is not accurate for AoA calculation due to phase offsets, thus we should reconstruct 3D-CSI based on the exact phase offset $\varphi^{ex}_{\Delta m,n,k}$ obtained in Eq. (5.6). From Eq. (5.4), we can observe that AoA calculation just depends on the phase difference between consecutive antennas. In this way, we can maintain accurate phase difference utilizing the following equation:

$$X(m, n, k) = C(1, n, k)e^{(m-1)\varphi^{ex}_{\Delta m,n,k}}, \tag{5.10}$$

where $X(m, n, k)$ is reconstructed 3D-CSI. For the nth subcarrier, $X(m, \cdot, k)$ is a $M \times K$ matrix, where $M = 2$, $K = 500$ in our system. In contrast, prior works utilize packet-level CSI to localize targets, where they can derive at most $M \times N$ CSI matrix from each packet. Thus, we can represent packet-level CSI as $X_{packet}(m, n)$. In OFDMA backscatter localization, subcarriers assigned to one certain tag are limited, which makes multi-subcarrier CSI unavailable. For one single subcarrier, $X_{packet}(m, \cdot)$ is reduced to a 2×1 matrix. In contrast, our fine-grained CSI can provide a 2×500 matrix $X(m, \cdot, k)$, which makes extended MUSIC scheme possible. Here, we show how to extend the dimension of $X(m, n, k)X^*(m, n, k)$ to improve the performance. In particular, supposing that we want to extend the number of rows with P times as shown in Fig. 5.8, we can utilize a P-size slide window to generate symbol-domain extension matrix, denoted by $X^{music}_n(i, j)$, where i is the index of the virtual antenna and j is start symbol of the slide window. For a given P, we can finally obtain a $MP \times (K - P + 1)$ matrix $X^{music}_n(i, j)$. In the following contents, we will introduce how to generate such a matrix.

Virtual Antenna Array To generate $X^{music}_n(i, j)$ mentioned, we first establish the virtual antenna array. As shown in Fig. 5.8, the basic idea is to pick CSI at different antennas in turn and then put it into a new matrix. For example, we first select $X(1, n, 1)$ and set it as the first antenna while setting $X(2, n, 1)$ as the second antenna. Then we simulate the 3rd and 4th virtual antennas utilizing $X(1, n, 2)$ and $X(2, n, 2)$ respectively. And so on, for remaining $X(m, n, k)$ in the current slide window. Mathematically, we have

$$X_n^{music}(i,1) = \begin{cases} X(1, n, \frac{i+1}{2}), \ i = 2p-1 \\ X(2, n, \frac{i}{2}), \quad i = 2p \end{cases}, \ p \in \mathbb{Z} \ and \ p < P, \tag{5.11}$$

where p denotes the current index. In this way, when the start pointer of the slide window moves to the jth symbol, we can summarize the relation between $X_n^{music}(i, j)$ and $X(m, n, k)$ as follows:

$$X_n^{music}(i,j) = \begin{cases} X\left(1, n, \frac{i+2j-1}{2}\right), \ i = 2p-1 \\ X\left(2, n, \frac{i+2j-2}{2}\right), \ i = 2p \end{cases}, \ p \in \mathbb{Z} \ and \ p < P. \tag{5.12}$$

According to this equation, we can acquire one $2P \times (K-P+1)$ matrix, where $2P$ denotes the number of virtual antennas while $(K - P + 1)$ denotes the number of samples. However, such a $X_n^{music}(i, j)$ cannot be directly utilized for AoA calculation. It is because we can just ensure that partial phase differences between consecutive virtual antennas are exactly correct. In particular, based on Eq. (5.10), we can just ensure accurate phase offset between $X(1, n, k)$ and $X(2, n, k)$ while the one between $X(2, n, k)$ and $X(1, n, k + 1)$ is inaccurate. Then we introduce how to revise this inaccurate phase difference.

Phase Offset Revision We employ an interpolation method to address this problem. The basic idea is to revise inaccurate phase difference utilizing adjacent accurate values. For example, for continuous 4 antennas in Fig. 5.8, we can employ { $X_n^{music}(1, j)$,$X_n^{music}(2, j)$,$X_n^{music}(3, j)$,$X_n^{music}(4, j)$ } to represent them. According to previous discussion, we know that phase difference between $X_n^{music}(2, j)$ and $X_n^{music}(3, j)$ is not accurate, denoted by $\varphi_{\Delta 23}$. In contrast, $\varphi_{\Delta 12}$ and $\varphi_{\Delta 34}$ are accurate, since they both come from the phase difference between $X(1, n, k)$ and $X(2, n, k)$. Therefore, we can employ the mean value of $\varphi_{\Delta 12}$ and $\varphi_{\Delta 34}$ to represent $\varphi_{\Delta 23}$, where $\varphi_{\Delta 23} = (\varphi_{\Delta 12} + \varphi_{\Delta 34})/2$. In this way, we can obtain a modified $X_n^{music}(i, j)$ to enable the MUSIC algorithm.

Symbol-Domain Matrix for MUSIC Previous processes allow us to generate the $2P \times (K - P + 1)$ matrix $X_n^{music}(i, j)$ to enable the MUSIC algorithm. Since $X_n^{music}(i, j)X_n^{music*}(i, j)$ is one $2P \times 2P$ matrix, compared with 2×2 matrix $X_{packet}(m, \cdot)X^*_{packet}(m, \cdot)$, the steering vectors calculation suffers less from noise. Meanwhile, since we establish $2P$ virtual antennas here, the steering vector can be written as $\mathbf{a}(\theta) = [1, \Phi(\theta), \ldots, \Phi(\theta)^{2P-1}]^T$ according to Eq. (5.9).

5.1.5.3 Multi-Domain Extension

In OFDMA backscatter systems, it is also possible to assign a group of subcarriers to one certain tag. For feasible multi-subcarrier CSI, SpotFi [87] propose the ToF sanitization algorithm to remove ***packet detection delay*** [227] via a linear fit of unwrapped multi-subcarrier phase angles. Since the accuracy of linear fit relies on the number of samples, it is difficult to obtain accurate results with limited subcarriers. In this section, we first reveal that the ToF sanitization algorithm is not necessary and cannot improve performance of localization systems. Then we show how to simultaneously perform SpotFi and our symbol-domain extension methods based on discontinuous allocated subcarriers for more accurate AoA measurement.

ToF Sanitization Algorithm Here we introduce the root reason why ToF sanitization proposed by SpotFi is not necessary in localization systems. Supposing that there are L-path signals arriving at receiver, ToF of lth path is denoted by t_l while packet detection delay is represented by t_{packet}. The receiver observes $s_r(n)$ of nth subcarrier, which consists of L-path signals with respective attenuation A_l and phase shift $e^{-j2\pi f t_l}$ caused by ToF t_l. Besides, there is also an extra phase offset $e^{-j2\pi f t_{packet}}$ caused by packet detection delay, thus we have

$$s_r(n) = \left(\sum_{l=1}^{L} A_l e^{-j2\pi f t_l}\right) e^{-j2\pi f t_{packet}}. \tag{5.13}$$

For nth subcarrier, its frequency could be denoted by $f_1 + (n-1)\Delta f$, where f_1 is the frequency of the first subcarrier and Δf is the frequency spacing between consecutive subcarriers. Therefore, we can denote the phase shift of nth subcarrier relative to the first subcarrier as follows:

$$f(n) = \underbrace{(\sum_{l=1}^{L} A_l e^{-j2\pi(n-1)\Delta f t_l})}_{\text{Part A}} \underbrace{(e^{-j2\pi(n-1)\Delta f t_{packet}})}_{\text{Part B}}. \tag{5.14}$$

As we know, the phase angle of complex exponential $e^{-j2\pi(n-1)\Delta f t_{packet}}$ could be denoted by $\phi(e^{-j2\pi(n-1)\Delta f t_{packet}}) = -j2\pi(n-1)\Delta f t_{packet}$. To remove t_{packet}, SpotFi tries to deal with the phase shift among different subcarriers via a linear fit. The basic idea is that they first find the best linear fit of $\phi(f(n))$ and then subtract the phase offset caused by packet detection delay. For example, supposing that n is the independent variable in equation $\phi(e^{-j2\pi(n-1)\Delta f t_{packet}}) = -j2\pi(n-1)\Delta f t_{packet}$, we can find the best linear fit with a slope $-j2\pi\Delta f t_{packet}$. Then we subtract $-j2\pi n\Delta f t_{packet}$ from $-j2\pi(n-1)\Delta f t_{packet}$ and therefore remove packet detection delay. However, we believe such a scheme is not necessary and cannot improve performance of system for the following two reasons:

First, proposed schemes would simultaneously remove parts of ToF information (i.e., $t_1, t_2, \ldots t_l$) and make ToF inaccurate, which has been verified in their own experiments [87]. The reason is that part A in Eq. (5.14) also contains linear phase angle components. One simple example is that when there is just one path signal,

$f(n)$ could be represented as $A_1 e^{-j2\pi(n-1)\Delta f(t_1+t_{packet})}$ and the corresponding phase angle is $\phi(n) = -j2\pi(n-1)\Delta f(t_1 + t_{packet})$. According to the ToF sanitization algorithm, we can obtain the best linear fit $-j2\pi\Delta f(t_1 + t_{packet})$ and therefore remove both t_1 and t_{packet}. Since ToF information t_1 has been removed, the ToF sanitization algorithm even leads to worse performance. Furthermore, when there are L paths, part A contains L signals with respective linear change. With all multipath signals summed, the phase angle is dominant by the strongest signal and shows linear-like result. Consequently, when removing packet detection delay of part B, the ToF sanitization algorithm still inevitably damages ToF information in part A.

Second, since ToF measured is not accurate, SpotFi just utilizes them to determine the direct path. The basic idea is that the smallest ToF is corresponding to the direct path. In particular, they remove packet detection delay in Eq. (5.14) and have

$$f'(n) = \left(\sum\nolimits_{l=1}^{L} A_l e^{-j2\pi(n-1)\Delta f t_l}\right) e^{-\varphi_{residual}} = \sum\nolimits_{l=1}^{L} A_l e^{-j2\pi(n-1)\Delta f t_l - \varphi_{residual}}, \tag{5.15}$$

where $\varphi_{residual}$ is the residual phase offset since SpotFi can never obtain the exact packet time delay. Then SpotFi employs the MUSIC scheme to parse a series of ToF, denoted by $\{t_1 + t_{residual}, t_2 + t_{residual}, \ldots, t_L + t_{residual}\}$, where we can find the direct path t_1 by finding the smallest ToF. In reality, without ToF sanitization algorithm, we can also obtain such a group of ToF, denoted by $\{t_1 + t_{packet}, t_2 + t_{packet}, \ldots, t_L + t_{packet}\}$, where we still can find the direct path based on the smallest ToF. In conclusion, the ToF sanitization algorithm is not necessary for clear multi-subcarrier information.

Multi-Domain Extension for MUSIC In previous section, we show how to establish the symbol-domain extension matrix for MUSIC algorithm. In reality, when a group of subcarriers are assigned to one certain tag, we can utilize both frequency-dimension and symbol-dimension CSI for further performance promotion. The basic idea is to further extend the symbol-domain matrix in Fig. 5.8 for multi-domain matrix generation. For this purpose, we first determine subcarrier IDs corresponding to current tag. For example, as shown in Fig. 5.9, supposing that subcarriers #1, #4, #6 are assigned to tag #2, we have $\mathcal{F}_2 = \{1, 4, 6\}$. Second, we determine the current start pointer and select corresponding column in the symbol-domain extension matrix. For example, we here select the first start symbol pointer of subcarrier #1, i.e., $X_1^{music}(i, 1)$. Third, for other subcarriers assigned to the same tag, we perform the second step for all assigned subcarriers and obtain subcarrier CSI group, i.e., $X_1^{music}(i, 1)$, $X_4^{music}(i, 1)$, $X_6^{music}(i, 1)$. Finally, we combine different $X_n^{music}(i, j)$ together utilizing matrix fusion shown in Fig. 5.9 and therefore obtain the multi-domain extension matrix for MUSIC algorithm.

For MUSIC algorithm, it is necessary to generate steering vector corresponding to the multi-domain extension matrix. In fact, based on subcarrier group and virtual

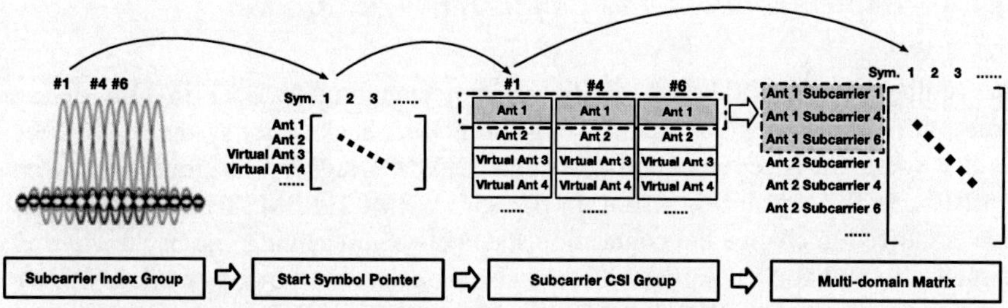

Fig. 5.9 Multi-domain extension

antennas, the steering vector generation is similar to SpotFi [87]. In contrast, our scheme does not rely on ToF sanitization algorithm for packet detection delay elimination, which has been proved unnecessary above. Meanwhile, our MUSIC extension scheme is based on limited subcarriers and virtual antennas generated by the symbol-domain extension scheme. Our steering vector, for $2P$ virtual antennas and N assigned subcarriers, can be written as

$$\mathbf{a}(\theta,\tau) = \Bigg[\underbrace{\Omega_T(1,\tau), \Omega_T(2,\tau), \ldots \Omega_T(N,\tau),}_{\text{Antenna 1}} \\ \underbrace{\Phi(\theta)\Omega_T(1,\tau), \ldots, \Phi(\theta)\Omega_T(N,\tau),}_{\text{Antenna 2}} \ldots, \\ \underbrace{\Phi(\theta)^{2P-1}\Omega_T(1,\tau), \ldots, \Phi(\theta)^{2P-1}\Omega_T(N,\tau)}_{\text{Virtual Antenna 2P}}\Bigg]^T, \tag{5.16}$$

where $\Omega_T(n,\tau) = e^{-j2\pi\mathcal{F}_T(n)f\tau}$ denotes the phase shift caused by different subcarriers while $\Phi(\theta)$ (Eq. (5.9)) denotes that caused by the antenna array. Inside, $\mathcal{F}_T(n)$ denotes nth element in $\mathcal{F}_T$. Therefore, we can determine possible τ and θ utilizing the MUSIC algorithm introduced by SpotFi.

Remarks Section 5.1.5 first utilizes symbol-dimension CSI to improve localization accuracy under single-subcarrier scenario and then utilize multi-dimension CSI (i.e., symbol-dimension and frequency-dimension CSI) to further improve localization accuracy under multi-subcarrier scenario. This section provides us with a new solution to address the challenge caused by limited antennas, subcarriers, and packets. Compared with previous works which employ multiple antennas, subcarriers, and packets for pure performance improvement, we utilize spatial-dimension CSI for phase offset elimination, frequency-dimension CSI for concurrent localization and symbol-dimension CSI to avoid massive packets transmission and maintain accuracy. Therefore, our system provides more practical schemes for concurrent low-power localization systems.

5.1.6 Implementation and Concurrent Localization

We utilize WARP v3 boards [236] to implement the transceiver. Our on-tag transmitter implementation refers to Wi-Fi OFDMA backscatter systems [118]. We utilize some COTS components to realize OFDMA tags by integrating them on an RF-4 PCB board in Fig. 5.10a. In particular, HMC190BMS8E SPDT switches are employed to change the connection state between receiving and backscattering circuits. To realize SSB signal, which is also described in previous work [292], we use a splitter/combiner BP2U+ [293], ADG902 SPST reflective switch [294] and a printed transmission line.

We utilize DIGILENT NEXYS4 FPGA development boards [295] to control tags communication while using Matlab 2016*a* to realize OFDM burst processing, phase offsets elimination and extended MUSIC schemes. For every tag localization, at least two AoAs with respect to different receivers are necessary. Since our excitation signal device (WARP v3 board) is full-duplex, it can act as both the transmitter and receiver without any hardware change. In this way, we can derive enough AoAs relative to different transceivers with only one pair of WARP v3 boards.

In this paper, we localize targets based on the following method. In particular, we define an error E to denote the deviation between AoAs that would be observed at each evaluated location and the corresponding AoAs that were actually observed as follows:

$$E = \sum\nolimits_{T_1}^{T_N} \sum\nolimits_{s=1}^{S} (\overline{\theta}_s(T_n) - \theta_s(T_n))^2, \tag{5.17}$$

where there are S receivers and T_N tags, $\theta_s(T_n)$ denotes AoA of tag T_n relative to the sth receiver while $\overline{\theta}_s(T_n)$ denotes geometrical AoA of tag T_n at the sth receiver corresponding to evaluated position. Finally, we can determine tag's position by finding the minimum E. Since OFDM subcarriers are orthonormal, each tag localization is independent from the others. Therefore, E defined in Eq. (5.17) takes all tags into consideration by directly summing evaluated error of multiple tags.

5.1.7 Experiments

We implement our design on WARP v3 boards and conduct comprehensive experiments in different scenarios to evaluate proposed mechanisms. In our system, each DIGILENT NEXYS4 FPGA development board [295] is utilized to control 4 different backscatter tags. During experiments, we employ $1 \sim 48$ tags (40 tags are shown in Fig. 5.10b) to verify system performance. The position of transceivers and tags can be found in Fig. 5.10c, where the TX-tag distance is $0 \sim 6$ m while the tag-RX distance is $0 \sim 6$ m. To localize backscatter tags, each WARP v3 board is equipped with $2 \sim 4$ antennas. We utilize the carton to build NLoS scenarios as

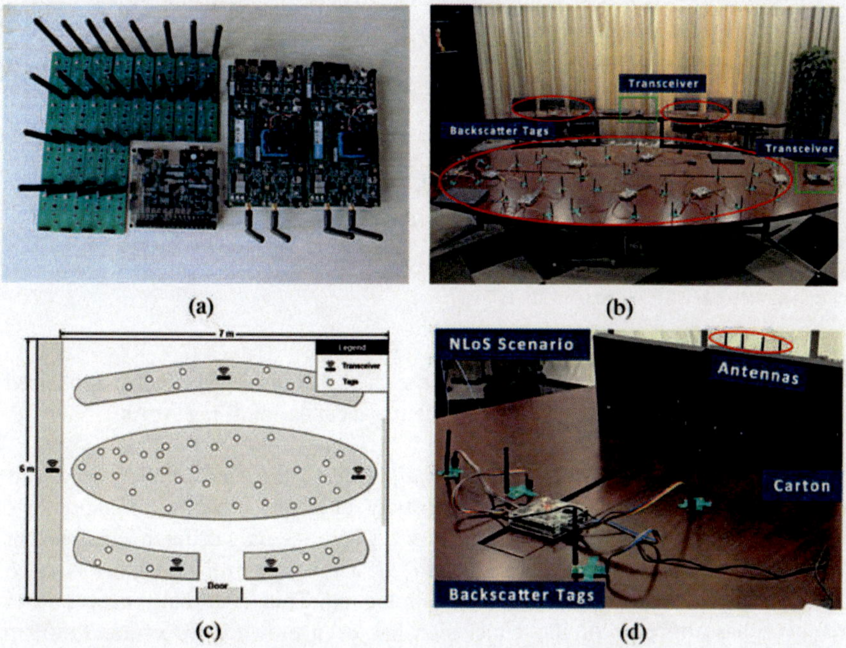

Fig. 5.10 Ground truth of the batch localization system. (**a**) shows experimental devices, where 24 tags, FPGA, and WARP v3 boards are presented; (**b**) shows environment of our experiment in one 6 m × 7 m conference room; (**c**) shows the layout of tags and transceivers; (**d**) shows how we build the NLoS scenario. (**a**) Experimental devices. (**b**) Experimental environment. (**c**) Experimental Layout. (**d**) NLoS Scenario

shown in Fig. 5.10d. For multi-tag localization, the transmitter broadcasts necessary wireless signals in channel 2.485 GHz.

5.1.7.1 Necessity of System Designs

We first verify the necessity of our system designs, which consist of three parts corresponding to three challenges. For each experiment, we conduct 500 measurements and compare the results with our proposed scheme.

Packet-Level CSI and 3D-CSI Now we are going to verify that 3D-CSI is necessary to address *challenge 1*. As explained in CSI tools [291], since packet-level CSI exactly comes from the preamble of OFDM bursts, we can utilize preamble CSI $C(m, n, pre)$ to denote packet-level CSI. As shown in Fig. 5.11a, we can observe that 80% error is within 65 degrees for packet-level CSI localization. Here, we analyze why packet-level CSI is unfeasible. Due to the synchronization mechanism in OFDMA backscatter systems, all OFDM burst preambles come from the transmitter, which denotes that packet-level CSI can just measure channel state

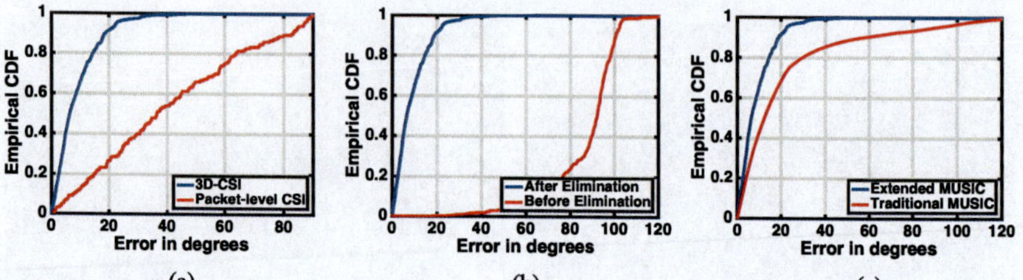

Fig. 5.11 Necessity of system designs. (**a**) Packet-level CSI & 3D-CSI. (**b**) Phase offsets elimination. (**c**) MUSIC & Extended MUSIC

between the transmitter and receiver. Since such packet-level CSI can just be utilized to localize the transmitter, we can never derive accurate multi-tag AoAs.

Phase Offsets Elimination To address *challenge 2*, we propose the phase offsets elimination scheme. To realize the necessity of phase offsets elimination, we localize multiple tags with the same data set, where the only difference is whether we remove phase offsets. It can be seen in Fig. 5.11b that error in degrees is much worse when there is not phase offsets elimination. Thus it is really necessary to remove phase offsets. Similar conclusion has been drawn by Ubicarse [180] in Fig. 9, and the underlying reason is as one kind of noise, phase offsets are so large that AoA calculation cannot work.

MUSIC and Extended MUSIC We explain the necessity of the extended MUSIC scheme addressing *challenge 3* by comparing the performance of traditional MUSIC and our scheme. As shown in Fig. 5.11c, our extended MUSIC scheme realizes half AoA error in degrees compared to the traditional MUSIC scheme. It is because that with limited antennas, subcarriers, and packets, existing MUSIC schemes introduced in Eq. (5.8) are reduced to pure AoA calculation in Eq. (5.4), which is easily interrupted by noise. For example, we can find that the maximum error is about 120 degrees in contrast to 40 degrees of our extended MUSIC scheme.

5.1.7.2 Localization Performance

After verifying the necessity of our system designs, we now comprehensively examine performance of the batch localization system, where the number of backscatter tags, subcarriers and transceivers, experimental scenarios, energy consumption are well considered. We conduct 10 experiments in each situation. Every experiment transmits 50 OFDM bursts and each burst contains 500 symbols. We measure AoA and position according to the information of each OFDM burst.

The Number of Tags To measure the relation between localization accuracy and the number of tags, we respectively employ different numbers of backscatter tags

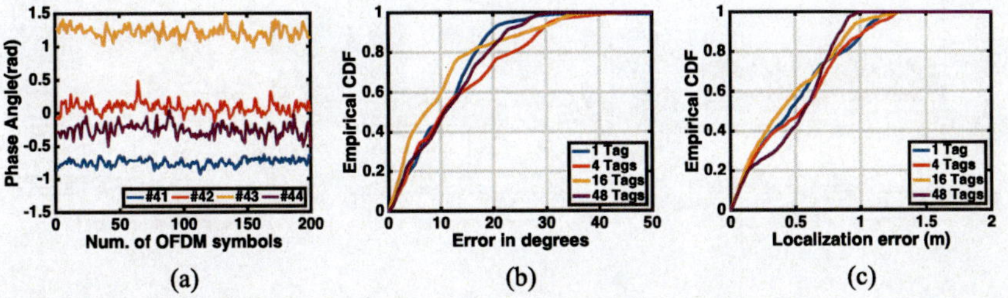

Fig. 5.12 Localization and Num. of tags. (**a**) Multi-tag separation. (**b**) AoA & Num. of tags. (**c**) Localization & Num. of tags

and conduct measurements for different scenarios. We first show how to separate multi-tag AoAs in Fig. 5.12a, where we plot $\varphi^{ex}_{\Delta m,n,k}$ (Eq. (5.6)) of different tags. Since $\varphi^{ex}_{\Delta m,n,k}$ denotes the exact phase difference relative to different AoAs, we can respectively derive AoA corresponding to every tag and therefore separate different tags. For distinct exhibition, we just select 4 different tags and plot $\varphi^{ex}_{\Delta m,n,k}$ in Fig. 5.12a, where we can obviously observe that different tags could be separated. Second, we select different numbers of tags (i.e., 1, 4, 16, and 48) to perform our concurrent localization method. The purpose of this experiment is to prove multi-tag localization is feasible in our system. From Fig. 5.12b, we can observe that with the increase of localized tags, AoA calculation is not interrupted, where the average errors are respectively 11.10, 13.34, 10.02 and 11.71 degrees, the median (90%) errors are respectively 11.10 (19.38), 10.35 (28.66), 7.25 (26.15)and 10.81 (22.34) degrees. Thus concurrent localization is feasible and does not significantly influence localization performance of tags. Third, we utilize derived AoAs to localize multiple tags and therefore obtain localization results shown in Fig. 5.12c, where we can observe average errors are respectively 0.49 m, 0.52 m, 0.45 m and 0.50 m, the median (90%) errors are respectively 0.46 m (1.03 m), 0.54 m (1.03 m), 0.37 m (0.91 m) and 0.56 m (0.85 m).

The Number of Subcarriers To verify our multi-domain extension scheme, we employ different subcarriers and conduct measurements for different scenarios. We first assign one certain tag multiple subcarriers and plot $\varphi^{ex}_{\Delta m,n,k}$ (Eq. (5.6)) assigned to this tag. As shown in Fig. 5.13a, different subcarriers have similar $\varphi^{ex}_{\Delta m,n,k}$. It is because that these subcarriers have been assigned to the same tag and are thus corresponding to the same AoA. Second, we select different numbers of subcarriers (i.e., 1, 2, 4, and 8) to perform our extended MUSIC method. As shown in Fig. 5.13b, with the increase of allocated subcarriers, AoA calculation is improved, where the average errors are respectively 11.10, 9.68, 7.66 and 7.13 degrees, the median (90%) errors are respectively 11.10 (19.38), 9.34 (17.95), 5.82 (17.18) and 5.41 (17.70) degrees. Third, we utilize derived AoAs to localize the tag and therefore obtain localization error shown in Fig. 5.12c, where we can observe average errors are respectively 0.49 m, 0.48 m, 0.39 m and 0.37m, the median

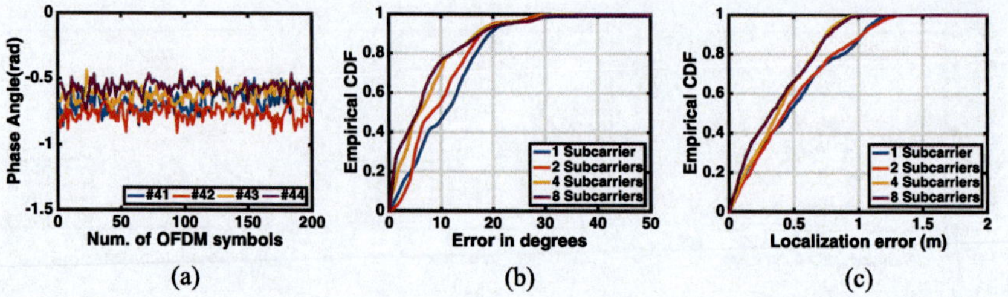

Fig. 5.13 Localization and Num. of subcarriers. (**a**) Multi-subcarrier phase angle. (**b**) AoA & Num. of subcarriers. (**c**) Localization & Num. of subcarriers

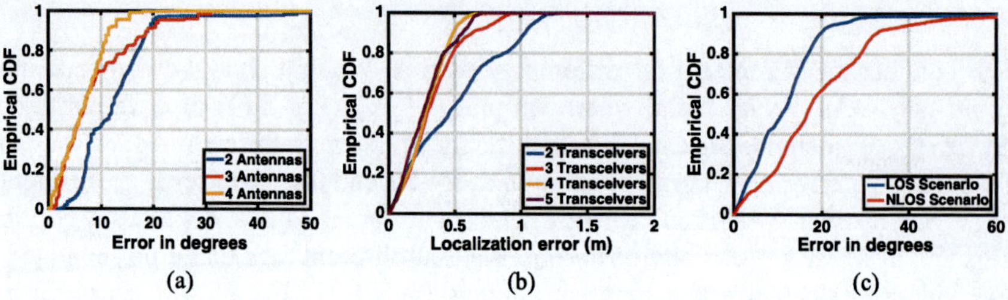

Fig. 5.14 Localization and Num. of transceivers. (**a**) AoA & Num. of Antennas. (**b**) Localization & Num. of transceivers. (**c**) LOS & NLOS scenarios

(90%) errors are respectively 0.46 m (1.03 m), 0.43 m (1.03 m), 0.38 m (0.77 m) and 0.33 m (0.75 m).

The Number of Antennas Since the performance of AoA measurement is dependent on the number of antennas, we conduct experiments with respect to different numbers of antennas. During experiments, we collect CSI and eliminate phase offset based on the method introduced in Sect. 5.1.4. With the increase of antennas, we can observe better performance as shown in Fig. 5.14a, where average errors are respectively 12.48, 8.74, and 6.65 degrees, the median (90%) errors are respectively 12.00 (19.00), 6.60 (19.00), and 6.50 (11.50) degrees.

The Number of Transceivers In batch localization systems, at least two transceivers are necessary to localize multiple tags. During our experiments, we employ different numbers of transceivers to realize batch localization. Since we determine multi-tag position based on AoAs relative to multiple transceivers via Eq. (5.17), with the increase of transceivers, we can observe better performance in localization as shown in Fig. 5.14b, where average errors are respectively 0.49 m, 0.33 m, 0.29 m and 0.28 m, the median (90%) errors are respectively 0.45 m (1.03 m), 0.30 m (0.63 m), 0.28 m (0.49 m) and 0.26 m (0.56 m).

LoS and NLoS To verify system performance under different scenarios, we respectively select LoS and NLoS scenarios to conduct experiments. We conduct measurements for LoS and NLoS scenarios, where average errors are respectively 11.10 and 18.78 degrees, the median (90%) errors are respectively 10.83 (19.38) and 16.65 (32.41) degrees.

Energy Consumption We employ the same tag design as Wi-Fi OFDMA backscatter systems [118], where the overall power consumption of backscatter tag is 55-81.3 μW. Here we briefly introduce the energy consumption which consists of the following three parts. The first part comes from the frequency synthesizer realized by All-Digital PLL whose power consumption is 0.47 μW/MHz. The second part is the digital part which consumes 17 μW. Third, we calculate the power consumption of the backscatter circuit by referring to parameters in the datasheets of electronic components. For different switching frequency, the power consumption is 36.9–54.4 μW.

5.1.8 Performance Comparison

In this section, we compare our batch localization system with previous methods to verify that we can realize system ***accuracy***, ***concurrency***, ***energy conservation*** and ***communication compatibility*** (Sect. 5.1.3). For our scheme, 3D-CSI introduced is available, which is one $2 \times 48 \times 500$ matrix. As a contrastive scheme, we calibrate phase offsets utilizing the scheme introduced by Phaser [94] and then localize tags one by one via SpotFi [87].

Accuracy To verify the accuracy of our system, we conduct 500 experiments to calculate AoA based on different kinds of CSI and different numbers of subcarriers. During our experiments, single-subcarrier AoA calculation without 3D-CSI is based on the traditional MUSIC scheme [185] while multi-subcarrier AoA calculation is based on the scheme introduced in SpotFi. In Fig. 5.15a, we can observe average errors are respectively 16.63, 12.66, 10.25 and 7.80 degrees, the median (90%) errors are respectively 13.20 (27.05), 11.35 (25.32), 7.00 (26.80) and 5.05 (17.84) degrees for different kinds of CSI. It denotes that based on one single packet (OFDM burst), we can achieve higher accuracy than existing systems.

Concurrency and Efficiency We conduct 500 experiments to examine the concurrency of our system, where we compare the number of valid localization results in our system with existing works. During our experiments, we respectively regard AoA calculation with error less than 10, 15, 20 degrees are valid. In Fig. 5.15b, our system shows about 50× efficiency than existing systems. There are two points to be discussed here. First, we can observe a high multiple at the beginning. It is because that existing works must utilize extra communication to calibrate down conversion phase offset and therefore localize zero tag. Since our system not only removes

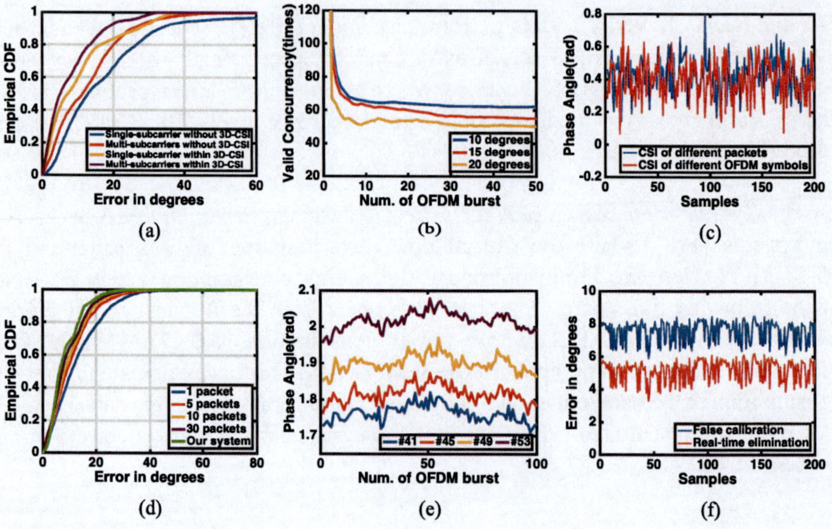

Fig. 5.15 System comparison. (**a**) CSI localization comparison. (**b**) Concurrency and effeciency. (**c**) CSI of different packets and symbols. (**d**) Energy and accuracy. (**e**) Phase offset calibration. (**f**) Calibration and accuracy

phase offsets in real time but also has derived multi-tag position, it can get an infinite multiplier here. Second, since we utilize 48 data subcarriers to localize multiple tags, we can realize 48× concurrency theoretically. In reality, according to Fig. 5.15b we can realize about 50–65 times efficiency, which is higher than theoretical 48 times. It is rooted in that existing methods show lower accuracy due to coarse-grained CSI shown in Fig. 5.15a, which denotes that we can derive more valid AoAs and thus realize higher efficiency.

Energy Conservation Our system does not require continuous massive packets for accuracy and therefore ensures energy conservation. Instead, we utilize symbol-dimension CSI and develop extended MUSIC scheme to ensure localization accuracy. During experiments, we first show the phase angle across different packets (OFDM bursts) and OFDM symbols in Fig. 5.15c, where we can observe the similar results. This experiment shows that it is feasible to replace massive-packet CSI with symbol-dimension CSI. Since we can derive 500 symbol-dimension CSI measurements based on one packet, we ensure energy conservation. Second, we compare energy consumption of our system and existing schemes by analyzing the results based on different numbers of packets (1, 5, 10, and 30). As shown in Fig. 5.15d, existing schemes require about 30 packets to obtain similar performance as our system, which consumes 30× energy consequently. Moreover, massive-packet localization proposed in previous works might reduce efficiency further. For example, SpotFi utilizes 10 packets to localize one single tag while our scheme utilizes one single packet to localize 48 tags. Thus our system can finish about 480 times localization while SpotFi just finishes one.

Communication Compatibility Communication compatibility denotes that we do not interrupt other devices' communication when localizing certain targets or calibrating phase offset. As shown in Fig. 5.13, we can exactly localize multiple tags concurrently. Since available CSI comes from valid communication, we therefore maintain concurrent communication when localizing certain targets. Moreover, our system removes phase offsets dynamically without special calibration process and therefore ensures communication compatibility. To stress the advantage of our phase offset elimination scheme, we calibrate phase offsets of different OFDM bursts and then plot the results in Fig. 5.15e. From this figure, we observe that phase offset calibration result is unsteady, which means previous work might derive inaccurate calibration results and therefore obtain inaccurate AoA. For example, in Fig. 5.15f, false calibration results lead to a higher error in contrast to our real-time calibration.

5.2 Autonomous Wi-Fi Device Map

Channel state information (CSI) based Wi-Fi localization can achieve admirable decimeter-level accuracy; however, such systems require labor-intensive site survey to calibrate the AP position and the antenna array direction, which hinders practical large-scale deployment. In this section, we reveal an interesting finding that the calibration efforts for deploying the CSI localization system can be significantly reduced by simply replacing the ordinary linear antenna layout of the AP with the non-linear layout. In particular, we first present an autonomous self-calibrating method to significantly facilitate site survey for deploying CSI localization systems. Then we propose a systematical evaluation mechanism to show the fundamental reason why linear antenna layout usually leads to serious errors and why non-linear antenna layout is better off. Finally, we build a testbed with COTS devices and conduct comprehensive experiments. Results show that triangular antenna layout can achieve 80% angle of arrival (AoA) measurement error within 9° for any direction in contrast to 16° based on linear antenna layout. Moreover, we can realize promising localization accuracy as previous works even without labor-intensive site survey, where 80% localization error is within 0.60 m.

5.2.1 *Motivation and Challenge*

In this section, we answer the following two questions: why self-calibration is necessary for localization applications and why designing such systems is challenging?

Fig. 5.16 Application scenarios. (**a**) Scenario A: Navigation. (**b**) Scenario B: Smart Home

5.2.1.1 Motivation

Indoor localization systems enable many intelligent applications such as smart home and indoor navigation. Self-calibrating mechanisms can facilitate deployability of these applications for the following reasons.

Figure 5.16a shows a navigation application in the mall, where numerous Wi-Fi access points are deployed. Compared with RSSI localization requiring prior knowledge about the location of the AP, CSI localization systems require that the location of the AP and the direction of the AP's antenna array are both perfectly known. The calibration of antenna array direction is, however, more difficult than the calibration of AP's position, especially when numerous APs are deployed. Therefore, such huge efforts will hinder the large-scale deployment of Wi-Fi localization systems.

Figure 5.16b shows smart home, where the system is able to adjust the light and temperature based on the user's position. To this end, we need to figure out the overall layout of these intelligent devices (light, access points and air condition) and then localize the user. For such a plug-and-play application, both consumers and providers prefer the system to accomplish the above tasks automatically. Thus, self-calibration can improve the user experience and reduce deployment costs of plug-and-play localization applications.

The proposed self-calibrating system can automatically figure out the overall layout of all Wi-Fi devices. Based on this layout, it is possible to determine the relative geometric relationship among these devices. Further, when navigate a user to one certain position in the map, we can attach the above overall layout to the map with a few calibrations.

5.2.1.2 Challenge

The basic idea of self-calibrating systems is to figure out the overall layout of all devices. To represent the geometric relationship between different devices, previous localization systems utilize distance derived from RSSI while CSI localization systems utilize AoA to realize decimeter-level accuracy. However, recently

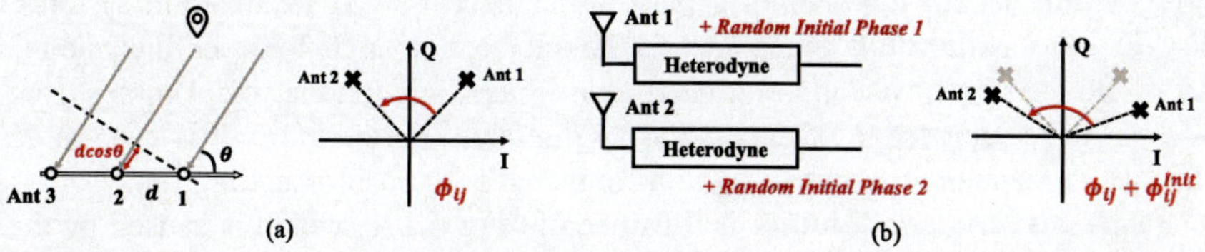

Fig. 5.17 We cannot derive valid AoA without manual calibration while AoAs are essential for self-calibration. (**a**) Basic Methodology: AoA Measurement. (**b**) The heterodyne leads to random initial phase distortion

proposed works [86, 178, 180] show the measurement of AoA suffers from the initial phase distortion that must be manually calibrated. Phaser [94] points out that the calibration of this phase distortion requires prior knowledge about the location of the AP and the direction of the AP's antenna array. Since overall layout construction requires accurate estimation of AoA while the measurement of AoA requires prior knowledge about the overall layout, self-calibration is a "chicken and egg" problem. This section will discuss this challenge based on theoretical analysis and experimental results.

The basic methodology of CSI localization is to first derive target's AoA with respect to different receivers and then localize the target by geometry. AoA is the intersection angle between the antenna array and the target. Supposing that AoA is denoted by θ, the process of AoA estimation could be illustrated in Fig. 5.17a. In particular, since distance between neighboring antennas is usually far less than the distance between antennas and the target, we can regard propagation paths arriving at antennas as a series of parallel lines. Then we can observe an extra flight distance of the signal at the jth antenna (labeled with the red line) that could be represented as $d_{ij}cos(\theta)$, where d_{ij} is the distance between ith and jth antennas. Such an extra flight distance leads to the corresponding signal phase shift denoted by $\phi_{ij}(\theta) = -2\pi d_{ij}cos(\theta)f/c$, where f is the frequency of signals and c is the speed of light. Since d_{ij} is fixed and can be known in advance, we can thus calculate AoA as follows:

$$\theta = arccos\left(-\frac{c\phi_{ij}}{2\pi d_{ij}f}\right), \tag{5.18}$$

where we can directly derive phase difference ϕ_{ij} between ith and jth antennas from collected CSI. Finally, we can localize the target by geometry.

Though CSI localization can realize decimeter-level accuracy, it is impractical to deploy such systems on a large scale due to labor-intensive site survey. The calibration in CSI localization consists of position calibration, direction calibration and phase distortion calibration. ***Position calibration*** means that we should have a prior knowledge about relative placement of APs, which is a common problem of CSI localization and RSSI localization. Though the position calibration issue can be addressed in RSSI localization [77], it can only achieve a 80% localization error within 2.15 m that cannot satisfy the demand of CSI localization systems. For ***direction calibration***, since AoA is the intersection angle between the antenna array and the target, we can never determine which direction the signal comes from without prior knowledge of antenna array direction. Furthermore, it is even more difficult to manually measure direction compared with position calibration.

Phase distortion calibration is illustrated in Fig. 5.17b, which is caused by the independent heterodyne for each antenna on the RF interface chip. In particular, while each heterodyne shares the same reference frequency produced by crystal oscillator, different heterodynes will finally introduce different phase distortion to the baseband signal due to the randomness of the steady state of the VCO PLL. This phase distortion is time-invariant when the system works stably. However, once the central frequency of communication changes, the VCO begins to work and finally achieve another random steady state. For a certain RF interface, different antennas would observe signals of different random phases after the heterodyne. Consequently, it incurs a random phase distortion denoted by ϕ_{ij}^{init} (labeled with red lines) between ith and jth antennas compared with the accurate result (labeled with gray lines). According to the above discussion, measured phase difference $\hat{\phi_{ij}}$ consists of two parts, i.e.,

$$\hat{\phi}_{ij}(\theta) = \phi_{ij}(\theta) + \phi_{ij}^{init} = -2\pi d_{ij} f cos(\theta)/c + \phi_{ij}^{init}. \tag{5.19}$$

Without removing ϕ_{ij}^{init}, we cannot derive accurate AoA since

$$\theta \neq arccos\left(-\frac{c(\phi_{ij}(\theta) + \phi_{ij}^{init})}{2\pi d_{ij} f}\right). \tag{5.20}$$

The above theoretical analysis motivates us to think whether the error caused by this phase distortion is tolerable and whether it is easy to calibrate this phase distortion. We conduct experiments in real environments to account for these two questions. Figure 5.18a shows the accuracy of AoA measurement w/o removing phase distortion. Without calibration, we can observe a 60° error of AoA estimation

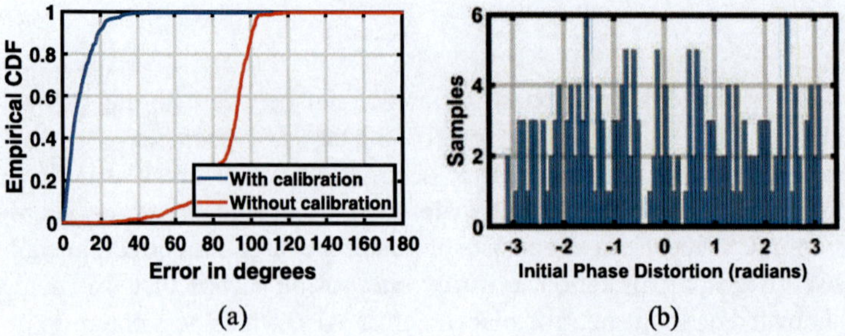

Fig. 5.18 Initial phase distortion. (**a**) AoA Measurement. (**b**) Phase Distortion Distribution

in average. Experiments in previous works [86, 94, 180] also reach the same conclusion, i.e., initial phase distortion cannot be ignored.

To remove this phase distortion, ArrayTrack [86] utilizes the transmission line to manually calibrate its value. It is, however, impractical because its value changes once the central frequency of communication changes. We conduct experiments to verify the feature of phase distortion and Fig. 5.18b shows the distribution of initial phase distortion, where we can see the initial phase distortion is exactly a random value. The other works [94, 178] require prior knowledge about AP's position and the direction of the antenna array for phase distortion calibration. Specifically, with perfect knowledge about the overall layout of APs, they can directly derive θ and the corresponding $\phi_{ij}(\theta)$. Then, they exchange messages among these devices to collect $\hat{\phi}_{ij}(\theta)$ and finally calculate phase distortion via $\phi_{ij}^{init} = \hat{\phi}_{ij}(\theta) - \phi_{ij}(\theta)$.

These experiments validate that it is necessary to master prior knowledge about the overall layout for phase distortion calibration and further AoA estimation. However, self-calibrating system exactly means this prior knowledge is unavailable and we must figure out the overall layout based on estimated AoAs. It is thus challenging to solve such a *"chicken and egg" problem*, where we want to estimate geometric relationship without initial phase distortion calibration.

5.2.2 System Overview

Our system is illustrated in Fig. 5.19, which consists of self-calibrating system designs and non-linear antenna array designs. The first part proposes that non-linear antenna array can address the "chicken and egg" problem while the second part shows how to select an appropriate non-linear antenna array for indoor localization.

Specifically, to address the "chicken and egg" problem, we utilize ADoA rather than AoA to represent the geometric relationship among different APs. However, in the presence of initial phase distortion, we cannot derive ADoA. Section 5.2.3

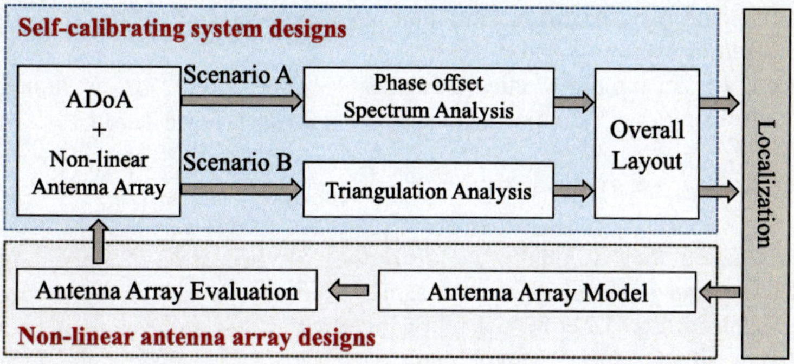

Fig. 5.19 System overview

reveals the fundamental reason and proposes that non-linear antenna layout can address this issue. Even so, it is still hard to derive a unique solution due to multipath effect and phase circularity. To this end, we design two different methods respectively suitable for two scenarios shown in Fig. 5.16. In scenario A, Wi-Fi APs are widely deployed, based on which we show how to derive ADoA with more APs involved. In scenario B, since there are usually limited intelligent devices in current smart home, we propose a new method to measure ADoA with just 3 devices. Finally, we utilize ADoA to figure out the overall layout and localize the target.

While the non-linear antenna array can address the "chicken and egg" problem, it is necessary to learn how to select a suitable non-linear antenna array. In Sect. 5.2.4, we model the antenna array to show how good the antenna array is for localization, based on which we reveal the fundamental reason why linear antenna layout usually leads to serious AoA measurement errors, and why non-linear layout is better off.

5.2.3 *Self-Calibrating System Designs*

5.2.3.1 Basic Idea

Angle Difference of Arrival The crux of self-calibrating systems is to solve the above "chicken and egg" problem. For this purpose, one possible method is to find another information such as time of flight (ToF) to represent geometric relation among existing APs; however, ToF is not always available for COTS Wi-Fi devices. For example, ToneTrack [95] requires strict synchronize mechanism to derive ToF and the other work [227] requires special frequency hopping mechanism. Another possible method is to construct the APs layout via RSSI [77]; however, it can only achieve a 80% localization error within 2.15 m that cannot satisfy the demand of CSI localization systems.

Recently proposed work [181] utilizing ADoA for indoor localization motivates us to think about whether it is possible to address this challenge via ADoA. In particular, as shown in Fig. 5.20a, the angle labeled with red line is defined as ADoA between B and C, which is denoted by $\Delta\theta_B$. Intuitively, we can subtract θ_{AC} from θ_{AB} to calculate $\Delta\theta_B$; however, the initial phase distortion ϕ_{ij}^{init} [86, 94] makes it impossible to measure θ_{AC} and θ_{AB}.

Instead, we can explore whether it is possible to measure ADoA by subtracting $\hat{\phi}_{ij}(\theta_{AC})$ from $\hat{\phi}_{ij}(\theta_{AB})$. Supposing that $K_{ij} = -2\pi d_{ij} f/c$, we have

$$\Delta\hat{\phi}_{ij} = \hat{\phi}_{ij}(\theta_{AB}) - \hat{\phi}_{ij}(\theta_{AC}) = K_{ij}(cos(\theta_{AB}) - cos(\theta_{AC})), \tag{5.21}$$

where we have removed initial phase distortion ϕ_{ij}^{init} because ϕ_{ij}^{init} corresponding to different propagation paths are the same. Even so, we cannot derive $\Delta\theta_B = \theta_{AB} - \theta_{AC}$ from Eq. (5.21) for the following obvious reason: we cannot calculate $(\theta_1 - \theta_2)$ based on $cos(\theta_1) - cos(\theta_2)$ only.

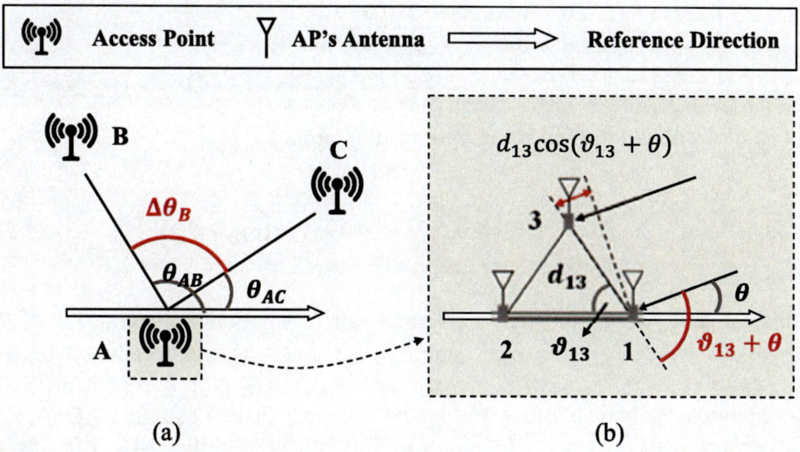

Fig. 5.20 ADoA measurement. (**a**) ADoA (Angle Difference of Arrival). (**b**) Non-linear Antenna Layout

Non-linear Antenna Layout Non-linear antenna layout can facilitate deployability of CSI indoor localization systems through deriving ADoA from phase difference in Eq. (5.21), which has not been found until now. In particular, for the triangular antenna layout shown in Fig. 5.20b, since all antennas are not in one straight line, we redefine AoA as the intersection angle between the reference direction and the arrived signal. Note that the "reference direction" of linear antenna layout is the same as antenna array direction. There are two critical parameters affecting phase difference between two antennas, i.e., distance between given two antennas and the intersection angle relative to reference direction respectively denoted by d_{ij} and ϑ_{ij}. In particular, to calculate ϕ_{13}, we can define the intersection angle between the antenna sub-array $\{1, 3\}$ and the target as $\vartheta_{13} + \theta$. Then, we can observe an extra flight distance $d_{13}cos(\vartheta_{13} + \theta)$ (labeled with the red line) by geometry and the corresponding phase shift $\phi_{13} = -2\pi d_{13}cos(\vartheta_{13} + \theta)f/c$. Finally, for any ϕ_{ij}, based on the same method, we have

$$\phi_{ij}(\theta) = 2\pi d_{ij}cos(\vartheta_{ij} + \theta)f/c. \tag{5.22}$$

Based on this new expression of $\phi_{ij}(\theta)$, we subtract $\hat{\phi}_{ij}(\theta_1)$ from $\hat{\phi}_{ij}(\theta_2)$ and have

$$\Delta\hat{\phi}_{ij}(\Delta\theta) = K_{ij}(cos(\vartheta_{ij} + \theta_2) - cos(\vartheta_{ij} + \theta_1)). \tag{5.23}$$

We take the equilateral triangular antenna layout in Fig. 5.20b as an example to show how to derive ADoA. In particular, we can select different antenna sub-arrays with respect to different ϑ_{ij} and have

$$\begin{cases} \Delta\hat{\phi}_{12} = K_{12}(cos(\theta_{AB}) - cos(\theta_{AC})) \\ \Delta\hat{\phi}_{13} = K_{13}(cos(60^\circ + \theta_{AB}) - cos(60^\circ + \theta_{AC})) \\ \Delta\hat{\phi}_{23} = K_{23}(cos(120^\circ + \theta_{AB}) - cos(120^\circ + \theta_{AC})). \end{cases} \tag{5.24}$$

It is equal to solving the following optimization problem

$$\arg\min_{(\theta_1,\theta_2)} \left(\sum_{i \neq j} |\Delta\hat{\phi}_{ij} + K_{ij}(cos(\vartheta_{ij} + \theta_1) - cos(\vartheta_{ij} + \theta_2))| \right). \tag{5.25}$$

Mathematically, we can always derive one pair of solutions satisfying Eq. (5.25) respectively denoted by $(\theta_{AB}, \theta_{AC})$ and $(\theta_{AC}+\pi, \theta_{AB}+\pi)$. It is because $cos(\theta_{AB}+\vartheta) - cos(\theta_{AC} + \vartheta) \equiv cos(\theta_{AC} + \pi + \vartheta) - cos(\theta_{AB} + \pi + \vartheta)$ for all ϑ. In fact, though these two solutions are not the same but share a common ADoA, i.e., $|\theta_{AB} - \theta_{AC}| = |(\theta_{AC} + \pi) - (\theta_{AB} + \pi)|$. This is why we utilize ADoA for layout construction.

Challenges for ADoA Measurement While proposed methods can measure ADoA under ideal situation, two critical challenges must be addressed in real scenarios.

Challenge 1: Multipath Effect For indoor localization scenarios, since there usually exists multipath effect, actual received signal is the combination of multipath signals rather than one single signal. Multipath effect would lead to an extra phase shift so that Eqs. (5.18) and (5.22) are no longer accurate even though the phase distortion has been calibrated. One effective method to classify multipath signals is the MUSIC algorithm employed in many recent works [86, 87, 94, 178–180, 192, 227]. However, the MUSIC algorithm also requires calibrated phase distortion and cannot be utilized to derive ADoA. It is thus necessary to propose a new method to resolve multipath.

Challenge 2: Phase Circularity Phase circularity means that signal's phase would always lie in $[0, 2\pi]$. For an actual phase $\phi > 2\pi$, we can only obtain $(\phi \mod 2\pi)$ instead of ϕ. It is because that the collected CSI is in the form of complex number, i.e., $\gamma e^{j\hat{\phi}}$, where γ denotes amplitude and $\hat{\phi}$ denotes phase. In the polar coordinate system, $\hat{\phi}$ is always in $[0, 2\pi]$. Due to phase circularity, the solution to Eq. (5.25) becomes non-unique. In particular, we can rewrite the phase subtraction (Eq. (5.23)) into the form of complex number and have

$$\frac{e^{j\hat{\phi}_{ij}(\theta_{AB})}}{e^{j\hat{\phi}_{ij}(\theta_{AC})}} = \frac{e^{jK_{ij}cos(\vartheta_{ij}+\theta_{AB})}}{e^{jK_{ij}cos(\vartheta_{ij}+\theta_{AC})}}. \tag{5.26}$$

In this way, Eq. (5.25) is equal to find the maximum $P(\theta_1, \theta_2)$, where

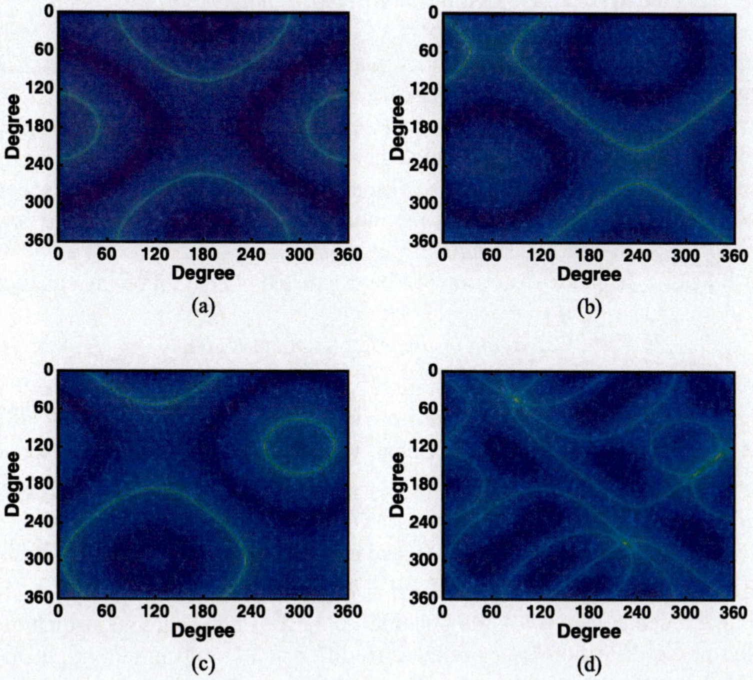

Fig. 5.21 We can derive finite but non-unique solutions via triangular antenna layout. (**a**) Linear Antenna Layout. (**b**) Antenna Sub-array {2,3}. (**c**) Antenna Sub-array {1,3}. (**d**) Triangular Antenna Layout

$$\frac{1}{P(\theta_1, \theta_2)} = \sum_{i \neq j} \left| \frac{e^{j\hat{\phi_{ij}}(\theta_1)} e^{-jK_{ij}cos(\vartheta_{ij}+\theta_1)}}{e^{j\hat{\phi_{ij}}(\theta_2)} e^{-jK_{ij}cos(\vartheta_{ij}+\theta_2)}} - 1 \right|. \tag{5.27}$$

Obviously, when $P(\theta_1, \theta_2) \to \infty$, we have $(\theta_1, \theta_2) \to (\theta_{AB}, \theta_{AC})$. To calculate (θ_1, θ_2), we can scan $(\theta_1, \theta_2) \in [0°, 360°] \times [0°, 360°]$ and derive a series of $P(\theta_1, \theta_2)$ shown in Fig. 5.21. The results of linear antenna layout are shown in Fig. 5.21a, where we can observe so many bright points that satisfy $(\theta_1, \theta_2) \to (\theta_{AB}, \theta_{AC})$. It is because that $\vartheta_{ij} \equiv 0$ for linear antenna layout, we can always obtain similar results based on different antenna sub-arrays. In contrast, for triangular antenna layout, we can respectively obtain the heat map as shown in Fig. 5.21a, b, and c by selecting different antenna sub-arrays {1,2}, {2,3} and {1,3}. Finally, we combine results of all antenna sub-arrays and determine the corresponding pair of AoAs in Fig. 5.21d, where the actual AoAs are (90°, 45°). However, we can still observe four different points including (90°, 45°), (270°, 225°), (340°, 125°) and (310°, 165°), which is caused by phase circularity and needs to be distinguished for accurate ADoA measurement.

5.2.3.2 Scenario A: Phase Distortion Spectrum Analysis

Multipath effect and phase circularity make it hard to derive accurate ADoA. Meanwhile, for desired results of Eq. (5.27), we must scan (θ_1, θ_2) in $[0, 360°] \times [0, 360°]$ space. When M APs are employed in the systems, the searching space would be extended to $[0, 360°]^M$, which inevitably leads to high time complexity. To address above issues, we here propose a method to realize phase distortion spectrum analysis. The basic idea is to overcome multipath effect by combining the signals with respect to different transmitters.

Specifically, the phase distortion can be calculated via the following equation:

$$e^{j\phi_{ij}^{init}(m,\theta)} = e^{j\hat{\phi}_{ij}(m)}/e^{j\phi_{ij}(\theta)} = e^{j(\hat{\phi}_{ij}(m)-\phi_{ij}(\theta))}, \tag{5.28}$$

where $e^{j\hat{\phi}_{ij}(m)}$ can be directly derived from CSI at mth transmitter and $e^{j\phi_{ij}(\theta)}$ is the phase difference caused by AoA. For all θ, we can always derive the corresponding phase distortion set denoted by $E_{init}(\theta, m) = \{e^{j\phi_{ij}^{init}}(m, \theta)|\forall i \neq j\}$ with respect to the mth transmitter. For example, if we scan $\theta \in [0°, 360°]$, we can derive a series of points $(\phi_{12}^{init}(m, \theta), \phi_{13}^{init}(m, \theta))$ that make up a phase distortion trace $\mathcal{T}_m$ as shown in Fig. 5.22a, where $e^{\phi_{12}^{init}(m,\theta)}, e^{\phi_{13}^{init}(m,\theta)} \in E_{init}(\theta, m)$. As verified in Fig. 5.21, the non-linear antenna layout has the potential to derive finite solutions, which means that the phase distortion traces relative to different M transmitters might not be completely overlapping with each other. To validate this guess, we give proofs as follows.

Theorem 17 *The traces $\mathcal{T}_{m_1}$ and $\mathcal{T}_{m_2}$ are completely overlapping only when $\theta_{m_1} = \theta_{m_2}$.*

Proof 56 For phase distortion trace, each θ is corresponding to one point in the trace. If there are completely overlapping traces $\mathcal{T}_{m_1}$ and $\mathcal{T}_{m_2}$, we can deduce that there are infinite pairs of (θ_1, θ_2) satisfying $(e^{j\phi_{12}^{init}(m_1,\theta_1)}, e^{j\phi_{23}^{init}(m_1,\theta_1)}) = (e^{j\phi_{12}^{init}(m_2,\theta_2)}, e^{j\phi_{23}^{init}(m_2,\theta_2)})$. According to Eq. (5.28), we can represent the initial

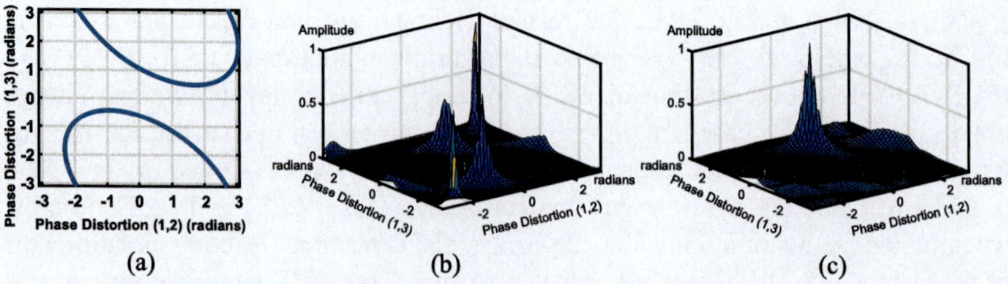

Fig. 5.22 Phase distortion spectrum analysis distinguishes non-unique solutions and improve accuracy. **(a)** Phase Distortion Trace. **(b)** $P(\phi_{12}^{init}, \phi_{13}^{init})$ Spectrum (2 Transmitters). **(c)** $P(\phi_{12}^{init}, \phi_{13}^{init})$ Spectrum (3 Transmitters)

phase distortion with $e^{j\hat{\phi}_{ij}(m)}$ and $e^{j\phi_{ij}(\theta)}$ and thus have

$$\begin{cases} e^{j(\phi_{12}(\theta_1)-\phi_{12}(\theta_2))} = e^{j(\hat{\phi}_{12}(m_1)-\hat{\phi}_{12}(m_2))} = e^{jC_1}, \\ e^{j(\phi_{23}(\theta_1)-\phi_{23}(\theta_2))} = e^{j(\hat{\phi}_{23}(m_1)-\hat{\phi}_{23}(m_2))} = e^{jC_2}. \end{cases} \tag{5.29}$$

As $e^{j\hat{\phi}_{ij}(m)}$ can be directly derived from CSI and is constant, we can simplify the results as e^{jC_1} and e^{jC_2}. If $\theta_1 \equiv \theta_2$, we have $C_1 \equiv C_2 \equiv 0$, and the traces $\mathcal{T}_{m_1}$ and $\mathcal{T}_{m_2}$ are completely overlapping. Under this situation, we have $\theta_{m_1} = \theta_{m_2}$ and the corresponding ADoA is 0. If $\theta_1 \neq \theta_2$, the above equation is similar to Eqs. (5.24) and (5.25), where the number of solutions is finite for non-linear antenna layouts. It means the intersection points of $\mathcal{T}_{m_1}$ and $\mathcal{T}_{m_2}$ are finite.

Since ϕ_{ij}^{init} relative to different propagation paths should be fixed without device rebooted [94], we can find the intersection point of these traces to find the actual value of phase distortion. Mathematically, we can define the $P(\phi_{12}^{init}, \phi_{13}^{init})$ as follows:

$$\frac{1}{P(\phi_{12}^{init}, \phi_{13}^{init})} = \sum_{m=1}^{M} |e^{j\mathcal{F}((\phi_{12}^{init}, \phi_{13}^{init}), \mathcal{T}_m)}|, \tag{5.30}$$

where $\mathcal{F}((\phi_{12}^{init}, \phi_{13}^{init}), \mathcal{T}_m)$ denotes the minimum distance between the point $(\phi_{12}^{init}, \phi_{13}^{init})$ and trace $\mathcal{T}_m$. Finding the peak of $P(\phi_{12}^{init}, \phi_{13}^{init})$ as shown in Fig. 5.22b, c, we can determine the actual value of phase distortion $(\hat{\phi}_{12}^{init}, \hat{\phi}_{13}^{init})$. Finally, we can perform the MUSIC algorithm with this phase distortion to estimate AoA and obtain corresponding ADoAs.

In order to improve the performance of phase distortion estimation, the principle of the above spectrum analysis is to take more APs into consideration. In particular, there are three AoA spectrums shown in Fig. 5.23, where the dot lines in different

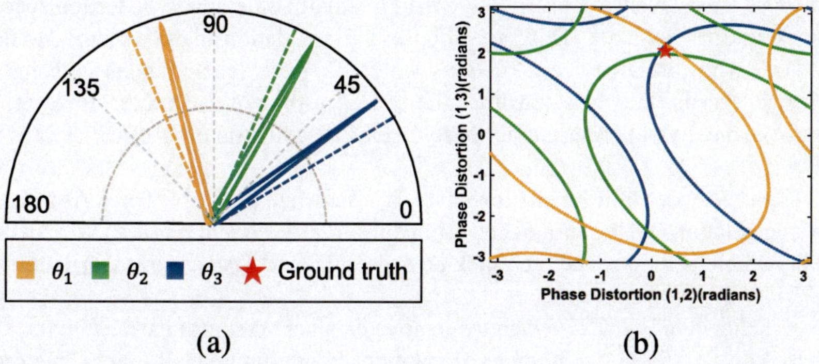

Fig. 5.23 The principle of spectrum analysis. (**a**) AoA Spectrum. (**b**) Phase Distortion Traces

colors represent the ground-truth AoA. Based on any subgroup of antennas, as the number of antennas is two, we can just derive a fused AoA $\hat{\theta}$ that suffers from multipath effect. When we generate the phase distortion trace with such CSI, we will observe a fixed phase shift of the trace denoted by $e^{j(\phi_{ij}(\hat{\theta})-\phi_{ij}(\overline{\theta}))}$, where $\overline{\theta}$ denotes ground-truth AoA. Consequently, we can obtain several 2D traces that are translated. Supposing that there is no multipath effect, these traces would completely intersect at the ground-truth point. To reduce the errors caused by multipath effect, we thus propose the spectrum analysis method to average the errors with respect to multiple APs.

Note that we can always obtain the solution by finding the maximum $P(\phi_{12}^{init}, \phi_{13}^{init})$. Moreover, as proved in Theorem 1, the solutions to Eq. (5.29) are finite unless $C_1 \equiv C_2 \equiv 0$, where we can directly deduce the corresponding ADoA is approximately equal to 0. To explore the minimum number of transmitters required for the phase distortion spectrum analysis, we conduct extensive experiments and the results are shown in Fig. 5.35d.

The advantage of proposed method is that we can analyze multiple signals without incurring high time complexity. In particular, since the generation of $E_{init}(\theta, m)$ relative to the mth transmitter is independent from the other transmitters, we can derive all $E_{init}(\theta, m)$ through scanning $[0°, 360°]$ space with M times rather than $[0°, 360°]^M$ space.

To summarize, as shown in Fig. 5.22a, it is feasible to derive the 2-D phase distortion traces with 3 antennas. Then, we can obtain finite solutions with non-linear antenna layouts as proved in Theorem 1. Finally, Fig. 5.22c shows we find the peak of $P(\phi_{12}^{init}, \phi_{13}^{init})$ with multiple transmitters. Consequently, this section can address the non-unique issue via 3 antennas, which coordinates with the number of antennas installed on commercial devices.

5.2.3.3 Scenario B: Triangulation Analysis

Phase distortion spectrum analysis is suitable for navigation scenarios, where numerous APs are deployed for localization. However, when a few intelligent devices are sparsely deployed in the mall or smart home, phase distortion spectrum analysis does not work. This part proposes a triangulation analysis method to derive ADoA with just 3 devices. Previous work [227] can locate targets with just one device, that is, they can construct the overall layout of 2 devices; however, they must manually calibrate initial phase distortion, which is impractical as proved in Fig. 5.18.

The basic idea of triangulation analysis is shown in Fig. 5.24. Three APs make up a triangulation and the sum of corresponding ADoAs should be 180°. We can derive all possible solutions and then pick an appropriate set $\{\Delta\theta_A, \Delta\theta_B, \Delta\theta_C\}$ satisfying $\Delta\theta_A + \Delta\theta_B + \Delta\theta_C = 180°$.

As shown in Fig. 5.21, when we remove the phase distortion and estimate ADoA based on Eq. (5.27), the number of solutions is non-unique but finite. This section

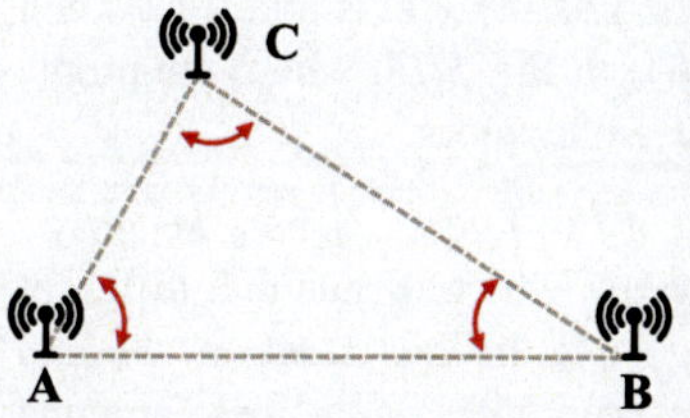

ADoA	$\angle A$	$\angle B$	$\angle C$
Solutions	$\angle A_1$	$\angle B_1$	$\angle C_1$
	$\angle A_2$	$\angle B_2$	$\angle C_2$
	...	...	...
s.t.	$\angle A + \angle B + \angle C = 180°$		

Fig. 5.24 The basic idea of triangulation analysis

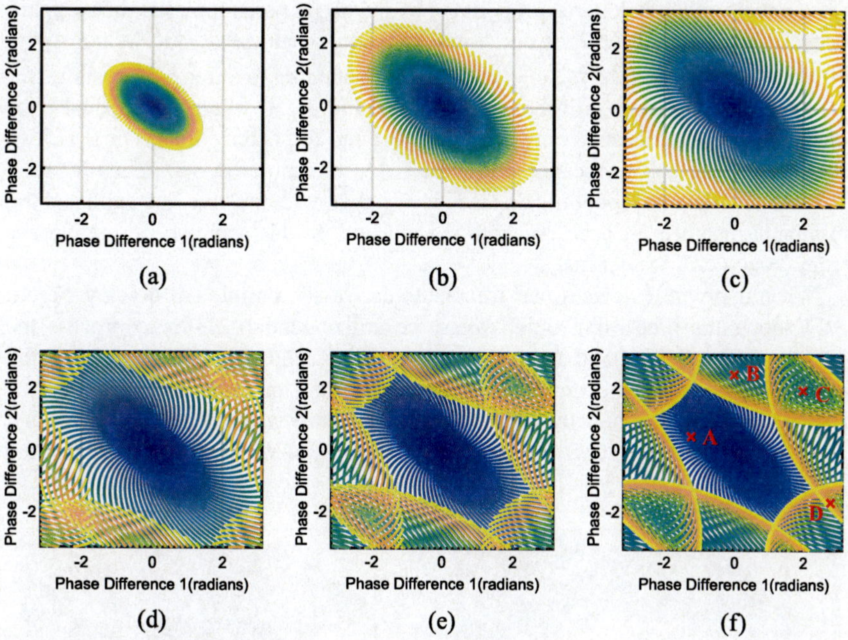

Fig. 5.25 The relationship between $\Delta\theta$ and $\Delta\hat{\phi}_{ij}$. **(a)** $\Delta\theta \in (0, 30)$. **(b)** $\Delta\theta \in (0, 60)$. **(c)** $\Delta\theta \in (0, 90)$. **(d)** $\Delta\theta \in (0, 120)$. **(e)** $\Delta\theta \in (0, 150)$. **(f)** $\Delta\theta \in (0, 180)$

mainly studies the number of solutions. For this purpose, we first analyze the mapping between $\Delta\theta$ and $\Delta\hat{\phi}_{ij}$, where $\Delta\hat{\phi}_{ij}$ in Eq. (5.23) denotes phase difference after initial phase distortion elimination. In particular, we utilize the following step to generate Fig. 5.25.

1. For $\Delta\theta = \theta_2 - \theta_1$, we can discuss all possible situations by scanning $(\theta_1, \Delta\theta)$ in $[0, 360°] \times [0, 180°]$.
2. For one certain $\Delta\theta$, we can obtain a series of $\Delta\hat{\phi}_{ij}(\Delta\theta)$ by searching θ_1 in $[0, 360°]$ space, where $e^{j\Delta\hat{\phi}_{ij}(\Delta\theta)} = e^{jK_{ij}(cos(\theta_1+\Delta\theta+\vartheta_{ij})-cos(\theta_1+\vartheta_{ij}))}$ can be deduced from Eq. (5.23).

3. We respectively choose $\Delta\hat{\phi}_{12}(\Delta\theta)$ and $\Delta\hat{\phi}_{23}(\Delta\theta)$ as the x axis and y axis and then plot different points $(\Delta\hat{\phi}_{12}(\Delta\theta), \Delta\hat{\phi}_{23}(\Delta\theta))$ in Fig. 5.25, where the points corresponding to different $\Delta\theta$ are filled with different colors.

We dynamically increase $\Delta\theta$ to obtain 6 different sub-figures. When $\Delta\theta$ grows to 90° (Fig. 5.25c), these points begin to move inward, which means that different ADoAs begin to incur a same $(\Delta\hat{\phi}_{12}(\Delta\theta), \Delta\hat{\phi}_{23}(\Delta\theta))$. In this situation, we can derive more than one solution based on $\Delta\hat{\phi}_{ij}(\Delta\theta)$ even with a non-linear antenna array. As $\Delta\theta$ increases, we observe increasing overlapping areas, where a smaller $\Delta\theta$ begins to be polluted by the bright parts. When $\Delta\theta$ grows to 180° (Fig. 5.25f), there remain some dark parts not polluted by the bright parts. It means the mapping between $\Delta\theta$ and $\Delta\hat{\phi}_{ij}(\Delta\theta)$ is one-to-one mapping when $\Delta\theta < 40°$. When $\Delta\theta > 40°$, one point $(\Delta\hat{\phi}_{12}(\Delta\theta), \Delta\hat{\phi}_{23}(\Delta\theta))$ is corresponding to at most 3 possible $\Delta\theta$, we respectively explore 4 different positions in Fig. 5.25f, where the residual error in Fig. 5.26 is the distance between this point and the other points. For θ_1, if we set the unit length of the scanning process is 1°, we can derive 360 points for each $\Delta\theta$. We plot the corresponding distance in Fig. 5.26 with the blue line and label the minimum distance for each $\Delta\theta$ with the red line. In this way, we can derive $\Delta\theta$ from $(\Delta\hat{\phi}_{12}(\Delta\theta), \Delta\hat{\phi}_{23}(\Delta\theta))$.

When there are 3 devices, we are able to derive all possible $\Delta\theta$ quickly. Figure 5.27 shows the processing time, where we utilize Matlab 2019a to process the data in DELL XPS 9570 (Intel Core i9-8950HK CPU@2.9 GHz). During the experiments, we set the unit length of the scanning process as 0.5°, 1°, 2°, 4°. Experimental results show that we can calculate ADoA within 9 ms, 5 ms, 2 ms and 1 ms for different unit lengths. After calculating all ADoAs, we can construct the

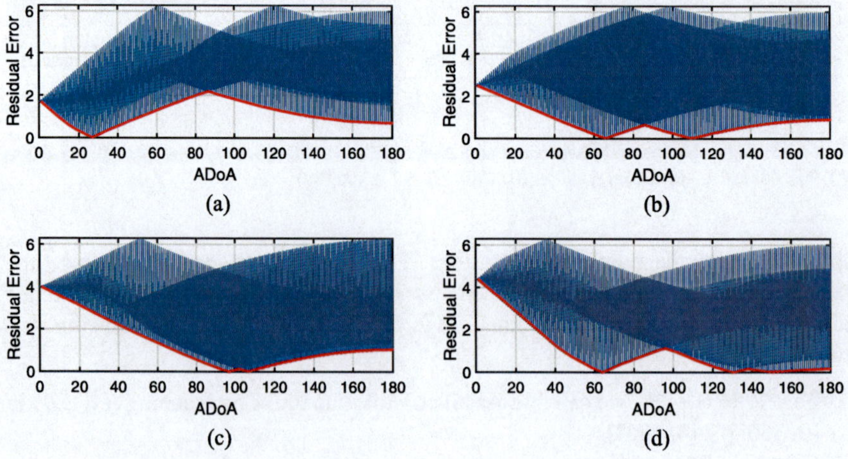

Fig. 5.26 ADoA measurement for positions in Fig. 5.25f. (**a**) Position A. (**b**) Position B. (**c**) Position C. (**d**) Position D

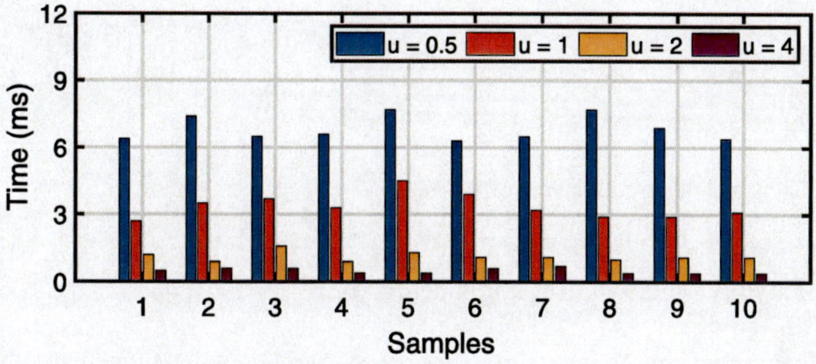

Fig. 5.27 Processing time for different unit lengths

overall layout among 3 APs through picking an appropriate set $\{\Delta\theta_A, \Delta\theta_B, \Delta\theta_C\}$ that satisfies $\Delta\theta_A + \Delta\theta_B + \Delta\theta_C = 180°$.

Algorithm 11 Triangulation analysis

Require: $E_{input} = \{\Delta\hat{\phi}_{ij}(\Delta\hat{\theta}) | \Delta\hat{\theta} = \Delta\theta_A, \Delta\theta_B, \Delta\theta_C\}$
Ensure: $\Delta\hat{\theta}_A, \Delta\hat{\theta}_B, \Delta\hat{\theta}_C$
  Part A: Generate all possible points;
  for each $\Delta\theta$ in [0, 180°] **do**
    Initialize the set $E(\Delta\theta)$;
    for each θ_1 in [0, 360°] **do**
      Calculate $\Delta\hat{\phi}_{ij}(\Delta\theta) \in E(\Delta\theta)$;
    end for
  end for
  Part B: Measure ADoA;
  for $\Delta\hat{\phi}_{ij}(\Delta\hat{\theta})$ in E_{input} **do**
    Calculate residual error $e = |\Delta\hat{\phi}_{ij}(\Delta\hat{\theta}) - \Delta\hat{\phi}_{ij}(\Delta\theta)|$;
    Find minimum three e and the corresponding three solutions $\Delta\theta_1, \Delta\theta_2, \Delta\theta_3$;
    if $\exists\Delta\theta_n < 40°$ **then**
      $\Delta\hat{\theta} = \Delta\theta_n$;
    else
      $\Delta\hat{\theta} = \{\Delta\theta_1, \Delta\theta_2, \Delta\theta_3\}$;
    end if
  end for
  Pick $\{\Delta\hat{\theta}_A, \Delta\hat{\theta}_B, \Delta\hat{\theta}_C\} = \arg\min_{(\Delta\theta_A, \Delta\theta_B, \Delta\theta_C)}(|\Delta\theta_A + \Delta\theta_B + \Delta\theta_C - 180°|)$;
  Part C: Revise ADoA;
  Find $\Delta\theta_{revise}$ that satisfying $\Delta\hat{\theta}_A + \Delta\hat{\theta}_B + \Delta\hat{\theta}_C + 3 \times \Delta\theta_{revise} = 180°$;
  $\{\Delta\hat{\theta}_A, \Delta\hat{\theta}_B, \Delta\hat{\theta}_C\} = \{\Delta\hat{\theta}_A, \Delta\hat{\theta}_B, \Delta\hat{\theta}_C\} + \Delta\theta_{revise}$;

We summarize the triangulation analysis in Algorithm 11, which consists of three parts. In part A, we generate all possible points in Fig. 5.25, which are fixed for a given antenna array; in part B, we present the method to measure ADoA, which can be quickly realized as shown in Fig. 5.27; in part C, we revise the ADoAs for layout construction.

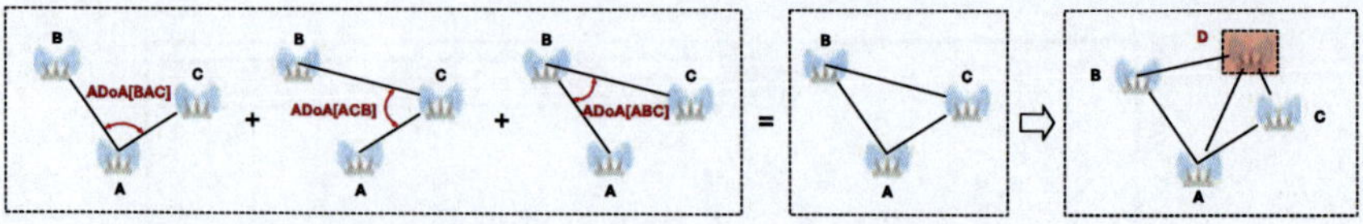

Fig. 5.28 Construct the overall layout of Wi-Fi APs

5.2.3.4 Layout Construction and Localization

Layout Construction We can figure out the overall layout of APs via the following method as shown in Fig. 5.28. First, we can select one AP group $\{A, B, C\}$ and construct the corresponding layout based on $\Delta\theta_{BAC}$, $\Delta\theta_{ACB}$, and $\Delta\theta_{ABC}$. Then, for any other AP such as D in Fig. 5.28, we can find all relevant ADoAs and derive a set E. For example, $E_D = \{ABD, ACD, ADB, ADC, BAD, BCD, BDA, BDC, CAD, CBD, CDA, CDB\}$. For each element $a \in E$, if we define ADoA derived from CSI as $\Delta\hat{\theta}_a$, we can derive the position of AP D as follows:

$$p = \arg\min_{p} \sum_{a \in E} (\Delta\theta_a(p) - \Delta\hat{\theta}_a)^2, \tag{5.31}$$

where $\Delta\theta_a(p)$ denotes the ADoA calculated by geometry when AP D is assumed to be in the estimated position p. Note that we can find the position of AP D via the subset of E such as $\{ABD, ACD, BAD, CAD\}$ in Fig. 5.28. Therefore, more available elements in E improve construction performance but it is not necessary to acquire all elements in E for overall layout construction.

Localization in Indoor Maps Now, we can construct the overall layout of intelligent devices to improve user experience in some plug-and-play applications such as smart home; however, when we want to attach these APs to indoor maps for navigation applications, it is also necessary to make extremely few calibration efforts. For example, for a practical scenario where N APs are deployed, we can determine the positions of all APs in indoor maps based on the positions of only 3 APs while previous works require N times position calibrations and the other N times antenna direction calibrations. When deployed on a large scale, we significantly reduce the calibration efforts to $3/2N$ of that in previous works. Therefore, triangular antenna layout can facilitate deployability of CSI indoor localization systems.

Since phase distortion can be calibrated in Fig. 5.22c and the layout construction has been realized in Fig. 5.28, we can localize targets based on ADoA. It is because that the localized target can be regarded as a movable AP. The advantage of ADoA localization is that we do not need related information about antenna array direction as illustrated in Fig. 5.28. Note that it is not necessary to acquire all ADoA for position measurement, our system is capable to localize regular devices that cannot retrieve CSI for AoA measurement such as mobile phones.

Remark Though ADoA has been utilized by previous work [181] for localization, **the innovation of our method is how to derive ADoA without any calibration** instead of how to figure out the overall layout via obtained ADoA. In particular, based on the definition of ADoA, it is obvious that $\Delta\theta_B$ is equal to $(\theta_{AB} - \theta_{AC})$ in Fig. 5.20. However, θ_{AB} and θ_{AC} are not available without manual calibration based on existing methods. Since it is impractical to manually calibrate hundreds of APs' positions and antenna array directions in indoor maps for large-scale employment, we must realize self-calibration that is a "chicken and egg" problem as discussed in Sect. 5.2.1.

5.2.4 *Non-Linear Antenna Array Designs*

Except for realizing self-calibrating mechanism, the non-linear antenna layout also outperforms the linear antenna layout during localization scenarios. Though advantages of non-linear antenna layout have been verified via experiments in previous work [183, 184], it is necessary to present a fundamental analysis to evaluate the performance of different antenna layouts. For example, why linear antenna layout usually leads to serious AoA measurement errors? Whether one certain antenna layout is free of these errors? How to choose an appropriate antenna layout for localization?

In this section, we will first reveal the fundamental reason why linear antenna layout usually leads to serious AoA measurement errors and then model the antenna array for localization applications to show how to design an appropriate antenna layout.

5.2.4.1 Limitations of Linear Antenna Array

Multipath effect affects the performance of AoA measurement. Actually, even without multipath effect, we can also not derive accurate AoA based on the linear antenna layout. It is because that AoA measurement of linear antenna layout suffers from another two kinds of problems consisting of symmetry [86] and circularity [179].

Symmetry can be illustrated in Fig. 5.29a, where signals with respect to two different AoAs (θ and $-\theta$) arrive at the antenna array. According to the relation between phase difference and AoA, we have $\varphi_{T1} = -2\pi dcos(\theta)f/c$ that is equal to $\varphi_{T2} = -2\pi dcos(-\theta)f/c$. It is thus difficult to judge which direction the signal comes from. Moreover, we can observe ***circularity*** in Fig. 5.29b. Supposing that there are two signals arriving at the receiver, their AoAs are respectively near to 0° and 180°. We thus have $\varphi_{T1} \approx -2\pi dcos(0°)f/c = -2\pi df/c$ and $\varphi_{T2} \approx -2\pi dcos(180°)f/c = 2\pi df/c$. Since previous works usually set d as half the wavelength, i.e., $d = \lambda/2$, where $\lambda = f/c$ is the wavelength of signal, we have $\varphi_{T1} \approx -\pi$ that approximates to $\varphi_{T2} \approx \pi$. It is therefore hard to derive accurate

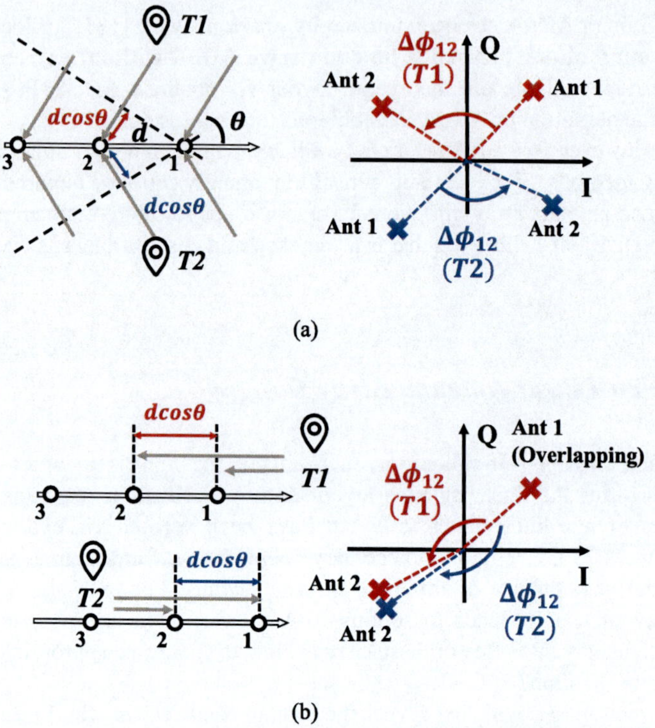

Fig. 5.29 The limitation of linear antenna array. (**a**) Limitation 1: Symmetry. (**b**) Limitation 2: Circularity

AoA near to 0° and 180°. Symmetry and circularity would lead to a serious AoA error that damages localization performance when APs are sparsely placed. Several previous works [86, 182, 192] claim that non-linear antenna layout can address the symmetry issue; however, there is still no systematic evaluation mechanism to show how good one given antenna layout is for AoA measurement. To this end, we model the antenna array for localization applications, based on which we show how to design an appropriate antenna layout.

To model the antenna array for localization, we first review the process in Fig. 5.17 and then find that the core information for AoA measurement is the phase difference. According to Eq. (5.22), the phase difference $\phi_{ij}(\theta)$ depends on two physical features d_{ij} and ϑ_{ij}. For AP equipped with M antennas, if any two antennas can make up the antenna sub-arrays, there are C_M^2 antenna sub-arrays which means C_M^2 different $\phi_{ij}(\theta)$. This section begins with the antenna sub-array, based on which we show how to model the antenna array for localization.

5.2.4.2 Model the Antenna Sub-Array

The basic idea of antenna sub-array modeling is to verify whether the mapping between $\phi(\theta)$ and θ is unique. Mathematically, for two sets Θ and Φ, $\mathcal{F} : \Theta \rightarrow \Phi$ is the mapping from set Θ to set Φ. We want to know whether $\mathcal{F}$ is one-to-one mapping. To this end, for any two AoAs denoted by θ_1 and θ_2, we calculate the distance $D(\theta_1, \theta_2)$ between $\phi(\theta_1)$ and $\phi(\theta_2)$, where $D(\theta_1, \theta_2) = (e^{j\phi(\theta_1)-j\phi(\theta_2)} - 1)^2$. We plot all $D(\theta_1, \theta_2)$ to generate the heat map shown in Fig. 5.30, where a dark area means a short distance and a bright area means a long distance.

Further, we change the values of d and ϑ to observe the impact on the heat map. In particular, (1) when we change ϑ, the heat map moves in the direction of the diagonal $(0°, 0°) \rightarrow (360°, 360°)$; (2) when we change d, the area around the diagonal becomes bright. However, we observe that new dark areas occur.

The heat map can visually represent the performance of localization. We begin with the linear antenna array in Fig. 5.30a. In this figure, we can observe the following three phenomena that are all verified by previous experiments [86, 94, 179]. First, since the value is zero at point $(\theta, 360° - \theta)$, we can never distinguish two symmetric AoAs denoted by θ and $-\theta$. Second, we can observe small values around points $(0°, 180°)$ and $(180°, 0°)$ and can hardly distinguish two different AoAs are respectively near to 0° and 180°. Third, the dark area seems to be wider around 0°

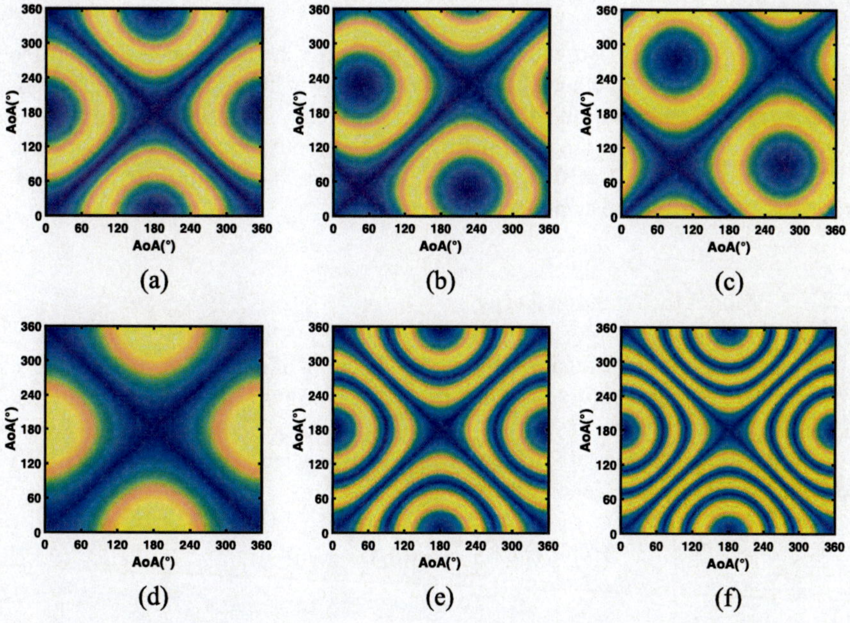

Fig. 5.30 Model of the antenna sub-array. **(a)** $d = 0.5\lambda$, $\vartheta = 0°$. **(b)** $d = 0.5\lambda$, $\vartheta = 45°$. **(c)** $d = 0.5\lambda$, $\vartheta = 90°$. **(d)** $d = 0.25\lambda$, $\vartheta = 0°$. **(e)** $d = 1\lambda$, $\vartheta = 0°$. **(f)** $d = 2\lambda$, $\vartheta = 0°$

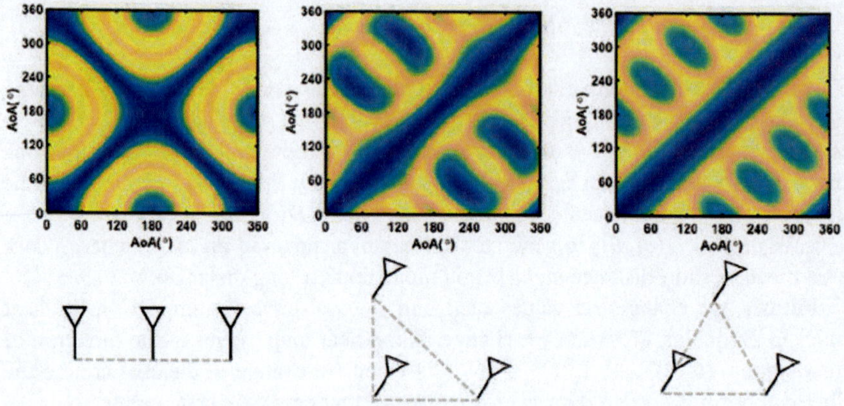

Fig. 5.31 Performance for the three-antenna array

than that around 90°, which means AoA measurement around 0° is more likely to be interrupted by noise.

When we adjust d, the dark area around the diagonal becomes wider (>0.5λ) or narrower (<0.5λ). It means that a longer distance between adjacent antennas leads to a higher accuracy but more ambiguities. The ambiguity means more than one θ is corresponding to the same phase difference $\phi(\theta)$ as shown in the new dark area in Fig. 5.30e, f.

When we adjust φ, we cannot resolve ambiguity because we just move the heat map without any other changes. But when we integrate all antenna sub-arrays to make up a non-linear antenna array, we can eliminate these dark areas (Sect. 5.2.4.3). Note that the diagonal $(0°, 0°) \rightarrow (360°, 360°)$ should be always dark, because $D(\theta_1, \theta_2) \equiv 0$ for $\theta_1 = \theta_2$. In contrast, the diagonal $(0°, 360°) \rightarrow (360°, 0°)$ represents the symmetry problem.

5.2.4.3 Model the Antenna Array

The antenna array can be regarded as the superposition of many antenna sub-arrays. While $\phi(\theta)$ makes up the set Φ for the antenna sub-arrays, $\phi_{ij}(\theta)$ makes up the set $\{\Phi_{ij} | i \neq j\}$ for antenna array.

Similar to the antenna sub-array, we need to verify whether $\mathcal{F} : \Theta \rightarrow \{\Phi_{ij} | i \neq j\}$ is one-to-one mapping. Generally, we have

$$D(\theta_1, \theta_2) = \sum_{i \neq j} (\phi_{ij}(\theta_1) - \phi_{ij}(\theta_2))^2. \tag{5.32}$$

Figure 5.31 shows the heat map for the three-antenna array, where d_{ij} is set as 0.45λ. It is because we can address the circularity problem by setting d_{ij} =

0.45λ, where the value around the point $(0°, 180°)$ becomes larger compared with Fig. 5.30a. Even so, the symmetry problem still exists. Previous works [86, 182] show that non-linear antenna array can resolve the symmetry issue. However, since non-linear antenna layout can be diverse, it is still a problem to select one appropriate layout. It is necessary to apply proposed mechanism to model antenna layout before practical deployment.

We verify that non-linear antenna layout can resolve symmetry and circularity issues by employing two different non-linear antenna arrays in Fig. 5.31. In this figure, we have (1) the antenna layout deployed in a right triangle can address the symmetry issue but incurs new dark areas around $(60°, 240°)$ and $(120°, 300°)$. Moreover, the dark area of the diagonal $(0°, 0°) \rightarrow (360°, 360°)$ is nonuniform, which means AoA measurement around 90° is more likely to be interrupted by noise; (2) in contrast, when the antenna layout is deployed in an equilateral triangle, the dark areas around $(60°, 240°)$ and $(120°, 300°)$ become brighter. We can only observe a small D around the point (θ, θ) and $(0°, 360°)$. In this scenario, when $\{\Phi_{ij} | i \neq j\}$ is slightly interrupted by noise, we can just misjudge θ as another value around θ instead of $-\theta$ (symmetry) or the other value with high error (circularity). Therefore, non-linear antenna array can help to address symmetry and circularity issues at the same time.

Further, we discuss the situation for more antennas in Fig. 5.32. With more antennas installed, the dark area around the diagonal becomes narrower, but the symmetry issue remains unsolved. We can remove these dark areas through placing the antenna into the regular polygon, which, however, leads to a wider dark area around the diagonal $(0°, 0°) \rightarrow (360°, 360°)$. In summary, a linear antenna layout can provide a very high accuracy for AoA measurement around 90° but lead to serious ambiguity. In contrast, when we deploy multiple antennas into the regular polygon, we can derive any AoA with a regular accuracy without ambiguity. When APs are densely deployed, since we can resolve symmetrical AoAs by geometry, it is better to select a linear antenna array. However, as proved in Sect. 5.2.3, when antennas are deployed non-linearly, especially as a regular polygon (Fig. 5.31), the system enables self-calibration and is more suitable for sparse deployment of APs. Since commercial Wi-Fi APs are often equipped with three antennas, we deploy these antennas in an equilateral triangle for localization.

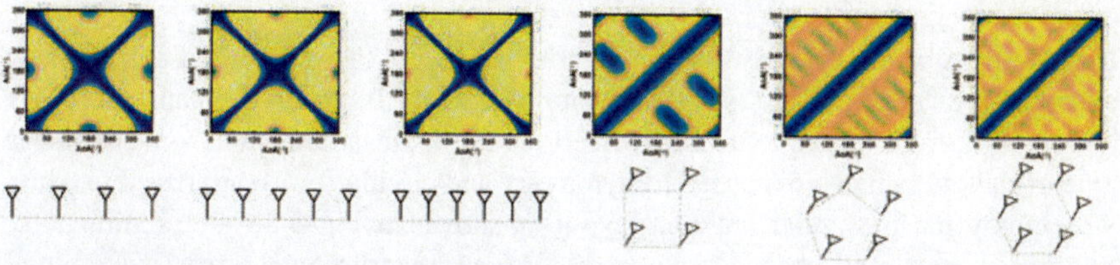

Fig. 5.32 Performance for the multi-antenna array

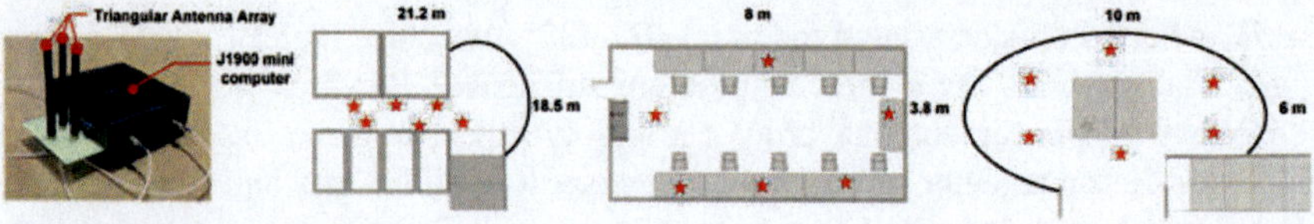

Fig. 5.33 Experimental setup

5.2.5 *Performance Evaluation*

5.2.5.1 Implementation

We implement the AP using Intel 5300 toolkit [97] based on J1900 minicomputer to collect CSI as shown in Fig. 5.33, where 6 clients have been deployed in different scenarios. The AP is equipped with 3 different antennas and can retrieve CSI of 30 subcarriers, which therefore enables 3×30 CSI matrix. The distance between neighboring antennas is 2.5 cm (about $0.45\times$ wavelength).

During experiments, we collect CSI at 10 Hz and process collected data using Matlab 2019a. The experiments are conducted in three different scenarios including corridors, indoor office and open space as previous works [87]. For linear antenna layout, we utilize the method introduced in SpotFi [87] to measure AoAs and localize targets as a comparison.

In this section, we first compare AoA measurement performance of different antenna layouts in Sect. 5.2.5.2, where we calibrate phase offsets via methods introduced in Phaser [94] and then calculate AoA based on calibrated results. Then, for each scenario, we select 5 different layouts to verify that self-calibration can be realized via triangular antenna layout in Sect. 5.2.5.3, where prior knowledge about the AP position, antenna array direction and initial phase distortion is not available (Fig. 5.33 shows one of the five layouts). Note that we avoid multiple APs arranged in a line in the experiment. Finally, we localize targets based on ADoA. During each experiment, if there is no special description, we all collect 200 packets for AoA/ADoA estimation, which costs 20 s.

5.2.5.2 Performance of Different Antenna Layouts

Symmetry and Circularity We conduct experiments to verify that the triangular antenna layout can address symmetry and circularity issues at the same time. As shown in Fig. 5.34a, we can observe serious AoA measurement errors for linear antenna layout, which are caused by symmetry and circularity. To address this issue, we employ the triangular antenna layout as shown in Fig. 5.33 and significantly improve system performance, where 80% AoA measurement error is within 9 degrees.

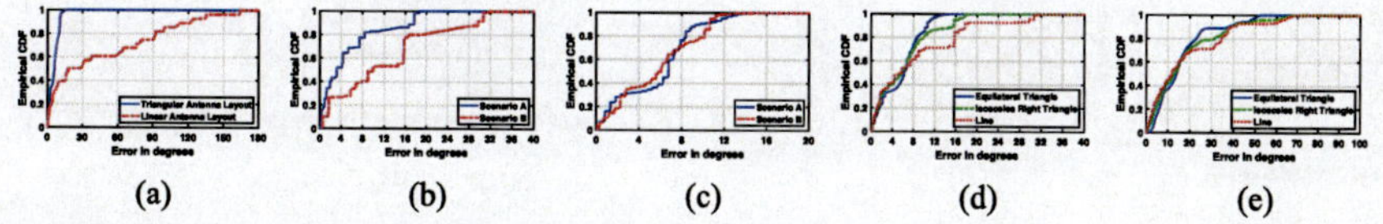

Fig. 5.34 Performance of different antenna layouts. (**a**) Symmetry and Circularity. (**b**) Linear Antenna Layout. (**c**) Triangular Antenna Layout. (**d**) Layout Comparison. (**e**) NLoS Scenarios

AoA Measurement for Any Direction Another advantage of the triangular antenna layout is that we can obtain satisfying accuracy with respect to any direction. To verify this advantage, we design two different scenarios A, B and conduct comprehensive experiments based on linear antenna layout and triangular antenna layout. In particular, we select 7 AoAs ({60°, 70°, 80°, 90°, 100°, 110°, 120°}) in scenario A and the other 7 AoAs ({0°, 10°, 20°, 30°, 40°, 50°, 60°}) in scenario B. Since symmetry and circularity will cause serious errors for linear antenna layout as shown in Fig. 5.34a, they are not considered here for a fair comparison. The results of linear antenna layout are shown in Fig. 5.34b, where we can observe that the performance of scenario B is obviously worse than that of scenario A. In contrast, we can derive similar results with respect to two different scenarios based on triangular antenna layout in Fig. 5.34c, where 80% AoA measurement errors are both within 10 degrees.

Impact of Antenna Layout To compare the performance of different antenna layouts, we select 3 different antenna layouts including line, equilateral triangle and isosceles right triangle and conduct comprehensive experiments for each antenna layout. The results are illustrated in Fig. 5.34d, e, where 80% errors are respectively less than 9°, 10°, and 16°. Therefore, equilateral triangular antenna layout outperforms the other two antenna layouts by relieving serious AoA measurement errors. We further explore the performance in NLoS scenario in Fig. 5.34e, where equilateral triangular antenna layout also outperforms the other two antenna layouts.

5.2.5.3 Construction Performance

Phase Distortion Spectrum Analysis We conduct experiments to examine ADoA measurement accuracy. During our experiments, we respectively calculate ADoA within and without calibration. In particular, we utilize the method introduced in Sect. 5.2.3 to measure ADoA without calibration. Meanwhile, we manually calibrate this phase offset and then calculate AoAs via SpotFi [87]. Based on known AoAs, we can obtain the corresponding ADoAs. The resulted CDFs of three scenarios are illustrated in Fig. 5.35a–c, where 80% errors are respectively less than 15.0°, 12.5°, and 12.0° without calibration. In contrast to results within calibration, where 80% errors are respectively less than 10.5°, 10° and 8°, we can derive accurate

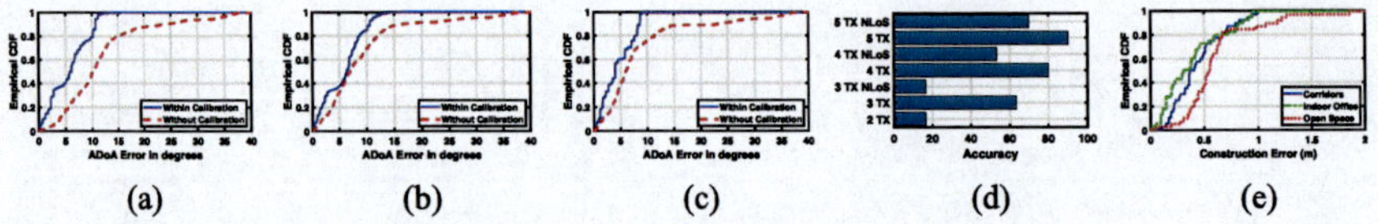

Fig. 5.35 Construction performance of phase distortion spectrum analysis. (**a**) Corridors. (**b**) Indoor Office. (**c**) Open Space. (**d**) Num. of APs. (**e**) Construction Accuracy

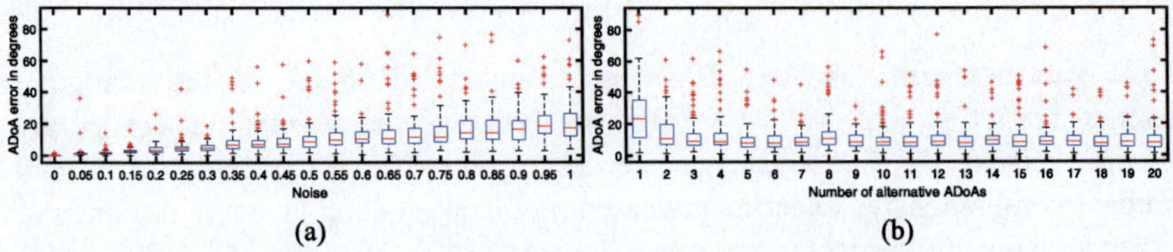

Fig. 5.36 Simulation results of triangulation analysis. (**a**) Noise (Simulation results). (**b**) The number of alternative ADoAs (Simulation results)

ADoAs at most time even without previous calibration. We further explore how the number of APs affects phase distortion spectrum analysis, where $2 \sim 5$ transmitter are chosen for comparison. When we find a correct intersection point and the initial phase distortion error is less than 0.2 radian, we define the result is accurate. For 30 different situations (5 layouts $\times$ 6 APs), the accuracy is 90% for 5 transmitters.

Triangulation Analysis We first simulate the proposed mechanism to choose suitable parameters for triangulation analysis. We conduct 100 simulations for each situation shown in Fig. 5.36. Figure 5.36a shows the relationship between accuracy of ADoA measurement and noise added on the phase difference $\Delta\hat{\phi_{ij}}$. With the increase of noise, the error of ADoA measurement is usually less than 20°. However, we can observe some anomalous points labeled with red markers, which implies that our algorithm has derived a wrong set of ADoA. In fact, we can change the number of alternative ADoAs (i.e., the number of alternative solutions $\Delta\theta_1$, $\Delta\theta_2$, $\Delta\theta_3$ in Algorithm 11) to promote our method. As shown in Fig. 5.36b, when the number of alternative ADoA is set as $5 \sim 9$ rather than 3 in Algorithm 11), we can observe less anomalous points. We further examine the time complexity of proposed methods and the simulation time for Fig. 5.36a, b is 11.24 s and 12.78 s respectively. Since there are 20 different situations and we conduct 100 simulations for each situation, we have calculate $20 \times 100 \times 3 = 6000$ ADoAs, where there are three angles in triangulation analysis. It means that each ADoA measurement costs about 2 ms. We then conduct real experiments to verify proposed methods. The resulted CDFs of three scenarios are illustrated in Fig. 5.37a, where 80% errors are respectively less than 20.0°, 20.5° and 15.5° without calibration.

Construction Accuracy Figure 5.35e shows construction performance of phase distortion spectrum analysis, where we conduct experiments under three different

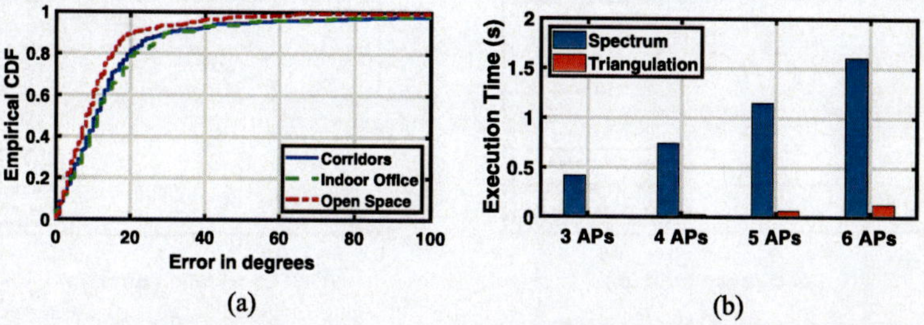

Fig. 5.37 Triangulation analysis and execution time. (**a**) Triangulation analysis. (**b**) Execution time

Table 5.1 Comparison of the proposed methods

Num.	Spectrum	Triangulation
3 APs	Not work	22.2°@80%
4 APs	25.5°@80%	21.3°@80%
5 APs	16.8°@80%	19.8°@80%
6 APs	12.5°@80%	20.0°@80%

scenarios including corridors, indoor office and open space shown in Fig. 5.33, where red stars means the actual APs' positions. Experiments show that we can realize 80% construction errors within 0.7 m, 0.6 m, and 0.68 m under different scenarios. Note that while we can derive more accurate ADoAs in open space, the open space covers a larger area than the other two scenarios and thus observes higher distance errors.

Execution Time We compare the execution time of the proposed methods in Fig. 5.37b. We can observe a linear increase in execution time of the phase distortion spectrum analysis. The execution times of phase distortion estimation are respectively 0.4197 s, 0.7468 s, 1.1525 s and 1.6080 s. In contrast, though it takes less time to perform the triangulation analysis, we can observe rapid growth of execution time as the number of APs increases, where the execution times are respectively 0.0061 s, 0.0252 s, 0.0601 s, and 0.1222 s. It is because that we need to calculate C_M^3 different triangles for the triangulation analysis and each triangle costs about $2\,\text{ms} \times 3 = 6\,\text{ms}$. As the spatial attributes of APs are steady, such execution time is acceptable. Moreover, the comparison of ADoA estimation is shown in Table 5.1. As the triangulation analysis just allows for three APs, there is no improvement as the number of APs increases. As a comparison, we would observe better performance with more APs involved based on the phase distortion spectrum.

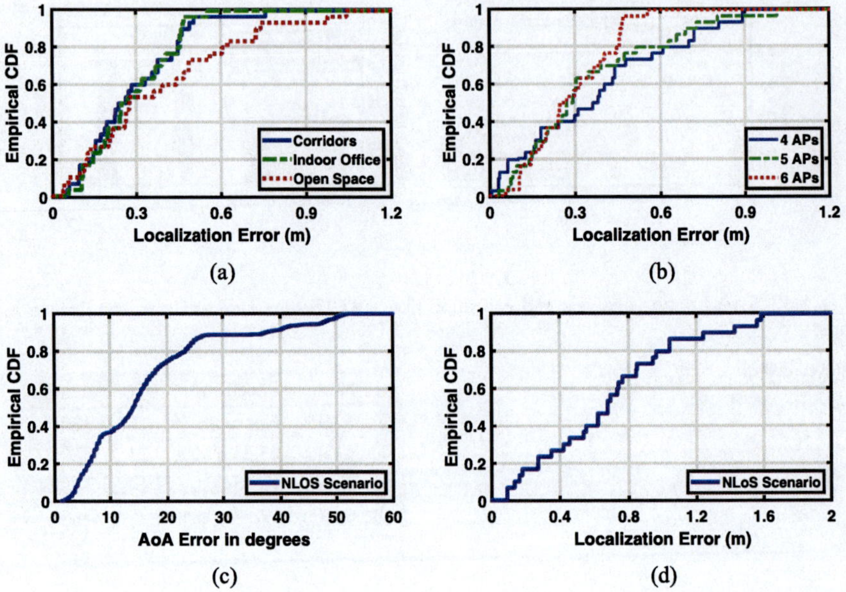

Fig. 5.38 Localization performance. (**a**) Calibration-free Localization. (**b**) Impact of AP Number. (**c**) AoA Measurement (NLoS). (**d**) Localization Performance (NLoS)

5.2.5.4 Localization Performance

Calibration-Free Localization Performance We conduct experiments to verify that triangular antenna layout can localize targets under calibration-free localization scenarios, which has been not realized via linear antenna layout. Figure 5.38a shows localization performance under three different scenarios including corridors, office room and open space, where 80% errors are respectively less than 0.45 m, 0.45 m and 0.60 m. Figure 5.38b shows the relation between localization performance and the number of APs, where we can observe a higher accuracy with more APs employed.

5.2.5.5 NLoS Scenarios

We conduct experiments to examine AoA measurement and localization accuracy of NLoS scenarios and then present the results in Fig. 5.38c, d, where 80% AoA measurement error is less than 24° while 80% localization error is less than 1.0 m.

Table 5.2 System Comparison

System name	80% Accuracy	Calibration-free?
Our system	0.70 m	Yes
SpotFi [87] and Phaser [94]	0.61 m	No
MonoLoco [192]	0.68 m	No
EZ [77]	2.15 m	Yes

5.2.5.6 System Comparison

We compare our system with SpotFi [87], Phaser [94], MonoLoco [192] and EZ [77] in Table 5.2. Among them, (1) EZ is the first work to realize self-calibration based on RSSI. Though this work published 9 years ago, it is still necessary to verify that CSI based self-calibrating systems outperform RSSI-based self-calibrating systems; (2) SpotFi is a classic work to realize decimeter-level localization accuracy and Phaser is the first work to calibrate phase distortion, which are respectively published in 2015 and 2014. We here utilize Phaser for calibration and SpotFi for localization; (3) MonoLoco is a recently proposed work published in 2018. Though MonoLoco can localize targets with just one AP, the phase distortion must be manually calibrated each time the device is rebooted. In this table, we can observe that despite the absence of any explicit pre-deployment calibration, our system yields a similar accuracy to existing works.

5.3 Calibration-Free CSI Fingerprints

Fingerprinting indoor localization systems exploit wireless signal propagation features to estimate the location of wireless devices, where the major challenge in practice is the all-consuming training process: it requires site survey to establish the mapping between the signal feature and the location where the feature is observed. In this paper, we present a Wi-Fi localization scheme based on channel state information (CSI) of wireless signals, which manages to relieve time-consuming site survey. In particular, we first propose how to automatically generate the theoretical fingerprints database based on the signal propagation model and geometric methods. Localization with the theoretical fingerprints database yields accuracy close to existing methods. Second, we improve localization accuracy by parsing the user's trajectory instead of restricting to the single spot, where human movement features introduce more information for localization. Third, we present an automatic update scheme for the theoretical fingerprints database to improve time efficiency for localization, which can save $94 - 98\%$ processing time for utilizing the CSI fingerprints database. We implement a prototype with COTS devices and conduct comprehensive experiments to verify proposed mechanisms. Results show that our design achieves 80% localization errors within 0.3 m, which is 3× accuracy compared with the state-of-the-art design leveraging CSI.

5.3.1 Analysis of Fingerprinting Localization

In the training phase of the fingerprinting localization system, we usually divide the localization service area into cells and collect wireless fingerprints on each cell to build up the fingerprints database. Performance of fingerprinting localization depends on the size of cells and the type of fingerprints.

5.3.1.1 The Size of Cells

If we choose a smaller cell, we expect higher localization accuracy; however, since the fingerprint of each cell needs to be collected manually, this also means more human efforts [270–274]. We summarize the deployment cost of fingerprinting systems in Table 5.3. For N APs deployed in the area of $(X \times Y)m^2$, we respectively choose $(1/P_1 \times 1/P_1)m^2$ and $(1/P_2 \times 1/P_2)m^2$ cells for RSSI and CSI fingerprints collection. For 2D plane, the deployment cost is $A \times B \times P_1^2$ in RSSI fingerprinting localization, where $1/P_1$ is usually $1m$. Similarly, the deployment cost is $A \times B \times P_2^2$ in CSI fingerprinting localization, where $1/P_2$ is usually $10 \sim 30$ cm. For example, when deployed in the area of (30×30) m^2, we respectively collect fingerprints each 1 m and 0.1 m for RSSI and CSI localization. Consequently, we can observe that CSI fingerprinting localization requires $O(10) \sim O(100)\times$ deployment cost compared with RSSI fingerprinting localization. To conclude, it is impractical to select a small cell for accurate CSI fingerprinting localization.

5.3.1.2 The Type of Fingerprints

Fingerprinting localization performs better using wireless fingerprints with the following favored characteristics: (1) the wireless fingerprint is stable at a fixed position; (2) wireless fingerprints at different locations can be easily distinguished.

RSSI and CSI fingerprints mentioned above can be denoted by r in dB and $\gamma e^{j\phi}$, respectively. In practice, we could retrieve CSI amplitueli2de γ and phase ϕ separately. Figure 5.39a shows the stability of the fingerprints, where we select 100

Table 5.3 Deployment of different localization systems

Type	Deployment cost	Examples
RSSI fingerprinting localization	$A \times B \times P_1^2$	900
CSI fingerprinting localization	$A \times B \times P_2^2$	90,000

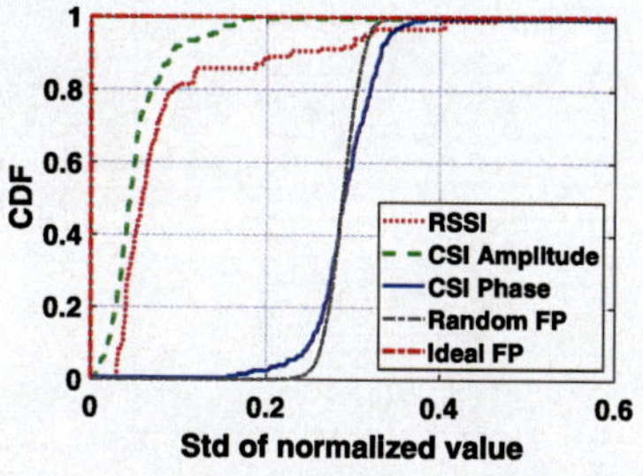

(a)

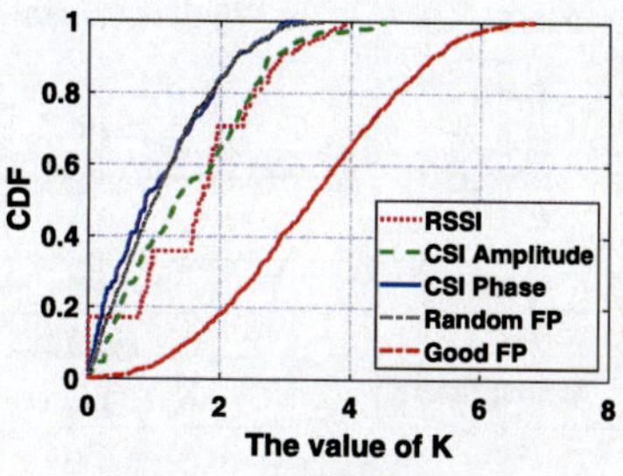

(b)

Fig. 5.39 The characteristics of RSSI and CSI fingerprints. (**a**) Stability of fingerprints. (**b**) Difference of fingerprints

different locations and calculate variance of those fingerprints at each location. The red line shows the ideal fingerprints (FP), where the standard deviation is always 0. The gray line shows the case where we generate random numbers between 0 and 1 as fingerprints, and the standard deviation is about 0.29.

Figure 5.39b shows the difference in fingerprints at different locations, where we select 100 pairs of locations with the inter-distance of one meter. We define K as

$$K = \frac{2 \times |f_{p_1} - f_{p_2}|}{std(f_{p_1}) + std(f_{p_2})}, \tag{5.33}$$

where f_p denotes the fingerprint in position p, $std(f_p)$ means the standard deviation of fingerprint f_p, and $|f_{p_1} - f_{p_2}|$ means the difference between f_{p_1} and f_{p_2}. The red line shows a "good" fingerprint that can provide localization accuracy within one meter, where the difference between fingerprints at different positions is greater than the standard deviation of the fingerprints at a fixed position. However, for RSSI and CSI fingerprints, we can observe that $K < 2$ most of the time, which means that the standard deviation of the fingerprints is in the same order of magnitude as the fingerprints difference at different positions. Thus we can see that it is hard to distinguish fingerprints at different positions.

This motivates us to think about why the CSI amplitude is so stable even at different locations? Why the CSI phase is so unstable even at the same location?

CSI Amplitude The reason for stable CSI amplitude is that the receiver has an automatic gain control (AGC) function [187]. Despite variation of the signal amplitude (RSSI) at the input, AGC can maintain a suitable signal amplitude at its output through dynamically adjusting the gain of the amplifiers. This means that the CSI amplitude is stable when the device moves. Since Wi-Fi devices are primarily for communication rather than localization, it is impractical to remove the AGC function that is helpful for reliable communication. Consequently, it is difficult to distinguish CSI amplitudes at different locations.

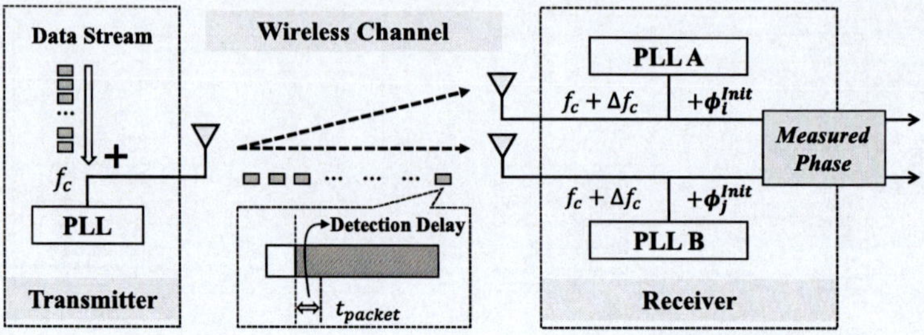

Fig. 5.40 CSI phase offset

CSI Phase Besides the influence of the communication channel, the CSI phase is also affected by the following three factors: carrier frequency offset (CFO) [91, 189, 227], packet detection delay [87, 227] and initial phase distortion [86, 94, 118, 120, 178, 227]. Note that these systems might use different names to define these factors.

For one pair of transceivers, we can simplify the communication process as shown in Fig. 5.40, where the transmitter generates a carrier f_c to transmit packets that will be detected and decoded by the receiver. However, there are small differences Δf_c between the carrier frequencies of the transmitter and receiver. This leads to a time-varying phase offset across each frequency band. Since phase is the integral of frequency over time, the phase offset $\phi_c(t)$ is

$$e^{j\phi_c(t)} = e^{-j\Delta f_c t} = e^{jf_c t} \times e^{-j(f_c+\Delta f_c)t}. \tag{5.34}$$

Moreover, Wi-Fi receivers detect the presence of a packet based on the energy of its first few time samples [227], which incurs the packet detection delay denoted by t_{packet}. For the signal at f_c, this time delay causes the following phase offset

$$e^{j\phi_p(t)} = e^{jf_c t_{packet}}, \tag{5.35}$$

where the packet detection delay $\phi_p(t)$ varies for different packets. Finally, the receiver uses a phase-locked loop (PLL) to generate the center frequency $f_c + \Delta f_c$ from the reference frequency of the crystal. The PLL will produce a random initial phase ϕ_{Init}. When the center frequency is fixed, the PLL will maintain the value of this initial phase. Since different antennas use different PLLs, for any pair of antennas, we can observe an initial phase distortion ϕ_{ij}^{Init} between ith and jth antennas, i.e.,

$$e^{j\phi_{ij}^{Init}} = e^{j(\phi_i^{Init} - \phi_j^{Init})}. \tag{5.36}$$

To conclude, carrier frequency offset and packet detection delay lead to time-varying CSI phase. Consequently, the CSI phase is instable even at a fixed position.

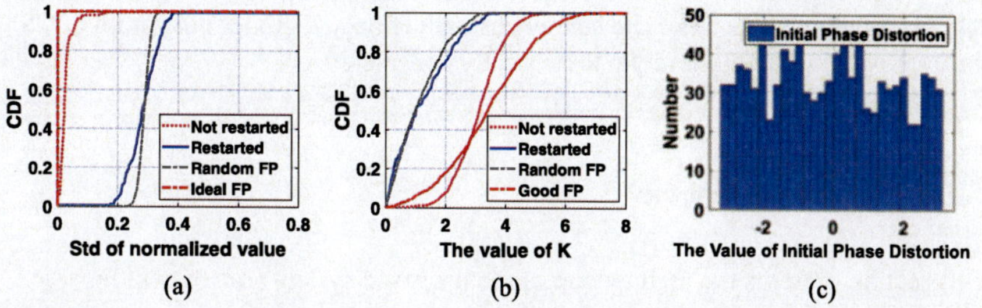

Fig. 5.41 The characteristics of CSI phase difference. **(a)** Stability. **(b)** Difference. **(c)** Phase distortion

Table 5.4 Fingerprints comparison

Fingerprints	$std(f_p)$	K
RSSI	0.117@80%	2.511@80%
CSI Amplitude	0.069@80%	2.476@80%
CSI Phase	0.320@80%	1.891@80%
Phase Difference (not restarted)	0.031@80%	3.730@80%
Phase Difference (restarted)	0.323@80%	2.175@80%

CSI Phase Difference Existing CSI model-based localization methods [86, 87, 94] employ the antenna array to remove these two phase offsets. This is because carrier frequency offset and packet detection delay are exactly the same at different antennas. We explore the characteristics of CSI phase difference and the experimental results show that CSI phase difference fingerprints outperforms RSSI, CSI amplitude and CSI phase fingerprints. Though the CSI phase difference suffers from the initial phase distortion ϕ_{ij}^{Init}, this initial phase distortion is not time-varying. As illustrated in Fig. 5.41a, b, when the device is not restarted, the fingerprint is stable at a fixed position and can be distinguished at different positions.

We compare different kinds of fingerprints in Table 5.4, where we can observe that the CSI phase difference fingerprint yields minimum $std(f_p)$ and maximum K. This means that CSI phase difference might be the most suitable fingerprint for indoor localization. However, once the device is restarted or the center frequency of communication changes, the PLL will generate a new random initial phase ϕ_{ij}^{Init}. This will make the existing fingerprints database no longer valid. It is worth noting that the localization error caused by this phase distortion cannot be ignored. The experimental results in Fig. 5.41c show the distribution of this initial phase distortion every time the device is restarted, where we can observe that this phase distortion can be arbitrary. Moreover, this phase distortion makes the fingerprint unstable even at a fixed position as illustrated in Fig. 5.41a, b.

Summary We can see that it is challenging to design a fingerprinting localization scheme leveraging CSI. The site survey cost reduction and proper utilization of CSI for improving accuracy have yet to be jointly considered.

5.3.2 System Overview

This section presents the architecture of the proposed system as shown in Fig. 5.42, where each component is briefly described.

Theoretical Fingerprints This is for generating a preliminary fingerprints database without site survey. The basic idea of fingerprinting localization as mentioned is to establish the mapping between signal features and physical locations. In our system, with respect to any Wi-Fi access point (AP), we associate the received signal's angle of arrival (AoA) with the CSI phase difference based on the theoretical signal propagation model. Then, if locations of APs are known in advance, we can transform AoAs of different APs into physical locations according to geometry principles. Even without prior knowledge of those APs' locations, we can still utilize the method introduced in [177, 190] to automatically obtain necessary information of APs. Consequently, we can establish the mapping between the inter-antenna phase difference and the physical locations.

Fingerprinting Localization The theoretical fingerprints database is actually not accurate enough. This is because the theoretical fingerprints database mentioned above is derived from the theoretical signal propagation model. Subject to the number and performance of sensors equipped in COTS devices, we might observe similar fingerprints at two locations that are far apart, which means a smaller K in Table 5.4. Instead of restricting to the single spot, we utilize the long short term memory (LSTM) network to realize the context-aware localization in order

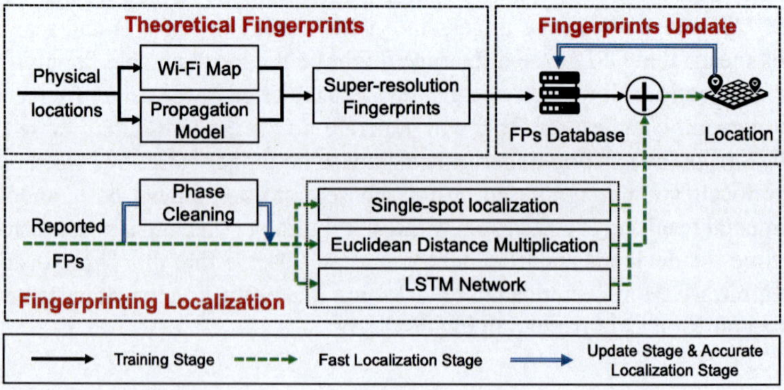

Fig. 5.42 System architecture

to track the user's trajectory over time. This idea stems from the following facts: (1) the continuity of the user's trajectory can help us eliminate the aforementioned ambiguous locations; (2) the wavelength of Wi-Fi signal is usually a few centimeters, which enables the fingerprint sequence to reflect the user's slight movement. Based on the fingerprints database, this paper not only integrates trajectory tracking [88, 91, 98, 127, 189] and localization methods [86, 87, 119, 227], but also takes human movement features into account.

Automatic Fingerprints Update Computing AoA for the theoretical fingerprints database requires time-consuming CSI processing to resolve multipath effect. This is because the theoretical fingerprints cannot take multipath effects into account like collected fingerprints. Benefiting from the context-aware localization method, we can provide accurate location labels for crowdsourcing data. This helps us accomplish the following two tasks: first, we can replace the theoretical fingerprints with the crowdsourcing fingerprints to realize automatic fingerprints update; second, the location can be utilized to label other types of crowdsourcing data such as RSSI, which enables the construction of RSSI fingerprints database.

Dual-Mode Localization We define dual-mode localization in the legend consists of fast and accurate localization stages. As accurate localization relies on time-consuming CSI processing (phase cleaning), we can remove phase cleaning stage to enable real-time localization. These reported fingerprints are temporarily stored. When the system is idle, we perform phase cleaning stage to estimate accurate positions for automatic fingerprints update. Moreover, if it is not necessary to provide real-time locations, we can directly perform accurate localization stage and update fingerprints database at the same time. One might wonder if it is possible to achieve fast and accurate localization. In reality, as the automatic fingerprints update stage runs, the fingerprints database becomes more accurate. Under this situation, we can obtain admirable accuracy even without time-consuming phase cleaning.

5.3.3 Theoretical Fingerprints Generation

Theoretical fingerprints generation is the inverse process of model-based localization. In particular, the model-based localization system first derives the AoA leveraging CSI phase difference and then transforms AoAs into locations according to geometric principles. Our method calculates AoAs with respective to APs at each location and then transform AoAs into CSI phase difference fingerprints. By contrast, the proposed method focuses more on the relationship among fingerprints in the overall localization area. For example, how do the fingerprints change when the user moves to a nearby location? Are there similar fingerprints at two remote locations? These global features will significantly improve localization performance. This section will first provide detailed processes for generating theoretical fingerprints. This brings some issues in practice such as initial phase

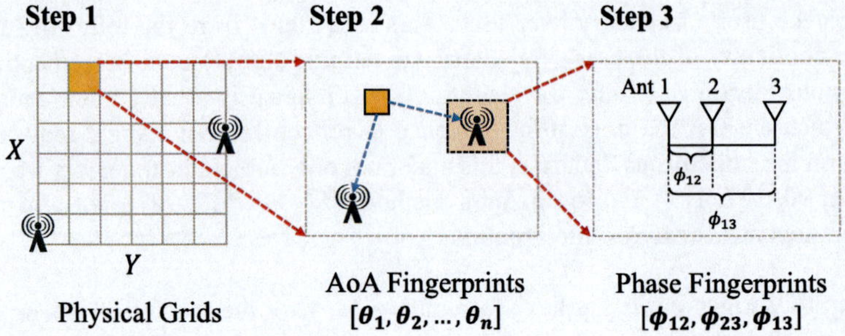

Fig. 5.43 Theoretical fingerprints database

distortion, calibration of AP position, and multipath effects, which are resolved in latter part.

5.3.3.1 Basic Idea and Challenges

We define the $(A \times B)$ m^2 localization area as area of interest (AoI). The theoretical fingerprints database can be built in three steps as illustrated in Fig. 5.43.

1. We first divide the AoI into $X \times Y$ blocks, with each block sized $(a \times b)$ m^2.
2. For each cell, such as the orange one, we can always obtain the corresponding AoA fingerprints $\boldsymbol{\Theta}$ by geometry when APs' positions are known.
3. For each AP equipped with M antennas, we can derive C_M^2 phase differences $\varphi(\theta, d)$ with different inter-antenna distances denoted by d. When N APs are deployed, these $N \times C_M^2$ phase differences make up the phase difference fingerprints $\boldsymbol{\Phi}$.

In the testing phase, the reported fingerprints f_t are compared within the database for estimating the user's location, where t denotes the time slot. Fingerprints matching can be done based on Euclidean distance, kNN and other methods [32, 38, 126, 188]. During theoretical fingerprints generation stage, the following two challenges must be addressed.

Challenge 1 preparation for fingerprints database generation. As analyzed in Sect. 5.3.1, the initial phase distortion exists for any two antennas, which makes the detected phase difference varies from $\varphi(\theta, d)$ as derived from the theoretical model. Such initial phase distortion must be calibrated. Moreover, the premise for computing the theoretical fingerprint is that we have prior knowledge of APs' positions. In fact, the antenna array direction of the AP is also required. For model-based localization systems, manual site survey is inevitable. A natural question to ask is: is it possible to implement a CSI fingerprinting localization system without site survey?

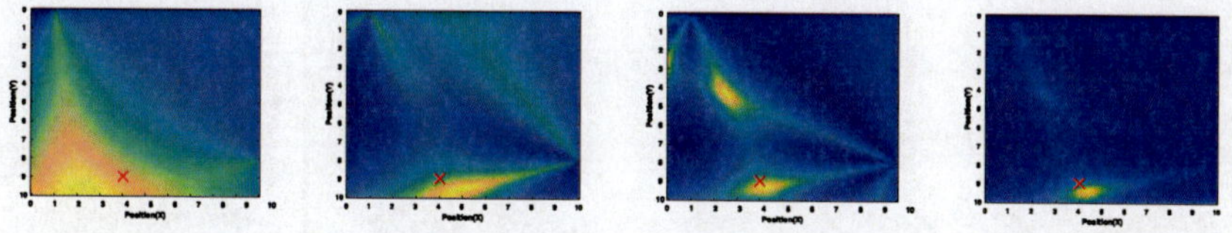

Fig. 5.44 Hybrid inter-antenna distance. (a) $d_1 = d_2 = 0.5\lambda$. (b) $d_1 = d_2 = 1.0\lambda$. (c) $d_1 = d_2 = 1.5\lambda$. (d) $d_1 = 0.5\lambda$ and $d_2 = 1.5\lambda$

Table 5.5 Calibration in CSI localization systems

System	[94]	[190]	[177]	[14]
Position Calibration	No	Yes	Yes	Yes
Direction Calibration	No	Yes	Yes	Yes
Phase distortion Calibration	Yes	No	Yes	No
Linear antenna array	Yes	Yes	No	Yes

Challenge 2 resolution and ambiguous locations. When we transform AoA into phase difference under various inter-antenna distance, there is a trade-off between resolution and the number of candidate locations. For example, in Fig. 5.44, we demonstrate values of Euclidean distance between f_t and the theoretical fingerprints $\boldsymbol{\Phi}$ at all positions, where two APs are placed at $(1, 0)$ and $(10, 8)$ respectively. The reported fingerprint is collected at position $(4, 9)$ labeled with the red mark cross. The bright yellow area means smaller Euclidean distance, which indicates possible spots where the device could be reporting the fingerprint. If we choose an inter-antenna distance 0.5λ as in Fig. 5.44a, though we do not observe ambiguous locations, the bright yellow area is so large that we cannot accurately estimate the device's position. When we increase the inter-antenna distance, we tend to observe smaller yellow areas at more candidate locations in Fig. 5.44b, c. By fusing fingerprints under different inter-antenna distances, we are able to eliminate both ambiguous locations and inaccurate measurements as shown in Fig. 5.44c, a respectively. We will reveal the potential of the hybrid inter-antenna distance for better performance through a fundamental analysis and then promote the localization system.

5.3.3.2 Mapping AoA into Phase Difference

We list existing self-calibrating approaches for localization systems in Table 5.5. When preparing for theoretical fingerprint generation, we can utilize these methods to calibrate AP or phase distortion. Note that our method can be applied to both linear [14, 94, 190] and non-linear [177] antenna layouts, because they both follow the theoretical signal model.

After the preparation process, we are able to obtain the AoA fingerprints $\boldsymbol{\Theta}$ by geometry. To transform $\boldsymbol{\Theta}$ into phase difference fingerprints that can be directly

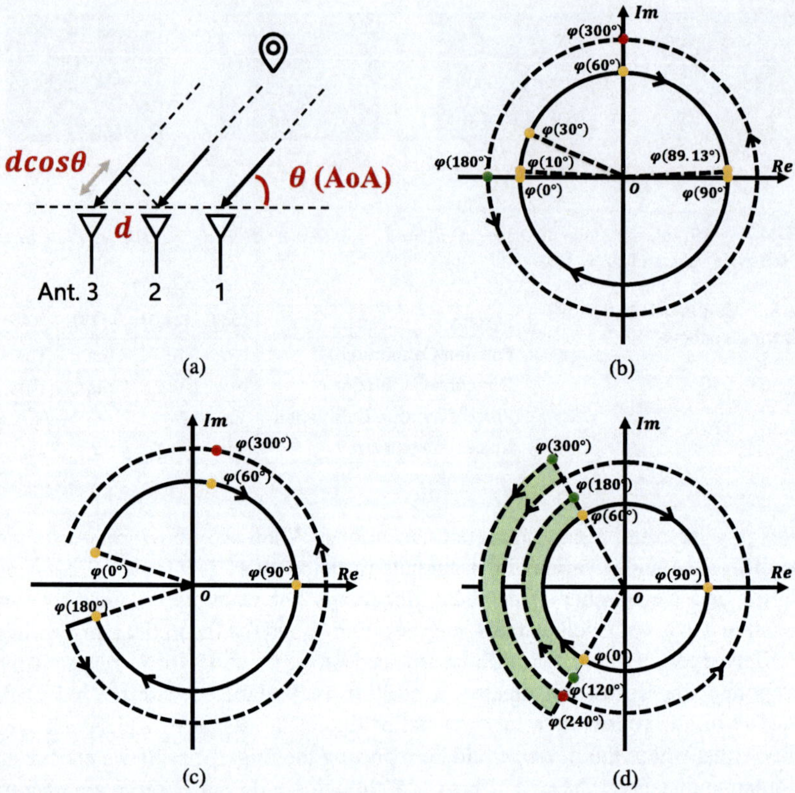

Fig. 5.45 AoA-fingerprints constellation diagram. **(a)** AoA localization. **(b)** $d = 0.5\lambda$. **(c)** $d = 0.45\lambda$. **(d)** $d = 2\lambda/3$

reported by devices, we leverage signal propagation model in Fig. 5.45a. When signals arrive at the antenna array of the receiver, since the inter-antenna distance d is much smaller than the distance between the transceivers, we can regard those signals observed by different antennas as a series of parallel lines. Thus we could observe an extra flight distance denoted by $dcos(\theta)$ between neighboring antennas as shown in the figure. This extra flight distance leads to an phase difference denoted by $\varphi(\theta, d)$. We have

$$\varphi(\theta, d) = 2\pi dcos(\theta)/\lambda, \tag{5.37}$$

where $\lambda = c/f$ is the wavelength of the electromagnetic wave, c is the speed of light and f is the frequency of the signal.

For a device with M antennas, there are at most C_M^2 inter-antenna distance d. We can use the constellation diagram to visualize the relationship between $\varphi(\theta, d)$ and θ. In particular, we can represent the phase as a complex number $e^{j\varphi(\theta,d)} =$

$cos(\varphi(\theta, d)) + jsin(\varphi(\theta, d))$. Then, we adjust the value of d and plot $e^{j\varphi(\theta,d)}$ on the complex plane to obtain Fig. 5.45b, c, d. Since we have chosen a fixed d in each figure, we can simplify $\varphi(\theta, d)$ to $\varphi(\theta)$ that means the inter-antenna phase difference caused by θ. In Fig. 5.45b, we can observe the following phenomenon: If $\theta = 0°$, the starting point is $\varphi(0°)$. As θ increases, $\varphi(\theta)$ rotates clockwise along the inner circle (the solid line); if $\theta > 180°$, it begins to rotate counterclockwise along the outer circle (the dot line). Note that the two circles in the figure are just for better understanding, they will completely coincide in the real scene.

Since the inter-antenna distance $d = 0.5\lambda$ is applied in most localization systems, we explore properties of the constellation diagram based on Fig. 5.45b to reveal the fundamental limitations of the existing setting.

1. When θ changes from 0 to 60 or from 60 to 90, the rotation angles of $\varphi(\theta)$ are both 90°. This means that corresponding points on the constellation diagram are unevenly distributed. Different AoAs have different tolerance to noise, we term this property as the **non-uniformity** of the phase difference. For example, when actual AoA is 0° and phase noise is $n = 0.01519\pi$, we may misjudge the AoA as 10 degrees ($\varphi(10°) = 0.98481\pi$). But when actual AoA is 90°, with the same noise, we just misjudge the AoA to 89.13 degrees ($\varphi(89.13°) = 0.01519\pi$). That is, AoAs around 0° is easier to be interrupted by the noise than those around 90°.
2. Since we can only derive the phase difference $\varphi(\theta) \in [0°, 360°]$, $\varphi(180°) = -\pi$ and $\varphi(0°) = \pi$ are located in the same point and cannot be distinguished in Fig. 5.45b. We term the property as **circularity** of phase difference.
3. Moreover, we have $\varphi(\theta) \equiv \varphi(360° - \theta)$, where we cannot distinguish **symmetrical** AoAs denoted by θ and $-\theta$.

As these properties will be affected by d, we explore the influence to show the advantage of different inter-antenna distances. First, Fig. 5.45c shows the situation where $d < 0.5\lambda$. For $\varphi(\theta, 0.45\lambda) = 0.9\pi cos\theta$, there is an unclosed ring as $\varphi(0°, 0.45\lambda) = 0.9\pi \neq \varphi(180°, 0.45\lambda) = -0.9\pi$, which eliminates the circularity issue. However, the distance between the neighboring points decreases in the figure, which implies that it is easier for the measured AoA to be impacted by the noise. Second, Fig. 5.45d shows the situation where $d > 0.5\lambda$. For $\varphi(\theta, 2/3\lambda) = (4/3)\pi cos\theta$, we can observe overlapping areas when $\theta \in [120°, 180°)$ and $\theta \in [0°, 60°)$ due to the circularity issue. Moreover, the distance between the neighboring points (e.g., $\varphi(60°)$ and $\varphi(90°)$) increases compared with Fig. 5.45c, which implies that it is harder for the measured AoA to be impacted by the noise. Consequently, larger inter-antenna distance causes more circularity issues, which mean more highly similar candidate locations, but improves anti-noise performance at each candidate location.

As the inter-antenna distance affects the relationship between the phase difference and AoA, then how does the distance affect the localization process? We define two types of errors denoted by E_1 and E_2 here. E_1 means the localization error caused by environmental noise while E_2 means the localization error caused by the circularity and symmetry. When d is small, we tend to observe a large E_1 due to

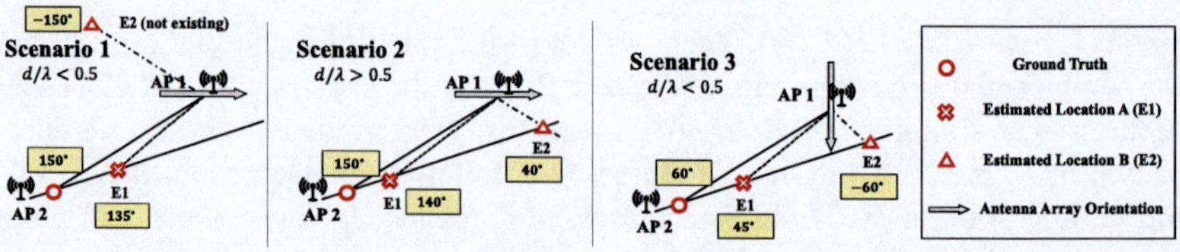

Fig. 5.46 The relationship between inter-antenna distance and localization. A small d relives E2 but causes serious E1

low resolution. When d is large, we tend to observe a large E_2 due to false candidate positions. We take the three scenarios as illustrated in Fig. 5.46 as examples.

1. In the first scenario, the inter-antenna distance d is less than 0.5λ, where the accurate AoA is 150°. We calculate a false $\theta_{E_1} = 135°$ caused by noise and a symmetrical $\theta_{E_2} = -150°$. Because E2 does not intersect with AoA of AP2, we just observe a localization error E1.
2. In the second scenario, if we change the inter-antenna distance to make $d > 0.5\lambda$, $\theta_{E_1} = 140°$ is closer to 150° due to the anti-noise performance. However, a larger inter-antenna distance makes circularity issues more serious. For example, $\theta = 150°$ and $\theta_{E_2} = 40°$ might be corresponding to the same phase difference. Consequently, we observe $E2 >> E1$ when we choose a wrong candidate position.
3. Note that when d is less than 0.5λ, we may also observe multiply candidate locations in the third scenario. It is because that when we change the antenna array direction, we cannot distinguish the symmetrical AoAs and therefore estimate two symmetrical positions simultaneously.

Figure 5.46 intuitively analyzes the relationship between d and localization. In the following section, we will clean the phase difference for super-resolution fingerprints generation.

5.3.3.3 Generate Super-Resolution Fingerprints

Mathematically, we define the AoA space $\boldsymbol{\Theta}$ as

$$\boldsymbol{\Theta} = \{\Theta_{x,y} | 1 < x < X, 1 < y < Y\}, \tag{5.38}$$

where $\Theta_{x,y} = [\theta_1, \theta_2, \cdots, \theta_N]$ denotes the AoA fingerprints with respect to N APs at location (x, y). Obviously, $\boldsymbol{\Theta}$ is an $X \times Y \times N$ matrix. Moreover, for transforming θ into C_M^2 different ϕ_{ij}, the obtained phase difference fingerprint $\boldsymbol{\Phi}$ will be an $X \times Y \times N \times C_M^2$ matrix. In contrast to RSSI fingerprints, where each AP reports just one RSSI measurement at each location, it takes more time to match high-dimensional CSI fingerprint with fingerprints database.

Phase difference cleaning this part proposes the fingerprint compression method that can resolve the multipath effect. In particular, after calibrating the initial phase distortion, we can employ the MUltiple SIgnal Classfication (MUSIC) [87, 185] to process the phase difference and then derive the AoA corresponding to the direct path. Finally, we transform the resulting AoA $\hat{\theta}$ into $[\hat{\phi}_{min}, \hat{\phi}_{max}]$ based on

$$\begin{cases} \hat{\phi}_{min} = \phi(\hat{\theta}, d_{min}) = 2\pi d_{min} cos(\hat{\theta})/\lambda, \\ \hat{\phi}_{max} = \phi(\hat{\theta}, d_{max}) = 2\pi d_{max} cos(\hat{\theta})/\lambda, \end{cases} \tag{5.39}$$

where d_{min} and d_{max} respectively denote the minimum and the maximum inter-antenna distances. We have

$$[c_1, c_2, \cdots, c_M] \xrightarrow{MUSIC} \hat{\theta} \xrightarrow{\phi(\theta,d)} [\hat{\phi}_{min}, \hat{\phi}_{max}], \tag{5.40}$$

where c_m denotes the CSI observed at mth antenna.

The fundamental reason for the above process is that MUSIC algorithm is a classic algorithm to resolve multipath signals with the antenna array. MUSIC can make full use of multiple antennas to derive accurate AoA results. The existing model-based localization system utilizes this AoA to localize the target directly. Instead, we convert this AoA into two phase differences $[\hat{\phi}_{min}, \hat{\phi}_{max}]$ based on the minimum and maximum inter-antenna distances. The phase difference fingerprints are better than AoA fingerprints for the following reason: when deploying large-scale localization systems, we cannot always guarantee that APs are always on the edge of AoI. This makes symmetric AoAs need to be discussed case by case. For example, due to the symmetry and circularity issues, a phase difference fingerprint is corresponding to at least two AoAs denoted by θ and $-\theta$. When N APs are deployed, there are 2^N different cases that need to be discussed. If we utilize phase difference fingerprints, a set of phase difference fingerprints can cover all the above cases.

Super-resolution fingerprints the phase difference fingerprint $\mathbf{\Phi}$ is a $X \times Y \times N \times 2$ matrix, i.e.,

$$\mathbf{\Phi} = \{\Phi_{x,y} | 1 < x < X, 1 < y < Y\}, \tag{5.41}$$

where $\Phi_{x,y} = [\boldsymbol{\phi}_1, \boldsymbol{\phi}_2, \cdots, \boldsymbol{\phi}_N]^T$ denotes the phase difference fingerprints with respect to N APs at location (x, y), $\boldsymbol{\phi}_n = [\hat{\phi}_{min}, \hat{\phi}_{max}]$ for the nth AP.

The proposed method outperforms model-based localization systems due to the following reasons.

(1) We can use $\hat{\phi}_{min}$ to remove as many ambiguous positions as possible to decrease $E2$ (candidate locations). Moreover, we do not have to worry about $E1$ caused by this small d, because the MUSIC algorithm has removed noise and obtained accurate AoAs through taking hybrid inter-antenna distance into consideration.

Algorithm 12 Theoretical fingerprint generation

Require: APs' position and antenna array direction [94]
Ensure: Theoretical fingerprints database $\boldsymbol{\Phi}$
  Construct Wi-Fi map and calibrate initial phase distortion;
  Divide the AoI into $X \times Y$ cells;
  Generate the AoA space $\boldsymbol{\Theta}$;
  for each θ in $\boldsymbol{\Theta}$ **do**
    Calculate $[\phi_{min}, \phi_{max}]^T$ as theoretical fingerprint $\boldsymbol{\phi}_n$;
  end for
  Generate theoretical fingerprints database $\boldsymbol{\Phi}$ based on $\boldsymbol{\phi}_n$.

(2) We can use $\hat{\phi}_{max}$ to magnify the effect of user movement on fingerprints, which will be deduced in the Eq. (5.42). In particular, as the device moves slightly, the AoA changes from θ to $(\theta + \Delta\theta)$. The variation of AoA can be reflected in phase difference, and we have

$$\begin{aligned}\Delta\phi &= \phi(\theta + \Delta\theta, d) - \phi(\theta, d)\\ &= -\frac{4\pi d}{\lambda} \times sin\left(\frac{2\theta + \Delta\theta}{2}\right) \times sin\left(\frac{\Delta\theta}{2}\right)\\ &\approx -\frac{4\pi d}{\lambda} \times sin(\theta) \times \frac{\Delta\theta}{2}, \text{ when } \Delta\theta \to 0.\end{aligned} \tag{5.42}$$

The person's movement in a short time will lead to a small $\Delta\theta$, so we can use $\Delta\theta/2$ to approximate $sin(\Delta\theta/2)$. Since $\Delta\phi \propto d$ for fixed θ and $\Delta\theta$, we can utilize the maximum inter-antenna distance d_{max} to enhance the effect of motion on fingerprints. After we magnify the impact of user movement on fingerprints, it is easier to utilize context-aware methods to track user movement. Algorithm 12 shows the method to generate the theoretical fingerprints database $\boldsymbol{\Phi}$.

5.3.4 *Fingerprinting Localization*

This section improves the accuracy based on the theoretical fingerprints database. The crux is to resolve ambiguous locations. In particular, we first present how to perform single-spot localization, where ambiguous locations have been removed with CSI amplitude. Furthermore, we realize the context-aware localization in order to track the user's trajectory over time for higher accuracy. This section explores two context-aware methods including the Euclidean distance multiplication and LSTM network. Euclidean distance multiplication is presented in our preliminary version of this article [186]. This version proposes to further promote the system by leveraging the LSTM network to involve user motion features such as walking speed.

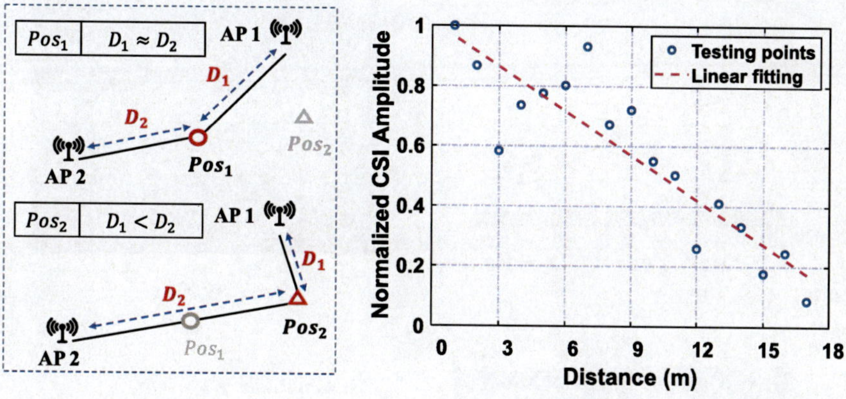

(a) (b)

Fig. 5.47 Remove ambiguous locations with CSI amplitude. **(a)** The basic principle. **(b)** The relationship between CSI amplitude and distance

5.3.4.1 Single-Spot Localization

While the hybrid inter-antenna distance can remove many ambiguous positions, the issue still exists if APs are sparsely deployed; we need to select the most accurate estimation among multiple highly similar candidate locations.

As illustrated in Fig. 5.47a (the third scenario in Fig. 5.46), even if we let $d < 0.5\lambda$, it is still possible to observe ambiguous positions Pos_1 and Pos_2. The location will affect distance between AP1 and AP2. For example, $D_1 \approx D_2$ in Pos_1 and $D_1 < D_2$ in Pos_2. To this end, we utilize the CSI amplitude to assist CSI phase difference. This is based on the observation that although the CSI amplitude will be adjusted by the AGC, it can still reflect the distance to some extent. Figure 5.47b shows the relationship between distance and CSI amplitude, which verifies the above assumption. Before finding the best matching fingerprint through traditional methods, we filter out ambiguous positions based on the following method.

(1) We filter out most fingerprints based on phase difference fingerprints and derive a series location $P = \{(x, y)||\hat{\Phi} - \Phi_{x,y}| < T_e\}$, where T_e denotes threshold.
(2) For each position in P, we can obtain the physical distances with respect to different APs, denoted by $\mathbf{D}_{x,y} = [D_1, D_2, \ldots, D_N]$.
(3) Based on the mapping relationship between the distance and CSI amplitude in Fig. 5.47b, we can translate distance $\mathbf{D}_{x,y} = [D_1, D_2, \ldots, D_N]$ into CSI amplitude fingerprints $\boldsymbol{\gamma}_{x,y} = [\gamma_1, \gamma_2, \ldots, \gamma_N]$.
(4) We normalize the value of $\boldsymbol{\gamma}_{x,y}$ and calculate difference between the normalized vectors and the measured CSI amplitude fingerprints $\boldsymbol{\gamma}$. By finding the 20% positions in set P with the smallest difference, we can obtain the subset $P_{sub} \subset P$. This step will help us filter out ambiguous positions.

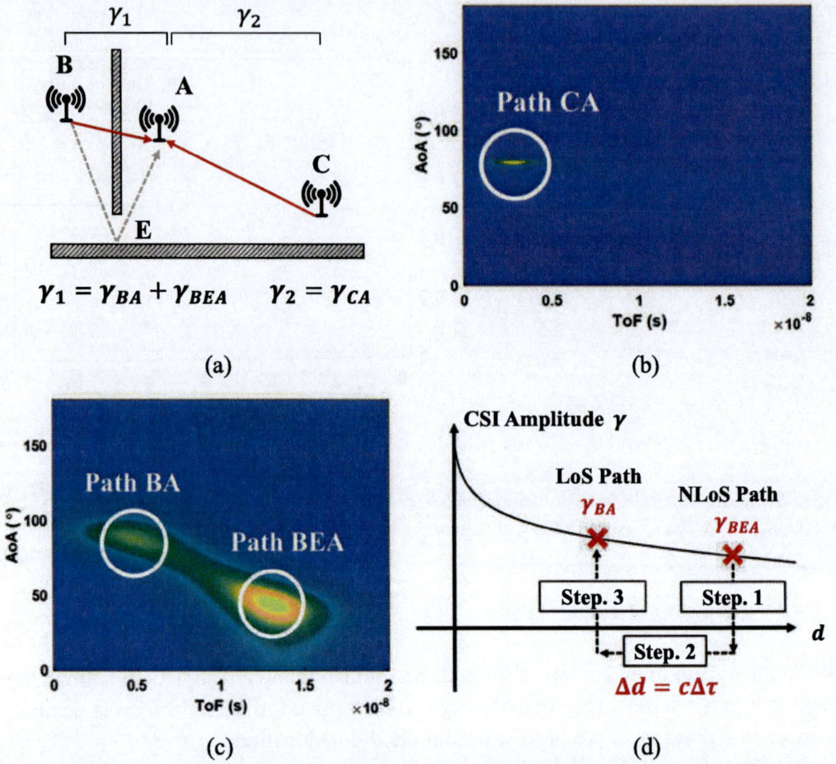

Fig. 5.48 Rescale CSI amplitude for NLoS scenario. (**a**) Scenario. (**b**) Path CA. (**c**) Path BA and BEA. (**d**) CSI amplitude rescaling

(5) For P_{sub}, we can derive the accurate position $p_t = (\hat{x}, \hat{y}) = \arg\min_{(x,y)\in P_{sub}}(|\hat{\Phi}-\Phi_{x,y}|)$ based on the minimum Euclidean distance.

The crux of this method is to identify non-line-of-sight (NLoS) scenario, which leads to a low CSI amplitude with respect to the direct path. Figure 5.48a shows that A receives signals from B and C, whose CSI amplitudes are respectively denoted by γ_1 and γ_2. As there is an obstacle between A and B, the CSI amplitude γ_1 is dominated by the NLoS signal.

MUSIC spectrum can be used to identify NLoS scenarios but cannot be used to rescale the CSI amplitude. In particular, we evaluate the MUSIC spectrum $P_{music}(\theta, \tau)$ based on θ and time of flight (ToF) τ [87, 208] in Fig. 5.48b, c. $P_{music}(\theta, \tau)$ is a two-dimensional matrix and its value in (θ, τ) represents the energy for the signal with θ and τ. In NLoS scenario, the direct path with the smallest τ_{direct} may not be corresponding to the maximum energy. Figure 5.48d shows the NLoS path $B \rightarrow E \rightarrow A$ dominates the energy of the signal, i.e., $\gamma_1 = \gamma_{BEA} + \gamma_{BA} \approx \gamma_{BEA}$. Since A-E-B is a reflected path and is longer than

Algorithm 13 Single-spot localization

Require: Channel state information $\boldsymbol{C} = [\boldsymbol{C}_1, \boldsymbol{C}_2, \cdots, \boldsymbol{C}_N]$
Ensure: Position $p_t = (x, y)$
for each $\boldsymbol{C}_n$ in $\boldsymbol{C}$ **do**
Utilize MUISC to derive direct path AoA $\hat{\theta}$;
Recognize NLoS scenario and rescale CSI amplitude γ;
Calculate $[\hat{\phi}_{min}, \hat{\phi}_{max}]^T$ as fingerprint $\hat{\boldsymbol{\phi}}_n$;
end for
All $\hat{\boldsymbol{\phi}}_n$ makes up phase difference fingerprint $\hat{\Phi}$;
Calculate a series location $P = \{(x, y)||\hat{\Phi} - \Phi_{x,y}| < T_e\}$;
Derive $P_{sub} \subset P$ based on CSI amplitude;
$p_t = (\hat{x}, \hat{y}) = \arg\min_{(x,y)\in P_{sub}}(|\hat{\Phi} - \Phi_{x,y}|)$.

the direct path A-C, it is obvious that $\gamma_1 \approx \gamma_{BEA} < \gamma_{AC} \approx \gamma_2$. Intuitively, the following equation can roughly describe the CSI amplitude of path A-B, i.e.,

$$\frac{\gamma_{BA}}{P_{music}(\theta_{BA}, \tau_{BA})} = \frac{\gamma_{BEA}}{P_{music}(\theta_{BEA}, \tau_{BEA})}. \tag{5.43}$$

However, it is useless because this path is blocked and cannot reflect the exact distance between A and B. In other words, as $\gamma_{BA} < \gamma_{BEA} < \gamma_{AC}$ based on Eq. (5.43), the direct path A-B will be considered to be longer than A-E-B and A-C.

To make γ_{BA} feasible to reflect D_{BA}, we should imagine such a scenario: assuming that the obstacle is removed, what the CSI amplitude should look like? To this end, our rescaling method is illustrated in Fig. 5.48d, which is based on the observation that although the MUSIC cannot provide accurate ToF measurement [87, 118], it provides an accurate time difference of flight (TDoF) denoted by $\Delta\tau = \tau_{BEA} - \tau_{BA}$ [192]. So we can first estimate the distance of the NLoS path, and then deduce the direct path D_{BA} via $c \times \Delta\tau$. Finally, we map the distance of direct path into rescaled CSI amplitude. Mathematically, we have

$$\gamma_1' = \gamma_{BA} = \mathcal{F}(\mathcal{F}^{-1}(\gamma_{BA}) - c\Delta\tau), \tag{5.44}$$

where $\mathcal{F}$ is the mapping function that transforms distance into CSI amplitude, c is the speed of the electromagnetic wave. Note that we just use CSI amplitude to filter out the ambiguous positions rather than localize targets, so coarse-grained translation with Fig. 5.47b is feasible. After coarse-grained filtering, we will use step 5 to derive fine-grained locations. The single-spot localization is summarized in Algorithm 13.

5.3.4.2 Euclidean Distance Multiplication

For reported fingerprint f_t at time t, the raw data can be denoted by $\boldsymbol{C} = [\boldsymbol{C}_1, \boldsymbol{C}_2, \cdots, \boldsymbol{C}_N]$, where $\boldsymbol{C}_i$ denotes the CSI at ith AP. As shown in Eq. (5.40), the MUSIC algorithm transforms the matrix $\boldsymbol{C}_n$ into estimated AoA denoted by $\hat{\theta}$, and we finally derive $\hat{\boldsymbol{\phi}}_n = [\hat{\phi}_{min}, \hat{\phi}_{max}]^T$ based on Eq. (5.39). These phase differences relative to different receivers make up the reported fingerprints

$$\hat{\Phi}_t = [\hat{\boldsymbol{\phi}}_1, \hat{\boldsymbol{\phi}}_2, \cdots, \hat{\boldsymbol{\phi}}_N]. \tag{5.45}$$

Algorithm 13 shows how to perform single-spot localization. The basic idea is to find the fingerprint that best matches $\hat{\Phi}_t$ from the fingerprints database $\boldsymbol{\Phi}$. Figure 5.44 shows the Euclidean distance between $\hat{\Phi}_t$ and the theoretical fingerprint $\Phi_{x,y}$ at each position. In the figure, the brightest point means the best matching fingerprint. We define the Euclidean distance matrix as $\Delta\boldsymbol{\Phi}$ and have

$$\Delta\boldsymbol{\Phi}_t = \{\Phi_{x,y} - \hat{\Phi}_t || 1 < x < X, 1 < y < Y\}. \tag{5.46}$$

As verified in SpotFi [87], the AoA of the direct path cannot always be found, because the number of antennas and communication bandwidth are both limited for commercial Wi-Fi devices. Moreover, because the AoA model is more reliable than the attenuation model, the CSI amplitude fingerprint can only be used to filter out the ambiguous positions. In fact, we can correct the AoA through considering a series of locations at the same time. For a series of reported fingerprints $S = [\hat{\Phi}_1, \hat{\Phi}_2, \ldots, \hat{\Phi}_t]$, one intuitive way is to multiply all Euclidean distance matrices together. We have $\Delta\boldsymbol{\Phi} = \prod_{t=1}^{T} \Delta\boldsymbol{\Phi}_t$.

Figure 5.49 shows several samples of $\Delta\boldsymbol{\Phi}_t$ and the final result $\Delta\boldsymbol{\Phi}$, where the sampling interval is $0.1s$. The bright yellow area means smaller Euclidean distance, which indicates possible spots where the device could be reporting the fingerprint. Figure 5.49b–d show the Euclidean distance matrix $\Delta\boldsymbol{\Phi}_t$ when $t = 0.8, 1.0, 0.1$ and Fig. 5.49e shows $\Delta\boldsymbol{\Phi} = \prod_{t=0s}^{1.0s} \Delta\boldsymbol{\Phi}_t$.. We filter Fig. 5.49c based on the result in Fig. 5.49e and finally obtain Fig. 5.49f, where the ambiguous location is removed.

In particular, in Fig. 5.49c, d, we can observe that the inaccurate AoA leads to a wrong position. When we multiply all Euclidean distance matrices together, we can obtain a matrix shown in Fig. 5.49e, where the wrong position can be removed through finding the brightest point. Then, we can utilize CSI amplitude to filter out the ambiguous positions and derive a reference position via

$$p = (\hat{x}, \hat{y}) = \arg \min_{(x,y) \in P_{sub}} |\Delta\boldsymbol{\Phi}|. \tag{5.47}$$

Considering the user's movement, we utilize p to revise $\Delta\boldsymbol{\Phi}_t$ in previous time slot. In particular, we consider the maximum human speed is 8 m/s and the window size is T_w. In this way, we can directly remove those locations $8T_w$ meter away from the position p. Figure 5.49f shows the heat map after the above filter, where the bad locations have been removed.

The basic idea of the above method is to first cluster the fingerprints over a period of time to roughly estimate the area where the user moves, and then filter out the ambiguous locations based on the results of the clustering.

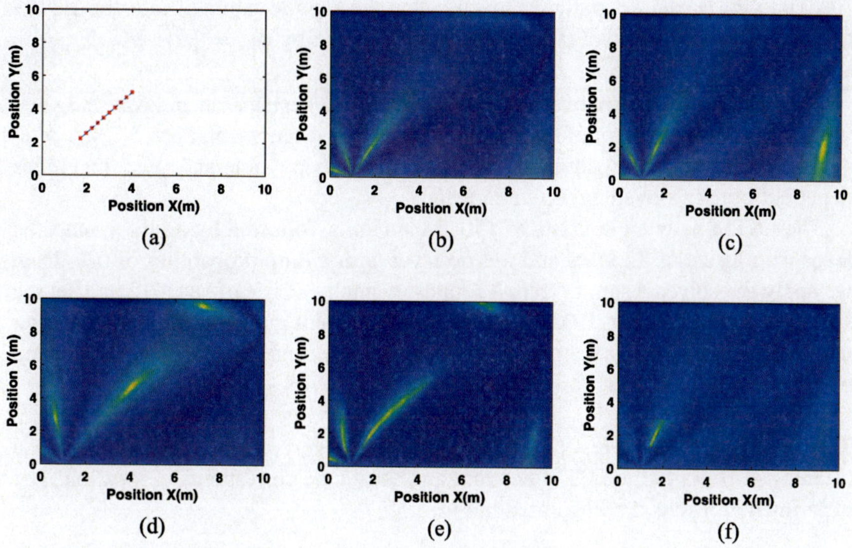

Fig. 5.49 Hybrid inter-antenna distance. (**a**) Ground truth. (**b**) Position A (t = 0.8 s). (**c**) Position B (t = 1.0 s). (**d**) Position C (t = 0.1 s). (**e**) Multiplication. (**f**) Revised position B

However, the pattern of object movement has not been taken into consideration in our system. For example, the targets tend to stand still or move in specific directions without frequently switching walking directions. In the next section, we will introduce how to ensure the above two factors when designing the LSTM network.

5.3.4.3 LSTM Network

Training Phase We simulate the pattern of user movement to generate 100000+ trajectories for training. It takes about 46 s to generate these trajectories. There are two types of trajectories as follows: (1) the user stands still at a random position, which accounts for 50% of the total training data; (2) the user moves in a random direction at a random speed, which accounts for 50% of the total training data. We use a normal distribution with an average of 1.2 m/s and a standard deviation of 0.2 m/s to determine the user's initial walking speed, i.e., $v \sim N(1.2\,\text{m/s}, (0.2\,\text{m/s})^2)$. For the acceleration of user motion, we let $a \sim N(0\,\text{m/s}^2, (0.3\,\text{m/s}^2)^2)$. Moreover, we choose the direction of motion from $[0°, 360°]$ with equal probability. Each training data contains 50 positions denoted by $\mathbf{p} = [p_1, p_2, \cdots, p_{50}]$, which is corresponding to 5 s in reality. Then, we convert the trajectories $\mathbf{p} = [p_1, p_2, \cdots, p_{50}]$ into phase difference fingerprints based on the theoretical fingerprints database. In particular, for position p_t, we first calculate the AoA fingerprint $[\theta_1, \theta_2, \theta_3, \theta_4]$ relative to the nearest 4 APs and transform each

AoA into $\boldsymbol{\phi}_n = [\phi_{min}, \phi_{max}]^T$. In this way, the sample inputted into the LSTM network at each time point is a 1×8 vector denoted by $\boldsymbol{\phi}_{in} = [\boldsymbol{\phi}_1^T, \boldsymbol{\phi}_2^T, \boldsymbol{\phi}_3^T, \boldsymbol{\phi}_4^T]$, where $\boldsymbol{\phi}_n^T = [\phi_{min}, \phi_{max}]$.

Note that the LSTM network is only a tool in our system, our innovation is that we design the input vector of the LSTM network. As discussed in Sect. 5.3.3.3, ϕ_{min} can help us remove as many ambiguous positions as possible and ϕ_{max} magnifies the effect of user movement on fingerprints.

The LSTM network consists of 150 hidden units, followed by a fully connected layer with a size of 32 units and a drop layer with a drop probability of 0.2. Then we apply the solver Adam to train 2 rounds in small batches of size 512 and set the specified learning rate as 0.01. Additionally, to prevent gradient explosions, we set the gradient threshold as 1. At each time, we input $\boldsymbol{\phi}_{in}$ and derive a position denoted by $\hat{p}_t$. Thus, the loss function can be written as $\mathcal{L} = ||p_t - \hat{p}_t||$.

Testing Phases In the test phase, we input the fingerprint after phase difference cleaning at 10 Hz to the LSTM network and obtain the corresponding location. The experimental results are illustrated in Fig. 5.55.

5.3.5 *Automatic Fingerprints Update*

Though MUSIC algorithm and LSTM network can be utilized to enable high accuracy, they all require frequent packet switching between transceivers. We are able to broadcast Wi-Fi packets using smart devices, and the packets will be captured simultaneously by several Intel 5300 NICs in monitor mode. Despite this, frequent packet switching for localization still does not conform to regular communication, where smart devices only exchange data with a single receiver. This section proposes an adaptive fingerprints update algorithm to make the system out of frequent packet exchanging, where we are able to locate targets with fewer packets.

The proposed method supports crowdsourcing in a non-participatory manner. In particular, the context-aware localization method provides an accurate location label for the crowdsourcing process. By matching it with the corresponding fingerprint, we can update the CSI fingerprints database or build RSSI fingerprints database. The RSSI fingerprints database can provide locations for most COTS devices.

CSI Fingerprints Update As shown in Fig. 5.50, when the user walks along the path $T1 - T2 - T3$, the crowdsourcing CSI fingerprints will be uploaded to the server. After localization, the server can automatically update the theoretical fingerprints database to the crowdsourcing fingerprints database. The advantage of the crowdsourcing fingerprints database is that it takes unpredictable environmental factors into consideration. For example, assume that there is a "bad" area in the environment due to NLoS or severe multipath effects, and the actual signal characteristics of the area do not match the theoretical fingerprint. We can infer

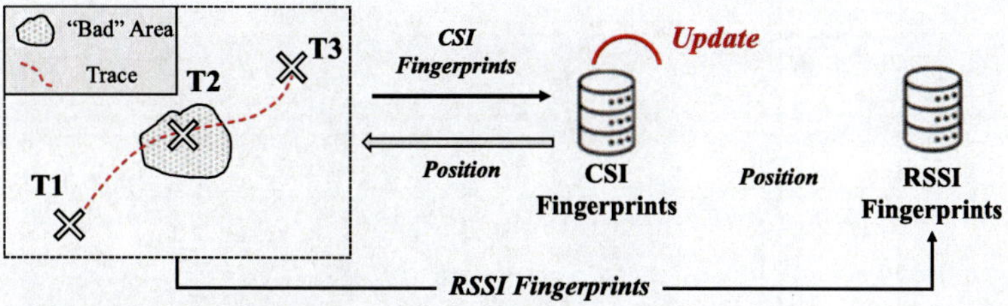

Fig. 5.50 Automatic fingerprints update process

these locations through the context-aware localization method, and then update the theoretical fingerprints database. Furthermore, even if frequent packets are no longer available, we can still obtain an accurate location by directly matching the reported fingerprint with the crowdsourcing fingerprints database.

RSSI Fingerprints Database The above method is also applicable to other crowdsourcing processes such as RSSI fingerprint construction. During experiments, we will construct the RSSI fingerprints database in a non-participatory manner and verify the localization performance.

5.3.6 Performance Evaluation

We implement APs and targets using J1900 minicomputer equipped with Intel Wi-Fi Link 5300 network interface controller (NIC). The device is off-the-shelf NIC equipped with three antennas that runs the CSI toolkit released by [97] that can retrieve the CSI as a 3×30 matrix at 10 Hz. We conduct experiments in two different scenarios as shown in Fig. 5.51a, b. Figure 5.51a shows the layout of the corridor and experimental environments, where there are 4 APs in $100\,\text{m}^2$ area. In this scenario, we verify the localization performance in the open indoor environment. Figure 5.51b shows the layout of the office building and experimental environments, where there are 5 APs in $224\,\text{m}^2$ area. In this scenario, we verify the localization performance in the complex indoor environment. We have explored different inter-antenna distances via external antenna bases. As shown in Fig. 5.53, we can realize the best performance based on the following setting: $d_{12} = 0.5\lambda$, $d_{23} = 2.0\lambda$, $d_{13} = 2.5\lambda$. Thus, it is set as the common setting for localization. During the training phase, we set the cell as 2.5 cm × 2.5 cm and generate the theoretical fingerprints database within 1 s. This cell size means $160{,}000 \sim 350{,}000$ different cells in these two scenarios. It is impractical to manually collect so many fingerprints. During the testing phase, we randomly select 50 positions and 10 trajectories to evaluate the localization performance.

As we know, most smartphones cannot retrieve CSI. Though Shadow Wi-Fi [191] presents how to teach smartphones to extract CSI, it is hard to obtain the

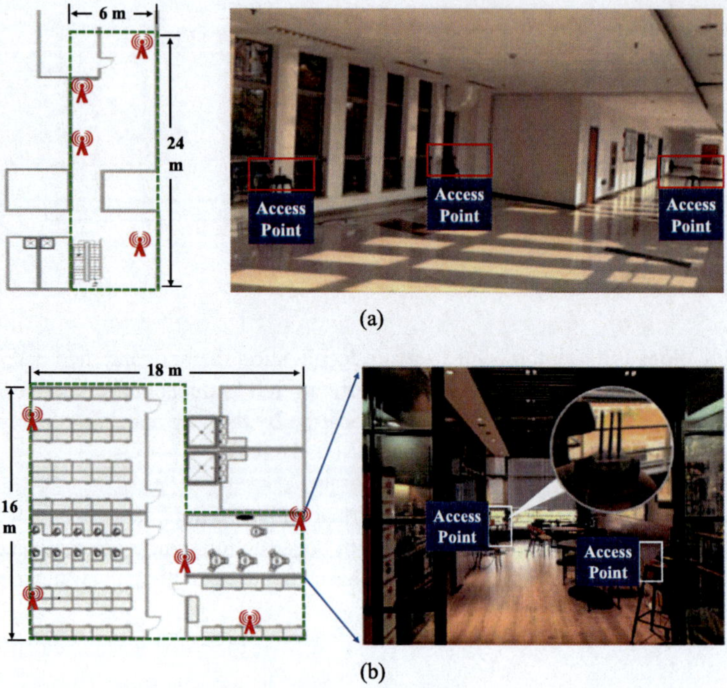

Fig. 5.51 Ground truth. (**a**) Corridor. (**b**) Research laboratory

antenna array direction of mobile phones due to diverse placements. To address this problem, the insights are as follows: (1) Shadow Wi-Fi can teach COTS mobile devices to transmit raw signals; (2) we can enable monitor mode in Wi-Fi APs [209] to extract the CSI of signals in the space, e.g., the above raw signals emitted by smartphones. As the performance of Shadow Wi-Fi has been verified, we do not implement their system here. During experiments, we enable monitor mode and HTTP service at each receiver and select one Wi-Fi AP to send packets. Finally, we use the *websave(·)* function in Matlab to obtain the CSI data files in real time.

5.3.6.1 Theoretical Fingerprints Database

Fingerprints Generation We choose different cell sizes to generate the theoretical fingerprints database. In the above two scenarios, the training time and testing time are shown in Fig. 5.52a. Based on the experimental results, we finally set the cell size as 2.5 cm. The corresponding training time is respectively 0.77 s (corridor) and 1.54 s (lab). Because the pre-training process is a one-time task and does not need to be completed in real time, such training time can be easily tolerated. Moreover, the processing time of the single-spot localization process shown in Algorithm 13 is

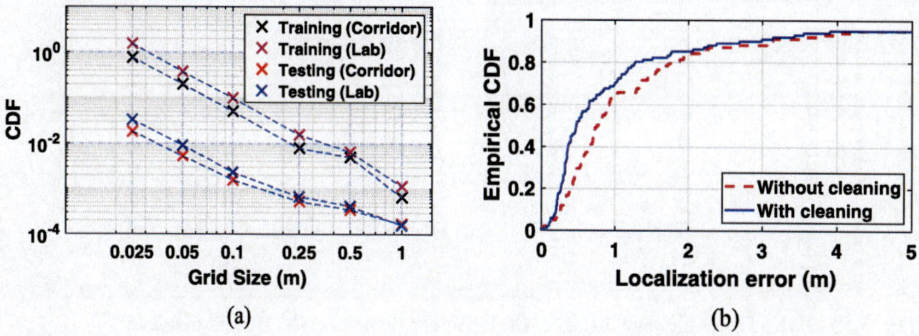

Fig. 5.52 Processing time and phase difference cleaning. (**a**) Processing time. (**b**) Phase difference cleaning

0.02 s (corridor) and 0.03 s (lab). Note that we consider the processing time of the MUSIC algorithm independently. It takes about 0.5 s to execute the SpotFi algorithm [87, 179, 208]. Since 0.03 s $\ll$ 0.5 s, this small cell size does not significantly increase the localization time of our system.

Phase Difference Cleaning The purpose of phase difference cleaning (Sect. 5.3.3.3) is to resist multipath effects. We compare localization performance w/o phase difference cleaning in Fig. 5.52b, where the 80% errors are respectively within 1.27 m (MUSIC + phase difference fingerprint) and 1.76 m (only phase difference fingerprint). Thus, the MUSIC algorithm can improve the accuracy but takes an extra 0.5 s to process the raw data.

Hybrid Distance The hybrid inter-antenna distance can help improve localization accuracy (Sect. 5.3.3.2). To this end, we first explore different inter-antenna distances in Fig. 5.53a. As the inter-antenna distance increases, we tend to get higher median accuracy. The median errors are respectively 1.20 m ($d = 0.5\lambda$), 0.72 m ($d = 1.0\lambda$), 0.55 m ($d = 1.5\lambda$) and 0.43 m ($d = 2.0\lambda$). This validates our analysis that a large inter-antenna distance can improve the anti-noise performance of the system. Meanwhile, the 80% localization errors are respectively within 2.56 m ($d = 0.5\lambda$), 2.05 m ($d = 1.0\lambda$), 3.28 m ($d = 1.5\lambda$) and 4.91 m ($d = 2.0\lambda$). This means that large inter-antenna distance can cause more $E2$. Figure 5.53b shows the localization accuracy of the hybrid inter-antenna distance, where the median errors are respectively 0.55 m ($0.5\lambda + 1.5\lambda$), 0.39 m ($0.5\lambda + 2.0\lambda$) and 0.42 m ($1.5\lambda+2.0\lambda$), and the 80% localization errors are respectively within 1.24 m, 1.00 m and 1.48 m. According to the experimental results, the hybrid inter-antenna distance can significantly improve localization performance, where we can realize the best performance based on the following setting: $d_{12} = 0.5\lambda$, $d_{23} = 2.0\lambda$, $d_{13} = 2.5\lambda$.

Hybrid Fingerprints We can utilize hybrid fingerprints consisting of CSI amplitude and CSI phase difference to filter out the ambiguous positions (Sect. 5.3.4.1). Figure 5.54a compares the localization performance w/wo CSI amplitude finger-

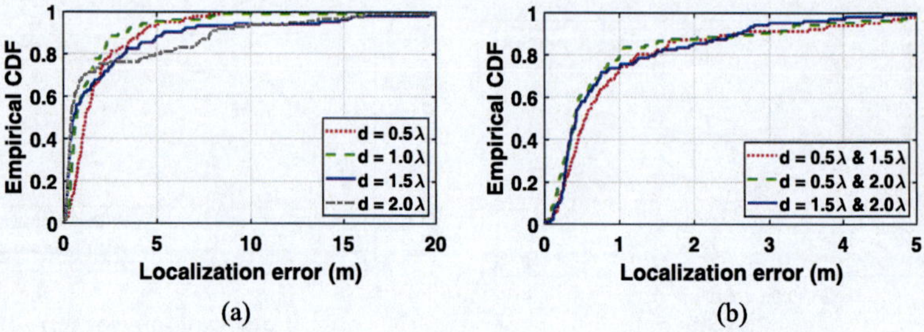

Fig. 5.53 Hybrid inter-antenna distance. (**a**) Different distances. (**b**) Hybrid distance

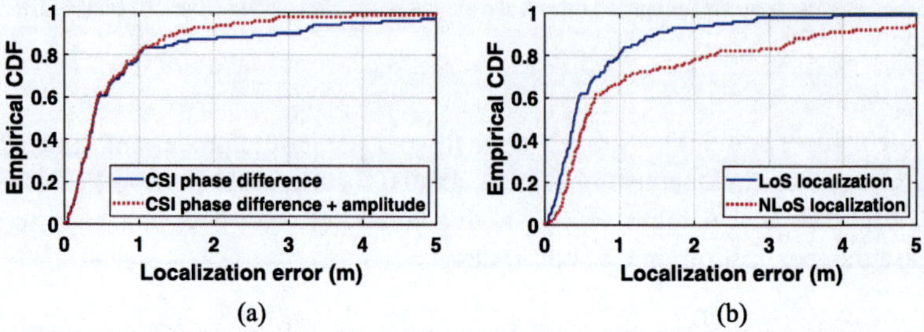

Fig. 5.54 Hybrid fingerprint and NLoS scenarios. (**a**) Hybrid CSI fingerprint. (**b**) LoS and NLoS

print, where the 90% localization errors are respectively within 1.51 m and 2.83 m. Since CSI amplitude fingerprint proves to be less reliable than CSI phase difference (Table 5.4), we just utilize the CSI amplitude to revise $E2$ rather than determine the target's position. This accounts for why they yield a similar 80% localization errors.

NLoS Performance Figure 5.54b shows the localization performance in NLoS scenarios, where the 80% localization errors are respectively within 0.98 m and 2.17 m. Taking the size of AoI into account, single-spot localization based on the theoretical fingerprints yields a similar accuracy to the existing localization methods such as SpotFi [87] and WiTag [179].

5.3.6.2 Fingerprinting Localization

Euclidean Distance Multiplication We can take a series of positions into consideration to improve localization performance. As illustrated in Fig. 5.55a, we can observe that the multiplication approach can achieve higher accuracy, where the 80% localization errors are respectively within 0.98 m, 0.74 m ($T_{window} = 0.3$ s), 0.67 m ($T_{window} = 0.5$ s) and 0.71 m ($T_{window} = 1.0$ s). However, the proposed

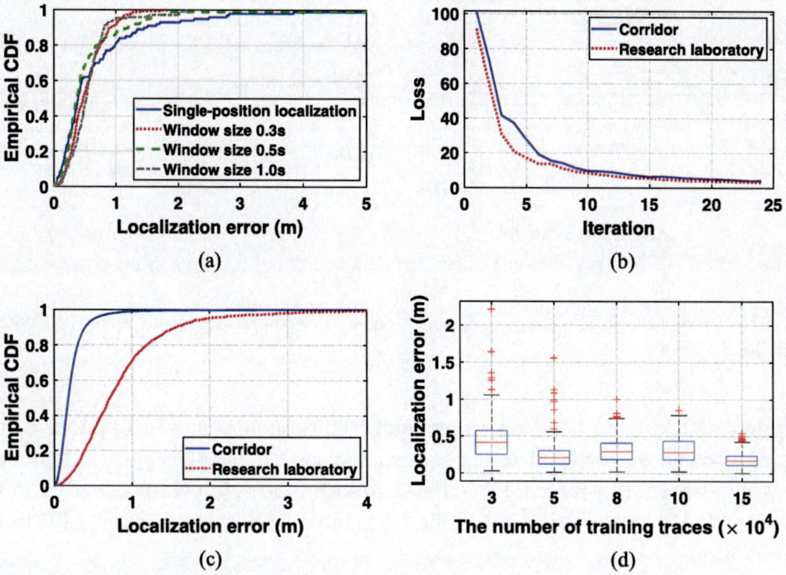

Fig. 5.55 Context-aware localization. (**a**) Multiplication approach. (**b**) Loss. (**c**) LSTM performance. (**d**) Number of traces

multiplication approach is a clustering method rather that a context-aware method. Thus, we use LSTM network to promote the system.

LSTM Network The loss curve of the LSTM network is shown in Fig. 5.55b, we can observe that the network can converge quickly. We utilize Matlab 2019a to train the LSTM network in DELL XPS 9570 (Intel Core i9-8950HK CPU@2.9 Hz, NVIDIA GeForce GTX 1050 Ti), and the experimental results show that the convergence time is within 60 s. In testing phase, it takes about 0.009 s to derive target's position each time. Figure 5.55c shows the localization performance under different scenarios, where the 80% localization errors are respectively within 0.30 m and 1.19 m. Moreover, we change the numbers of trajectories to verify the localization performance in Fig. 5.55d. As the training trajectory increases, we will observe fewer ambiguous locations labeled with the red marker. As discussion in Sect. 5.3.4.3, it takes 62 s to generate 150,000 trajectories.

5.3.6.3 Automatic Fingerprints Update

CSI Fingerprinting Localization We update the CSI fingerprints database and localize the target based on the direct fingerprints matching method. Note that the phase difference cleaning method proposed in Sect. 5.3.3.3 does not need to be executed for crowdsourcing fingerprints here. This is because the crowdsourcing fin-

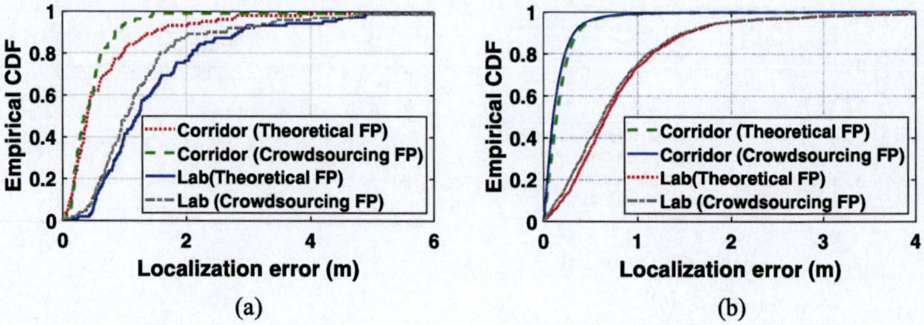

Fig. 5.56 Crowdsourcing fingerprints database. (**a**) CSI fingerprinting localization. (**b**) Context-aware localization

gerprints database has taken environmental factors such as multipath into account. Figure 5.56a shows the localization performance, where the 80% localization errors are within 0.98 m (theoretical FP in the corridor), 0.60 m (crowdsourcing FP in the corridor), 2.10 m (theoretical FP in the lab), and 1.56 m (crowdsourcing FP in the lab).

Context-Aware CSI Fingerprinting Localization Based on the crowdsourcing fingerprints database, we can still utilize the LSTM network to optimize the localization performance. Although we need to re-simulate user trajectories to train the LSTM network, this is just a one-time job and does not take much time. Similar to the results in the previous section, it takes just 0.3 ms for localization in testing phase. As shown in Fig. 5.56a, b, we can summarize the accuracy relationship as follows: context-aware localization based on crowdsourcing FP $\approx$ context-aware localization based on theoretical FP > crowdsourcing FP > theoretical FP. The root cause of the above relationship is that crowdsourcing fingerprint is dependent on the location labels from the context-aware localization process, and it is difficult to further improve the localization accuracy based on the context-aware method.

Fingerprints Update and Human Orientation Figure 5.57a shows the difference between theoretical fingerprints database and the crowdsourcing fingerprints database, where we calculate $\frac{\|f_{theoretical}-f_{crowd}\|_2}{N}$ for each updated position. The sum of the update positions is about 10 Hz $\times$ 10 traces $\times$5 seconds as illustrated in Sect. 5.3.4.3. For $d_{12} = 0.5\lambda$, $d_{13} = 2.5\lambda$, we can observe a minor variation for $d_{12} = 0.5\lambda$ that determines the global positions. Though the fingerprints difference with respect to $d_{13} = 2.5\lambda$ is large, we just utilize its variation over time to magnify the impact of user movement on fingerprints as discussed in Sect. 5.3.3.3. That is to say, we pay more attention to its variation over time rather than its own value. Furthermore, we explore the impact of user orientation in Fig. 5.57b. When user's orientation changes, most phase differences remain stable, and the 80% localization errors are within 0.4 m.

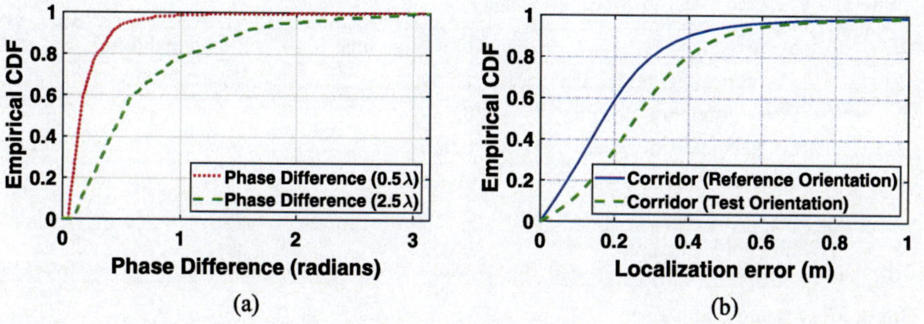

Fig. 5.57 Fingerprints update and human orientation. (**a**) Fingerprints update. (**b**) Human orientation

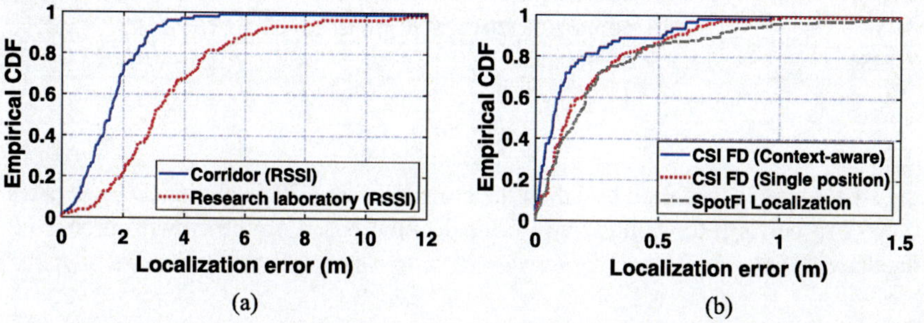

Fig. 5.58 RSSI localization and the size of AoI. (**a**) RSSI figerprint localization. (**b**) Localization in a small area

RSSI Fingerprinting Localization The proposed context-aware localization method can also provide the location labels for the other crowdsourcing process. Figure 5.58a shows the localization performance of the crowdsourcing RSSI fingerprinting, where the 80% localization errors are within 2.44 m and 4.68 m. Since localization accuracy is similar to existing RSSI fingerprinting localization systems [126, 187], the proposed method enables crowdsourcing RSSI fingerprints collection.

The Size of AoI In CSI localization systems, the size of AoI is an important factor affecting localization accuracy. We select a 4 m × 8 m area to conduct experiments, and the results show that the 80% localization errors are within 0.20 m, 0.34 m and 0.41 m for different methods.

Time Complexity Table 5.6 summarizes the execution time of different stages. It takes 1.54 s to train the theoretical FP database (ID 1). It takes 0.53 s to perform the single-spot localization based on the theoretical FP database (ID 2,3). It takes 70 s ∼ 120 s to train the LSTM network (ID 4,5). It takes 0.5 s to perform the context-aware localization based on the theoretical FP database (ID 2,6). It takes 0.03 s to perform

Table 5.6 The execution time of different stages

ID	Stage	Execution time	Run in real time?
1	Theoretical FP generation	1.54 s	No
2	Phase difference cleaning	0.5 s	Yes
3	Single-spot localization	0.03 s	Yes
4	Trajectories generation	about 62 s	No
5	LSTM training	$5 \sim 60$ s	No
6	LSTM localization	0.009 s	Yes

Table 5.7 System Comparison

System	Localization accuracy	Num. of antennas
Our System	0.30 m@80%	3
SpotFi	1.00 m@80%	3
RIM	0.50 m@80% (20 Hz) 0.13 m@80% (40 Hz)	6
CRISLoc	1.35 m@80%	1
Triangular antenna array	1.18 m@80%	3

the single-spot localization based on the crowdsourcing FP database (ID 3). It takes 0.009 s to perform the context-aware localization based on the crowdsourcing FP database (ID 6).

5.3.6.4 System Comparison

We compare our system with the following four systems. **SpotFi [87]** is proposed in 2015. It is a classic CSI model-based localization method, where they can achieve decimeter-level localization accuracy based on the receivers equipped with just 3 antennas; **RIM [98]** is proposed in 2019. It is the latest device-based tracking system. However, they need more antennas and data collection at high frequencies; **CRISLoc [187]** is proposed in 2019. It is the latest CSI fingerprinting localization system. This system uses the CSI amplitude as the fingerprint and designs a fingerprint update algorithm for moved APs; **Triangular Antenna Array [177]** is proposed in 2019. The system realizes calibration-free model-based localization. It is necessary to compare our method with the model-based localization system.

As shown in the Table 5.7, our system outperforms SpotFi [87] and [177]. It is because that we utilize the context-aware localization method to promote the system. Moreover, since CSI phase difference fingerprint is better CSI amplitude fingerprint, our system yields a higher accuracy than CRISLoc [187]. Finally, our system can achieve a similar accuracy as RIM when we collect CSI data at 20 Hz. However, they require the device equipped with 6 antennas, which is not always available.

Chapter 6
Conclusions

In this part, we summarize this book and give a direction toward future work.

6.1 Research Summary

Localizing mobile terminal is the basic prerequisite for many Internet-of-Things (IoT) applications. Although many terminals have installed Global Navigation Satellite System (GNSS, such as GPS) functional modules for localization, the actual application scenarios are complex and the environment is diverse, which make satellite systems unable to meet the needs under all circumstances. For example, environments such as indoors or the "urban canyon" with high-rise buildings, and the walls of buildings will block satellite signals and make the terminal unable to be localized.

Wireless localization is a kind of technique to estimate the wireless terminal's location by utilizing characteristics of wireless communication signals, such as Wi-Fi, LTE, and BLE. The technique has been attracting interests from both academia and industrial in the past decides; however, most of the research work is conducted using the methodology of "empirical study," that is, observing experimental results or phenomena and then coming up with a new idea, and then verifying the new idea with experiments again. A systematic theoretical study on the topic has been long needing but lacking. Answers to some fundamental issues of wireless localization technique are still unknown.

In this book, we present a systematic theoretical study of the wireless localization technique. Guided by the theoretical results, we will provide design approaches for improving performance of the localization system and making deployment of the system more convenient.

X. Tian et al., *Wireless Localization Techniques*, Wireless Networks,
https://doi.org/10.1007/978-3-031-21178-2_6

6.2 Future Work

The wireless indoor localization technique continues to develop from diverse perspectives. We summarize some interesting and attractive points as follows.

Environmental Heterogeneity It is obvious that the settings of home and office scenes are considerably different. The different settings of the scenes can also cause an impact on the signal, such as the placement of obstacles and the number of electronic devices. In detail, the signals are reflected on the surface of a static object, and the reflection path varies from location to location. What is more, the electromagnetic waves emitted by the electronic devices also affect the received signals. Nevertheless, most current sensing systems are based on static physical models without taking the effects of scene setup into account. Therefore, the implementation of dynamic adaptive algorithms is quite necessary.

Open CSI Interface CSI tool makes it easier for commercial devices to obtain CSI data. However, the type of NIC devices in most state-of-the-arts is currently limited, such as Intel 5300, Atheros, etc. There is still a significant gap between this type of NIC and commercial mainstream NICs in terms of protocol, rate, and bandwidth. If the scope of CSI-adapted NICs cannot be extended, the applications such as Wi-Fi sensing or localization will be still some distance away from large-scale commercialization. Furthermore, even if CSI collection is commercially available at scale, the physical data collection is inevitably challenged by the heterogeneity of NICs between different models. Therefore, how to provide an open and transparent CSI interface is crucial for the commercialization of Wi-Fi applications.

Location Technology Based on Multi-Modal Information In the field of smart IoT, multi-modal information fusion has become more and more popular in recent works. Multi-modal information can better describe the scenes and human behaviors from diverse dimensions. The RF signals, computer vision, acoustic, and other information can provide opportunities for localizing. For example, combining both footsteps and wireless signals will help to track the position of the moving user. However, due to the different modalities of information, it is still challenging to find a unified processing paradigm for fusing multi-modal information effectively.

Seamless Outdoor/Indoor Localization Localization applications have attracted considerable attention from researchers. GPS technology is the current mainstream technology for outdoor localization. However, GPS technology cannot work indoors due to weak indoor signals. There are many indoor localization technologies. However, it cannot achieve seamless outdoor and indoor localization. Hence, it is necessary to integrate with outdoor and indoor localization systems to form a seamless whole scene location service, allowing the system to track a user on the street and in a mall, on different floors, and even underground parking garage, the in-time logistics of a package, etc.

Acknowledgments The publication of the book is supported by the National Key Research and Development Program of China 2020YFB1708700 and National Natural Science Foundation of China (No. 61922055, 61872233, 62102282, 61572319). The authors would like to thank those who made technical contributions to this book (alphabetical order by the first name): Binyao Jiang, Dong Xu, Duowen Liu, Fengyuan Zhu, Hao Li, Jinyu Shi, Kaikai Sheng, Ke Liu, Mei Wang, Qi Zhang, Qianru Li, Ruofei Shen, Sijie Xiong, Sujie Zhu, Tuo Yu, Wencan Zhang, Wenxin Li, Xinyu Wu, Xuanqi Meng, Xueyu Mao, Yang Wan, Yang Zhang, Yingling Mao, Yucheng Yang, Yutian Wen, Zhehui Zhang, Zhenyu Song, Zhicheng Gu.

References

1. J. Wang, D. Fang, Z. Yang, H. Jiang, X. Chen, T. Xing, L. Cai, E-HIPA: An energy-efficient framework for high-precision multi-target-adaptive device-free localization. IEEE Trans. Mob. Comput. **16**(3), 716–729 (2016)
2. I. Sabek, M. Youssef, A.V. Vasilakos, ACE: An accurate and efficient multi-entity device-free WLAN localization system. IEEE Trans. Mob. Comput. **14**(2), 261–273 (2014)
3. C. Liu, D. Fang, Z. Yang, H. Jiang, X. Chen, W. Wang, L. Cai, RSS distribution-based passive localization and its application in sensor networks. IEEE Trans. Wirel. Commun. **15**(4), 2883–2895 (2015)
4. C. Wu, J. Xu, Z. Yang, N.D. Lane, Z. Yin, Gain without pain: Accurate WiFi-based localization using fingerprint spatial gradient. Proc. ACM IMWUT **1**(2), 1–19 (2017)
5. I. Kashiwagi, T. Taga, T. Imai, Time-varying path-shadowing model for indoor populated environments. IEEE Trans. Veh. Technol. **59**(1), 16–28 (2009)
6. S.P. Kuo, Y.C. Tseng, A scrambling method for fingerprint positioning based on temporal diversity and spatial dependency. IEEE Trans. Knowl. Data Eng. **20**(5), 678–684 (2008)
7. R. Sharma, V. Badarla, Geometrical optimization of a novel beacon placement strategy for 3D indoor localization, in *Proceedings of the IEEE ANTS* (2018), pp. 1–6
8. S.P. Chepuri, G. Leus, A.J. van der Veen, Sparsity-exploiting anchor placement for localization in sensor networks, in *Proceedings of the IEEE EUSIPCO* (2013), pp. 1–5
9. Y. Chen, J.A. Francisco, W. Trappe, R.P. Martin, A practical approach to landmark deployment for indoor localization, in *Proceedings of the IEEE SECON*, vol. 1 (2006), pp. 365–373
10. Y. Li, Z. He, Z. Gao, Y. Zhuang, C. Shi, N. El-Sheimy, Toward robust crowdsourcing-based localization: A fingerprinting accuracy indicator enhanced wireless/magnetic/inertial integration approach. IEEE Internet Things J. **6**(2), 3585–3600 (2018)
11. J. Xu, H. Chen, K. Qian, E. Dong, M. Sun, C. Wu, Z. Yang, ivr: Integrated vision and radio localization with zero human effort. Proc. ACM IMWUT **3**(3), 1–22 (2019)
12. Z. Li, X. Zhao, F. Hu, Z. Zhao, J.L.C. Villacrés, T. Braun, SoiCP: A seamless outdoor–indoor crowdsensing positioning system. IEEE Internet Things J. **6**(5), 8626–8644 (2019)
13. U.M. Qureshi, Z. Umair, G.P. Hancke, Problem of wireless fingerprint duplication in fingerprinting based indoor localization, in *Proceedings of the IEEE ISIE* (2018), pp. 1270–1275
14. R. Ayyalasomayajula, A. Arun, C. Wu, A. Shaikh, S. Rajagopalan, Y. Hu, I.K. Jain, LocAP: Autonomous Millimeter Accurate Mapping of WiFi Infrastructure, in *Proceedings of the NSDI* (2020), pp. 1115–1129

X. Tian et al., *Wireless Localization Techniques*, Wireless Networks,
https://doi.org/10.1007/978-3-031-21178-2

15. Y. Gao, J. Niu, R. Zhou, G. Xing, ZiFind: Exploiting cross-technology interference signatures for energy-efficient indoor localization, in *Proceedings of the IEEE INFOCOM* (2013), pp. 2940–2948
16. R. Nandakumar, K.K. Chintalapudi, V.N. Padmanabhan, Centaur: Locating devices in an office environment, in *Proceedings of the ACM MobiCom* (2012), pp. 281–292
17. S. Papaioannou, H. Wen, Z. Xiao, A. Markham, N. Trigoni, Accurate positioning via cross-modality training, in *Proceedings of the ACM Sensys* (2015), pp. 239–251
18. M. Ibrahim, H. Liu, M. Jawahar, V. Nguyen, M. Gruteser, R. Howard, F. Bai, Verification: Accuracy evaluation of WiFi fine time measurements on an open platform, in *Proceedings of the ACM MobiSys* (2018), pp. 417–427
19. M. Ibrahim, A. Rostami, B. Yu, H. Liu, M. Jawahar, V. Nguyen, R. Howard, Wi-go: accurate and scalable vehicle positioning using wifi fine timing measurement, in *Proceedings of the ACM MobiSys* (2020), pp. 312–324
20. K. Jiokeng, G. Jakllari, A. Tchana, A.L. Beylot, When FTM Discovered MUSIC: Accurate WiFi-based Ranging in the Presence of Multipath, in *Proceedings of the IEEE INFOCOM* (2020), pp. 1857–1866
21. S. Nikitaki, G. Tsagkatakis, P. Tsakalides, Efficient multi-channel signal strength based localization via matrix completion and Bayesian sparse learning. IEEE Trans. Mob. Comput. **14**(11), 2244–2256 (2015)
22. Y. Shu, P. Cheng, Y. Gu, J. Chen, T. He, TOC: Localizing wireless rechargeable sensors with time of charge. ACM Trans. Sens. Netw. **11**(3), 1–22 (2015)
23. B. Huang, M. Liu, Z. Xu, B. Jia, On the performance analysis of wifi based localization, in *Proceedings of the IEEE ICASSP* (2018), pp. 4369–4373
24. K.S.V. Prasad, E. Hossain, V.K. Bhargava, Machine learning methods for RSS-based user positioning in distributed massive MIMO. IEEE Trans. Wirel. Commun. **17**(12), 8402–8417 (2018)
25. L. Huo, Wang, A target tracking algorithm using Grey Model predicting Kalman Filter in wireless sensor networks, in *Proceedings of the IEEE iThings* (2017), pp. 604–610.
26. A. Behboodi, F. Lemic, A. Wolisz, Hypothesis testing based model for fingerprinting localization algorithms, in *Proceedings of the IEEE VTC Spring* (2017), pp. 1–6
27. S. He, S.H.G. Chan, Tilejunction: Mitigating signal noise for fingerprint-based indoor localization. IEEE Trans. Mob. Comput. **15**(6), 1554–1568 (2015)
28. J.F. Huang, G.Y. Chang, G.H. Chen, A historical-beacon-aided localization algorithm for mobile sensor networks. IEEE Trans. Mob. Comput. **14**(6), 1109–1122 (2014)
29. S. He, S.H.G. Chan, L. Yu, N. Liu, Fusing noisy fingerprints with distance bounds for indoor localization, in *Proceedings of the IEEE INFOCOM* (2015), pp. 2506–2514
30. P.H. Tseng, K.T. Feng, Geometry-assisted localization algorithms for wireless networks. IEEE Trans. Mob. Comput. **12**(4), 774–789 (2012)
31. K. Kaemarungsi, P. Krishnamurthy, Properties of indoor received signal strength for WLAN location fingerprinting, in *IEEE MOBIQUITOUS* (2004), pp. 14–23
32. P. Bahl, V.N. Padmanabhan, Radar: An in-building RF-based user location and tracking system, in *Proceedings of the IEEE INFOCOM* (2000), pp. 775–784
33. P. Castro, P. Chiu, T. Kremenek, R. Muntz, A probabilistic room location service for wireless networked environments, in *Proceedings of the ACM UbiComp* (2001), pp. 18–34
34. A. Haeberlen, E. Flannery, A.M. Ladd, A. Rudys, D.S. Wallach, L.E. Kavraki, Practical robust localization over large-scale 802.11 wireless networks, in *Proceedings of the ACM MobiCom* (2004), pp. 70–84
35. M.A. Youssef, A. Agrawala, A. Udaya Shankar, Wlan location determination via clustering and probability distributions, in *Proceedings of the IEEE PerCom* (2003), pp. 143–150
36. M.A. Youssef, A. Agrawala, Handling samples correlation in the Horus system, in *Proceedings of the IEEE INFOCOM* (2004)
37. M. Brunato, R. Battiti, 'Statistical learning theory for location fingerprinting in wireless lans. Comput. Netw. **47**(6), 825–845 (2005)

38. M.A. Youssef, A. Agrawala, The horus wlan location determination system, in *Proceedings of the ACM MobiSys* (2005), pp. 205–218
39. H. Hashemi, Impulse response modeling of indoor radio propagation channels. IEEE J. Sel. Areas Commun. **11**(7), 967–978 (1993)
40. G. Chandrasekaran, M.A. Ergin, J. Yang, S Liu, Y. Chen, M. Gruteser, R.P. Martin, Empirical evaluation of the limits on localization using signal strength, in *Proceedings of the IEEE SECON* (2009), pp. 1–9
41. T.V. Haute, E.D. Poorter, I. Moerman, F. Lemic, V. Handziski, A. Wolisz, N. Wirström, T. Voigt, Comparability of RF-based indoor localisation solutions in heterogeneous environments: an experimental study, *Int. J. Ad Hoc Ubiquitous Comput.* **23**(1–2), 92–114 (2016)
42. S. He, D.H. Shin, J. Zhang, J. Chen, Toward optimal allocation of location dependent tasks in crowdsensing, in *IEEE INFOCOM* (2014), pp. 745–753
43. Y. Wen, J. Shi, Q. Zhang, X. Tian, Z. Huang, H. Yu, Y. Cheng, X. Shen, Quality-driven auction-based incentive mechanism for mobile crowd sensing. IEEE Trans. Veh. Technol. **64**(9), 4203–4214
44. X. Tian, W. Zhang, J. Wang, W. Li, S. Li, X. Wu, Y. Yang, Online pricing crowdsensed fingerprints for accurate indoor localization, in *IEEE VTC-Fall* (2017), pp. 1–5
45. S. Yang, P. Dessai, M. Verma, M. Gerla, FreeLoc: Calibration-free crowdsourced indoor localization, in *Proceedings of the IEEE INFOCOM* (2013), pp. 2481–2489
46. S.H. Jung, B.C. Moon, D. Han, Performance evaluation of radio map construction methods for Wi-Fi positioning systems. IEEE Trans. Intell. Transp. Syst. **18**(4), 880–889 (2016)
47. W. Yu, J. Zhang, C. Wang, An on-demand approach for indoor localization based on crowdsourced Wi-Fi fingerprints, in *Proceedings of the IEEE ICCSNT*, vol. 1 (2015), pp. 1226–1229
48. L. Li, X. Guo, N. Ansari, H. Li, A hybrid fingerprint quality evaluation model for wifi localization. IEEE Internet Things J. **6**(6), 9829–9840 (2019)
49. K. Han, H. Huang, J. Luo, Quality-aware pricing for mobile crowdsensing. IEEE/ACM Trans. Networking **26**(4), 1728–1741 (2018)
50. X. Tian, W. Zhang, Y. Yang, X. Wu, Y. Peng, X. Wang, Toward a quality-aware online pricing mechanism for crowdsensed wireless fingerprints. IEEE Trans. Veh. Technol. **67**(7), 5953–5964 (2018)
51. S. Yang, K. Han, Z. Zheng, S. Tang, F. Wu, Towards personalized task matching in mobile crowdsensing via fine-grained user profiling, in *Proceedings of the IEEE INFOCOM* (2018), pp. 2411–2419
52. Y. Jiang, H. Qiu, M. McCartney, G. Sukhatme, M. Gruteser, F. Bai, R. Govindan, Carloc: Precise positioning of automobiles, in *Proceedings of the IEEE Sensys* (2015), pp. 253–265
53. A. Rai, K.K. Chintalapudi, V.N. Padmanabhan, R. Sen, Zee: zero-effort crowdsourcing for indoor localization, in *Proceedings of the ACM MobiCom* (2012), pp. 293–304
54. H. Wang, S. Sen, A. Elgohary, M. Farid, M. Youssef, R.R. Choudhury, No need to war-drive:unsupervised indoor localization, in *Proceedings of the ACM MobiSys* (2012), pp. 197–210
55. Z. Yang, C. Wu, Y. Liu, Locating in fingerprint space: wireless indoor localization with little human intervention, in *Proceedings of the ACM MobiCom* (2012), pp. 269–280
56. C. Wu, Z. Yang, Y. Liu, W. Xi, Will: wireless indoor localization without site survey, in *IEEE Transactions on Parallel and Distributed Systems* (2013), pp. 839–848
57. C. Wu, Z. Yang, Y. Liu, Smartphones based Crowdsourcing for Indoor Localization. IEEE Trans. Mob. Comput. **13**(10), 2199–2214 (2013)
58. T. Higuchi, S. Fujii, H. Yamaguchi, T. Higashino, Mobile node localization focusing on stop-and-go behavior of indoor pedestrians. IEEE Trans. Mob. Comput. **13**(7), 1564–1578 (2013)
59. C. Meng, H. Xiao, L. Su, Y. Cheng, Tackling the redundancy and sparsity in crowd sensing applications, in *Proceedings of the ACM Sensys* (2016), pp. 150–163
60. X. Guo, N. Ansari, L. Li, H. Li, Indoor localization by fusing a group of fingerprints based on random forests. IEEE Internet Things J. **5**(6), 4686–4698 (2018)

61. R. Margolies, A. Sridharan, V. Aggarwal, R. Jana, N.K. Shankaranarayanan, V.A. Vaishampayan, G. Zussman, Exploiting mobility in proportional fair cellular scheduling: Measurements and algorithms. IEEE/ACM Trans. Networking **24**(1), 355–367 (2014)
62. A. Chakraborty, L.E. Ortiz, S.R. Das, Network-side positioning of cellular-band devices with minimal effort, in *Proceedings of the IEEE INFOCOM* (2015), pp. 2767–2775
63. A. Ray, S. Deb, P. Monogioudis, Location of LTE measurement records with missing information, in *Proceedings of the IEEE INFOCOM* (2016)
64. S.C. Ergen, H.S. Tetikol, M. Kontik, R. Sevlian, R. Rajagopal. P. Varaiya, RSSI-fingerprinting-based mobile phone localization with route constraints. IEEE Trans. Veh. Technol. **63**(1), 423–428 (2013)
65. K. Majeed, S. Sorour, T.Y. Al-Naffouri, S. Valaee, Indoor localization and radio map estimation using unsupervised manifold alignment with geometry perturbation. IEEE Trans. Mob. Comput. **15**(11), 2794–2808 (2015)
66. H. Zheng, M. Gao, Z. Chen, X.Y. Liu, X. Feng, An adaptive sampling scheme via approximate volume sampling for fingerprint-based indoor localization. IEEE Internet Things J. **6**(2), 2338–2353 (2019)
67. K. Dong, Z. Ling, X. Xia, H. Ye, W. Wu, M. Yang, Dealing with insufficient location fingerprints in Wi-Fi based indoor location fingerprinting, in *Wireless Communications and Mobile Computing* (2017)
68. S. Yoon, K. Lee, Y. Yun, I. Rhee, ACMI: FM-based indoor localization via autonomous fingerprinting. IEEE Trans. Mob. Comput. **15**(6), 1318–1332 (2015)
69. X. Liu, S. Aeron, V. Aggarwal, X. Wang, M. Wu, Adaptive Sampling of RF Fingerprints for Fine-Grained Indoor Localization. IEEE Trans. Mob. Comput. **15**(10), 2411–2423 (2015)
70. G. Shen, Z. Chen, P. Zhang, T. Moscibroda, Y. Zhang, Walkie-Markie: Indoor pathway mapping made easy, in *Proceedings of the USENIX NSDI* (2013), pp. 85–98
71. C. Luo, H. Hong, C.M. Chan, PiLoc: a self-calibrating participatory indoor localization system, in *Proceedings of the 13th International Symposium on IPSN-14* (2014), pp. 143–153
72. C. Luo, H. Hong, M.C. Chan, J. Li, X. Zhang, Z. Ming, MPiLoc: Self-calibrating multi-floor indoor localization exploiting participatory sensing, in *IEEE Transactions on Mobile Computing* (2017), pp. 141–154
73. X. Tong, K. Liu, X. Tian, L. Fu, X. Wang, FineLoc: A Fine-grained Self-calibrating Wireless Indoor Localization System. IEEE Trans. Mob. Comput. **18**(9), 2077–2090 (2018)
74. Y. He, J. Liang, Y. Liu, Pervasive Floorplan Generation Based on Only Inertial Sensing: Feasibility, Design, and Implementation. IEEE J. Sel. Areas Commun. **35**(5), 1132–1140 (2017)
75. K. Vu, R. Zheng, Geometric algorithms for target localization and tracking under location uncertainties in wireless sensor networks, in *Proceedings of the IEEE INFOCOM* (2012), pp. 1835–1843
76. S.H. Fang, T. Lin, Principal component localization in indoor WLAN environments. IEEE Trans. Mob. Comput. **11**(1), 100–110 (2011)
77. K. Chintalapudi, A. Padmanabha Iyer, V.N. Padmanabhan, Indoor localization without the pain, in *Proceedings of the ACM MobiCom* (2010), pp. 173–184
78. H. Liu, Y. Gan, J. Yang, S. Sidhom, Y. Wang, Y. Chen, F. Ye, Push the Limit of WiFi based Localization for Smartphones, in *Proceedings of the ACM MobiCom* (2012), pp. 305–316
79. S. Jana, S.K. Kasera, On fast and accurate detection of unauthorized wireless access points using clock skews. IEEE Trans. Mob. Comput. **9**(3), 449–462 (2008)
80. D. Zhang, Y. Liu, X. Guo, M. Gao, L.M. Ni, On distinguishing the multiple radio paths in rss-based ranging, in *Proceedings of the IEEE INFOCOM* (2012), pp. 2201–2209
81. F. Gu, J. Niu, L. Duan, WAIPO: A fusion-based collaborative indoor localization system on smartphones. IEEE/ACM Trans. Networking **25**(4), 2267–2280 (2017)
82. H. Chenji, R. Stoleru, Toward accurate mobile sensor network localization in noisy environments. IEEE Trans. Mob. Comput. **12**(6), 1094–1106 (2012)

83. S. Hilsenbeck, D. Bobkov, G. Schroth, R. Huitl, E. Steinbach, Graph-based data fusion of pedometer and WiFi measurements for mobile indoor positioning, in *Proceedings of the ACM UbiComp* (2014), pp. 147–158
84. Y. Kim, H. Shin, H. Cha, Smartphone-based Wi-Fi pedestriantracking system tolerating the RSS variance problem, in *Proceedings of the IEEE PerCom* (2012), pp. 11–19
85. Z. Yang, Z. Zhou, Y. Liu, From RSSI to CSI: Indoor localization via channel response. ACM Comput. Surv. **46**(2), 1–32 (2013)
86. J. Xiong, K. Jamieson, ArrayTrack: A fine-grained indoor location system, in *Proceedings of the NSDI* (2013), pp. 71–84
87. M. Kotaru, K. Joshi, D. Bharadia, S. Katti, SpotFi: Decimeter Level Localization Using WiFi, in *Proceedings of the ACM SIGCOMM* (2015), pp. 269–282
88. K. Qian, C. Wu, Y. Zhang, G. Zhang, Z. Yang, Y. Liu, Widar2. 0: Passive human tracking with a single wi-fi link, in *Proceedings of the ACM MobiSys* (2018)
89. J. Wang, H. Jiang, J. Xiong, K. Jamieson, X. Chen, D. Fang, B. Xie, LiFS: Low Human-effort, Device-free Localization with Fine-grained Subcarrier Information, in *Proceedings of the ACM MobiCom* (2016), pp. 243–256
90. T.C. Tai, K.C.J. Lin, Y.C. Tseng, Toward reliable localization by unequal AoA tracking, in *Proceedings of the ACM MobiSys* (2019), pp. 444–456
91. X. Li, D. Zhang, Q. Lv, J. Xiong, S. Li, Y. Zhang, H. Mei, IndoTrack: Device-Free Indoor Human Tracking with Commodity Wi-Fi, in *Proceedings of the ACM Ubicomp* (2017), pp. 1–22
92. X. Li, S. Li, D. Zhang, J. Xiong, Y. Wang, H. Mei, Dynamic-MUSIC: Accurate Device-free Indoor Localization, in *Proceedings of the ACM UbiComp* (2016), pp. 196–207
93. X. Tong, H. Li, X. Tian, X. Wang, Wi-fi localization enabling self-calibration. IEEE/ACM Trans. Networking **29**(2), 904–917 (2021)
94. J. Gjengset, J. Xiong, G. McPhillips, K. Jamieson, Phaser: Enabling phased array signal processing on commodity WiFi access points, in *Proceedings of the ACM MobiCom* (2014), pp. 153–164
95. J. Xiong, K. Sundaresan, K. Jamieson, ToneTrack: Leveraging Frequency-Agile Radios for Time-Based Indoor Wireless Localization, in *Proceedings of the ACM MobiCom* (2015), pp. 537–549
96. R.H. Venkatnarayan, M. Shahzad, S. Yun, C. Vlachou, K.H. Kim, Leveraging Polarization of WiFi Signals to Simultaneously Track Multiple People. Proc. ACM IMWUT **4**(2), 1–24 (2020)
97. D. Halperin, W. Hu, A. Sheth, D. Wetherall, Tool release: gathering 802.11 n traces with channel state information, in *Proceedings of the ACM SIGCOMM* (2011), pp. 53–53
98. S. Sen, J. Lee, K.H. Kim, P. Congdon, Avoiding multipath to revive inbuilding WiFi localization in *Proceedings of the ACM MobiSys* (2013), pp. 249–262
99. Y. Zeng, P.H. Pathak, P. Mohapatra, WiWho: wifi-based person identification in smart spaces, in *Proceedings of the IEEE IPSN* (2016)
100. E. Elnahrawy, X. Li, R.P. Martin, The limits of localization using signal strength: A comparative study, in *Proceedings of the IEEE SECON* (2004), pp. 406–414
101. M. Youssef, M. Abdallah, A. Agrawala, Multivariate analysis for probabilistic WLAN location determination systems, in *Proceedings of the the Second Annual International Conference on Mobile and Ubiquitous Systems: Networking and Services* (2005)
102. F. Graziosi, F. Santucci, A general correlation model for shadow fading in mobile radio systems. IEEE Commun. Lett. **6**(3), 102–104 (2002)
103. R. Battiti, M. Brunato, A. Delai, Optimal wireless access point placement for location-dependent services, Technical Report (2003). http://eprints.biblio.unitn.it/489/1/DIT-03-052-withCover.pdf
104. M.A. Youssef, A. Agrawala, *On the Optimality of Wlan Location Determination Systems* (2003). http://www.cs.umd.edu/~moustafa/papers/cnds04.pdf
105. H. Liu, J. Yang, S. Sidhom, Y. Wang, Y. Chen, F. Ye, Accurate WiFi Based Localization for Smartphones Using Peer Assistance. IEEE Trans. Mob. Comput. **13**(10), 2199–2214 (2013)

106. N. Patwari, J. Ash, S. Kyperountas, I. Hero, A.O., R. Moses, N. Correal, Locating the nodes: cooperative localization in wireless sensor networks. IEEE Signal Process. Mag. **22**(4), 55–69 (2005)
107. N. Patwari, A.O.H., M. Perkins, N.S. Correal, R.J. O'Dea, Relative location estimation in wireless sensor networks. IEEE Trans. Signal Process. **51**(8), 2137–C2148 (2003)
108. M. Angjelichinoski, D. Denkovski, V. Atanasovski, L. Gavrilovska, Cramér-Rao Lower Bounds of RSS-Based Localization with Anchor Position Uncertainty. IEEE Trans. Inf. Theory **61**(5), 2807–2834 (2015)
109. International Telecommunication Union (ITU), Guidelines for evaluation of radio interfacce technologies for IMT-Advanced, Report ITU-R M. 2135–1 (2009) pp. 30–33. https://www.itu.int/dms_pub/itu-r/opb/rep/R-REP-M.2135-1-2009-PDF-E.pdf
110. WINNER II, WINNER II Channel Models, IST-4-027756 (2007), pp. 44–45. http://www.cept.org/files/1050/documents/winner2%20-%20final%20report.pdf
111. Z. Zhang, D. Liu, S. Zhu, S. Chen, X. Tian, Squeeze More from Fingerprints Reporting Strategy for Indoor Localization, in *Proceedings of the IEEE SECON* (2016)
112. T.V. Haute, E.D. Pooter, F. Lemic, V. Handziski, N. Wirström, T. Voigt, A. Wolisz, I. Moerman, Platform for benchmarking of RF-based indoor localization solutions. IEEE Commun. Mag. **53**(9), 126–133 (2015)
113. T.V. Haute, E.D. Pooter, I. Moerman, F. Lemic, V. Handziski, A. Wolisz, N. Wirström, T. Voigt, Comparability of RF-based Indoor Localization Solutions in Heterogeneous Environments: An Experimental Study. Int. J. Ad Hoc and Ubiquitous Computing **23**(1/2), 92–114 (2016)
114. F. Lemic, A. Behboodi, V. Handziski, V. Wolisz, Experimental decomposition of the performance of fingerprinting-based localization algorithms, in *Proceedings of the International Conference on Indoor Positioning and Indoor Navigation* (2016), pp. 695–704
115. S. He, S.H.G. Chan, Wi-Fi fingerprint-based indoor positioning: Recent advances and comparisons, in *IEEE Communications Surveys and Tutorials* (2016), pp. 466–490
116. Y. Wen, X. Tian, X. Wang, S. Lu, Fundamental limits of RSS fingerprinting based indoor localization, in *Proceedings of the IEEE INFOCOM* (2015)
117. X. Tian, R. Shen, D. Liu, Y. Wen, X. Wang, Performance Analysis of RSS Fingerprinting Based Indoor Localization. IEEE Trans. Mob. Comput. (2016). http://ieeexplore.ieee.org/stamp/stamp.jsp?arnumber=7797481
118. X. Tong, F. Zhu, Y. Wan, X. Tian, X. Wang, Batch Localization Based on OFDMA Backscatter. Proc. ACM Ubicomp **3**(1), 25 (2019)
119. Y. Xie, J. Xiong, M. Li, K. Jamieson, mD-Track: Leveraging multi-dimensionality for passive indoor Wi-Fi tracking, in *Proceedings of the ACM MobiCom* (2019)
120. C. Wu, F. Zhang, Y. Fan, K.J. Liu, RF-based Inertial Measurement, in *Proceedings of the ACM SIGCOMM*, pp. 117–129 (2019)
121. Y. Xie, Z. Li, M. Li, Precise power delay profiling with commodity wifi, in *Proceedings of the ACM Mobicom* (2015), pp. 53–64
122. S. Obayashi, J. Zander, A body-shadowing model for indoor radio communication environments View Document. IEEE Trans. Antennas Propag. **46**(6), 920–927 (1998)
123. M. Ayadi, A. Zineb, Body Shadowing and Furniture Effects for Accuracy Improvement of Indoor Wave Propagation Models. IEEE Trans. Wireless Commun. **13**(11), 5999–6006 (2014)
124. I. Kashiwagi, T. Taga, T. Imai, Time-varying path-shadowing model for indoor populated environments view document. IEEE Trans. Veh. Technol. **59**(1), 16–28 (2010)
125. J. Wang, D. Fang, Z. Yang, H. Jiang, X. Chen, T. Xing, L. Cai, E-HIPA: An Energy-Efficient Framework for High-Precision Multi-Target-Adaptive Device-free localization. IEEE Trans. Mob. Comput. **16**(3), 716–729 (2017). http://ieeexplore.ieee.org/stamp/stamp.jsp?arnumber=7469386
126. J. Wang, H. Jiang, J. Xiong, K. Jamieson, X. Chen, D. Fang, B. Xie, LiFS: Low Human Effort, Device-Free Localization with Fine-Grained Subcarrier Information, in *Proceedings of the ACM MobiCom* (2016)

127. K. Qian, C. Wu, Z. Yang, Y. Liu, K. Jamieson, Widar: Decimeter-Level Passive Tracking via Velocity Monitoring with Commodity Wi-Fi, in *Proceedings of the ACM MobiHoc* (2017)
128. H. Wang et al., No need to war-drive: Unsupervised indoor localization, in *Proceedings of the ACM MobiSys* (2012), pp. 197–210
129. T. Lin, S. Fang, W. Tseng, C. Lee, J. Hsieh, A group-discrimination-based access point selection for WLAN fingerprinting localization. IEEE Trans. Veh. Technol. **63**(8), 3967–3976 (2014)
130. K. Nagano, Y. Kawahara, K. Aihara, Size-constrained submodular minimization through minimum norm base, in *Proceedings of the 28th International Conference on Machine Learning* (2011)
131. Z. Svitkina, L. Fleischer, Submodular approximation: Sampling-based algorithms and lower bounds. https://arxiv.org/pdf/0805.1071.pdf (2010)
132. H. Liu, J. Yang, S. Sidhom, Y. Wang, Y. Chen, F. Ye, Accurate wifi based localization for smartphones using peer assistance, IEEE Trans. Mob. Comput. **13**(10), 2199–2214 (2014)
133. X. Zheng, H. Liu, J. Yang, Y. Chen, R.P. Martin, X. Li, A study of localization accuracy using multiple frequencies and powers. IEEE Trans. Parallel Distrib. Syst. **25**(8), 1955–1965 (2013)
134. X. Bai, Z. Yun, D. Xuan, T.H. Lai, W. Jia, Deploying four-connectivity and full-coverage wireless sensor networks, in *Proceedings of the IEEE INFOCOM* (2008)
135. R. Kershner, The number of circles covering a set. Am. J. Math. **61**(3), 665–671 (1939)
136. M. Wang, Z. Zhang, X. Tian, X. Wang, Temporal correlation of the RSS improves accuracy of fingerprinting localization, in *Proceedings of the IEEE INFOCOM* (2016)
137. C. Fischione, F. Graziosi, F. Santucci, Approximation for a sum of on-off lognormal processes with wireless applications. IEEE Trans. Commun. **55**(10), 1984–1993 (2007)
138. C. Oestges, N. Czink, B. Bandemer, P. Castiglione, F. Kaltenberger, J. Paulraj, Experimental Characterization and Modeling of Outdoor-to-Indoor and Indoor-to-Indoor Distributed Channels. IEEE Trans. Veh. Technol. **59**(5), 2253–2265 (2010)
139. S. Szyszkowicz, H. Yanikomeroglu, J. Thompson, On the Feasibility of Wireless Shadowing Correlation Models. IEEE Trans. Veh. Technol. **59**(9), 4222–4236 (2010)
140. K. Kaemarungsi, P. Krishnamurthy, Properties of indoor received signal strength for WLAN location fingerprinting, in *IEEE Mobile and Ubiquitous Systems: Networking and Services* (2004), pp. 14–23
141. U. Grenander, G. Szegö, Toeplitz forms and their applications, in *University of California Press*, vol. 321 (2001)
142. L.D. Branges, The Stone-Weierstrass theorem. Proc. Am. Math. Soc. **5**(4), 822–824 (1959)
143. D. Halperin, W. Hu, A. Sheth, D. Wetherall, Predictable 802.11 packet delivery from wireless channel measurements, in *Proceedings of the ACM SIGCOMM* (2010), pp. 159–170
144. S.M. Kay, *Fundamentals of Statistical Signal Processing: Estimation Theory* (Prentice-Hall, Englewood Cliffs, 1993)
145. Y. Shen, M. Z. Win, Fundamental limits of wideband localization part I: A general framework. IEEE Trans. Inf. Theory **56**(10), 4956–4980 (2010)
146. O. Tekdas, V. Isler, Sensor placement for triangulation-based localization. IEEE Trans. Autom. Sci. Eng. **7**(3), 681–685 (2010)
147. M.A. Spirito, On the accuracy of cellular mobile station location estimation. IEEE Trans. Veh. Technol. **50**(3), 674–685 (2001)
148. N. Salman, H.K. Maheshwari, A.H. Kemp, M. Ghogho, Effects of anchor placement on mean-CRB for localization, in *Proceedings of the The 10th IFIP Annual Mediterranean Ad Hoc Networking Workshop* (2011), pp. 115–118
149. K. Ligett, A. Roth, Take It or Leave It: Running a survey when privacy comes at a cost, in *Proceedings of the International Conference on Internet and Network Economics* (2012), pp. 378–391
150. R. Cummings, K. Ligett, A. Roth, Z. Wu, J. Ziani, Accuracy for sale: Aggregating data with a variance constraint, in *Proceedings of the Conference on Innovations in Theoretical Computer Science* (2015), pp. 317–324

151. S. Shalev-Shwartz, Online learning and online convex optimization. Foundations and Trends in Machine Learning **4**(2), 107–194 (2012)
152. N. Cesa-Bianchi, A. Conconi, C. Gentile, On the generalization ability of on-line learning algorithms, in *IEEE Transactions on Information Theory* (2004)
153. E. Dudin, Y.G. Smetanin, A review of cloud computing. Sci. Tech. Inf. Process. **38**(4), 280–284 (2011)
154. M. Nabaee, A. Pooyafard, A. Olfat, Enhanced object tracking with received signal strength using Kalman filter in sensor networks, in *Proceedings of the IEEE International Symposium on Telecommunications (ISTEL)* (2008), pp. 318–323
155. R. Faragher, R. Harle, Location fingerprinting with Bluetooth low energy beacons. IEEE J. Sel. Areas Commun. **33**(11), 2418–2428 (2015)
156. Y. Shu, K.G. Shin, T. He, J. Chen, Last-Mile Navigation Using Smartphones, in *Proceedings of the ACM MobiCom* (2015), pp. 512–524
157. N. Nakajima, S. Tang, T. Ogishi, S. Obana, Improving Precision of BLE-based Indoor Positioning by Using Multiple Wearable Devices, in *Ajunct Proceedings of the ACM MOBIQUITUS* (2016), pp. 118–123
158. G.G. Anagnostopoulos, M. Deriaz, Accuracy Enhancements in Indoor Localization with the Weighted Average Technique, in *Proceedings of the SENSORCOMM* (2014)
159. D.K. Cho, M. Mun, U. Lee, W.J. Kaiser, M. Gerla, Autogait: A mobile platform that accurately estimates the distance walked, in *PerCom* (2010), pp. 116–124
160. F. Li, C. Zhao, G. Ding, C. Liu, F. Zhao, A reliable and accurate indoor localization method using phone inertial sensors, in *Proceedings of the ACM Ubicomp* (2012), pp. 421–430
161. S. Chen, M. Li, K. Ren, C. Qiao, Crowd map: Accurate reconstruction of indoor floor plans from crowdsourced sensor-rich videos, in *IEEE ICDCS* (2015), pp. 1–10
162. D. Milioris, G. Tzagkarakis, A. Papakonstantinou, Low dimensional signal-strength fingerprint-based positioning in wireless LANs. Ad Hoc Netw. **12**, 100–114 (2014)
163. S. Nikitaki, G. Tsagkatakis, P. Tsakalides, Efficient multichannel signal strength based localization via matrix completion and Bayesian sparse learning. IEEE Trans. Mobile Comput. **14**(11), 2244–2256 (2015)
164. D. Milioris, M. Bradonjic, P. Muhlethaler, Building complete training maps for indoor location estimation, in *Proceedings of the IEEE INFOCOM* (2015), pp. 75–76
165. A. Edelman, T.A. Arias, S.T. Smith, The geometry of algorithms with orthogonality constraints. SIAM J. Matrix Anal. Appl. **20**(2), 303–353 (1998)
166. P. Jain, P. Netrapalli, *Fast Exact Matrix Completion with Finite Samples*. https://arxiv.org/pdf/1411.1087.pdf
167. D. Lazzaro, A nonconvex approach to low-rank matrix completion using convex optimization. Numer. Linear Algebra Appl. **23**(5), 801–824 (2016)
168. K. Mohan, M. Fazel, Iterative reweighted algorithms for matrix rank minimization. J. Mach. Learn. Res. **13**(1), 3441–3473 (2012)
169. Z. Gu, Z. Chen, Y. Zhang, Y. Zhu, M. Lu, A. Chen, Reducing fingerprint collection for indoor localization. Comput. Commun. **83**, 56–63 (2016)
170. P. Jain, P. Netrapalli, S. Sanghavi, Low-rank matrix completion using alternating minimization, in *Proceedings of the forty-fifth annual ACM symposium on Theory of computing* (2013), pp. 665–674
171. Z. Lin, M. Chen, Y. Ma, *The Augmented Lagrange Multiplier Method for Exact Recovery of Corrupted Low-rank Matrices*. https://arxiv.org/pdf/1009.5055.pdf.
172. L. Balzano, R. Nowak, B. Recht, Online identification and tracking of subspaces from highly incomplete information, in *Communication, Control, and Computing* (2010), pp. 704–711
173. L. Balzano, S.J. Wright, Local convergence of an algorithm for subspace identification from partial data. Found. Comput. Math. **15**(5), 1279–1314 (2015)
174. D. Zhang, L. Balzano, Global convergence of a Grassmannian gradient descent algorithm for subspace estimation. https://arxiv.org/pdf/1506.07405.pdf
175. E.J. Candes, J. Romberg, Sparsity and Incoherence in Compressive Sampling. Inverse Prob. **23**(3), 969–985 (2007)

176. S. Fisk, A Note on Weyl's Inequality. Am. Math. Mon. **104**(5), 257–258 (1997)
177. X. Tong, H. Li, X. Tian, X. Wang, Triangular Antenna Layout Facilitates Deployability of CSI Indoor Localization Systems, in *Proceedings of the IEEE SECON* (2019)
178. Z. Chen, Z. Li, X. Zhang, G. Zhu, Y. Xu, J. Xiong, X. Wang, AWL: Turning Spatial Aliasing From Foe to Friend for Accurate WiFi Localization, in *Proceedings of the ACM CoNEXT*, pp. 238–250 (2017)
179. M. Kotaru, P. Zhang, S. Katti, Localizing Low-power Backscatter Tags Using Commodity WiFi, in *Proceedings of the ACM CoNEXT* (2017), pp. 251–262
180. S. Kumar, S. Gil, D. Katabi, D. Rus, Accurate indoor localization with zero start-up cost, in *Proceedings of the ACM MobiCom*, pp. 483–494 (2014)
181. A. Olivier, G. Bielsa, I. Tejado, M. Zorzi, J. Widmer, P. Casari, Lightweight Indoor Localization for 60-GHz Millimeter Wave Systems, in *Proceedings of the IEEE SECON* (2016)
182. O. Abari, D. Vasisht, D. Katabi, A. Chandrakasan, Caraoke: An E-toll transponder network for smart cities, in *Proceedings of the ACM SIGCOMM* (2015)
183. Y. Xie, Y. Zhang, J.C. Liando, M. Li, SWAN: Stitched Wi-Fi ANtennas, in *Proceedings of the ACM MobiCom*, pp. 51–66 (2018)
184. C. Zhang, F. Li, J. Luo, Y. He, iLocScan: Harnessing multipath for simultaneous indoor source localization and space scanning, in *Proceedings of the ACM Conference on Embedded Network Sensor Systems* (2014), pp. 91–104
185. R. Schmidt, Multiple emitter location and signal parameter estimation. IEEE Trans. Antennas Propag. **34**(3), 276–280 (1986)
186. Y. Wan, X. Tong, Y. Peng, X. Tian, Calibration-Free Localization Based on fingerprint of Channel State Information, in *Proceedings of the of IEEE 5th International Conference on Computer and Communications* (2019)
187. Z. Gao, Y. Gao, S. Wang, D. Li, Y. Xu, H. Jiang, CRISLoc: Reconstructable CSI fingerprint for Indoor Smartphone Localization, arXiv preprint arXiv:1910.06895 (2020)
188. S. Sen, B. Radunovic, R.R. Choudhury, T. Minka, You are facing the Mona Lisa: Spot localization using PHY layer information, in *Proceedings of the ACM MobiSys*, pp. 183–196 (2012)
189. M. Kotaru, S. Katti, Position tracking for virtual reality using commodity wifi, in *Proceedings of the IEEE CVPR* (2017), pp. 68–78
190. Z. Chen, X. Tong, X. Tian, Deriving AP Position and Antenna Array Orientation for Wi-Fi Localization, in *Proceedings of the IEEE WSCP* (2019)
191. M. Schulz, J. Link, F. Gringoli, M. Hollick, Shadow Wi-Fi: Teaching smartphones to transmit raw signals and to extract channel state information to implement practical covert channels over Wi-Fi, in *Proceedings of the MobiSys* (2018), pp. 256–268
192. E. Soltanaghaei, A. Kalyanaraman, K. Whitehouse, Multipath Triangulation: Decimeter-level Wi-Fi Localization and Orientation with a Single Unaided Receiver, in *Proceedings of the ACM MobiSys* (2018)
193. *EVARILOS Testbed*. http://evarilos.intec.ugent.be/
194. Technical report. https://www.dropbox.com/s/nuu7i6ggd9pmjqy/IEEE-TMC-Best%20Strategy-TechReport.pdf?dl=0 (2016)
195. *A Power Monitor for Android-Based Mobile Platforms*. http://ziyang.eecs.umich.edu/projects/powertutor/index.html (2011)
196. Qualcomm developer network, *Trepn Power Profiler*. https://developer.qualcomm.com/software/trepn-power-profiler (2017)
197. IEEE 802.11 Working Group et al., IEEE standard for information technology–telecommunications and information exchange between systems–local and metropolitan area networks–specific requirements–part 11: Wireless lan medium access control (mac) and physical layer (phy) specifications amendment 6: Wireless access in vehicular environments. *IEEE Std*, vol. 802 (2010), p. 11p
198. Vijay Kumar Lab demonstration. http://www.kumarrobotics.org/videos/
199. NEW ATLAS. http://newatlas.com/darpa-drone-autonomous-45-mph/41810/

200. C. Birkner, MLB's iBeacon Technology is A Marketing Home Run. https://www.ama.org/publications/MarketingNews/Pages/mlb-ibeacons.aspx
201. T. Reddy, 15 Companies From Airports to Retail Already Using Beacon Technology. https://www.umbel.com/blog/mobile/15-companies-using-beacon-technology/
202. iBeacon.io, http://ibeacon.com.hk/
203. WeChat mini program. https://mp.weixin.qq.com/debug/wxadoc/dev/api/iBeacon.html#wxstopbeacondiscoveryobject
204. Minewtech E5. https://en.minewtech.com/ibeacon/E5.html
205. Cell ID based localization. https://combain.com/about/about-positioning/cell-id-positioning/
206. Wikipedia. https://en.wikipedia.org/wiki/Low-rank_approximation
207. Technical Report. https://www.dropbox.com/s/nq72vr7x3xvbpzk/subspace.pdf?dl=0
208. SpotFi source code. https://bitbucket.org/mkotaru/spotfimusicaoaestimation/src/master/ (2019)
209. http://dhalperi.github.io/linux-80211n-csitool/old_installation.html, accessed 2020.06
210. A. Jimenez, F. Seco, C. Prieto, J. Guevara, A comparison of pedestrian dead-reckoning algorithms using a low-cost mems imu, in *IEEE WISP* (2009), pp. 37–42
211. G. Fei, J. Niu, L. Duan, P^3-LOC: A Privacy-Preserving Paradigm-Driven Framework for Indoor Localization. IEEE/ACM Trans. Netw. **26**(6), 2856–2869 (2018)
212. G. Fei, J. Niu, L. Duan, WAIPO: A fusion-based collaborative indoor localization system on smartphones.IEEE/ACM Trans. Netw. **25**(4), 2267–2280 (2017)
213. I. Bisio, F. Lavagetto, M. Marchese, A. Sciarrone, Smart Probabilistic Fingerprinting for WiFi-based Indoor Positioning with Mobile Devices. Elsevier Pervasive and Mobile Computing **31**, 107–123 (2016). http://dx.doi.org/10.1016/j.pmcj.2016.02.001
214. A. Haeberlen, E. Flannery, A.M. Ladd, A. Rudys, D.S. Wallach, L.E. Kavraki, Practical Robust Localization over Large-Scale 802.11 Wireless Networks, in *Proceedings of the ACM MobiCom* (2004)
215. I. Bisio, F. Lavagetto, A. Sciarrone, S. Yiu, A $Smart^2$ Gaussian Process Approach for Indoor Localization with RSSI Fingerprints, in *Proceedings of the IEEE ICC* (2017)
216. Z. Yang, C. Wu, Z. Zhou, X. Zhang, Y. Wang, Y. Liu, Mobility Increases Localizability: A survey on wireless indoor localization using inertial sensors. ACM Comput. Surv. **47**(3), Article 54 (2015)
217. A.M. Hossain, W.S. Soh, Cramer-rao bound analysis of localization using signal strength difference as location fingerprint, in *Proceedings of the IEEE INFOCOM* (2010), pp. 1–9
218. WINNER, WINNER Channel Models. IST-4-027756, pp. 44–45 (2007). http://www.cept.org/files/1050/documents/winner2
219. C. Wu, Z. Yang, Y. Liu, Smartphones based crowdsourcing for indoor localization. IEEE Trans. Mob. Comput. **14**(2), 444–457 (2014)
220. W. Sun, J. Liu, C. Wu, Z. Yang, X. Zhang, Y. Liu, MoLoc: On distinguishing fingerprint twins, in *Proceedings of the IEEE ICDCS* (2013), pp. 226–235
221. S. He, S.H.G. Chan, Wi-Fi fingerprint-based indoor positioning: Recent advances and comparisons. IEEE Commun. Surv. Tutorials **18**(1), first quarter (2016)
222. B. Iyer, S. Jegelka, J. Bilmes, Fast semidifferential-based submodular function optimization, in *Proceedings of the 30th International Conference on Machine Learning* (2013)
223. D. Zhang, K. Lu, R. Mao, Y. Feng, Y. Liu, Z. Ming, L.M. Ni, Fine-Grained localization for multiple transceiver-free objects by using RF-Based technologies. IEEE Trans. Parallel Distrib. Syst. **25**(6), 1464–1475 (2014)
224. K. Sheng, Z. Gu, X. Mao, X. Tian, W. Wu, X. Gan, X. Wang, The collocation of measurement points in large open indoor environment, in *Proceedings of the IEEE INFOCOM* (2015)
225. S. Fang, B. Lu, Y. Hsu, Learning location from sequential signal strength based on GSM experimental data. IEEE Trans. Veh. Technol. **61**(2), 726–736 (2012)
226. S. Fang, T. Lin, A dynamic system approach for radio location fingerprinting in wireless local area networks. IEEE Trans. Commun. **58**(4), 1020–1025 (2010)
227. D. Vasisht, S. Kumar, D. Katabi, Decimeter-Level location with a single WiFi access point, in *Proceedings of the NSDI* (2016), pp. 165–178

228. M. Win, A. Conti, S. Mazuelas, Y. Shen, W. Gifford, D. Dardari, M. Chiani, Network localization and navigation via cooperation. IEEE Commun. Mag. **49**(5), 56–62 (2011)
229. W. Dai, Y. Shen, M. Win, Energy-Efficient Network Navigation Algorithms. IEEE J. Sel. Areas Commun. **33**(7), 1418–1430 (2015)
230. H. Wymeersch, J. Lien, M. Win, Cooperative Localization in Wireless Network. Proc. IEEE **97**(2), 427–450 (2009)
231. K. Pahlavan, P. Krishnamurthy, Y. Geng, Localization challenges for the emergence of the smart world. IEEE Access **3**, 3058–3067 (2015)
232. Y. Gwon, R. Jain, Error characteristics and calibration-free techniques for wireless lan-based location estimation, in *Proceedings of the Second International Workshop on Mobility Management and Wireless Access Protocols* (2004), pp. 2–9
233. H. Lim, L.-C. Kung, J. Hou, H. Luo, Zero-configuration, robust indoor localization: Theory and experimentation, in *Proceedings of the IEEE INFOCOM* (2005)
234. P. Bahl, V.N. Padmanabhan, A. Balachandran, Enhancements to the RADAR user location and tracking system, technical report, Microsoft Research, Tech. Rep. (2000)
235. T. Roos, P. Myllymäki, H. Tirri, P. Misikangas, J. Sievänen, A probabilistic approach to wlan user location estimation. Int. J. Wireless Inf. Networks **9**(3), 155–164 (2002)
236. Warp project. http://warpproject.org
237. Atheros CSI Tool. http://pdcc.ntu.edu.sg/wands/Atheros/#cite
238. Statistics of CSI localization experiments. https://www.dropbox.com/s/0aect8y9m1pzkae/CSI-Exp%20Statistics.pdf?dl=0
239. H. Li, K. Ota, M. Dong, M. Guo, Mobile Crowdsensing in Software Defined Opportunistic Networks. IEEE Commun. Mag. **55**(6), 140–145 (2017)
240. T. Kumrai, K. Ota, M. Dong, P. Champrasert, An incentive-based evolutionary algorithm for participatory sensing, in *Proceedings of the IEEE GLOBECOM 2014* (2014)
241. T. Kumrai, K. Ota, M. Dong, P. Champrasert, Mobile agent-based energy-aware and user-centric data collection in wireless sensor networks. Comput. Netw. **74**(Part B), 58–70 (2014)
242. R.K. Ganti, F. Ye, H. Lei, Mobile crowdsensing: current state and future challenges. IEEE Commun. Mag. **49**(11), 32–39 (2011)
243. K. Han, H. Huang, J. Luo, Posted pricing for robust crowdsensing, in *Proceedings of the ACM MobiHoc* (2016), pp. 261–270
244. R. Kawajiri, M. Shimosaka, H. Kashima, Steered crowdsensing: Incentive design towards quality-oriented place-centric crowdsensing, in *Proceedings of the ACM UbiComp* (2014), pp. 691–701
245. M.H. Cheung, R. Southwell, F. Hou, J. Huang, Distributed time-sensitive task selection in mobile crowdsensing, in *Proceedings of the ACM MobiHoc* (2015), pp. 157–166
246. S. Reddy, D. Estrin, M. Srivastava, Recruitment framework for participatory sensing data collections, in *International Conference on Pervasive Computing* (2010), pp. 138–155
247. H. Jin, L. Su, D. Chen, K. Nahrstedt, J. Xu, Quality of information aware incentive mechanisms for mobile crowd sensing systems, in *Proceedings of the ACM MobiHoc* (2015), pp. 167–176
248. D. Peng, F. Wu, G. Chen, Pay as how well you do: A quality based incentive mechanism for crowdsensing, in *Proceedings of the ACM MobiHoc* (2015), pp. 177–186
249. L. Gao, F. Hou, J. Huang, Providing long-term participation incentive in participatory sensing, in *2015 IEEE Conference on Computer Communications (INFOCOM)* (2015), pp. 2803–2811
250. Q. Zhang, Y. Wen, X. Tian, X. Gan, X. Wang, Incentivize crowd labeling under budget constraint, in *2015 IEEE Conference on Computer Communications (INFOCOM)* (2015), pp. 2812–2820
251. C.K. Tham, T. Luo, Quality of contributed service and market equilibrium for participatory sensing. IEEE Trans. Mob. Comput. **14**(4), 829–842 (2014)
252. D. Zhao, X. Li, H. Ma, Budget-feasible online incentive mechanisms for crowdsourcing tasks truthfully. ACM/IEEE Trans. Netw. **24**(2), 647–661 (2014)
253. M. Altini, D. Brunelli, E. Farella, L. Benini, Bluetooth indoor localization with multiple neural networks, in *IEEE International Symposium on Wireless Pervasive Computing (ISWPC)* (2010), pp. 295–300

254. S.S. Chawathe, Low-latency indoor localization using bluetooth beacons, in *Proceedings of the IEEE Conference on Inteliigent Transportation Systems (ITSC)* (2009), pp. 1–7
255. F. Evennou, F. Marx, E. Novakov, Map-aided indoor mobile positioning system using particle filter, in *IEEE Wireless Communications and Networking Conference, 2005*, vol. 4 (2005), pp. 2490–2494
256. H. Wang, H. Lenz, A. Szabo, J. Bamberger, U.D. Hanebeck, WLAN-based pedestrian tracking using particle filters and low-cost MEMS sensors, in *2007 4th Workshop on Positioning, Navigation and Communication* (2007), pp. 1–7
257. A.S. Paul, E.A. Wan, RSSI-based indoor localization and tracking using sigma-point kalman smoothers. IEEE J. Sel. Top. Sign. Proces. **3**(5), 860–873 (2009)
258. J.L. Rullán-Lara, S. Salazar, R. Lozano, Real-time localization of an UAV using Kalman filter and a Wireless Sensor Network. J. Intell. Robot. Syst. **65**(1–4), 283–293 (2012)
259. X. Wu, R. Shen, X. Tian, P. Liu, X. Wang, iBILL: Using iBeacon and Inertial Sensors for Accurate Indoor Localization in Large Open Areas. IEEE Access **5**, 14589–14599 (2017)
260. Enhanced 911—Wireless Services. https://www.fcc.gov/general/enhanced-9-1-1-wireless-services
261. Ericsson White Paper, Postioning with LTE. http://www.sharetechnote.com/Docs/WP-LTE-positioning.pdf (2011)
262. European Global Navigation Satellite Systems Agency How to Enable Better Location for Emergency Calls: Galileo and 112. https://www.gsa.europa.eu/news/how-enable-better-location-emergency-calls-galileo-and-112
263. Rohde and Schwarz, LTE Location Based Services Technology Introduction White paper (2013). http://www.rohde-schwarz-wireless.com/documents/LTELBSWhitePaper_RohdeSchwarz.pdf
264. Spirent, White Paper: An Overview of LTE Positioning (2012). https://www.spirent.com/~/media/White%20Papers/Mobile/LTE_LBS_White_Paper_2012.pdf
265. 3GPP Release 8. http://www.3gpp.org/specifications/releases/72-release-8
266. 3GPP Release 9. http://www.3gpp.org/specifications/releases/71-release-9
267. Huawei, The second phase of LTE-Advanced LTE-B: 30-fold capacity boosting to LTE. www.huawei.com/ilink/en/download/HW_259010 (2013)
268. R. Margolies, R. Becker, S. Byers, S. Deb, R. Jana, S. Urbanek, C. Volinsky, Can you find me now? Evaluation of network-based localization in a 4G LTE network, in *Proceedings of the IEEE INFOCOM* (2017)
269. P. Nurmi, S. Bhattacharya, J. Kukkonen, A grid-based algorithm for on-device GSM positioning, in *Proceedings of the ACM UbiComp* (2010)
270. X. Wang, L. Gao, S. Mao, S. Pandey, CSI-based fingerprint for indoor localization: A deep learning approach. IEEE Trans. Veh. Technol. **66**(1), 763–776 (2016)
271. X. Wang, L. Gao, S. Mao, PhaseFi: Phase fingerprint for indoor localization with a deep learning approach, in *Proceedings of the IEEE GLOBECOM* (2015), pp. 1–6
272. T.F. Sanam, H. Godrich, A Multi-View discriminant learning approach for indoor localization using amplitude and phase features of CSI. IEEE Access **8**, 59947–59959 (2020)
273. X. Wang, X. Wang, S. Mao, CiFi: Deep convolutional neural networks for indoor localization with 5 GHz Wi-Fi, in *2017 IEEE International Conference on Communications (ICC)* (2017), pp. 1–6
274. T.F. Sanam, H. Godrich, FuseLoc: A CCA Based Information Fusion for Indoor Localization Using CSI Phase and Amplitude of Wifi Signals, in *ICASSP 2019–2019 IEEE International Conference on Acoustics, Speech and Signal Processing (ICASSP)* (IEEE, New York 2019), pp. 7565–7569
275. Y. Zhuang, J. Yang, Y. Li, L. Qi, N. El-Sheimy, Smartphone-based indoor localization with bluetooth low energy beacons. Sensors **16**(5), 596 (2016)
276. P. Pannuto, B. Kempke, P. Dutta, Slocalization: sub-μW ultra wideband backscatter localization, in *2018 17th ACM/IEEE International Conference on Information Processing in Sensor Networks (IPSN)* (2018)

277. Y.C. Tung, K.G. Shin, EchoTag: Accurate infrastructure-free indoor location tagging with smart-phones, in *Proceedings of the ACM MobiCom* (2015), pp. 525–536
278. R. Nandakumar, V. Iyer, S. Gollakota, 3D Localization for Sub-Centimeter Sized Devices, in *Proceedings of the ACM Sensys* (2018), pp. 108–119
279. V. Iyer, R. Nandakumar, A. Wang, S.B. Fuller, S. Gollakota, Living IoT: A Flying Wireless Platform on Live Insects, in *Proceedings of the ACM MobiCom* (2019), pp. 1–15
280. K. Kaemarungsi, P. Krishnamurthy, Modeling of indoor positioning systems based on location fingerprinting, in *Proceedings of the IEEE INFOCOM*, vol. 2 (2004), pp. 1012–1022
281. K. Sheng, Z. Gu, X. Mao, X. Tian, W. Wu, X. Gan, X. Wang, The Collocation of Measurement Points in Large Open Indoor Environment, Tech. Rep. (2014). http://iwct.sjtu.edu.cn/Personal/xwang8/paper/INFOCOM2015-collocation-Tech.pdf
282. H.-C. Chang, L.-C. Wang, A simple proof of Thue's Theorem on circle packing, arXiv preprint arXiv:1009.4322 (2010)
283. N. Cressie, Statistics for spatial data. Terra Nova **4**(5), 613–617 (1992)
284. C. Bettstetter, On the minimum node degree and connectivity of a wireless multihop network, in *Proceedings of the ACM MobiHoc* (2002), pp. 80–91
285. Technical report (2016). https://www.dropbox.com/s/0jxjf5nalrnz8vn/IEEE%20TMC-TemporalTechnical%20report.pdf?dl=0
286. J.F. Cai, E.J. Cands, Z. Shen, A singular value thresholding algorithm for matrix completion. SIAM J. Optimization **20**(4), 1956–1982 (2010)
287. G. Shen, Z. Chen, P. Zhang, T. Moscibroda, Y. Zhang, WalkieMarkie: Indoor pathway mapping made easy, in *Proceedings of the 10th USENIX Conference Network System Design Implementation* (2013), pp. 85–98
288. B. Kellogg, V. Talla, S. Gollakota, J.R. Smith, Passive Wi-Fi: Bringing Low Power to Wi-Fi Transmissions, in *13th USENIX Symposium on Networked Systems Design and Implementation* (2016), pp. 151–164
289. L.X. Chuo, Z. Luo, D. Sylvester, D. Blaauw, H.S. Kim, Rf-echo: A non-line-of-sight indoor localization system using a low-power active rf reflector asic tag, in *Proceedings of the 23rd Annual International Conference on Mobile Computing and Networking* (2017), pp. 222–234
290. P. Zhang, C. Josephson, D. Bharadia, S. Katti, Freerider: Backscatter communication using commodity radios, in *Proceedings of the 13th International Conference on emerging Networking EXperiments and Technologies* (2017), pp. 389–401
291. D. Halperin, W. Hu, A. Sheth, D. Wetherall, Linux 802.11 n CSI tool (2010)
292. P. Zhang, D. Bharadia, K. Joshi, S. Katti, Hitchhike: Practical backscatter using commodity wifi, in *Proceedings of the 14th ACM Conference on Embedded Network Sensor Systems CD-ROM* (2016), pp. 259–271
293. Power Splitter/Combiner—Mini Circuits. Retrieved May 3, 2018 from https://www.minicircuits.com/pdfs/BP2U+.pdf
294. SPST reflective switch (2017), ADG902 by ADI. Retrieved May 3, 2018 from https://www.analog.com/media/en/technical-documentation/data-sheets/ADG901902.pdf
295. FPGA development board, NEXYS4 by DIGILENT (2017). Retrieved May 3, 2018 from https://reference.digilentinc.com/media/reference/programmable-logic/nexys-4-ddr/nexys4ddrrm.pdf